37 Springer Series in Chemical Physics

Edited by Robert Gomer

Springer Series in Chemical Physics

Editors: Vitalii I. Goldanskii   Fritz P. Schäfer   J. Peter Toennies

Managing Editor: H.K.V. Lotsch

Volumes 1–39 are listed on the back inside cover

A. Meisel　G. Leonhardt　R. Szargan

X-Ray Spectra and Chemical Binding

With 230 Figures

Springer-Verlag Berlin Heidelberg New York
London Paris Tokyo Hong Kong

Professor Dr. Armin Meisel
Dozent Dr. Rüdiger Szargan
Karl-Marx-Universität, Sektion Chemie, Talstrasse 35, DDR-7010 Leipzig

Professor Dr. Gunter Leonhardt
Technische Universität Karl-Marx-Stadt, Sektion Chemie und Werkstofftechnik,
Strasse der Nationen 62, DDR-9010 Karl-Marx-Stadt

Series Editors

Professor Dr. Fritz Peter Schäfer
Max-Planck-Institut für
Biophysikalische Chemie
D-3400 Göttingen-Nikolausberg
Fed. Rep. of Germany

Professor Vitalii I. Goldanskii
Institute of Chemical Physics
Academy of Sciences
Kosygin Street 4
Moscow V-334, USSR

Professor Dr. J. Peter Toennies
Max-Planck-Institut für Strömungsforschung
Böttingerstrasse 6–8
D-3400 Göttingen
Fed. Rep. of Germany

Guest Editor: Professor Robert Gomer
The James Franck Institute
The University of Chicago
5640 Ellis Avenue, Chicago, IL 60637, USA

Managing Editor: Dr. Helmut K. V. Lotsch
Springer-Verlag, Tiergartenstrasse 17
D-6900 Heidelberg, Fed. Rep. of Germany

ISBN-13: 978-3-642-82264-3   e-ISBN-13: 978-3-642-82262-9
DOI: 10.1007/978-3-642-82262-9

Library of Congress Cataloging-in-Publication Data. Meisel, Armin. [Röntgenspektren und chemische Bindung. English] X-ray spectra and chemical binding / A. Meisel, G. Leonhardt, G. Szargan ; translated and edited by Elisabeth Källne, Richard D. Deslattes. p. cm.—(Springer series in chemical physics ; v. 37) Translation of: Röntgenspektren und chemische Bindung. Bibliography: p. Includes index. ISBN-13: 978-3-642-82264-31. X-ray spectroscopy. I. Leonhardt, Gunter. II. Szargan, Rüdiger. III. Title. IV. Series. QD96.X2M4413 1989 543'.08586—dc20 89-11589

This text was prepared using the PS™ Technical Word Processor
2154/3150-543210 – Printed on acid-free paper

Preface

In the 12 years that have passed since the publication of *Röntgenspektren und chemische Bindung*, the original German work on which this text is based, several aspects of high-resolution X-ray spectroscopy have developed rapidly. After accepting the suggestion of Dr. R.D. Deslattes that a translation should be prepared, we thoroughly revised the book in order to include these recent developments in theory, equipment and applications. The latest research results and publications that had appeared by 1987/1988 have been taken into account in the analysis. However, as the general treatment of the German edition is still valid today, the organization of the contents did not have to be modified. The present status of the field is adequately described just by the addition of supplementary material in the English edition. We have merely rearranged what was previously Chapter 4 into tabular form and placed it as an appendix. This presentation produces greater clarity and increases the ease with which the information can be referred to.

We thank Dr. E. Källne for undertaking the translation. We are grateful to Dr. R.D. Deslattes and Dr. H. Lotsch for their conscientious and critical checking of the translation. Chapters 3 and 6 were updated by our colleague Dr. H. Sommer, to whom special thanks are due. It is our hope that this translation and revision will make our text available to a larger section of the scientific community.

A. Meisel
G. Leonhardt
R. Szargan

Leipzig,
January 1989

Contents

1. Introduction

X rays have found wide application in science and technology, e.g. in qualitative and quantitative chemical analysis, in investigations of the geometric and electronic structure of matter, and in defectoscopy as well as in medicine for screening and therapy. The diagnostic possibilities of X rays were recognized very soon after their discovery in 1895. However, it required the historic work of M. v.Laue, W. Friedrich, and P. Knipping (1912), the predictions of W.H. and W.L. Bragg (1912), and the investigations by H.G.J. Moseley (1913) for the systematic development of X-ray diffraction and spectroscopic research to become possible.

The cornerstone is naturally the Bragg relation for monochromatic X rays

$$2d\sin\theta = n\lambda , \tag{1.1}$$

where d is the lattice constant, θ is the Bragg or glancing angle, n is the order of reflection (an integer > 1), and λ is the wavelength of the X rays. Determination of d values provides information on crystal structure, the magnitude of the lattice parameter, crystal perfection, and the size of crystallites (by means of X-ray diffraction analysis). When a crystal with known lattice constants is used, the wavelength of the diffracted X rays in emission or absorption is determined by (1.1) (X-ray spectroscopic analysis). In this way a qualitative analysis is performed using Moseley's law. By also considering relative intensities of the X-ray spectra, quantitative chemical analysis can be performed. This has been applied in industry with increasing frequency using refined, automatic and process-controlling analysis.

In this book other applications of X-ray spectroscopic research are discussed: the investigations of fine structure of X-ray emission and absorption spectra and the dependence of this fine structure on the chemical bond of the studied atom and the insights obtained from these results concerning the electronic structure of matter. Detailed studies on the fine structure of X-ray spectra were begun by *M. Siegbahn* [1.1]; J. Bergengren, A.E. Lindh, and O. Lundqvist (see [1.2]) first observed the influence of chemical binding on X-ray spectra. The research field was further developed in the 1930s and 1940s, in particular by various schools in England (H.W.B. Skinner), France (Y. Cauchois), Germany (A. Faessler), India (G.B. Deodhar), Japan (T. Hayashi, M. Sawada), the Netherlands

(D. Coster), USA (J.A. Bearden, L.G. Parratt), and the USSR (M.A. Blokhin, I.B. Borovskii). High-resolution spectroscopy applied today from the hard to the ultrasoft X-ray region seems to be one of the most reliable methods for experimental determination of chemical bonds in molecules and solids. This has been achieved through a highly developed sample technique in conjunction with modern spectral-analysis methods, as well as by combining experimental data with results from molecular-orbital (MO) and band-structure calculations. Coordinated studies have also been possible with the newer but already widespread method of photoelectron spectroscopy. Finally, one cannot fail to note the important new dimension provided through the application of synchrotron radiation.

In addition to [1.1,2], the physical basis and possible applications of high-resolution X-ray spectroscopy, as defined above, have been treated in monographs by *Compton* and *Allison* [1.3], *Cauchois* [1.4] and *Blokhin* [1.5,6] and in some handbooks [1.7-9]. Later, monographs appeared by *Azaroff* [1.10], *Agarwal* [1.11], and *Bonnelle* and *Mande* [1.12] on X-ray spectroscopy, and by *Barinskii* and *Nefedov* [1.13,14], *Zimkina* and *Fomichev* [1.15], *Nemoshkalenko* and *Aleshin* [1.17], *Mazalov* and co-workers [1.18], *Karazija* [1.19], and *Teo* and *Joy* [1.20] on particular X-ray spectroscopic subjects. Reviews of previously studied X-ray spectra are contained in the contribution by *Faessler* to the Landolt-Börnstein tables [1.21], in two volumes from the National Bureau of Standards, USA [1.22, 23], and in papers from *Meisel* [1.24] and *Hitchcock* [1.25]. The most recent values of X-ray wavelengths and atomic energy levels have been summarized by *Bearden* [1.26], and *Blokhin* and *Shveitser* [1.27].

Information on particular applications of X-ray spectroscopy has recently appeared in reviews, e.g. by *Faessler* [1.28] on X-ray spectroscopic studies of chemical binding in solids; by *Parratt* [1.29] on band structure of solids; by *Fabian* et al. [1.30,31] on emission bands, in particular from metals and alloys; by *Gohshi* [1.32] on the influence of chemical bonds on emission lines; and by *Barinskii* [1.33] on X-ray spectroscopic studies of chemical structural problems. Further, *Cauchois* and *Bonnelle* [1.34], and *Srivastava* and *Nigam* [1.35] summarized the role of X-ray spectroscopy in coordination chemistry. *Nefedov* [1.36] treated the delineation of electronic structure for free molecules and isolated groups in crystals; *Schwarz* [1.37] the X-ray absorption spectra of free molecules; and *White* [1.38] and *Grasserbauer* et al. [1.39] the influence of chemical binding on soft X-ray spectra studied with microprobes. *Nefedov* [1.40] and *Sumbaev* [1.41] summarized chemical effects in X-ray spectroscopy; *Urch* [1.42] X-ray emission spectroscopy; *Nordgren* and *Agren* [1.43] ultra-soft X-ray emission spectra; *Pendry* [1.44] X-ray absorption near-edge fine structure; and *Rabe* and *Haensel* [1.45], *Lee* and colleagues [1.46], *Garner* and *Hasnain* [1.47], *Hayes* and *Boyce* [1.48] the application of extended absorption fine structure (EXAFS) for structural determinations. In books edited by *Kunz* [1.49], *Winick* and *Doniach* [1.50], and *Koch* [1.51] some chapters summarize X-ray spectra obtained using synchrotron radiation. *McGlynn* et al. [1.52] edited a book on investigations in the vacuum ultraviolet region. *Prins* and *Koningsberger* [1.53] edited a book on X-ray absorption spectroscopy. Finally, two reviews by *Cauchois* should be mentioned: one gives an excellent survey on experimental techniques [1.54] and the other gives a historical overview of the development of X-ray spectroscopy [1.55].

International conferences on X-ray spectroscopy have been held since 1965 in following places:

i)   Ithaca, NY, in 1965 [1.56]
ii)   Leipzig in 1965 [1.57]
iii)   Kiev in 1968 [1.58]
iv)   Paris in 1970 [1.59]
v)   Munich in 1972 [1.60]
vi)   Helsinki in 1974 [1.61]
vii)   Gaithersburg, MD, in 1976 [1.62]
viii) Sendai in 1978 [1.63] (for the first time held jointly with the conference on inner-shell ionization)
ix)   Stirling in 1980 [1.64]
x)   Eugene, OR, in 1982 [1.65]
xi)   Leipzig in 1984 [1.66]
xii)   Paris in 1987 [1.67]

The next conference will be held in USA in 1990. Furthermore, the following conferences have been held, partly with international participation, during the last few years: In the USSR conferences on X-ray spectroscopy in 1963 in Erevan [1.68]; 1966 in Apatity [1.69]; 1971 in Ivano-Frankovsk [1.70]; 1973 in Alma-Ata [1.71]; 1975 in Rostov (Don) [1.72]; 1978 in Leningrad, 1981 in Lvov [1.73], 1985 in Irkutsk [1.74], and 1988 in Leningrad. A smaller conference on X rays and X-ray photoelectron spectra and chemical bonding was held in 1974 in Novosibirsk [1.75]. In the United States, annual conferences have been held in Denver, CO, on applications of X-ray analysis [1.76, 77], of which [1.76] deals particularly with X-ray spectroscopic studies. In Great Britain two conferences were held in Glasgow in 1967 and 1971, which concentrated mostly on X-ray emission bands from metals and alloys [1.78, 79], and another in Daresbury in 1982 on the structure of non-crystalline materials [1.80]. In Poland an international seminar on X-ray and electron spectroscopy was held in Jablonna in 1981 [1.81]. International conferences on EXAFS and near-edge structure were held in Frascati (1982) [1.82], in Stanford, CA (1984) [1.83], in Fontevraud (1986) [1.84], and in Seattle, WA (1988).

Several more recent X-ray spectroscopic studies were presented at the following conferences, although they did not particularly concentrate on X-ray spectroscopy: in Tokyo 1971 [1.85], Hamburg 1974 [1.86], Montpellier 1977 [1.87], Charlottesville, VA 1980, Jerusalem 1983 [1.88], and Lund 1986 [1.89] at conferences on vacuum ultraviolet radiation physics;

in Minsk 1962, 1967, and 1971 at conferences on chemical bonding in semiconductors [1.90-94]; in Kiev 1968, 1970, 1971, 1972, and 1977 at conferences on the electronic structure of transition metals and their compounds [1.95-100]; in Gaithersburg, MD 1969 at a conference on the electronic density of states of solids [1.101]; and in Atlanta, GA 1972 and Freiburg (FRG) 1976 at conferences on inner-shell ionization processes [1.102], and in Hamburg 1982 in synchrotron radiation instrumentation [1.103]. Since 1980 there have been synchrotron radiation conferences in Novosibirsk every two years and a conference on progress in X-ray studies by synchrotron radiation was held in Strasbourg 1985.

Here, published monographs, conference proceedings and papers in scientific journals up to 1987 have been considered, but because of space limitations not every individual study has been cited. In Chap.2 the atomic processes connected with emission and absorption of X rays are treated systematically using mainly one-electron theory. Chapter 3 contains a discussion of the principal components of an X-ray spectrometer, a comparison of different instruments, in particular with respect to luminosity and resolving power, together with a discussion of analysis of experimental data. Chapter 4 discusses X-ray spectroscopic determination of oxidation states, ionicity of binding, geometrical structure and effective atomic charges. After a summary of information deduced from X-ray emission band and absorption spectra concerning energy levels of molecules and complex compounds, Chap.5 gives some selected examples with interpretation of the corresponding X-ray spectra and thereby an overview of the results achieved so far. In the last chapter, the information content of X-ray spectra from solids is considered and an attempt is made to give general conclusions on the structure of certain substances as suggested by the spectra. Experimentally and theoretically studied X-ray spectra of pure elements and their compounds, published mainly since 1965, are listed in the Appendix; surveys of older publications can be found elsewhere [1.21-24].

2. Theoretical Basis

At the beginning of this chapter a general classification and characterization of X-ray emission and absorption processes is given (Sect.2.1). In the following subsection (Sect.2.2) models to explain the chemical shift and multiplet structure of diagram lines are reviewed, illustrating the theoretical approach to the energy of diagram lines on different levels of sophistication (frozen orbital and adiabatic approximation, transition operator method). Relationships between the intensity profile of X-ray emission bands of solids, molecules, and metal-ligand clusters, respectively, and the density of states and the transition probability are explained (Sect.2.3). The application of theoretical models MO LCAO, DV-Xα, and SW-Xα for the interpretation of energy separation and relative intensities of spectral features is discussed. In Sect.2.4 the origin of non-diagram lines or satellites is treated in detail including multiplet splitting, cross transitions, multiple ionization, and the radiative Auger effect. The influence of chemical bond on the energy and intensity of different types of satellites is studied. Section 2.5 deals with the elementary processes causing the emission of bremsstrahlung and explains the two techniques of recording bremsstrahlung and isochromat spectra. The relationship between the intensity profile of a bremsstrahlung spectrum and the density of states in the unoccupied part of the band, the absorption features due to the electron interaction in the target and the transition probability is represented. The dependence of absorption-edge energy and the absorption coefficient on the atomic number and the wavelength, respectively, is given in Sect. 2.6.1. The absorption cross section is discussed in terms of the one-electron model including potential barrier effects. The origin of the X-ray absorption near-edge structure (XANES) and the extended X-ray absorption fine structure (EXAFS) is explained in Sects.2.6.2 and 2.6.3, respectively. The influence of effective charge and potential barriers on the absorption features is discussed, applying the (Z+1) analogy.

2.1 General Survey

Spectroscopic methods can be used to study many different elementary processes which lead to emission of X rays or are initiated by X rays.

This monograph focusses on information regarding electronic structure of matter deduced from characteristic X-ray emission spectra, bremsstrahlung spectra and X-ray absorption spectra.

Characteristic X-ray emission spectra consist of intense diagram lines, mostly weak satellites and emission bands. These are produced in the interaction of electrons (primary excitation), photons (secondary or fluorescence excitation) or ions with the atoms in the target. The target can be solid, liquid or gaseous. The requirement for emission is ionization of an atom in an inner shell.

Continuous X-ray spectra or bremsstrahlung spectra are produced by stopping electrons in the anode material of an X-ray tube and are superimposed on the characteristic spectra of the target material. The short-wavelength limit of the bremsstrahlung spectrum is determined by the maximum kinetic energy of the exciting electrons.

The process resulting in X-ray emission can be studied not only by measuring the wavelength dependence of the intensity, but also by measuring intensity and its dependence on the kinetic energy of the exciting electrons, when characteristic isochromats or bremsstrahlung isochromats are obtained.

The X-ray absorption spectrum reflects the wavelength-dependent intensity reduction produced by inelastic interaction of X-ray photons with electrons in inner shells of atoms of the material transmitting the radiation (true absorption) and displays prominent X-ray absorption edges with accompanying manifolds of fine-structure features.

The spectrum of scattered radiation produced by monochromatic X rays consists of the Rayleigh line, the Compton band and Raman lines. These latter reflect different photon energies produced by inelastic scattering from bound electrons of the scattering material. A survey of the most important X-ray spectroscopic methods is shown in Fig.2.1, which also includes a description of the above-mentioned processes. The figure displays the energy levels participating in each transition in a simplified manner.

The various X-ray spectra depend not only on the kind of elementary process taking place in the active atom but also on the surrounding atoms, so that by choosing special methods, information can be obtained on the electronic structure in chemical compounds. We now proceed to survey the mechanism of the elementary processes producing these X-ray spectra.

Characteristic X rays originate from transitions between different energy states of an atom, which are determined by the nuclear charge and the number and distribution of electrons in different possible electronic configurations. The energy E_i of an X-ray level may be defined as the energy of a state where the atom has lost an electron in an inner shell i. The transition $i \rightarrow f$ from the state i into the state f with smaller energy leads to emission of an X-ray photon or to alternative nonradiative processes.

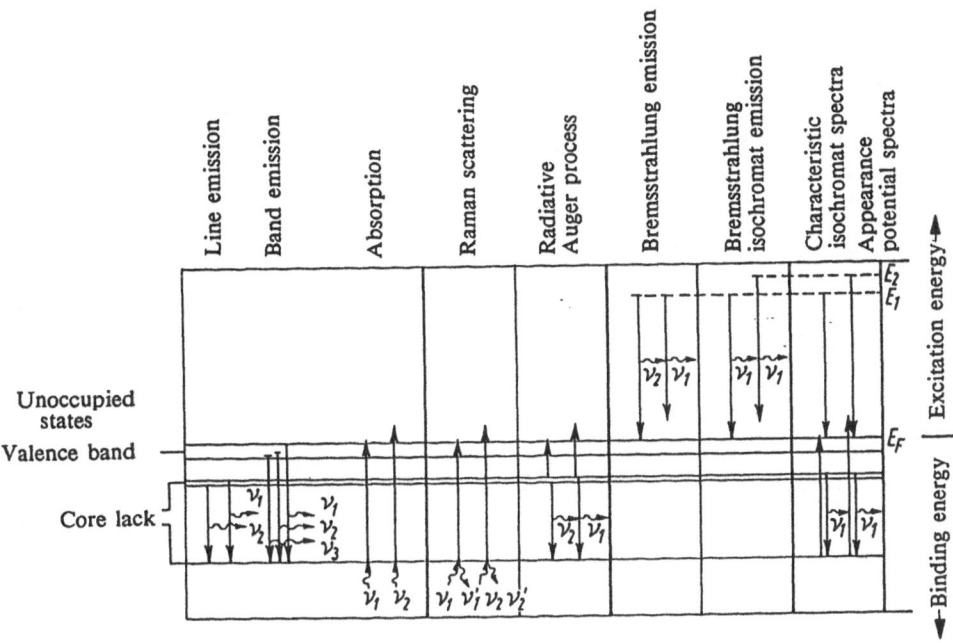

Fig.2.1. Schematic representation of X-ray spectroscopic methods [2.1]

Theoretical estimation of the X-ray transition energy requires calculating the total energy of the system before and after the transition with the corresponding initial and final electron configurations. In the simplest approximation, the transition is associated with a quantum-number change of only one electron, while the remaining electrons are assumed to be unperturbed. The transition $i \rightarrow f$ is accordingly equivalent to a single-electron transition in the opposite direction, i.e., a transition of an electron from level f into a vacancy in level i. The energy difference E_{if} between initial and final states determines the energy of the emitted X-ray photon except that differences between E_{if} and the emitted energy $h\nu$ can occur due to significant differences in ionization times, reorganization times of the orbitals due to the changed ionization potential (relaxation time) and the decay time of the excited state. Figure 2.2 shows the dependence of the ionization time (the time for an electron with a certain kinetic energy to leave the atom with a mean radius of 0.2 nm), the relaxation time (orbital period for an electron in a circle corresponding to the maximum in the electron-probability distribution) and the lifetime of the X-ray state and its dependence on the kinetic energy of the electron [2.2]. In spite of possible errors in these estimates, it can be seen that the relaxation times for electrons in outer shells, for a given element, are larger than those for electrons in inner shells. Since the lifetime of an X-ray state is smaller than the relaxation time for electrons in outer shells for atoms of higher atomic number, an X-ray transition can take place before

7

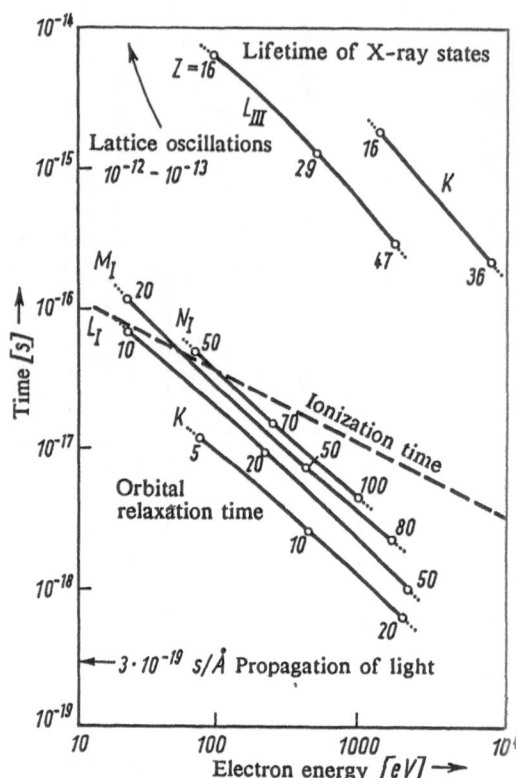

Fig.2.2. Ionization, relaxation and decay times for X-ray states and their dependence on the electron's kinetic energy [2.2]

the complete relaxation of the corresponding orbital occurs, so that excited states occur with energies higher than those of normal X-ray states [2.3].

The expected complete orbital relaxation before the decay of the X-ray state (Fig.2.2) for K, L, and M levels of the transition elements of the fourth period is evident from the result that a ^{55}Fe source emits Mn Kα, and Kβ radiation following K capture [2.4].

The X-ray levels of an atom, i.e., the energy levels of an atom with a hole in an inner closed shell can, to a first approximation, be deduced from the optical energy levels of a hydrogenlike atom if the effect of the electrons on the field of the atom is described by the constants of total and internal screening σ_t and σ_i. The energy of an X-ray level can be calculated in the following way

$$E(n, \ell, j) = Rhc \frac{M_z}{M_z + m} \left[\frac{(Z - \sigma_t)^2}{n^2} + \frac{\alpha^2 (Z - \sigma_i)^4}{n^4} \left(\frac{n}{j+1/2} - \frac{3}{4} \right) \right]. \quad (2.1)$$

Here the first term shows the main part of the energy, i.e., the sum of kinetic energy and potential energy of the electron. If the second term, which describes the interaction between the magnetic moments of a given electron and the remaining electrons, is neglected, it follows that the square root of the level energy varies linearly with Z. From this also follows that the square root of the photon energy of a transition between two levels is also linearly dependent on Z (Moseley's law for X-ray lines). If two levels are characterized by the same quantum numbers n and ℓ but different spin orientations, i.e., by different j values, it describes a spin doublet whose energy difference is determined from the difference in interaction energy between the magnetic moments. In this way spin doublets of X-ray lines always occur when the initial or final state of a transition is a spin doublet. For example, the $L\ell,\eta$ doublet ($L_{II,III} \rightarrow M_I$) should be considered as a spin doublet of the first kind (initial state) (Fig.2.3), while the $K\alpha_{1,2}$ doublet ($K \rightarrow L_{II,III}$) should be referred to the second category (final state).

The same symbols employed to classify X-ray states are used to classify those shells with the X-ray vacancy, thus the index I describes a removal of an s electron, and the indices II and III the removal of a p electron with j = 1/2 or 3/2, respectively. The X-ray terms K, L_{II} and M_{II} then correspond to states with the configurations $1sA(^2S)$, $1s^2 2s^2 2p^5 A(^2P_{1/2})$ and $1s^2 2s^2 2p^6 3s^2 3p^5 A(^2P_{3/2})$, where A corresponds to the outer closed shell. The optical term levels are written in parentheses. Transitions represented by $K \rightarrow L_{II,III}$ and $K \rightarrow M_{II,III}$ correspond to the electron jumps $2p \rightarrow 1s$ and $3p \rightarrow 1s$, respectively. Here, the Roman numbers are retained in view of their wide-spread historical usage although they are evidently superfluous.

The shapes and widths of X-ray lines, emitted in transitions between inner atomic levels, depend on the shapes and widths of the participating levels (neglecting instrumental broadening). If the probability for an atomic transition from a state i into any state f is P_i, then the mean lifetime of an atom in state i (the effective lifetime of this state) is $\tau_i = 1/P_i$. According to the uncertainty principle as regards energy and lifetime, the energy E_i of a state is not exactly determined, but has a spread $\Delta E_i \propto 1/\tau_i = P_i$. As the number of possible transitions from state i into state f is large for holes in the inner shells, it might be supposed that the width of the K state of an atom would be largest. Although this expectation is generally realized, there are important exceptions, such as those encountered in the super Coster-Kronig transitions. The widths of the valence levels are thus small (corresponding to their large lifetimes), but are commonly broadened by interaction with neighboring atoms.

Besides radiative transitions, there are also nonradiative transitions where the energy released from filling an inner hole in level i excites (or ionizes) an electron from an outer shell. This (Auger) process shortens the lifetime of an excited state so that the natural width of a level contains both radiative and Auger contributions. If the Z dependence of the level width ΔE were determined only by radiative transitions, then $\Delta E \propto Z^4$. This relation is not true for a large number of cases (Sect.2.2).

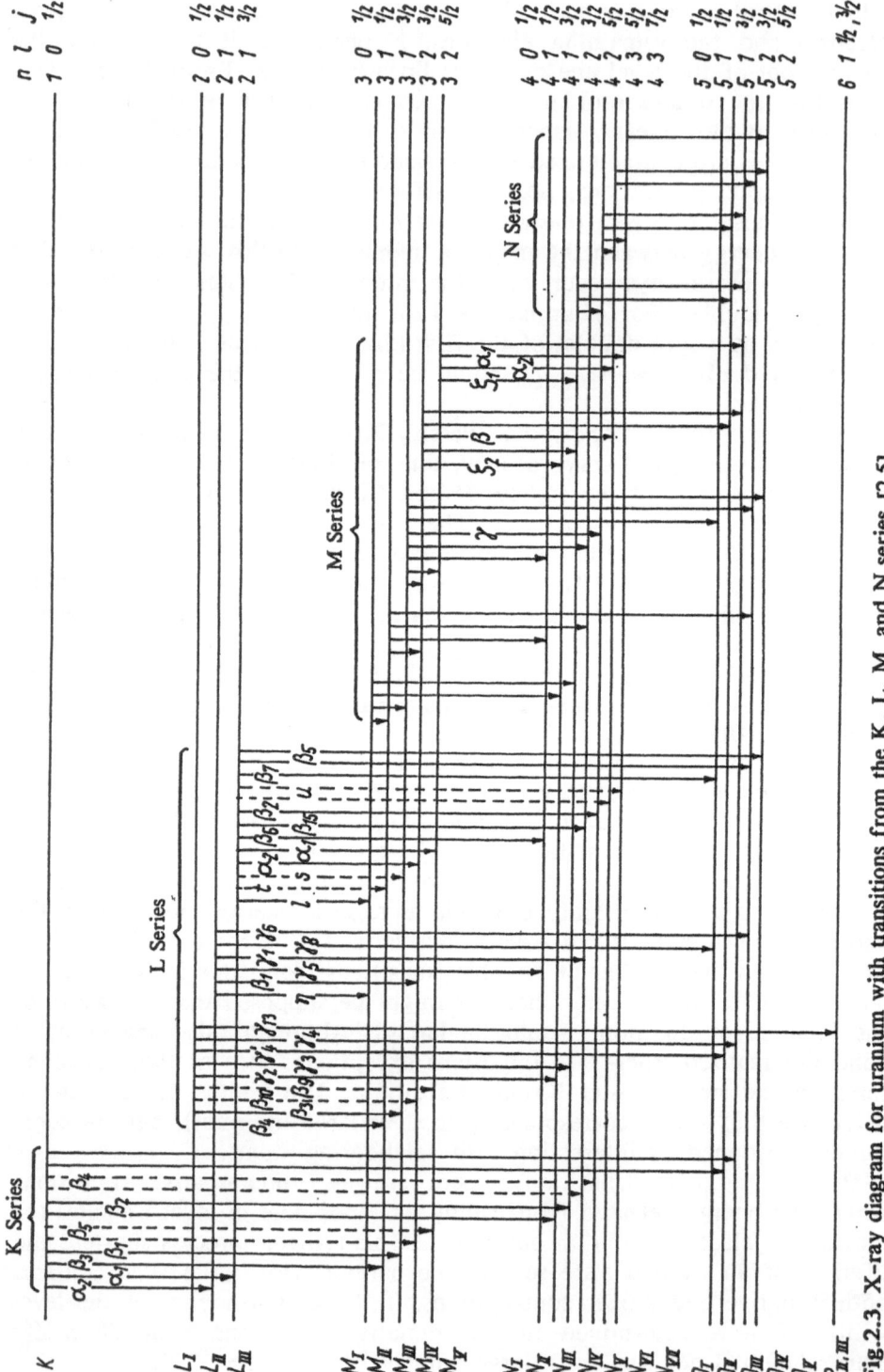

Fig. 2.3. X-ray diagram for uranium with transitions from the K, L, M, and N series [2.5]

In an elementary approximation, the shape of an X-ray line is described by the Lorentzian

$$I(\nu) = I_{\nu_0} \frac{1}{1 + [(\nu-\nu_0)/b]^2} \qquad (2.2)$$

which follows from any time-independent decay probability. In this, the spectral intensity distribution $I(\nu)$ is distributed symmetrically around ν_0. The parameter b is the half width at half height (2b: full width at half maximum).

For a very thin anode, the intensity of X-ray lines is proportional to the ionization probability F_i of the inner level, the statistical weight of this level g_i, the transition probability P_{if} for a transition from the state i into state f, and the energy of the transition $h\nu_{if}$. The influences of fluorescence yield, anode material and excitation methods are not considered here, but in Chap. 3.

F_i denotes the probability for excitation of the complete line series, and P_{if} is a measure of the relative intensity of the lines which occur from transitions from level i into level f. Quantum theory gives for the number of photons emitted from a target during a unit time

$$N_{if}(\omega, j) = \frac{e^2 \omega}{2hc^3 \pi} |D_{if}|^2 \qquad \text{with} \qquad (2.3)$$

$$D_{if} = \int \psi_i^* \, \omega r_j \, e^{ikr} \psi_f \, d\tau$$

(r_j: components of r in the polarization direction j).

If the exponential term is expanded in a power series, i.e.,

$$e^{ikr} = 1 + ikr + (ikr)^2/2! + \ldots \qquad (2.4)$$

it is immediately obvious from the so-called dipole approximation that with $r \simeq 10^{-8}$ cm for valence electrons and $k \simeq 6 \cdot 10^6$ cm^{-1} for an X ray with a wavelength of 10 nm, the terms following 1 can be neglected in (2.4), i.e., the dipole approximation is adequate. For radiation of shorter wavelengths, which includes the inner electrons, with r of a magnitude of 10^{-9} cm in the transition, the dipole approximation is not sufficient. Then the contribution from the second-order term in (2.4), the quadrupole radiation, must be included.

If the constant frequency in (2.3) is taken out of the integral, then the transition probability in the dipole approximation is given by

$$|P_{if}| \propto \nu^3 |r_{if}|^2 , \qquad \text{with} \qquad (2.5)$$

$$r_{if} = \int \psi_i^* r_j \psi_f \, d\tau \quad \text{and} \quad \nu = \omega/2\pi .$$

The integral r_{if} is the customary matrix element for the transition probability. The intensity of a line is derived from (2.5) through a summation over all states i and f with the same energy difference $h\nu$ as

$$I(\nu) \simeq \sum_{f} \sum_{i} \delta(E_i - E_f - h\nu) \nu^4 |r_{if}|^2 \qquad (2.6)$$

where δ denotes the Dirac delta function.

From the matrix element r_{if} the selection rules for the strongest lines, the electric dipole or diagram lines, are given by

$$|\Delta\ell| = 1 , \quad |\Delta j| = 0 \text{ or } 1 \quad (\Delta j = 0 \text{ only for } j \neq 0) , \quad \Delta n \neq 0 .$$

For the (mostly weak) quadrupole lines the selection rules give $|\Delta\ell| = 0,2$ and $|\Delta j| = 0,1,2$. Furthermore, lines with the selection rules $|\Delta\ell| = 0$ and $|\Delta j| = 0,1$ (magnetic dipole lines) have been observed for some heavier elements. These selection rules give the scheme shown in Fig.2.3 for the X-ray transitions. The arrows denote electric dipole transitions.

Lines which are emitted from an atom with a vacancy in the K shell are assigned to the K series, similarly all lines which occur through filling of a vacancy in the L shell are assigned to the L series, etc. The component with shortest wavelength of each series, i.e., transitions from valence electrons, are broadened and to some extent significantly structured for bound atoms. These transitions are called emission bands.

Besides the diagram lines, other generally weak lines or satellites exist, which cannot be explained within the framework of a single-vacancy energy diagram. They are caused by transitions in multiply ionized atoms, by many-electron processes, by multiplet splitting of X-ray terms and modifications of the atomic energy levels due to chemical bonding.

The relation between electron configuration of an atom and X-ray emission and absorption spectra can be seen in Table 2.1 in conjunction with Fig.2.3. Table 2.1 gives the electron distribution in free atoms, i.e., undisturbed by chemical bonding. From this and the selection rules, it is possible to derive expected emission lines and absorption spectra for monoatomic gases. The profiles of X-ray emission bands of metals and compounds cannot be predicted in all cases from these electron distributions. Thus, the K emission bands of Li, Be and Na can be observed, whereas occupancy of the participating valence levels in free atoms begins only with boron and aluminum, respectively. The reason for this discrepancy will be treated in Sect.2.3.

The attenuation of an X-ray beam upon penetrating a layer of material is proportional to the path length. The intensity decreases according to $dI_x/I_x = -\mu dx$. The proportionality factor μ is the linear absorption coefficient and is dependent on the absorbing material and the X-ray wavelength. The mass-absorption coefficient μ_m is defined as $\mu_m = \mu/\delta$ (δ:

Table 2.1a. Electron distributions in free atoms of the elements 1 through 36

	Level							
Element	K 1s	L_I 2s	$L_{II,III}$ 2p	M_I 3s	$M_{II,III}$ 3p	$M_{IV,V}$ 3d	N_I 4s	$N_{II,III}$ 4p
1 H	1	–	–	–	–	–	–	–
2 He	2	–	–	–	–	–	–	–
3 Li	2	1	–	–	–	–	–	–
4 Be	2	2	–	–	–	–	–	–
5 B	2	2	1	–	–	–	–	–
6 C	2	2	2	–	–	–	–	–
7 N	2	2	3	–	–	–	–	–
8 O	2	2	4	–	–	–	–	–
9 F	2	2	5	–	–	–	–	–
10 Ne	2	2	6	–	–	–	–	–
11 Na				1	–	–	–	–
12 Mg				2	–	–	–	–
13 Al				2	1	–	–	–
14 Si		[Ne]		2	2	–	–	–
15 P				2	3	–	–	–
16 S				2	4	–	–	–
17 Cl				2	5	–	–	–
18 Ar				2	6	–	–	–
19 K						–	1	–
20 Ca						–	2	–
21 Sc						1	2	–
22 Ti						2	2	–
23 V						3	2	–
24 Cr						5	1	–
25 Mn						5	2	–
26 Fe						6	2	–
27 Co		[Ar]				7	2	–
28 Ni						8	2	–
29 Cu						10	1	–
30 Zn						10	2	–
31 Ga						10	2	1
32 Ge						10	2	2
33 As						10	2	3
34 Se						10	2	4
35 Br						10	2	5
36 Kr						10	2	6

Table 2.1b. Electron distributions in free atoms of the elements 37 through 71

Element		$N_{IV,V}$ 4d	$N_{VI,VII}$ 4f	O_I 5s	$O_{II,III}$ 5p	$O_{IV,V}$ 5d	$O_{VI,VII}$ 5f	P_I 6s
37 Rb		–	–	1	–	–	–	–
38 Sr		–	–	2	–	–	–	–
39 Y		1	–	2	–	–	–	–
40 Zr		2	–	2	–	–	–	–
41 Nb		4	–	1	–	–	–	–
42 Mo		5	–	1	–	–	–	–
43 Tc		6	–	1	–	–	–	–
44 Ru		7	–	1	–	–	–	–
45 Rh		8	–	1	–	–	–	–
46 Pd	[Kr]	10	–	–	–	–	–	–
47 Ag		10	–	1	–	–	–	–
48 Cd		10	–	2	–	–	–	–
49 In		10	–	2	1	–	–	–
50 Sn		10	–	2	2	–	–	–
51 Sb		10	–	2	3	–	–	–
52 Te		10	–	2	4	–	–	–
53 I		10	–	2	5	–	–	–
54 Xe		10	–	2	6	–	–	–
55 Cs		10	–	2	6			1
56 Ba		10	–	2	6	–	–	2
57 La		10	–	2	6	1	–	2
58 Ce		10	1	2	6	1	–	2
59 Pr		10	3	2	6	–	–	2
60 Nd		10	4	2	6	–	–	2
61 Pm		10	5	2	6	–	–	2
62 Sm		10	6	2	6	–	–	2
63 Eu	[Kr]	10	7	2	6	–	–	2
64 Gd		10	7	2	6	1	–	2
65 Tb		10	9	2	6	–	–	2
66 Dy		10	10	2	6	–	–	2
67 Ho		10	11	2	6	–	–	2
68 Er		10	12	2	6	–	–	2
69 Tm		10	13	2	6	–	–	2
70 Yb		10	14	2	6	–	–	2
71 Lu		10	14	2	6	1	–	2

density of the irradiated material), and yields the relative intensity decrease when multiplied by the absorber's mass per unit area. The relative change in intensity of an X-ray beam with 1 cm² cross-section, caused by one atom, is the atomic absorption coefficient calculated from $\mu_a =$

Table 2.1c. Electron distributions in free atoms of the elements 72 through 104

Element		$N_{IV,V}$	$N_{VI,VII}$	O_I	$O_{II,III}$	$O_{IV,V}$	$O_{VI,VII}$	P_I	$P_{II,III}$	$P_{IV,V}$	Q_I
		4d	4f	5s	5p	5d	5f	6s	6p	6d	7s
72 Hf		10	14	2	6	2	–	2	–	–	–
73 Ta		10	14	2	6	3	–	2	–	–	–
74 W		10	14	2	6	4	–	2	–	–	–
75 Re		10	14	2	6	5	–	2	–	–	–
76 Os		10	14	2	6	6	–	2	–	–	–
77 Ir		10	14	2	6	7	–	2	–	–	–
78 Pt		10	14	2	6	9	–	1	–	–	–
79 Au	[Kr]	10	14	2	6	10	–	1	–	–	–
80 Hg		10	14	2	6	10	–	2	–	–	–
81 Tl		10	14	2	6	10	–	2	1	–	–
82 Pb		10	14	2	6	10	–	2	2	–	–
83 Bi		10	14	2	6	10	–	2	3	–	–
84 Po		10	14	2	6	10	–	2	4	–	–
85 At		10	14	2	6	10	–	2	5	–	–
86 Rn		10	14	2	6	10	–	2	6	–	–
87 Fr		10	14	2	6	10	–	2	6	–	1
88 Ra		10	14	2	6	10	–	2	6	–	2
89 Ac		10	14	2	6	10	–	2	6	1	2
90 Th		10	14	2	6	10	–	2	6	2	2
91 Pa		10	14	2	6	10	2	2	6	1	2
92 U		10	14	2	6	10	3	2	6	1	2
93 Np		10	14	2	6	10	4	2	6	1	2
94 Pu		10	14	2	6	10	6	2	6	–	2
95 Am	[Kr]	10	14	2	6	10	7	2	6	–	2
96 Cm		10	14	2	6	10	7	2	6	1	2
97 Bk		10	14	2	6	10	9	2	6	–	2
98 Cf		10	14	2	6	10	10	2	6	–	2
99 Es		10	14	2	6	10	11	2	6	–	2
100 Fm		10	14	2	6	10	12	2	6	–	2
101 Md		10	14	2	6	10	13	2	6	–	2
102 No		10	14	2	6	10	14	2	6	–	2
103 Lr		10	14	2	6	10	14	2	6	1	2
104 Ku		10	14	2	6	10	14	2	6	2	2

$m_a \mu / \delta \cdot N_A$, which is identical to the atomic cross-section for absorption or scattering (m_a: relative atomic mass, N_A: Avogadro's constant).

On the assumption that absorption and scattering are independent, the absorption coefficient can be written as the sum of the coefficient for

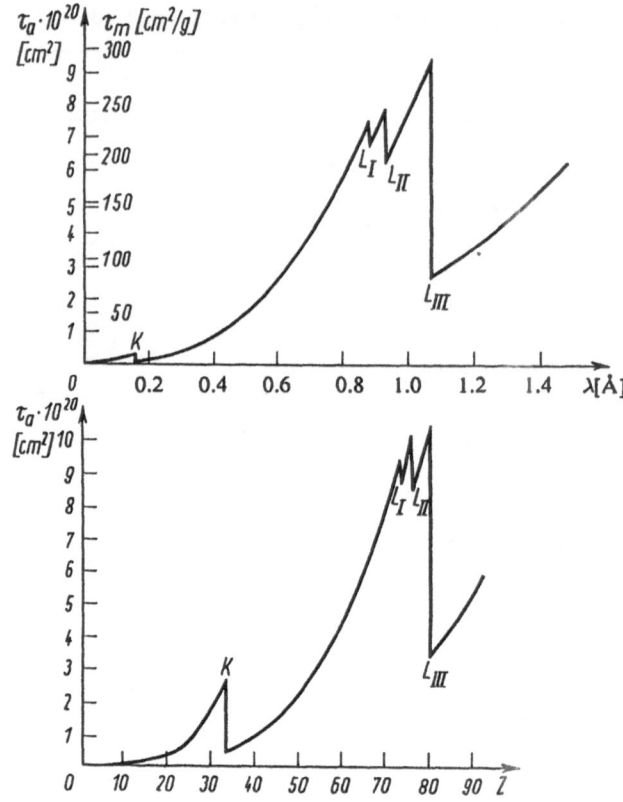

Fig.2.4. Wavelength dependence of the atomic absorption cross-section τ_a and mass absorption coefficient τ_m for platinum (top), and the dependence of τ_a at 0.1 nm on the atomic number of the absorbing element (below) [2.5]

true absorption τ and scattering σ, i.e., $\mu = \tau + \sigma$. True absorption increases rapidly with increasing wavelength and increasing atomic number so that the scattering contribution to the absorption of heavier elements ($Z > 24$) and longer wavelengths ($\lambda > 0.1$ nm) is so small that it can be disregarded.

The true absorption coefficient characterizes a process where the excitation of an atom following interaction with X rays results in the ejection of an electron from an inner shell. If the photon energy exceeds the binding energy E_i of the electron, then the excess energy appears as kinetic energy of the electron.

Photons with less energy than the energy of the corresponding X-ray state are not absorbed, so that the absorption starts at $E_i = h\nu$ in the form of an absorption edge. The energy of the edge corresponds to the energy of the corresponding X-ray level, so that (2.1) gives approximately a quadratic dependence of the edge energy on Z. The variation of the true absorption coefficient of Pt with wavelength is shown in the upper part of Fig.2.4 while the variation of the atomic cross-section with Z is indicated in the lower part.

If the spectra are recorded with instruments with higher resolution, absorption curves differing from those given in Fig.2.4 are obtained. Firstly, the absorption coefficient does not rise sharply at the edge but usually follows an arctan curve. Furthermore, on the long-wavelength side more or less pronounced structures can be observed, which often appear superimposed on the arctan function. Secondly, clear maxima and minima can be observed in the absorption coefficient on the short-wavelength side of the edge over a few hundreds eV. The near-edge fine structure, extending over a few eV on the absorption curve, is generally understandable in terms of transitions from inner states of the ejected electron during the absorption process into bound states, while the extended fine structure is, for the most part, interpretable by considering scattering of the outgoing electron in the potential field of the atom, molecule or crystal.

The differences between X-ray spectra from free atoms and atoms in molecules and crystals discussed below are specific to each observed spectrum. While only relatively small differences in wavelength, shift and shape occur for X-ray diagram lines, the changes in shape and energy of X-ray emission bands and absorption spectra are sometimes very strong. In the following sections of the present chapter several reasons for the effects caused by chemical bonding are discussed theoretically. The information revealed by X-ray spectra is clearly evident in the last chapters.

2.2 X-Ray Diagram Lines

X-ray diagram lines are emitted following transitions between terms of many-electron configurations which, in an elementary picture, can be described by the different electronic populations of the inner shells of an atom. Even if these shells do not directly participate in chemical bonding, the energy of the X-ray term is sensitive to the distribution of valence electrons between different atoms of a molecule, so that wavelengths and shapes of X-ray diagram lines from atoms in different surroundings differ to a varying extent. The influence of chemical binding on the $K\alpha$ doublet of the elements of the third period is evident from the shifts of the lines towards shorter wavelengths, arising when the positive charge of the atom under study increases [2.6-9]. Besides the oxidation state, the electronegativity of the surrounding atoms and the coordination [2.10] affect the wavelength and, in some instances, the shape of the (unresolved) $K\alpha_{1,2}$ doublet [2.9]. The influence of chemical bonding on the X-ray diagram lines of elements in the third period is mostly evident through line shifts of the X-ray transitions while the change of line shape (e.g., linewidth) is less pronounced. A complex dependence of the $K\alpha_{1,2}$ and $K\beta_{1,3}$ emission spectra on the bonding has been observed in compounds of the 3d elements [2.11-14]. Both long- and short-wavelength shifts of the line peaks occur and the above-mentioned relation between oxidation stage and wavelength is not evident. Besides a change of the doublet separation with bonding character (in particular, depending on the number of unpaired d electrons), a divergence of the shape of the X-ray line from the dispersion profile is observed as clear asymmetries occur (Table 2.2). The asymmetry index α (ratio between the long-wavelength to the short-wavelength part of the linewidth) of the $K\alpha_1$ line is usually larger than 1 and increases (linearly in a first approximation) with the number of unpaired 3d electrons. On the contrary, α for the $K\alpha_2$ line is often smaller than 1. The widths of the $K\alpha$ lines do not follow the Z^4 dependence (Fig.2.5) and increase with an increasing number of unpaired 3d electrons. A similar influence on the width of the $K\beta_{1,3}$ doublet has been noted for the 4d elements [2.16,17]. In these cases the absolute values of the linewidth changes are of the same order of magnitude as the line shifts.

Recently chemical effects have also been studied on $K\beta/K\alpha$ intensity ratios of transition-metal compounds [2.18-22].

The long-known relation for the 3d elements between the shape of the $K\alpha$ lines and the number of unpaired valence electrons [2.23-25] has led to a multiplet theory (see below) which explains the line shape satisfactorily. The validity of this theory has been suggested by several calculations [2.15,26-33]. An experimental demonstration favoring the multiplet-splitting picture was given, for example, by photoelectron spectroscopy of the M shells from Mn and Fe compounds [2.34]. Calculations of chemical effects, line shifts and line splittings are done in two ways. The more exact calculation proceeds by computing total energies for the emitting molecule in the initial i and final f states. A simpler approach describes the redistribution of electrons due to chemical binding as a change of the effective number of outer electrons or a change in number of unpaired electrons in a hypothetical isolated atom, i.e., the calculations are performed for free ions.

In the first case, the energy of an X-ray state E_i is given by the difference in the energy of a molecule between the ground state ($\mathscr{E}$) and an ionized state ($\mathscr{E}_i$) with a hole in an inner shell of the atoms

$$E_i = \mathscr{E}_i - \mathscr{E} . \tag{2.7}$$

The energies $\mathscr{E}$ and $\mathscr{E}_i$ for molecules with closed or open shells can be calculated using the Hartree-Fock-SCF-MO theory [2.35,36]. For systems with closed shells, in which all electrons are paired with those of opposite spin, the total energy can be approximated by the sum of one-electron energies ϵ_i^0 (the energy that each electron would have in the field of the existing nuclei), the Coulomb term J_{ij} (interaction between charges) for each pair of electrons, and the exchange energy K_{ij} for each electron pair which has the same spin

Table 2.2. Change of linewidth ΔE, line asymmetry $\Delta\alpha$ and doublet separation $E_{1,2}$ of the $K\alpha_{1,2}$ lines in compounds of 3d elements and their dependence on number of unpaired electrons n in free ions [2.15]

Element	Formal charge[1]	n	ΔE [eV]	$\Delta E/n$ [eV]	$\Delta\alpha$	$\Delta\alpha/n$	$E_{1,2}$ [eV]	$E_{1,2}/n$ [eV]
Ti	4+	0	0		0		0	
	3+	1	+0.38	+0.38	+0.19	+0.19	+0.25	+0.25
	2+	2	+0.19	+0.10	+0.37	+0.19	+0.43	+0.22
	Metal		-0.06		+0.24		+0.40	
V	5+	0	0		0			
	4+	1	+0.33	+0.33	+0.12	+0.12		
	3+	2	+0.57	+0.29	+0.23	+0.12		
	Metal		-0.11		+0.24			
Cr	6+	0	0		0		0	
	3+	3	+0.96	+0.32	+0.29	+0.10	+0.56	+0.19
	2+	4	+1.16	+0.29	+0.32	+0.08	+0.68	+0.17
	Metal		+0.09		+0.23		+0.32	
Mn	7+	0	0		0		0	
	6+(2+)	1	-0.09		+0.09	+0.09	+0.18	
	(3+)	2	+0.19	+0.10	+0.16	+0.08	+0.19	+0.10
	4+	3	+0.37	+0.12	+0.25	+0.08	+0.48	+0.16
	3+	4	+0.53	+0.13	+0.39	+0.10	+0.39	+0.10
	2+	5	+0.66	+0.13	+0.44	+0.09	+0.53	+0.11
	Metal		-0.16		+0.37		+0.14	
Fe	(2+)	0	0		0		0	
	(3+)	1	+0.31	+0.31	+0.04	+0.04	+0.20	+0.20
	6+	2	+0.87	+0.44	+0.30	+0.15		
	2+	4	+1.37	+0.34	+0.43	+0.11	+0.55	+0.14
	3+	5	+1.42	+0.28	+0.46	+0.09	+0.53	+0.11
	Metal		+0.71		+0.40		+0.20	
Co	(2+, 3+)	0	0		0		0	
	(2+)	1-2	+0.39	+0.26	+0.11	+0.07	+0.06	+0.04
	2+	3	+1.08	+0.36	+0.22	+0.07	+0.31	+0.10
	Metal		+0.13		+0.18		0.00	
Ni	(2+)	0	0		0		0	
	2+	2	+0.54	+0.27	+0.14	+0.07	+0.15	+0.08
	Metal		-0.06		-0.04		-0.01	

[1]) Values in parenthesis: Complex compounds

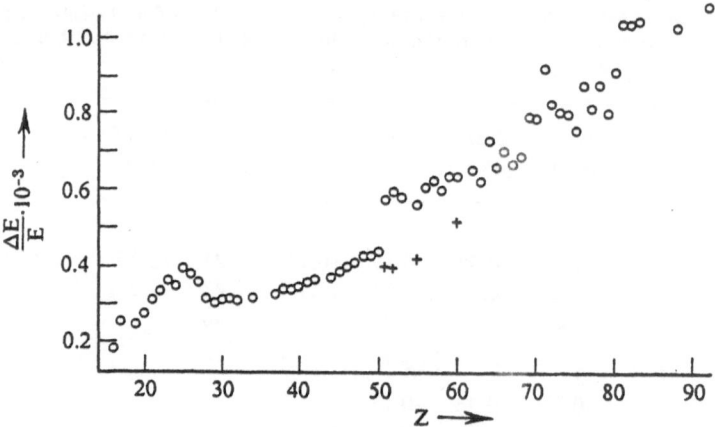

Fig.2.5. Relative $K\alpha_1$ linewidth as a function of atomic number [2.15]

$$\mathcal{E} = \langle \Psi | H | \Psi \rangle = \sum_i 2\epsilon_i^0 + \sum_{i,j} (2J_{ij} - K_{ij}), \tag{2.8}$$

where Ψ are antisymmetric, normalized, many-electron wave functions (Slater determinants) built up by the spin orbitals ϕ_i,

$$H = \sum_i \left[-\frac{1}{2} \nabla_i^2 \right] - \sum_{i\mu} \frac{Z_\mu}{r_{i\mu}} + \sum_{i<j} \frac{1}{r_{ij}} ,$$

$$\epsilon_i^0 = \langle \phi_i \left| -\frac{1}{2} \nabla_i^2 - \sum_\mu \frac{Z_\mu}{r_{i\mu}} \right| \phi_i \rangle ,$$

$$J_{ij} = \langle \phi_i \phi_j \left| \frac{1}{r_{12}} \right| \phi_i \phi_j \rangle ,$$

$$K_{ij} = \langle \phi_i \phi_j \left| \frac{1}{r_{12}} \right| \phi_j \phi_i \rangle .$$

The orbital energies ϵ_i^{SCF} can be calculated from the Hartree-Fock equations of the operator H^{SCF} as the eigenvalues of the spin orbitals ϕ_i

$$\epsilon_i^{SCF} = \langle \phi_i | H^{SCF} | \phi_i \rangle = \epsilon_i^0 + \sum_j (2J_{ij} - K_{ij}) . \tag{2.9}$$

Thus, following (2.8) the total energy is given by the sum of the orbital energies and the one-electron energies, viz.,

$$\mathcal{E} = \sum_i (\epsilon_i^0 + \epsilon_i^{SCF}) . \qquad (2.10)$$

Analogous calculations can be performed for molecules with open shells, where the total energy $\mathcal{E}_i$ and the orbital energy is obtained for ions with a hole in the shell i. The total energy E, of course, differs from the Hartree-Fock limit $\mathcal{E}^{HF}$ (the best solution of the Hartree-Fock equations corresponding to the energy obtained using an infinite basis set) due to relativistic and correlation effects, i.e.,

$$E = \mathcal{E}^{HF} + \mathcal{E}_{rel} + \mathcal{E}_{corr} . \qquad (2.11)$$

Nevertheless, the total energy obtained from ab-initio SCF calculations is rather better than might be expected, since in the subtraction of energies of initial and final states the influences of relativistic and correlation terms tend to cancel. Good agreement between theoretical and experimental data is to be expected only when differences due to the correlation energies are small. The influence of electron correlations can be calculated through configuration interaction (CI), i.e., through mixing of wave functions of different states.

If it is assumed that the orbitals change only slightly upon ejection of an electron from an atom (frozen-orbital approximation), then most of the terms in the equation are the same for $\mathcal{E}$ and $\mathcal{E}_i$. The energy of an ion thus differs from the energy of a neutral atom or molecule only by the value of the integrals which include the excited electron i. Accordingly,

$$- E_i \simeq \epsilon_i^0 + \sum_j (2J_{ij} - K_{ij}) = \epsilon_i . \qquad (2.12)$$

Corresponding to this equation, known as Koopmans' theorem, the orbital energies ϵ_i can be used to associate electronic configurations with experimental ionization potentials. It must be emphasized, however, that the usefulness of the theorem is strongly limited when a redistribution of the electrons follows electron excitation, i.e., a reorganization to the new potential (orbital relaxation). As is seen from Table 2.3, the values of ionization potentials, determined from Koopmans' theorem, are too high since the redistribution of electron orbitals following removal of an electron decreases the ionization energy.

To simplify the ΔSCF method of calculating ionization energies from (2.7) using separate calculations of the energies for the initial and final states, a so-called transition-state model has been developed which is

Table 2.3. Comparison of ionization energies (in atomic units) of neon, calculated using Koopmans' theorem (E^{Ko}), Slater's method of transition states (E^{TS}), the method of transition operator (E^{TO}), and the adiabatic or ΔSCF method ($E^{\Delta SCF}$) with experimental values [2.37]

	E^{Ko}	E^{TS}	E^{TO}	$E^{\Delta SCF}$	E^{exp}
E_{1s}	32.7725	31.9037	31.9239	31.9218	31.98
E_{2s}	1.9304	1.8084	1.8133	1.8127	1.78

characterized by noninteger occupation numbers in the participating orbitals [2.38]. The results of these calculations show that, in general, a relaxation error can be neglected (Table 2.3). The same result is obtained using a simultaneous optimization of initial and final states for a translation by minimizing the mean of the expectation value of the corresponding determinant. Through this method [2.37] a transition operator is introduced, which figuratively speaking, suggests that "half" of an electron is in one orbital and the other "half" is in another orbital. The ionization energies are estimated in this method by calculating eigenvalues of the transition operator:

$$- E_i = \epsilon_i^{TO} \; . \tag{2.13}$$

Comparison with ionization potentials calculated using this method with experimental values shows that only Koopmans' approximation introduces appreciable relaxation error (Table 2.3).

The energy of an X-ray transition E_{ij} is given from (2.7) by the difference between the total energies of two molecular ions

$$E_{if} = E_i - E_f = \mathscr{E}_i - \mathscr{E}_f \; , \tag{2.14}$$

or using (2.12 or 13) as the difference between two orbital energies

$$E_{if} = \epsilon_f - \epsilon_i \; . \tag{2.15}$$

Due to the complexity of these relations, energy calculations of X-ray terms and transitions have been performed using ab-initio SCF-MO method in the adiabatic approximation (ΔSCF method) during the last twenty years [2.39-44]. A considerable reduction in computational time is achieved if a free-ion model is used to calculate the position of center of gravity of the X-ray terms and the change in splittings within the X-ray multiplet for different compounds, when methods developed for optical spectra of atoms can be used [2.45, 46]. The average energy of the Coulomb and exchange interaction of pairs with equivalent electrons from (2.8) (using centro-symmetric atomic potentials) is given by

$$\left[\langle ij|\frac{1}{r_{12}}|ij\rangle - \langle ij|\frac{1}{r_{12}}|ji\rangle\right]_m$$

$$= F^0(\ell;\ell) - \frac{1}{4\ell+1}[c^2(\ell0;\ell0)F^2(\ell;\ell) + c^4(\ell0;\ell0)F^4(\ell;\ell) +...] \ . \tag{2.16}$$

The average interaction energy for nonequivalent electrons is given by

$$\left[\langle ij|\frac{1}{r_{12}}|ij\rangle - \langle ij|\frac{1}{r_{12}}|ji\rangle\right]_m$$

$$= F^0(\ell_1;\ell_2) - \frac{1}{2}[(2\ell_1+1)(2\ell_2+1)]^{-1/2}\sum_k c^k(\ell_1 0;\ell_2 0)G^k(\ell_1;\ell_2) \tag{2.17}$$

with

$$F^k(n_i\ell_i;n_j\ell_j) = R^k(ij;ij)$$

$$= \int_0^\infty\int_0^\infty R^*_{n_i\ell_i}(r_1)R^*_{n_j\ell_j}(r_2)R_{n_i\ell_i}(r_1)R_{n_j\ell_j}(r_2)\frac{2r^k(a)}{r^{k+1}(b)} r_1^2 r_2^2 dr_1 dr_2$$

and

$$G^k(n_i\ell_i;n_j\ell_j) = R^k(ij;ji) \ .$$

The total average energy of the electron configuration is then given by a relation similar to (2.8) with analogous dependences, as given in (2.16, 17) for the energy of the electron-electron interaction. X-ray emission lines, in most cases, do not appear as single, resolved maxima, rather they are produced from multiplet-multiplet transitions where spin-orbit splittings arising mainly from electrostatic exchange energies calculated from (2.16, 17) can be used to determine the center of gravity of the X-ray lines. The relative positions of the center of gravity of the multiplet of the energy levels of atoms in compounds can be approximated by the differences of the average energies of the ions whose charge resembles the real configuration of the compound. Calculation of the theoretical line shift $\delta E_{K\alpha}$ from the energies of the X-ray states [2.47-50] proceeds from either the adiabatic approximation (2.14)

$$\delta E_{K\alpha} = (\mathcal{E}_{1s} - \mathcal{E}_{2p})_{ion} - (\mathcal{E}_{1s} - \mathcal{E}_{2p})_{atom} \tag{2.18}$$

or from the orbital energies [2.51-54] using

23

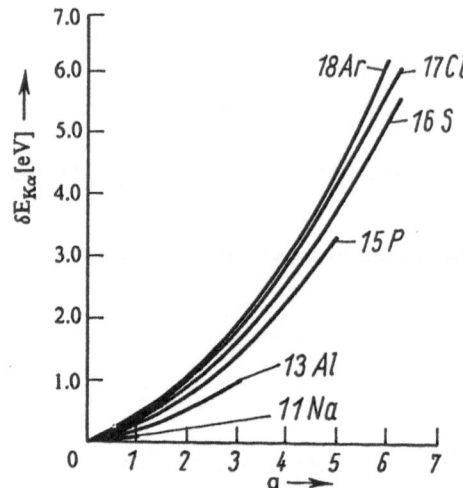

Fig.2.6. Kα line shifts for elements of the third period as a function of atomic charge [2.55]

$$\delta E_{K\alpha} = (\epsilon_{2p} - \epsilon_{1s})_{ion} - (\epsilon_{2p} - \epsilon_{1s})_{atom} \tag{2.19}$$

together with Koopmans' theorem (2.15). In either case, effective atomic charges can be determined from a comparison of the calculated and experimentally obtained shifts. Figure 2.6 shows line shifts $\delta E_{K\alpha}$ obtained theoretically as described above for elements of the third period and their dependence on atomic charges.

Line-shift calculations can be simplified if it is assumed that the wave functions in the integrals (2.16, 17) do not change when an atom changes to an ion that emits a valence electron. Then the ionization energies E_i of the atom and ion with an ejected $n\ell$ electron differ only through the missing Coulomb and exchange integrals so that the shift of the Kα lines can be estimated using wave functions of the neutral atoms [2.26]

$$\delta E_{K\alpha} = [F^0(1sn\ell) - G^k(1sn\ell)] - [F^0(2pn\ell) - G^k(2pn\ell)] . \tag{2.20}$$

Whilst the results thus obtained agree well with experimental values in some cases, they give only order-of-magnitude agreement if a further (attractive) simplification of the model calculation is introduced. In this model the level shift δE_i associated with a replacement of an atom by an ion is associated with a change of screening due to the missing valence electron, i.e., the difference in the energy levels E_i of the atoms of the atomic numbers Z and Z+1 is multiplied by part of the electron density from the valence electron in the vicinity of the i shell

$$\delta E_i = [E_i(Z+1) - E_i(Z)]\sigma_{in\ell} , \qquad \text{with}$$

$$\sigma_{in\ell} = \int_0^{r_i(max)} R_{n\ell}{}^2(r) \, dr . \qquad (2.21)$$

The partial screening $\sigma_{in\ell}$ can be calculated using hydrogenic wave functions [2.56]. No satisfactory agreement with experimental shifts in the region of the third period could be obtained with this method even when Hartree-Fock wave functions were used [2.57].

As mentioned at the beginning of this section, changes in line width and shape as well as shifts of X-ray lines occur in response to changes in the environment of the emitting atom; such changes can sometimes result in noticeable splitting of the lines. These effects can be understood in terms of multiplet splitting of the levels caused by spin-orbit and electrostatic exchange of the inner electrons with the unpaired outer electrons.

The splitting appears even in the simplest case, i.e., for a closed-shell ion (naturally excepting the inner shell with the electron hole), as a doublet splitting with the spin-orbit interaction parameter ξ given by

$$E = - (\xi/2)[J(J+1) - L(L+1) - S(S+1)] \qquad (2.22)$$

(E: term position relative to the center of gravity of the multiplet energy).

If more electrons are missing in the ion, additional splittings occur, which can be accurately calculated using ab-initio SCF-MO or solid-state calculations only. However, as for line shifts, it is also possible here to predict the multiplet splitting using a free-ion model, whereby calculations using both ΔSCF methods [2.34,58] and a Koopmans' theorem approach are possible. Several calculations have been performed which, for strong spin-orbit interaction ($\xi \gg a$, a being the exchange interaction parameter), use of j-j coupling with the electrostatic interaction as perturbation [2.15,30], whilst at stronger electrostatic interaction, the spin-orbit interaction can be taken as a perturbation using an LS coupling as the basis [2.27-29,32,33]. Intermediate coupling is needed for $\xi \simeq a$ [2.26].

The limit of the j-j coupling is generally realized when the interacting electrons belong to different shells, so that, for example, the multiplet structure of the $K\alpha_{1,2}$ doublet of the 3d transition elements, which arises from transitions $1s3d^n \rightarrow 2p^5 3d^n$ (only the unfilled shells are given), is readily understood in this representation. In this case, calculation of the energy term of the electron configuration $2p^5 3d^n$ is sufficient as the splitting of the K level due to the small exchange interaction between the 1s and 3d electrons can be neglected ($\delta E \simeq 0.04$ eV) [2.27,28]. Because of spin-orbit interaction the $2p^5$ configuration produces the states $^2P_{3/2}$ and $^2P_{1/2}$ with a distance of $(3/2)\xi_{2p}$, see (2.22), neglecting interaction with unpaired electrons. If this interaction is included, then the $L_{II,III}$ terms are split and the magnitude of the splitting can be obtained from

$$E_{L_{III}} = - (1/3)a[J(J+1) - S(S+1) - j(j+1)] ,$$

$$E_{L_{II}} = + (1/3)a[J(J+1) - S(S+1) - j(j+1)] \qquad (2.23)$$

with $J = j + S$; $\quad j_{L_{III}} = 3/2$; $\quad j_{L_{II}} = 1/2$; $\quad S = n/2$,

and with the exchange parameter

$$a = \alpha^2 \left[\frac{2}{15}G^1(2p3d) + \frac{3}{35}G^3(2p3d) \right].$$

The field of the ligands can be included using the factor α^2 (n: number of unpaired 3d electrons) [2.26].

The general features of the $K\alpha_{1,2}$ doublet from 3d transition elements (see below) are well represented by (2.23). A further improvement in the agreement of the splitting can be obtained using intermediate coupling.

The LS coupling limit is usually invoked when the interacting electrons occupy the same shell; the multiplet structure of the $K\beta$ ($1s3d^n \rightarrow 3p^5 3d^n$) lines of the 3d transition elements can be explained [2.27-30].

With the above methods, several observed phenomena can be understood. Thus, for example, the asymmetry index of the $K\alpha_1$ line of the 3d elements is larger than one and increases with the number of unpaired 3d electrons since the sublevel with the largest J value (proportional to the number of unpaired 3d electrons) is strongly shifted towards the short-wavelength side. (The contribution of the individual multiplet components to the intensity of the line follows from the statistical weight $g = 2J+1$). For the $K\alpha_2$ line the arrangement of the components is reversed, with the result that a shift of the component with smallest J value towards the short-wavelength side gives an asymmetry index smaller than (or close to) one. Corresponding to the shift of the multiplet components the maximum of the $K\alpha_1$ line shifts towards shorter wavelengths and the maximum of the $K\alpha_2$ towards longer wavelengths. Thus the increase in doublet separation is proportional to the increase of the number of unpaired 3d electrons.

Asymmetry of the lines corresponding to the splitting of the $L_{II,III}$ levels also leads to increase in width when the number of unpaired electrons increases in the bonding. A summary of these effects is given in Table 2.2. The multiplet structure of the $K\beta$ lines is exemplified in Fig. 2.7 by the Fe $K\beta_1\beta'$ lines. In agreement with theoretical predictions [2.27, 28, 32, 33], two lines occur where the shorter-wavelength one is the more intensive. With a theoretical intensity ratio $I_{K\beta'}/I_{K\beta_1} = n/(n+2)$ for $K\beta_1$ (level ^{n+2}P) and $K\beta'$ (level ^nP), the increase in intensity of the long-wavelength component can be explained by an increase in number of unpaired 3d electrons n.

Limitations of the free-ion model are evident from discrepancies in predicting unperturbed line widths and departures from dispersion profiles

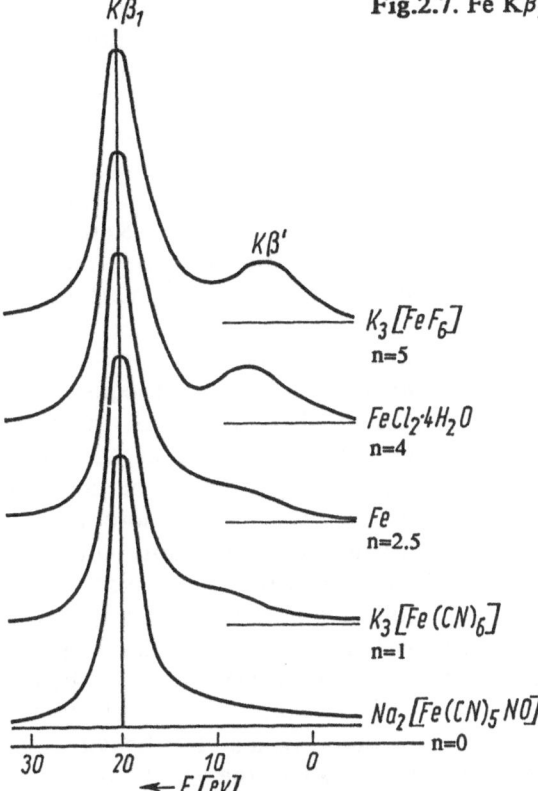

Fig.2.7. Fe $K\beta_1\beta'$ lines in various compounds [2.59]

by lines from diamagnetic compounds, which cannot be explained by interactions with unpaired d electrons. Evidently more factors must be included in the theory, factors which influence the emission lines (strong covalent bonding, 3d electron pairs). The multiplet model described here should, in favorable cases, correspond to conditions for compounds with mostly ionic bonding character.

2.3 X-Ray Emission Bands

Investigations of X-ray emission bands were started in the 1930s by *Siegbahn* [2.60], *Skinner* [2.61], and *Farineau* [2.62]. In recent years, due to improved instrumentation, these have been applied to a large number of metals, alloys, semiconductors, insulators, and molecular gases, so that numerous spectra (Chaps.5 and 6) are available with which one might hope to test the validity of molecular and solid-state theories.

In contrast with the deeper-inner-shell transitions which reflect changes of the surroundings of the atom through shifts and (mostly small) changes in shape, the X-ray emission bands from an element and its various compounds often show totally different intensity distributions. The energy levels of the valence states in the material under study must be determined, and attention paid to the vacancy production mechanism to explain the intensity profiles of X-ray emission bands. This can be done with the electron theory for solids developed by A. Sommerfeld and F. Bloch or with the MO theory developed by F. Hund and R.S. Mulliken for molecular gases and isolated groups of molecules in solids. Relativistic effects in the region of valence electrons corresponding to (2.11) must be included when the emission bands of compounds with heavy elements (Z > 50) are calculated [2.63].

In contrast to isolated atoms, the motion of valence electrons in molecules and crystals is perturbed by the surrounding as a consequence of comparable orbital dimensions and interatomic distances. Thus, in molecules the degeneracy of the valence states vanishes because of additional interaction terms in the Hamilton operator. An MO-level diagram (Fig.2.8) illustrates molecules obtained in this way with more or less densely packed energy levels. For solid targets which in reality contain a great number of valence electrons, the discrete energy levels are so closely packed that we can talk about a nearly continuous band of energy levels (Fig.2.9). The energy width of the band is determined from the coupling. Since the wave functions of the different deep inner electrons hardly overlap, the energy width of an inner level in a solid of a light element is very small compared to the width of the outer bands (for heavier elements the widths of the inner levels are large due to the small lifetime of the corresponding states).

Electronic states in a solid are determined, after calculating the energy levels of the electrons moving in a crystal with fixed nuclei, according to the dispersion relation, i.e., the dependence of the energy on the wave vector k of the electrons. The energy values obtained form bands which are separated from each other by forbidden zones.

Several approximate methods (e.g., the OPW, APW, pseudopotential, kp, and cellular methods) have been developed to perform the relatively complicated calculations in crystals [2.64,65], which are all based on the Bloch wave function for an electron in a periodic potential

$$\Psi_k(r) = u_k(r) e^{ikr} . \tag{2.24}$$

The periodic functions $u_k(r)$ depend on the propagation directions of the electrons and on the periodicity of the lattice. If it is assumed that the modulating function is constant, then $\Psi_k(r)$ describe plane waves (Sommerfeld's model for free electrons) and the energy is a quadratic function of k. If the periodic functions $u_k(r)$ are included, discontinuities in the energies at certain critical values of k are produced, which lie on the surface of the Brillouin zone in k space. The level density increases in the vicinity of these discontinuities. If one collects all k values whose Bloch

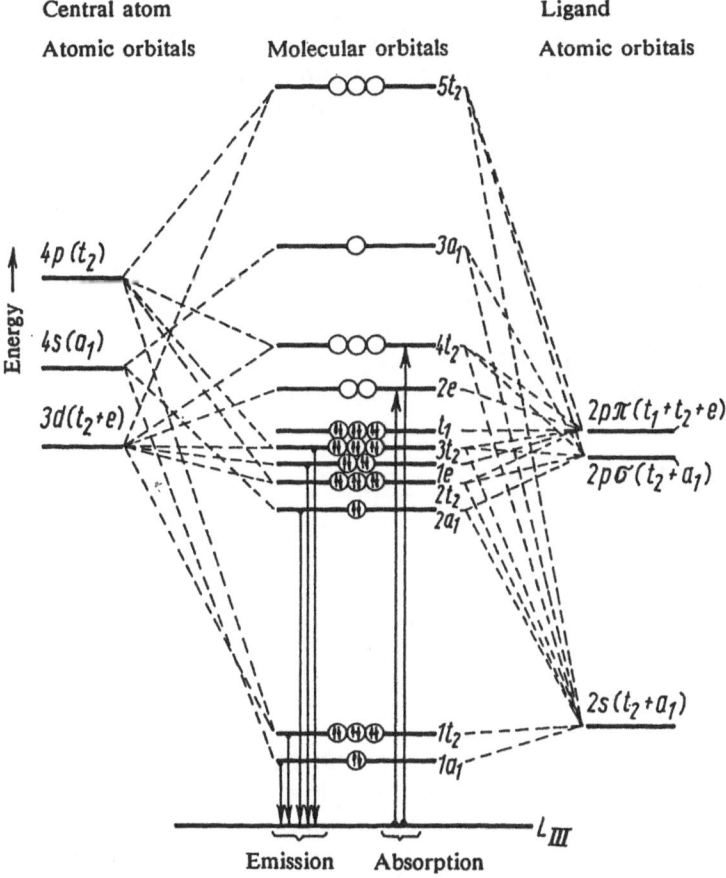

Central atom
Atomic orbitals

Molecular orbitals

Ligand
Atomic orbitals

Emission   Absorption

Fig.2.8. MO energy level diagram for tetrahedrally coordinated 3d metal atoms

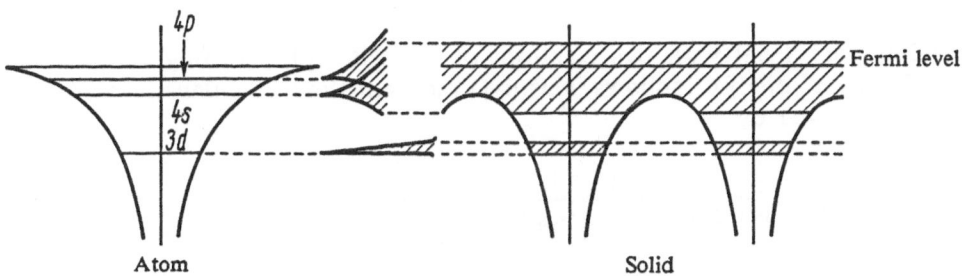

Fig.2.9. Schematic band diagram for metals belonging to the fourth period

functions have a specific energy, an isoenergetic surface results. It is called a Fermi surface if the specific energy corresponds to the Fermi energy. The Fermi energy is that energy which at absolute zero has the highest occupied states. As the numbers of the k values in a given energy region are different, the density of states $N(E)dE$ is introduced to de-

29

scribe the level density which corresponds to the number of levels in the unit volume of the solid in the energy region E and E+dE. The density of states in k space can be determined from a summation of all points in k space between the surface E and E+dE, i.e., through integration in the space between the isoenergetic surfaces:

$$N(E)dE \propto \int d^3k .$$ (2.25)

The distance between the surfaces E and E+dE in k space, measured perpendicular to the surfaces, i.e., in direction of the vector $\nabla_k E$, is given by dk = $1/|\nabla_k E|$dE. If the surface element d^2k is replaced by dS, then the density of states is obtained [after integration over the isoenergetic surface S(E) and after summation over all bands j] from

$$N(E) = \frac{V}{8\pi^3} \sum_j \int_{S(E)} \frac{dS}{|\nabla_k E_j|} ,$$ (2.26)

$$N(E) = \frac{V}{8\pi^3} \sum_j \int_{Bz} \delta(E_k^j - E)d^3k .$$ (2.27)

The second equation describes the summation over all states with energy E (δ: Dirac delta function) by integration over the volume of the Brillouin zone ($V/8\pi^3$: number of states within the unit volume).

Using the quadratic dependence of k on energy for free electrons, the surfaces of constant energy are spherical shells with radii k and k+dk. The density of states can then be written as

$$N(E) \propto \sqrt{E} .$$ (2.28)

Filling each level with two electrons of opposite spin, starting at the level with lowest energy, follows Fermi-Dirac statistics. The density of filled states is given by

$$N_e(E) = \frac{N(E)}{1 + \exp[(E-E_F)/k_B T]} ,$$ (2.29)

where k_B is the Boltzmann constant. At absolute zero, the occupancy ends abruptly at the Fermi level E_F (Fig.2.10).

The intensity of an emitted X-ray band produced by atomic transitions in a solid from an initial state with a vacancy in an inner shell, described by $\phi_{n\ell}$, to a valence state ϕ_k is calculated by a summation over all states and multiplying with a k-dependent transition probability P(E, k):

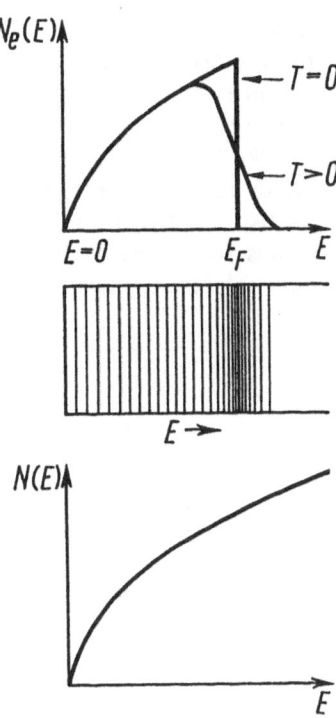

Fig.2.10. Density of states N(E) and density of occupied states $N_e(E)$ of free electrons

$$I_{ik}(E) \propto \nu^3 \sum_j \int_{S(E)} \frac{P(E,k)}{|\nabla_k E_j|} \, dS \ . \tag{2.30}$$

Calculating the intensity of X-ray transitions using (2.30) requires one-electron states which satisfy a Schrödinger equation. There are two alternatives for the assumed potential. For an X-ray emission process, there is firstly a final-state potential obtained by neglecting the core hole, and secondly, an initial-state potential in which the presence of a completely screened core hole is assumed.

Static single-particle calculations of the intensity distribution for the initial- and final-state potentials of the emission process, i.e., using single-particle wave functions calculated with and without a core hole, give rather different results for the Na $L_{II,III}$ spectra, as an example. If these are compared with experimental spectra there is obviously a better agreement for the final-state calculation [2.66]. A dynamical calculation including interaction between valence electrons and the core vacancy, and also allowing for the fact that the core hole is switched on or off during the X-ray emission process [2.67], shows a shape similar to that obtained from a static final-state calculation [2.66]. If, as in the case of the $L_{II,III}$ absorption spectra, the static final-state spectra, i.e., in this case the result of a calculation using single-particle wave functions including the core

31

vacancy, agree with the predictions from a dynamical calculation, then the following "final-state rule" for X-ray spectra of metals can be given: Single-particle calculations of emission and absorption spectra are valid if single-particle wave functions of the final states are used [2.66].

If it is assumed that the matrix element $\langle \phi_{n\ell} | r | \Psi_{kj} \rangle$ is not dependent on k, then $I(E) \propto \nu^3 P(E) N(E)$ gives an approximate relation for the intensity, which is useful for qualitative comparisons, as different wave functions Ψ_k and $\Psi_{k'}$ can give the same energy values for states with different symmetries. A series expansion of the valence-electron wave function Ψ_k in the form

$$\Psi_{kj}(r) = \sum_{\ell m} C_{\ell m}(jk) \, R_\ell(E,r) \, Y_{\ell m}(\Theta, \Phi) \tag{2.31}$$

makes it possible to describe the total density of states N(E) as composed of partial density of states $N_\ell(E)$ which expresses the number of states of a certain symmetry, i.e.,

$$N_\ell(E) = \sum_j \int_{S(E)} \frac{|C_\ell(jk)|^2}{|\nabla_k E_j|} \, dS \, . \tag{2.32}$$

$C_\ell(jk)$ describes the part of the functions with ℓ symmetry in the state Ψ_k. The intensity of the emission band is then given as a sum of the partial density of states multiplied with the corresponding transition probability

$$I_\ell(E) \propto \nu^3 \left[a N_{\ell-1}(E) \, P_{\ell,\ell-1}(E) + b N_{\ell+1}(E) \, P_{\ell,\ell+1}(E) \right] \, . \tag{2.33}$$

The index ℓ refers to the symmetry of the initial state. Due to the dipole selection rule only final states with $\ell \pm 1$ need to be considered. The expression

$$P_{i\ell}(E) = \left| \int_0^\infty R_i(E,r) \, R_\ell(r) \, r \, dr \right|^2 \tag{2.34}$$

gives the radial factor, a and b being the angular parts of the transition probability. If one assumes that the intensity distributions of the K and L emission bands are given by

$$I_K(E) \propto \nu^3 \, N_p(E) \, P_{sp}(E) \, ,$$

$$I_L(E) \propto \nu^3 \left[N_s(E)\, P_{ps}(E) + N_d(E)\, P_{pd}(E) \right] \qquad (2.35)$$

(the contribution of s → p transitions in the L emission bands is often neglected in the transition elements), then a complete picture of the density of states of occupied bands in solids can be obtained from an investigation of emission bands with initial states of different symmetries.

When the lattice points in a solid are occupied by different atoms, then the symmetry of valence electron functions Ψ_k for different kinds of atoms can be different. This can be taken into account by determining the local partial density of states which is obtained from an integration of the wave function in a part of space surrounding the lattice points, for example within the APW spheres. For a binary compound such as TiNi, for example, the total density of states is given by the sum of the local partial density of states in the Ti and Ni spheres, and the contributions of plane waves between the spheres (Fig.2.11).

Compared with the more exact equation, (2.30), applications of the simplified equations, (2.33,35), yield satisfying results for several cases. Neglecting the k dependence of the transition probability which can be deducted from a k-dependent radial part of the wave function (2.31) always offers at least one possible reason for discrepancies in comparisons between experimental curves and those calculated from (2.33).

Molecular gases and solids whose structural constituents can be considered as isolated groups, e.g. complexes, emit X-ray emission bands

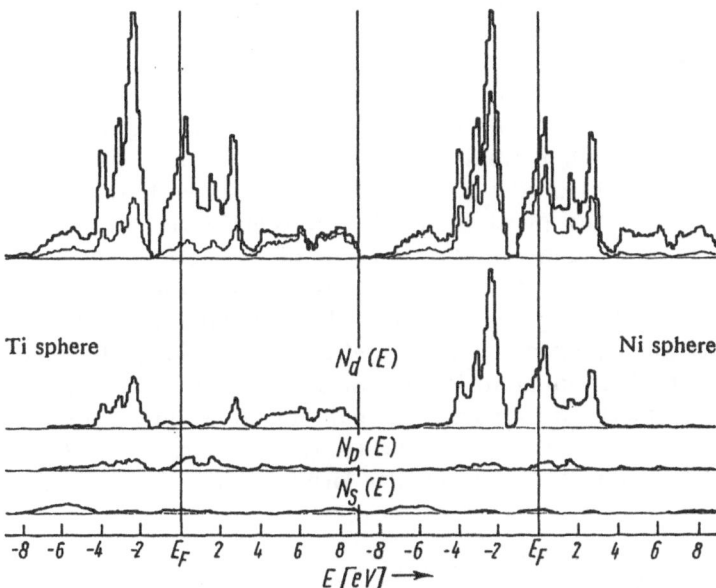

Fig.2.11. Total density of states (top) and local partial density of states (below) in Ni and Ti spheres of the intermetallic compound TiNi [2.68]

whose energy and intensity distributions can be interpreted using the MO theory. In contrast with the corresponding spectra of free atoms, X-ray emission bands of molecular gases consist of several bands which can be distinguished as separate maxima caused by the large splitting of the MO levels. The energies of the individual transitions from a singly ionized state with a vacancy in an inner shell to a final state with a vacancy in one of the available molecular orbitals differ by the possible values of the energy differences of the final states. Thus, the measured energy values of the X-ray transitions give the relative positions of the molecular levels (Chap. 5).

Since interpretation of X-ray emission bands is often complicated by superposition of satellites produced by transitions between multiply ionized states, calculated ionization potentials (binding energies) must often be compared to identify individual maxima. The difference between two ionization energies gives the energy difference corresponding to the X-ray transition.

Production of inner or outer vacancies in molecules leaves ions which differ from the ground state by modified equilibrium distances and bonding angles. Therefore, vibrational structures can be observed in X-ray spectra, which are useful for studying the energy surfaces of molecules with a core-hole state. Such experimental [2.69] and also theoretical studies [2.70-72] have been presented for small molecules. Using a Franck-Condon analysis, the equilibrium geometry and force constants for the core-hole state could be deduced from the resolved vibrational structures.

Extension of the method to larger molecules is difficult due to overlapping vibrational progressions and the relatively complicated energy surfaces. Nevertheless, good results were obtained, for example, for the NH_3 molecule using a combination of ab-initio results for the force constant and equilibrium geometry, and experimental results [2.70].

The observation of a vibrational structure with splitting of some tenths of eV requires a small natural width of the levels participating in the transition. As the widths of the K levels for elements with $Z > 10$ do not fulfill this condition, the observation of vibrational structure is easy only for K spectra from elements of the first row of the periodic table.

The intensity distribution of X-ray emission bands from molecular gases is determined from energy separations, the relative intensities and widths of different superposed components, which are produced by electronic transitions from individual molecular orbitals to an inner vacancy of an atom. The intensity of a particular component is given by the matrix element of the transition probability (2.5). Using the dipole approximation and expanding the molecular orbitals into their components, viz.,

$$\Psi = \sum_{A,\ell} C_{A\ell}\, \phi_{n\ell}(A) , \qquad (2.36)$$

one obtains the following terms

$$\langle \Psi_f |\mathbf{r}| \Psi_i \rangle = \sum_{AA'\ell\ell'} C_{f,A\ell} \langle \phi_f^{n\ell}(A) |\mathbf{r}| \phi_i^{n'\ell'}(A') \rangle C_{i,A'\ell'} . \tag{2.37}$$

If the origin of the coordinate system is taken to coincide with the position of the atom A' with the inner vacancy and the wave function of the initial state is taken as an atomic function ($\Psi_i \simeq \phi_i$), then calculation using (2.37) gives

$$\langle \Psi_f |\mathbf{r}| \phi_i \rangle \simeq \sum_{A\ell} C_{f,A\ell} \langle \phi_f^{n\ell}(A) |z_{A'}| \phi_i(A') \rangle . \tag{2.38}$$

Especially suited to the calculation of the dipole matrix obtained from (2.38) is a discrete numerical integration scheme in which all integrals are relatively easy to calculate.

Contributions to the matrix element and thereby to the intensity of a component in an emission spectrum arise from the corresponding structures of individual terms in (2.38) through hybridization and delocalization ($A=A'$, $\ell=\ell'\pm1$) and through crossing ($A\neq A'$). Terms $A = A'$ and $\ell = \ell', \ell'\pm2,...$ are equal to zero following the dipole selection rule. The importance of individual contributions is seen, for example, from a calculation for the structural unit SiO (Fig.2.12). Contrary to the more usual assumptions, it is found that the matrix elements of the transitions $5\sigma \rightarrow \sigma$ (Si 1s) and $5\sigma \rightarrow 2\sigma$ (O 1s) are mostly determined by delocalization [3p(Si) $\rightarrow$ 1s(Si)] and/or hybridization [2p(O) $\rightarrow$ 1s(O)], whilst the cross term $\langle \phi_{Si} |z| \phi_O \rangle$ contributes much less.

This result for SiO cannot be generalized into a rule since, for example, results from discrete variational Xα calculations (DV-Xα) of the X-ray emission bands for the anions XO_4^{n-}, XO_3^{n-}, and XO_2^{n-} (X = P, S, and Cl) have shown that an error of 50% in the relative intensities of individual components can occur if the cross transitions are ignored [2.73]. If the cross term is completely neglected, then for the intensity of a component of the K emission band (2.6) gives

$$I_1 \propto \nu^4 C_{1,A'p}^2 \left[\langle \phi_1^{np}(A') |z_{A'}| \phi^{1s}(A') \rangle \right]^2 . \tag{2.39}$$

The intensity I_2 of a transition with the final state Ψ_2 is given by this one-center model in an analogous way. Thus, the relative intensities of the components of an X-ray emission band in a molecule are determined from the relation of the LCAO coefficients (i.e., their corresponding squares) when the atomic orbitals in (2.36) for all molecular orbitals Ψ_f can be considered equal, i.e.

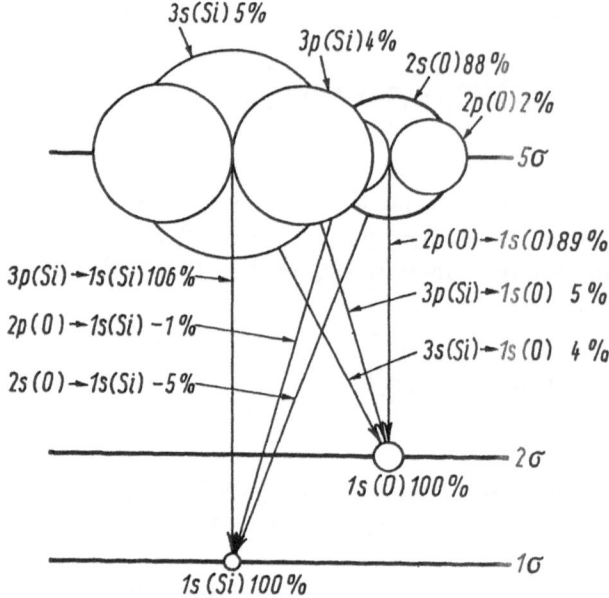

Fig.2.12. Components of the transitions $5\sigma \rightarrow 1\sigma$ (part of the Si Kβ emission band) and $5\sigma \rightarrow 2\sigma$ (part of the O K emission band). Numbers above the 5σ level give the relative atomic populations [2.42]

$$I_1/I_2 = C_1{}^2/C_2{}^2 . \tag{2.40}$$

Different degrees of degeneracy of the levels 1 and 2 determine the corresponding coefficients. Superposition of the calculated components and the addition of assumed Lorentzian profiles leads to theoretical emission bands which, for some cases, are in excellent agreement with those determined experimentally. If the above-described one-center model is applied to calculate the L emission bands, components occur both from s $\rightarrow$ p and d $\rightarrow$ p transitions according to the dipole selection rules. In particular, when d orbitals are important, agreement is often less satisfactory [2.74]. An extensive theoretical analysis of the S L$_{\text{II,III}}$ emission spectra of SF$_6$ shows that improved results are obtained when the transition moments are calculated including both the d orbitals and two-center terms [2.72].

Earlier studies [2.42] have shown that an important assumption for the applicability of (2.40), namely, same shape of the atomic orbitals $\phi_f{}^{n\ell}$ in the MO's, need not be fulfilled, so that all predictions obtained in this way may be dubious. Figure 2.13 shows the change of the Si 3p atomic orbitals in different molecular orbitals of the system SiO compared to the shape of the 3p orbitals of the free atom. The contraction seen in this figure results in a differential enhancement of the amplitudes in the region of the K and L shells by a factor of three. The importance of this phenomenon for the dipole matrix element of the transition $\Psi_f \rightarrow \phi_{1s}$ can

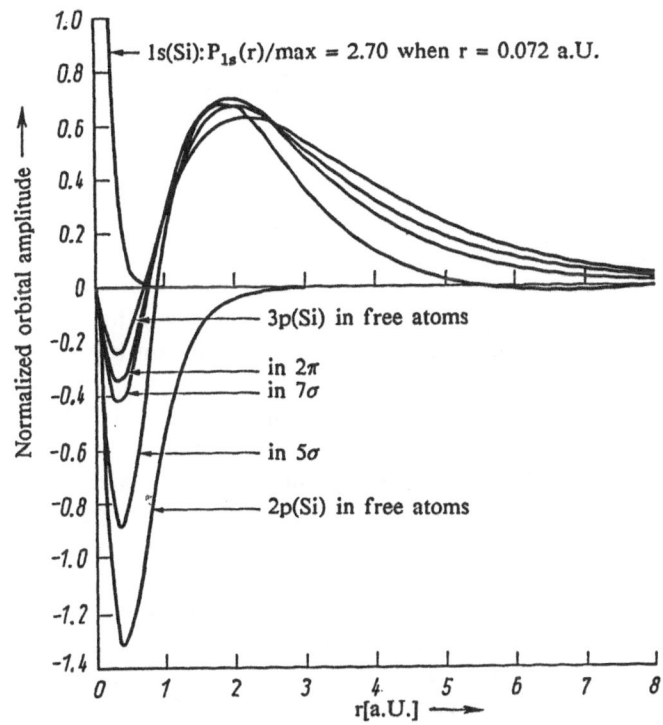

 shows a graph with the y-axis labeled "Normalized orbital amplitude" ranging from -1.4 to 1.0, and the x-axis labeled "r[a.U.]" ranging from 0 to 8. The following labels appear on the graph:

- 1s(Si): $P_{1s}(r)/\mathrm{max} = 2.70$ when $r = 0.072$ a.U.
- 3p(Si) in free atoms
- in 2π
- in 7σ
- in 5σ
- 2p(Si) in free atoms

Fig.2.13. Modified Si 3p atomic orbitals in various molecular orbitals of SiO [2.42]

be seen from Table 2.4, where the dipole matrix element of the modified Si 3p orbital is compared to the matrix element of the unmodified orbital.

Better results can be obtained from (2.39) if the approximation of (2.40) is abandoned and the matrix elements are included in the calculation using the scattered wave $X\alpha$ method in the muffin-tin approximation [2.75] for calculating molecules. In this case the wave functions of individual spherical atomic shells of a molecule are described by a partial-wave expansion. Because of the small extension of the core-wave function, it is sufficient to calculate the matrix element within the atomic sphere of in-

Table 2.4. Comparison of a few dipole matrix elements with modified and unmodified atomic orbitals (in atomic units) [2.42]

	$\langle \phi_r^{n\ell}(\mathrm{Si})|z|\phi^{1s}(\mathrm{Si})\rangle$				
$\phi_r^{n\ell}$:	ϕ^{2p}	ϕ^{3p}	$\phi_{5\sigma}^{3p}$	$\phi_{2\pi}^{3p_x}$	$\phi_{7\sigma}^{3p}$
	0.0429	−0.0081	−0.0293	−0.0115	−0.0139

terest using a relation analogous to (2.39), whereby the cross-transition terms disappear [2.63]. The muffin-tin approximation and the partial-wave expansion lead to a partition of the valence electrons which can be described by

$$n_f = q_{in}{}^f + q_{out}{}^f + \sum_{A=1}^{N} \sum_{\ell} q_{A\ell}{}^f \, ,$$

where n_f is the degeneracy of state f, $q_{in}{}^f$ the charge in the interatomic region, $q_{out}{}^f$ is the charge outside the outer sphere, N is the number of a-tomic centers in the molecule, and $q_{A\ell}{}^f$ is the ℓ-like charge in the A sphere. The magnitude of the charge q_A is useful for qualitatively inter-preting the emission bands. It reproduces the intensity distribution satis-factorily in many cases [2.76-79]. The relatively cumbersome calculation of the intensity distribution of X-ray emission bands of solids, especially for complicated crystal structures, can be avoided by considering a small region or cluster. Since X-ray spectra represent the local density of states near an atom, they can be accurately described by the electronic structure of the cluster with discrete energy levels. It is thus possible to obtain very good theoretical emission bands using the well-known LCAO, SW Xα, and other calculations, as may be inferred from (2.38-40) [2.80,81].

A number of peculiarities found in X-ray emission bands, e.g., long-wavelength tails and most satellites, cannot be explained within the one-electron model discussed above since it does not consider collective elec-tron motions occurring with the emission process. Some of these many-electron effects are discussed in Chap.6.

2.4 X-Ray Satellites

X-ray transitions deduced from the hydrogenic model (Fig.2.3) are not sufficient to explain all the structure found in emission spectra from ele-ments and their compounds. Those (mostly weak) structural features ob-served close to diagram lines whose photon energy cannot be calculated from energy differences in a simple atomic model are generally referred to as nondiagram lines or satellites.

Just as diagram lines are influenced by the environment of the emit-ting atom, so are the X-ray satellites, too. Relative intensities and energies of the satellites appear to depend more strongly on chemical bonding than is the case for the diagram lines. Since there are several different possibil-ities for the origin of the X-ray satellites, knowing this origin is an evi-dent prerequisite for explaining bonding effects.

Satellites can originate [2.5] from transitions between multiply ionized states [2.82-84], from simultaneous (multiple) electronic transitions [2.85] and from electron transitions from normally unoccupied outer orbitals to inner vacancies [2.86, 87]. In most cases the satellites are expected to appear on the short-wavelength side of the corresponding main line. To explain the well-known existence of long-wavelength satellites, three processes are recognized: (i) electron transitions from the outer levels of an ion into the K level of another ion in a crystal (cross-transition) [2.87]; (ii) two-electron processes, whereby the photon energy is decreased through a simultaneous excitation of an outer electron into the continuum (radiative Auger) [2.20, 88]; and (iii) the (occasionally strong) influence of multiplet splittings, as discussed below.

Earlier studies made it possible to classify more thoroughly the different sorts of satellites. In these studies all of the above explanations were considered and developed further. The following scheme introduced by Åberg [2.88] is limited to dipole transitions and structures which are caused by photon or electron excitation and does not treat the spectra caused by ion-atom collisions [2.89-92].

The explanation of diagram lines given by the hydrogenic model describes isolated atoms with closed shells (X-ray vacancy included) and does not consider the relaxation of the electron orbitals that accompanies creation of the X-ray vacancy, i.e., it considers the electrons before and after the X-ray transition in the same (frozen) orbitals. Incremental improvements to the hydrogenic model suggest mechanisms for satellites arising from the several processes mentioned above. Figure 2.14 shows the scheme published by Åberg [2.88]. It begins with the frozen-orbital approximation for atoms with closed shells, for atoms with partially filled outer shells, and for molecules and solids. It predicts the diagram lines and, through multiplet splitting and cross-transitions, predicts the corresponding satellite structure. In this way, it is natural to consider transitions between singly ionized configurations, so that the distinction of diagram and satellite lines in the spectrum seems somewhat formal.

If the frozen-orbital approach is abandoned and relaxation processes are considered, then through direct and indirect multiple ionization the subsequent deexcitation transitions show up as short-wavelength satellites while (through many-electron processes) long-wavelength satellites also appear.

2.4.1 Satellites from Multiplet Splitting

The multiplet structures of X-ray emission lines depicted in Fig.2.2 can appear as fully resolved splitting through a sufficiently strong exchange interaction between unfilled subshells and X-ray vacancies [2.93]. Thus, the long-wavelength satellites of lines that are emitted from atoms with unfilled d or f subshells can be explained as direct multiplet splitting

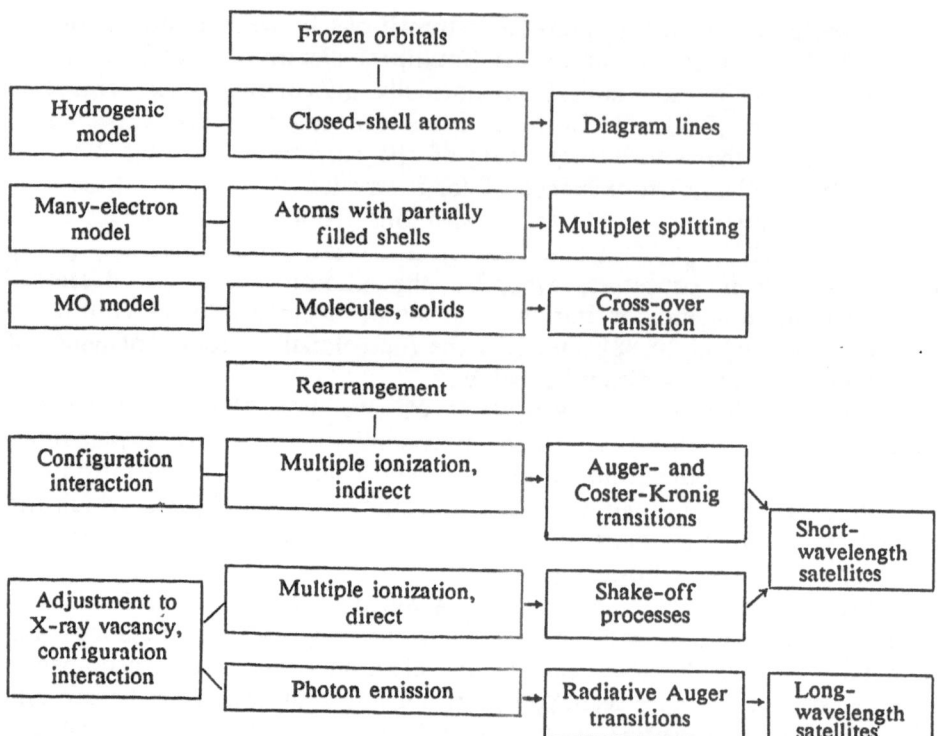

Fig.2.14. Schematic overview of the origins and classification of X-ray diagram lines and satellites [2.88]

[2.26]. Thus, for example, the well-known $K\beta'$ satellite in the $K\beta_{1,3}$ emission spectrum from the 3d transition elements can be explained as a multiplet structure of the $3p^5 d^n$ states (Fig.2.7).

Figure 2.15 shows an analogous interpretation of the V $K\beta_{1,3}\beta'$ spectrum. As the terms with the highest multiplicity have the lowest energy and the intensity from a multiplet component is proportional to the multiplicity, the allowed transitions $K\beta_{1,3}$ are accompanied by a weaker long-wavelength maximum $K\beta'$. Similar investigations of long-wavelength structures have been reported for the $L\gamma_{2,3}$ spectra ($L_I \rightarrow N_{II,III}$) of the 4d elements [2.27,28] and the $L\gamma_{2,3}$, $L\gamma_1$ ($L_{II} \rightarrow N_{IV}$) and $L\beta_{2,15}$ spectra ($L_{III} \rightarrow N_{IV,V}$) in the rare-earth metals [2.94-96].

This explanation of the satellite structure in terms of multiplet splitting is supported by observed splittings in photoabsorption spectra [2.97], in photoelectron spectra [2.98] and in Auger spectra [2.99] and also through the observation that the excitation mechanism has little influence [2.27,28,32,33,93,100].

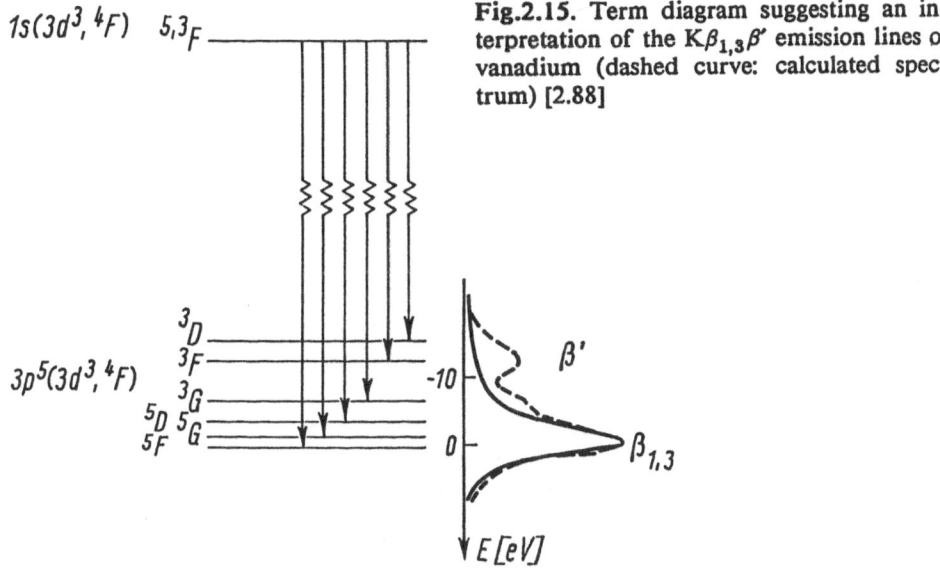

Fig.2.15. Term diagram suggesting an interpretation of the $K\beta_{1,3}\beta'$ emission lines of vanadium (dashed curve: calculated spectrum) [2.88]

2.4.2 Satellites from Cross-Transitions

Several of the weaker maxima near the main emission bands in molecules and solids originate from mixing valence orbitals of the emitting atom with orbitals from the surrounding atoms. The first observations of such satellites in the K spectra of KCl and NaCl [2.87] were explained specifically as transitions from valence electrons of one ion in the crystal into the K vacancy of the other and for this reason the transition was called a cross-transition. Following this observation, corresponding structures were found in many spectra where the participation of the valence electrons of molecules, complex ions in crystals and solids as carbides, nitrides and oxides leads to such structures. As is clear from the discussion in Sect.2.3, it is possible to calculate the intensity of the components in the emission band of a molecule within the framework of an MO model from the dipole matrix elements using the MO-LCAO function. From this it follows that the weaker maxima occurring close to the main peaks in the emission band from molecules and isolated groups in crystals are, similar to the multiplet-type satellites, components of diagram lines. Therefore, the relative intensities should be substantially independent of the excitation mode.

In several investigations the K emission bands from carbides, nitrides, oxides and fluorides of the elements from the second, third and fourth periods were studied and assignments made for accompanying long-wavelength satellites $K\alpha'$, $K\beta'$, and $K\beta''$ [2.101-103]. These can be interpreted in the following way: Because of the mixing of the 2s orbitals of the ligand X (C, N, O, and F) with the valence orbitals of the metal, on the long-wavelength side of the emission band K → V (V: valence

41

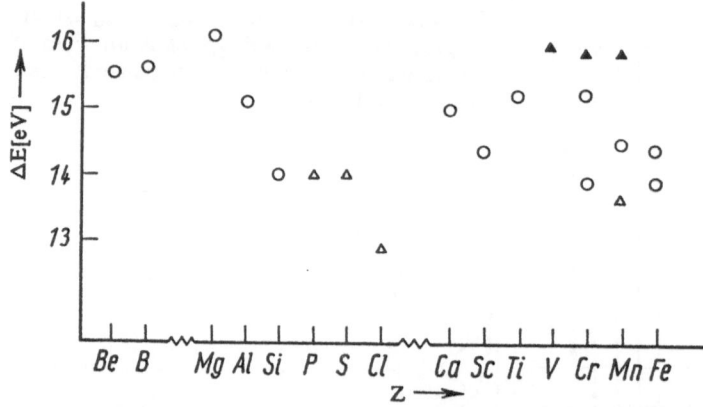

Fig.2.16. Energy difference ΔE between main peak and long-wavelength structure of the $K\alpha$, $K\beta_{1,3}$ and $K\beta_5$ spectra of oxides (circles) and oxygen-containing complex ions (triangles) for elements from the 1st, 2nd and 3rd periods [2.88]

orbitals) of the metal a satellite appears, which can be associated with the transition $K \rightarrow X(2s)$, where $X(2s)$ is the molecular orbital formed by 2s ligand orbital and valence orbitals of the metal. As the valence orbitals of the metal are mixed with the 2p orbitals of the ligand the distance of the emission band to the satellite can, to a first approximation, be derived from

$$E_{K\beta_1} - E_{K\beta'} = E_{2s}(X^-) - E_{2p}(X^-) . \qquad (2.41)$$

Here $E(X^-)$ is the average $n\ell$-orbital energy of the free ion X^-. Figure 2.16 shows a compilation of the different values for the distance of the satellites $K\alpha'$, $K\beta'$, and $K\beta''$ to the emission band for different oxygen compounds. A more detailed discussion of the structure in the spectra of the above-mentioned compounds is given in Chaps. 5 and 6.

2.4.3 Satellites from Transitions Between Multiply Excited or Ionized States

If an additional electron is missing in any shell during a transition that leads to an X-ray emission, the energy of the emitted photon is, in general, larger than the initial diagram line. Thus corresponding to each process known to occur with high probability, weaker short-wavelength satellites appear [2.83, 84]. To label transitions between doubly ionized states, a nomenclature corresponding to that of diagram lines is used. Thus, for example, in association with the main line $K\beta_{1,3}$ the satellite $KL_{III} \rightarrow L_{III}M_{II,III}$ is called $K\beta'''$. The initial state is, in this case, an atom ionized in the K and L_{III} shell. The group of $K\alpha$ satellites for the elements $8 \leq Z \leq 18$ ($\alpha', \alpha'', \alpha_3, \alpha_3'$ and α_4) is given by the five $KL \rightarrow L^2$ transitions pre-

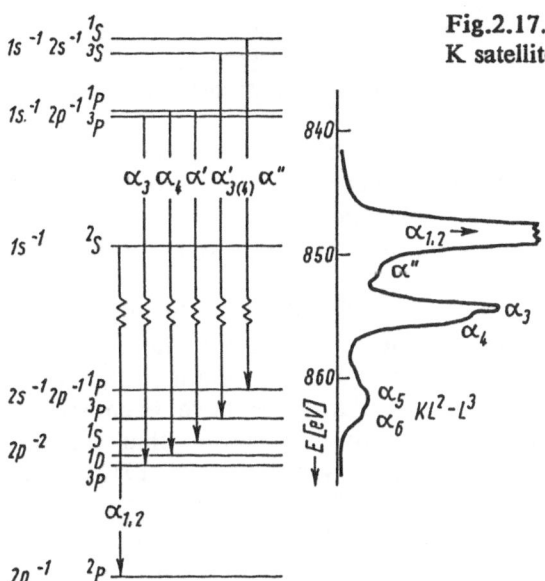

Fig.2.17. Level diagram for interpreting the K satellites of neon [2.88]

sent in an LS coupling scheme whose assignment can be seen from Fig. 2.17 (α_5 and α_6 are provisionally described to transitions between triply ionized states).

Known processes leading to multiply ionized atoms include successive ionization, multiple ionization by radiationless (Auger) transitions and multiple ionization through electron and photon collisions. Of these, only the indirect multiple ionization through Auger transitions [2.82] and the direct multiple ionization (shake-off) [2.104, 105] need be considered because of the short lifetime of the X-ray vacancy.

From anomalously small intensities of L_I emission spectra, the possibility of multiple ionization through internal conversion was discovered [2.82]. Specifically, an L_I level vacancy created by ionization may be filled by an L_{III} electron; the available energy is sufficient to ionize the $M_{IV,V}$ levels for elements with $Z < 50$. In this way, a nonradiative transition $L_I \rightarrow L_{III}M_{IV,V}$ (Auger effect) can follow. whereby the atom is left in a doubly ionized state. Auger transitions of the type $X_i \rightarrow X_j Y$, where X_i and X_j belong to different subshells of the same shell, are called Coster-Kronig transitions and are important in the production of multiply ionized states. The above type of description has been helpful in the classification and calculation of the L satellites, for example for Ag [2.106], Au [2.107], several heavy elements [2.108] and for the elements with $37 < Z < 56$ [2.109]. Coster-Kronig satellites have been analyzed and accurately separated by means of threshold excitation of Cu L_{III} X-ray emission spectrum using synchrotron radiation [2.110].

Satellites in the K series cannot be explained by indirect multiple ionization since multiply ionized states with a K vacancy cannot be created through the Auger effect. On the other hand, calculation of the direct KL double ionization from electron collisions [2.105] leads to good agreement for KL and K states and the ratio of the measured $K\alpha'$, α_3, α_4, and $K\alpha_1$ intensities.

Ealier papers [2.104, 111, 112] have shown that multiple ionization with appreciable strength is possible and a theory of X-ray satellites based on the sudden approximation has been developed [2.113, 114]. A survey of other aspects of direct multiple ionization and multiple excitation has been given by *Krause* [2.115].

Most intensity from satellite transitions for the doubly ionized systems falls a few eV on the high-energy side of the parent line and causes high-energy tails, broadening and energy shifts of the main peaks. These effects on X-ray emission bands of molecular gases have been studied recently [2.116]. For some cases it is possible, however, that satellites appear superimposed on the corresponding diagram line [2.117] or even appear on the long-wavelength side [2.112, 118]. These cases arise when the additional vacancy is present in an outer shell. Figure 2.18 shows the different cases of satellites from multiply ionized states of the type LM, LN, LMM, and LNN. From calculations of the energies of initial and final states it follows that due to the widths of the satellite lines, only some of them are observable as separated from the diagram line. Satellites emitted from initial states LM, LMM or LMN appear on the high-energy side generally within a few eV.

Those with the initial state LN or LNN tend to lie very close to the main line. Figure 2.19 shows the superposition of the diagram line Zr $L\beta_1$ with photons which originate from the decay of $L_{II}N$ (reached through direct and indirect multiple ionization) and triply ionized states $L_{II}NN$

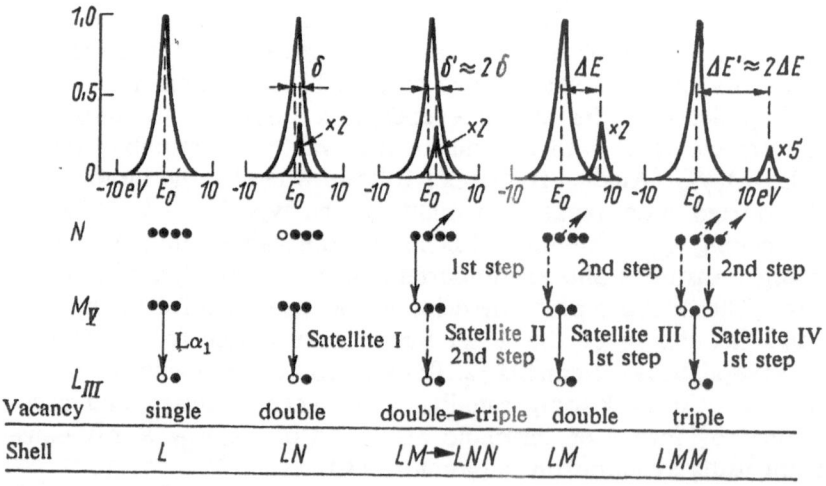

Fig.2.18. Production mechanism of various satellites in the L spectrum of zirconium [2.117]

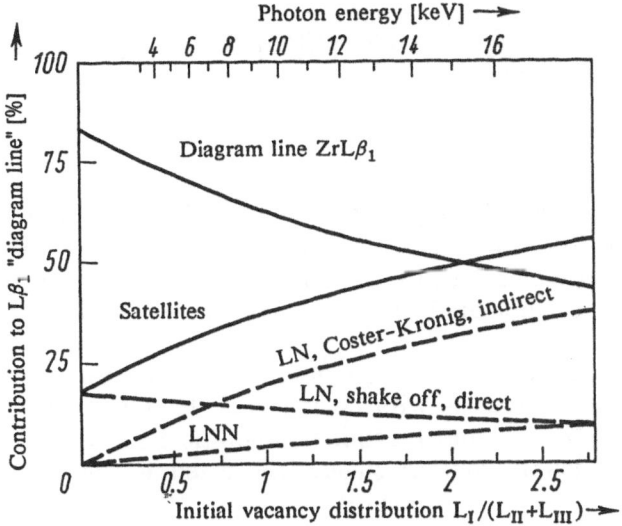

Fig.2.19. Components of the Zr $L\beta_1$ "diagram line" as a function of excitation energy and the initial vacancy distribution [2.117]

(reached through an Auger transition in a doubly ionized atom in the state LM or through a combination of direct or indirect multiple ionization of the N shell). Since with increasing energy of the exciting photons the ionization cross section of the 2s shell of Zr decreases more slowly than the corresponding 2p cross section [2.119], the vacancy ratio $L_I/(L_{II}+L_{III})$ increases so that the subsequent increase of probability for Coster-Kronig transitions decreases the intensity of the main line. Even with small excitation energies there is appreciable disturbance of the main line by satellites unless an excitation energy can be used that is only a few electron volts above the singly ionized state. Since the influence of the chemical bonding on the shape and position of the main line and satellite is different, the relative changes could be observed for different compounds using comparative measurements of line intensity, shape, and width.

To calculate the photon energy of the satellite lines, one requires the energies of multiply ionized states. A crude estimation is possible using

$$E_{1s}^{2s}(Z) = E_{1s}(Z) + E_{2s}(Z+1) . \tag{2.42}$$

Here the vacancy in the K shell of an atom is included by changing the screening by one unit. In this way the ionization energy of an electron in the L shell for an atom with charge Z+1 is used as an approximation for the corresponding energy of an atom with charge Z with an additional vacancy in the K shell.

More accurate values for the energy of the doubly ionized states can be extracted from experimental Auger energies [2.120]. The energy E_j^k of the atom ionized in the j and k shells is given by

$$E_i - E_j^k = E_{kin}^e + E_{kin}^a .\qquad (2.43)$$

If the recoil energy of the atom E_{kin}^a is neglected, one obtains E_j^k from the kinetic energy E_{kin}^e of the Auger electrons in the transition $i \rightarrow jk$ from the spectra of the solid element (kinetic energy in vacuum added to the work function) and from tabulated values of the energy E_i of the singly ionized state i.

Theoretical calculation of the energy of satellite E_s that is emitted in a transition from an atom or a molecule with holes in the shells i and k to states with vacancies in the shells j and k requires ab-initio SCF calculations for the system with doubly ionized states. The satellite energy is then given by the differences in the total energy $\mathcal{E}$:

$$E_s = \mathcal{E}_i^k - \mathcal{E}_j^k .\qquad (2.44)$$

In most calculations with open shells the individual orbital energies give approximate values for the ionized energies of the corresponding orbitals. Therefore, a good approximation for the energy of the doubly ionized system results if the energy ϵ^k of the k orbital is subtracted from the total energy $\mathcal{E}$ of the singly ionized atom or molecule. The energy of the satellite is then given by

$$E_s = \mathcal{E}_i - \epsilon_i^k - \mathcal{E}_j + \epsilon_j^k .\qquad (2.45)$$

The distance of the satellite to the main line follows from the difference between the orbital energies

$$E_s - E_{ij} = \epsilon_j^k - \epsilon_i^k .\qquad (2.46)$$

The relative intensities of X-ray satellites (i.e., the relative probability for transitions involving singly and multiply ionized states) follows from the distribution of atoms over the possible initial states. If it is assumed that the system studied is in dynamical equilibrium and that the numbers of atoms in different states of simultaneously excited configurations are equal, then it is possible to calculate the number of photons $n(f \rightarrow ij)$ from transitions between all states of the configuration j with a vacancy i to the configuration f through [2.88]

$$n(f|ij) = \frac{\Gamma^R(f|ij)}{\Gamma(ij)} \left[P(ij)N_i + \sum_{k\ell > ij} W(ij|k\ell)N'(k\ell) \right] ,\qquad (2.47)$$

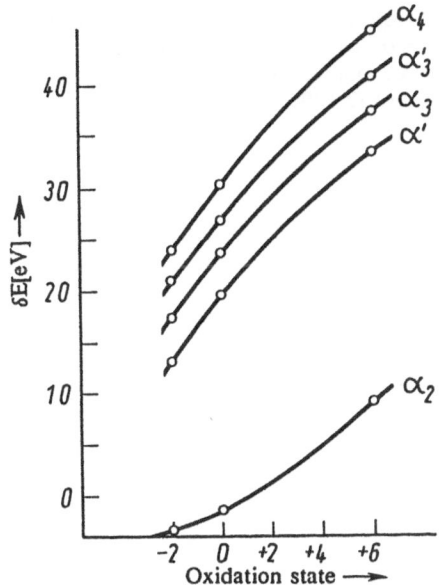

Fig.2.20. Chemical shift δE of the $K\alpha_2$ line and the $K\alpha$ satellites of sulfur as a function of oxidation state [2.127]

where $\Gamma(ij)$ and $\Gamma^R(f|ij)$ denote the total decay rate and radiative decay rate, respectively, of the configuration ij. Further, P(ij) means the relative probability of promoting atoms to the configuration ij, N_i is the total number of atoms created with a vacancy i. The sum in parentheses describes the excitation of atoms in configuration ij from kℓ whereby the relative decay rate W(ij|kℓ) gives the number of atoms of configuration kℓ of all nonradiative and radiative transitions which take the kℓ into the ij. Applications of the above equation to obtain relative satellite intensities have been reported in [2.121, 122].

As found for the X-ray diagram lines, for multiple-ionization satellites there is also a dependence of energy and intensity on the oxidation state of the emitting atom [2.123-126]. Theoretical interpretation of chemical shifts of the satellites requires calculation of the energies of multiply ionized states participating in the transition. Thus, an approximation analogous to that used to calculate chemical shifts of main lines (Sect.2.2) can be applied. Calculations using the free-ion model in an LS coupling scheme for different oxidation stages of sulfur (Fig.2.20) [2.127] showed that the chemical shift of all the $K\alpha$ satellites increases with the positive charge of the emitting atom and is larger than the shift of the $K\alpha_{1,2}$ lines.

By calculating shifts of satellites for ions compared to neutral atoms it is also possible to determine effective charges of the atoms from the experimentally determined shifts of the X-ray satellites. *Chun* et al. [2.128] thus obtained effective charges for several compounds of the elements from the third period from semiempirical and ab-initio calculations of X-ray transitions of the type KL → LL for the elements and for some of

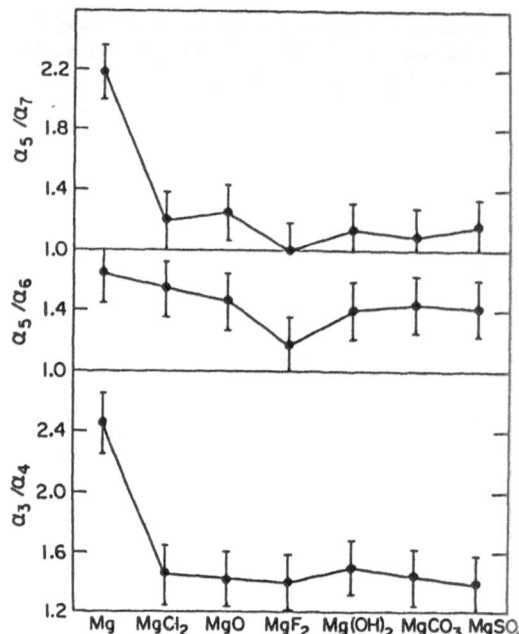

Fig.2.21. Intensity distribution of the $K\alpha$ satellites for several magnesium compounds excited with 5.4 MeV He ions [2.130]

their ions, which agreed relatively as well as absolutely with the charges determined from the $K\alpha_{1,2}$ shifts.

Influence of chemical binding on the relative intensity of the $K\alpha_{3,4}$ satellites (excited with photons or electrons) has been discussed earlier [2.126, 129]. The intensity ratio $K\alpha_3/K\alpha_4$ changes sharply when comparing spectra from metal and oxide excited with helium ions [2.130]. On the other hand, the intensity differences are small for a considerable range of magnesium compounds (Fig.2.21). The explanation of this is that only in the metal is it possible to have Coster-Kronig transitions with participation of the M shell electrons [2.126]. The strong dependence of the $K\alpha_3/K\alpha_4$ intensity ratio on the mass number of the projectile in the region $1 \le N \le 16$, when using heavy particles for excitation, can be explained by the relative population of the $1s^{-1}2p^{-1}$ ^1P and ^3P initial states which can be influenced by electronic rearrangement before $K\alpha$ emission [2.131, 132]. High-resolution ion-excited $K\alpha$ emission spectra with the initial states KL^n (number of vacancies: $0 < n < 6$) from Al, Si, S, and Cl compounds show strong influence of chemical binding on intensity ratios of satellite components (Fig.2.22). The reason for these effects could be interatomic processes, which are known to contribute to the deexcitation of the multiply ionized systems [2.132]. Recently numerous studies of chemical effects in satellite emission have been reported [2.133-137].

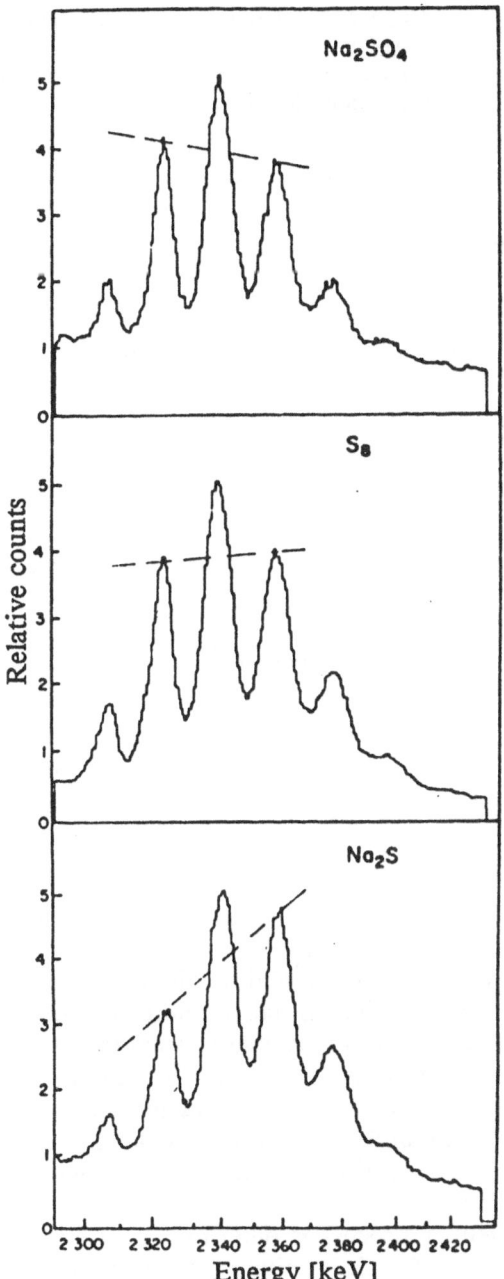

Fig.2.22. Sulfur Kα spectra from various substances excited with 2.0 MeV oxygen ions [2.132]

2.4.4 Satellites from Radiative Auger Transitions

Deexcitation of an atom with a vacancy in an inner shell to a state with less energy can proceed not only through a nonradiative transition accom-

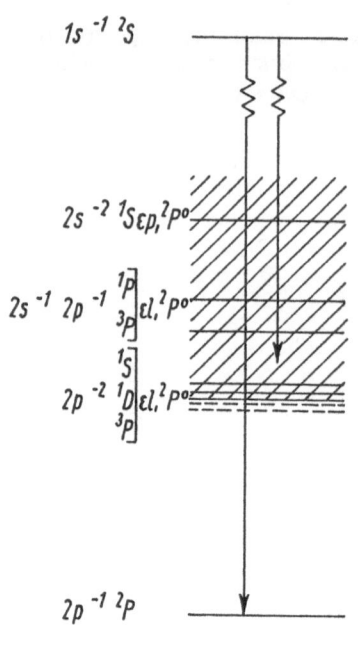

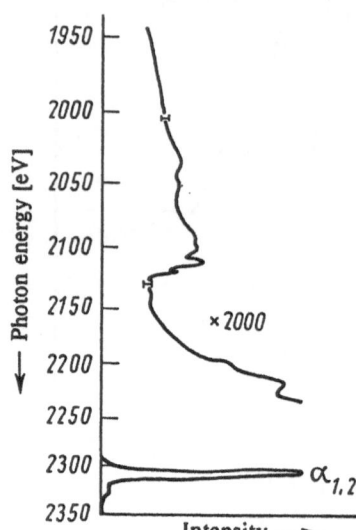

Fig.2.23. Satellite structures on the long-wavelength side of the $K\alpha_{1,2}$ doublet of sulfur and a level diagram suggesting their origins [2.133]

panied by electron emission or a radiative process, but is also possible through a simultaneous emission of an electron and a photon. The energy of the emitted photon in this case is given by

$$h\nu = E_i - E_{ff'} - E_e .\qquad(2.48)$$

The energy difference between the singly ionized initial state E_i and the doubly ionized final state $E_{ff'}$ corresponds to the nonradiative Auger transition i → ff'. The energies E_e of the emitted and excited electrons, respectively, extend from discrete values, which correspond to bound states, to the zero energy value, and to a continuous energy spectrum with positive values and a maximum value determined by the kinetic energy E_e = E_i-$E_{ff'}$. Correspondingly, the emitted radiation is characterized by a strong cut-off on the high-energy side and is followed by a structure which on the long-wavelength side has a continuously decreasing intensity (Fig.2.23).

The probability of a radiative Auger process is determined by configuration interaction between single- and double-vacancy configurations in the final state and by the change in the potential which affects the electron when the vacancy in the atom changes location. The influence of these two factors of the emitted radiation has been discussed by *Åberg* [2.139].

Since the final states of the satellite emission are also used to describe the shake-up structure in the photoelectron spectra, there is a consequent relation between the relative energies and intensities of the XES and XPS satellites. If a satellite state (shake-up) is ascribed to each singly ionized initial and final state of the X-ray emission, four possible transitions are obtained whose relative intensities can be predicted through the following approximation [2.140]: For core-core XES there is an interference effect so that transitions from an initial satellite state to a singly ionized final state and semi-Auger transitions from a singly ionized initial state into a shake-up state are very weak (if at all observable) even if the corresponding shake-up structure in the photoelectron spectrum is very strong. This situation is especially evident in the $K\alpha$ spectrum of Ni and in the Ni 2p XPS [2.141]. Transitions between satellite states have almost the same transition moments as those between singly ionized states. As the satellite transitions nearly coincide in energy with those from singly ionized states, their experimental demonstration is difficult. On the other hand, the relative intensity of the semi-Auger process in the case of a core-valence XES, i.e., for a transition from a singly ionized core state to satellite state with vacancies in the valence shell, is the same as for the shake-up process in XPS, as both are governed by the same component of the frozen-orbital configuration in the satellite state.

Examples of observed weak satellite structures on the long-wavelength side of the $K\alpha$ spectra of the elements Mg, Al, Si, and S are given in Fig.2.23. In agreement with the above-described mechanism the structures coincide in the low-energy region with the KLL Auger energies while the maxima in the edge region arise from bound states.

Satellites of the type K $\rightarrow$ LL have been observed on the long-wavelength side of the $K\alpha$ spectra for the elements $3 < Z < 22$. Similar structures were found for atomic gases (Ne [2.142]), molecular gases (N_2 [2.143]), for certain metals (Li [2.144], Ti [2.145]) and for compounds with similar structures.

Long-wavelength satellites of the type K $\rightarrow$ MM have been observed for the elements $15 < Z < 22$ [2.146]. Figure 2.24 shows the corresponding spectrum for gaseous argon. The positions of doubly ionized final states for the particular multiplets in LS coupling are marked. For the configuration $3s^2 3p^4$, discrete states also appear close to the edge of the continuous emission. The indicated energies are based on optical data and LMM Auger energies. The possibility that $K\beta$ satellites appear as transitions between doubly ionized states corresponding to $KL_{II,III} \rightarrow L_I M_I$ [2.147] was ruled out from the results of measurements of the satellites with excitation energies below the $KL_{II,III}$ energy [2.148].

In contrast to K emission, where one has semi-Auger satellites with only small intensity (around 1% of the parent line), the L emission spectra are found to have long-wavelength satellites of the same origin but with considerably higher intensity (more than 10% of the parent line). One prominent structure on the long-wavelength side of the Ar $L_{II,III}$ ($3s \rightarrow 2p$) diagram lines [2.149] was identified as a collection of several discrete lines in an experiment with improved resolution [2.150, 151]. In accordance

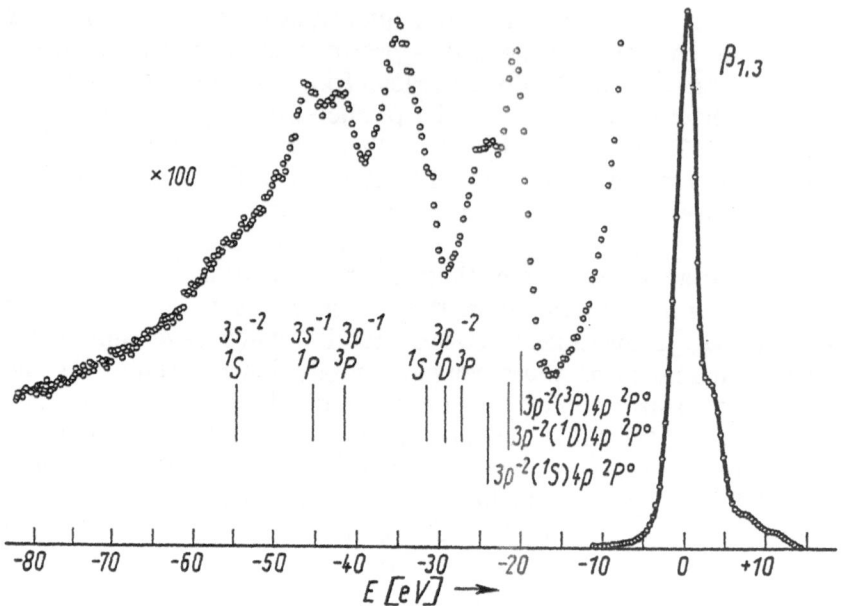

Fig.2.24. Satellite structures on the long-wavelength side of the $K\beta_{1,3}$ doublet of argon and level diagram for interpretation [2.88]

with an earlier interpretation, the structure was assigned to states with final configuration $3s^2 3p^4 nd(^2S)$ with n = 3,4 (Fig.2.25). Analogous structures occur for example in Cl L spectra of KCl and of halogenated methanes [2.149, 152].

Influence of chemical bonding on radiative Auger transitions should proceed similarly to the case of absorption and Auger electron spectra. Since the excited electron is coupled to two holes, in contrast to the photo-absorption process, excitation can occur in the satellite spectra even

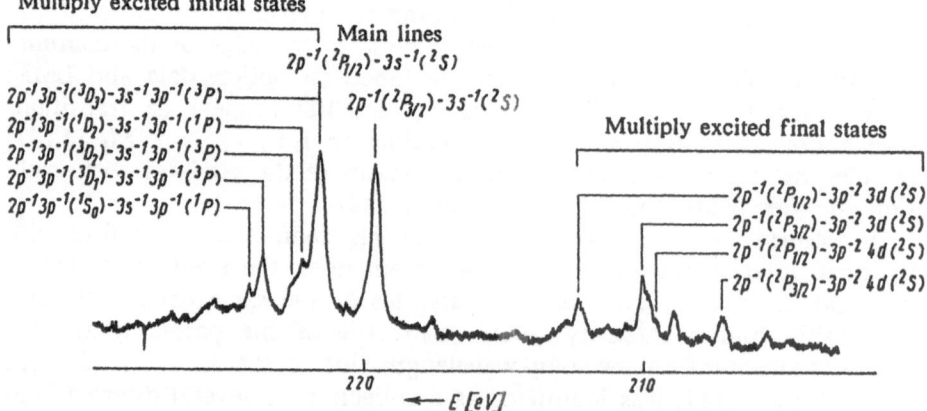

Fig.2.25. Interpretation of the $L\ell,\eta$ emission spectrum of argon [2.150, 151]

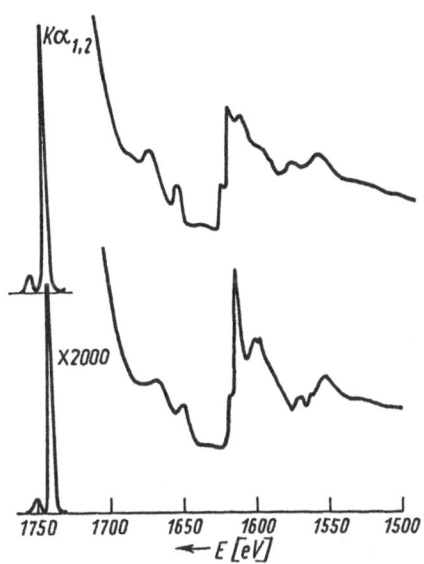

Fig.2.26. Satellite structure on the long-wavelength side of the Si $K\alpha_{1,2}$ doublet in silicon (top) and SiC (bottom) [2.153]; for SiO_2 (Fig.6.25)

if the absorption spectrum does not show any similar maximum. Chemical shifts should occur in opposite directions to those found in absorption spectra. Figure 2.26 shows the influence of chemical bonding on the semi-Auger transition $K \rightarrow LL$ in Si. The mean value of the shifts of maxima near 1600 eV is a measure of the shift of the total $K \rightarrow L_{II,III}^2$ tail of the spectrum assuming that these peaks can be assigned to double transitions of the type $K \rightarrow LL$. On the other hand, this shift should equal the change in the $L_{II,III}$ energy but with opposite sign. For SiO_2 there is a good agreement between the shift of the $K \rightarrow LL$ satellites ($-7eV\pm1eV$) and the value of the $L_{II,III}$ shift ($5.6eV\pm0.2eV$). Based on the free-ion model the energies of the $K \rightarrow L_{II,III}^2$ transitions give the effective charges of silicon atoms in compounds [2.153].

2.5 Bremsstrahlung and X-Ray Isochromat Spectra

Transitions of free electrons of a specified kinetic energy into the unfilled conduction band of a solid lead to bremsstrahlung, assuming that there is no interaction with the core electrons of the atoms in the target. If the energy E_0 of the electrons bombarding the target is held constant, i.e. by keeping the X-ray tube voltage constant, then the bremsstrahlung spectrum in the region of the short-wavelength limit can be recorded by changing the Bragg angle setting of a spectrometer. From Fig.2.27 it is clear that with increasing wavelength, final states are reached through transitions with increasing energy differences from the Fermi level so that

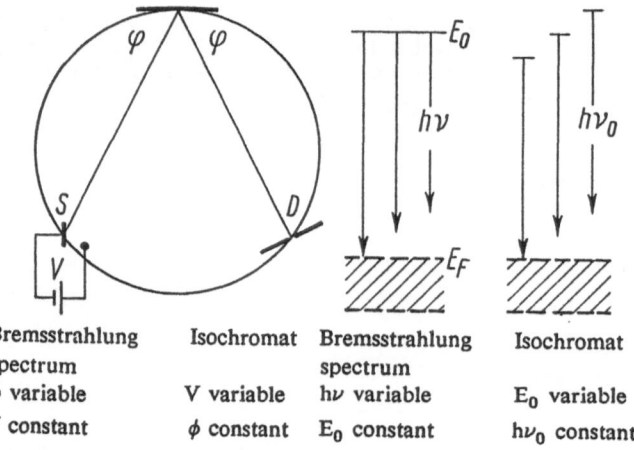

Bremsstrahlung spectrum	Isochromat	Bremsstrahlung spectrum	Isochromat
ϕ variable	V variable	$h\nu$ variable	E_0 variable
V constant	ϕ constant	E_0 constant	$h\nu_0$ constant

Fig.2.27. Basis of bremsstrahlung spectroscopy and bremsstrahlung-isochromat spectroscopy [2.154]. (S: source, D: detector)

in this way a spectrum with the short-wavelength limit of $h\nu = E_0 - E_F$ is recorded.

An analogous result is obtained if the intensity at a fixed wavelength is recorded when the energy of the incoming electrons is changed stepwise by corresponding changes of X-ray tube voltage. The recorded bremsstrahlung "isochromat" spectrum represents the intensity of a certain photon energy as a function at accelerating electron voltage. As the bremsstrahlung spectrum in the region of the short-wavelength limit and the bremsstrahlung isochromat spectrum were produced by the same elementary process, the information from the two recordings is equivalent and the spectra have similar shape (Fig.2.28). The differences revealed from the spectra recorded by the two techniques can be ascribed to effects of wavelength-dependent self-absorption, and to variations of the spectrometer's efficiency.

The intensity of the bremsstrahlung spectrum reflects the density of states in the unoccupied part of the band and absorption features due to the electron-transport process in the target and to an energy-dependent transition probability matrix. If, in the first approximation, the influence of absorption on the bremsstrahlung-intensity distribution is neglected (that means an omission of the absorption factor $\exp(-\mu x \sec\psi)$ with μ being the linear absorption coefficient and ψ the exit angle), then the intensity distribution determined by the above-mentioned factors can be written as [2.155]

$$I(E_0, h\nu) = \int_0^{\Delta E} f(E_0 - h\nu - \epsilon) \, P(h\nu + \epsilon) \, N(\epsilon) \, d\epsilon \ . \tag{2.49}$$

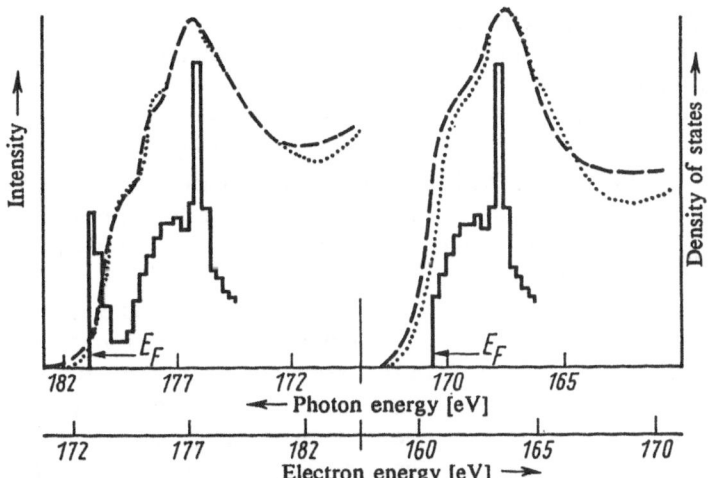

Fig.2.28. Comparison of bremsstrahlung spectra at the short-wavelength limit (dashed), and bremsstrahlung isochromat (dotted) from tantalum (left) and tungsten (right) with densities of unoccupied states (solid) [2.154]

The relative number of electrons that emit a photon of energy $h\nu$ after an energy loss during transport in the target through a transition to a state of energy ϵ in the conduction band can be expressed by the distribution function $f(E_0 - h\nu - \epsilon)$. The relation between the energy E of the electrons in the target, the energy of the bombarding electrons E_0, the emitted photon energy $h\nu$, the energy loss δE_0 and the maximum possible energy loss $\Delta E = E_0 - h\nu$ can be seen from Fig.2.29. Here $P(h\nu + \epsilon)$ gives the probability of the transition and $N(\epsilon)$ gives the number of possible final states. Under the assumption that discrete energy losses dominate over the continuous slowing-down of electrons in the target, the distribution function close to the short-wavelength limit can be eliminated, i.e., under the influence of the characteristic energy losses the intensity distributions of

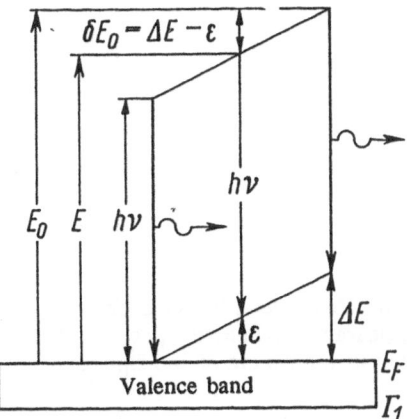

Fig.2.29. Energy diagram to illustrate calculation of a bremsstrahlung spectrum (for explanation, see text) [2.155]

both bremsstrahlung and bremsstrahlung isochromat are proportional to the density of states (under the assumption of a negligible energy-dependent transition probability).

The extent of agreement between the theoretical density of states and the spectra is suggested in Fig.2.28. The experimental bremsstrahlung isochromat spectra are shown for Ta and W, and compared to the calculated density of states for tantalum, whereby the short-wavelength limits of the spectra are aligned with the Fermi edge from the density-of-states curve, as given by the rigid-band model.

If the energy of the incoming electrons is enough to excite a core electron of the target atoms into the conduction band, a vacancy is created in the core with a transition probability showing an energy dependence on the kinetic energy of the incoming electrons. Investigation of the intensity of the characteristic X-ray emission spectrum as a funtion of incoming electron energy gives a characteristic X-ray isochromat or excitation spectrum [2.156].

The nondispersive recording of X rays emitted from a target as a function of electron energy leads to an excitation or Appearance Potential Spectrum (APS) [2.157]. This gives, in general, a recording similar to the corresponding characteristic isochromat, since the signal is created by characteristic radiation and branching ratios are approximately independent of energy (Fig.2.30).

The intensity of a characteristic isochromat is determined by the probability of creating the inner vacancy which constitutes the initial state in the recorded X-ray transition. If energy loss by the incoming electrons (energy E_0) is neglected, one has to sum over all possible transitions in the region ΔE between the threshold for core electrons of binding energy E_i to the Fermi level and $E_0 - E_i$. As seen from Fig.2.31 the states ϵ and

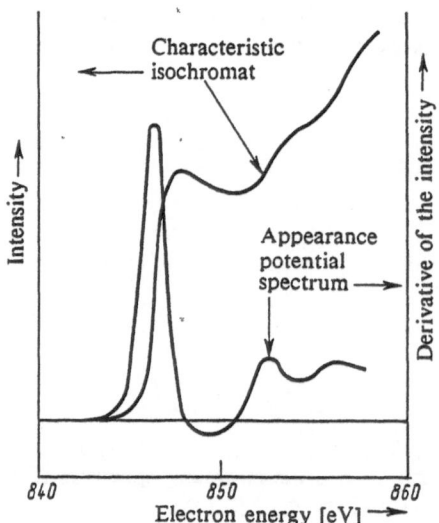

Fig.2.30. Characteristic isochromat [2.156] and appearance potential spectrum [2.158] of nickel in the region of the L_{III} edge

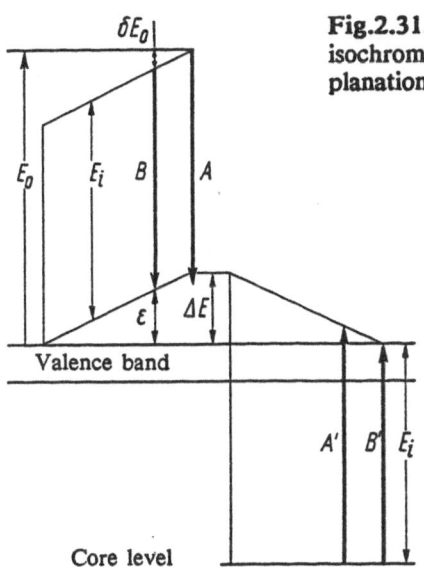

Fig.2.31. Energy diagram for calculating characteristic isochromat and appearance potential spectra (for an explanation, see text) [2.155]

ϵ', where the incoming and excited core electrons are found after an excitation process are located in the unfilled conduction band. They are related by $\epsilon' = \Delta E - \epsilon$ with $\Delta E = E_0 - E_i$ so that the intensity of a characteristic isochromat can be written as

$$I(E_0) = \int_0^{\Delta E} f(0)\, F_i(E_0)\, N(\epsilon)\, N'(\Delta E - \epsilon)\, d\epsilon \,, \qquad (2.50)$$

where F_i is the ionization cross-section [2.155].

If energy losses are included, transitions of type B are added to type A accounted for in (2.50). In this case an incoming electron will reach the final state in the conduction band after an energy loss δE_0. To determine the intensity of the characteristic radiation it is necessary to sum over possible energy losses δE_0 and possible final states ϵ:

$$I(E_0) = \int_0^{\Delta E} N(\epsilon) \left[\int_0^{\Delta E - \epsilon} f(\delta E_0) F_i(E_0 - \delta E_0) N(\Delta E - \epsilon - \delta E_0)\, d\delta E_0 \right] d\epsilon \,. \qquad (2.51)$$

Thus, when discrete energy losses dominate continuous electron-bremsstrahlung losses, $f(\delta E_0) = \delta(0)$. The intensity formula shows that the spectra are characterized by an autoconvolution of the density of states so that the proportionality to the density of states is generally lost. The same is true for the band-structure information that is revealed in an appearance potential spectrum.

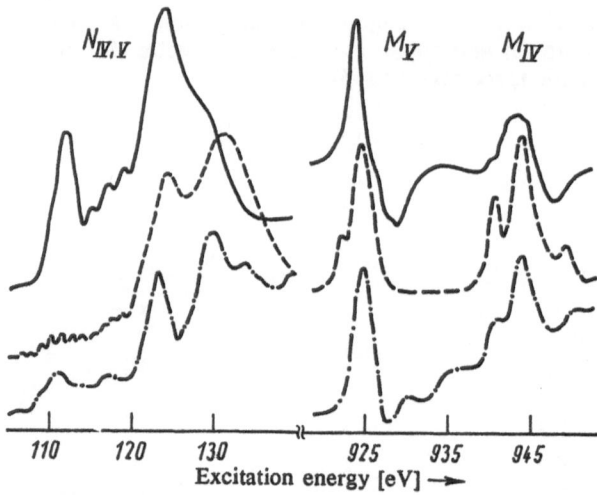

Fig.2.32. Comparison of X-ray (solid) and Auger electron appearance potential spectra (dashed-dotted) in the region of the $N_{IV,V}$ and $M_{IV,V}$ edges of praseodymium with the corresponding absorption spectra (dashed) [2.159]

However, it is possible in some cases to discern a similarity in the detailed structure of the appearance potential spectrum and the X-ray absorption spectrum, as seen from the $N_{IV,V}$ and $M_{IV,V}$ spectra of praseodymium (Fig.2.32). Besides the already described spectra, the figure includes the Auger electron appearance potential spectra [2.161] which are very similar to the intensity distribution of the absorption spectrum. The obvious agreement between the two suggests an analogy between absorption spectrum and quantum yield spectrum (total electron yield spectrum). This analogy was already discovered more than twenty years ago [2.161-166].

Investigations of the bremsstrahlung and the isochromat spectra thus give some information on the density of states in the unfilled part of the band in a region of 10 to 20 eV above the Fermi level [2.167- 174]. Information about negative ions from BIS studies of gas phase atoms or molecules are also available [2.175]. At higher energies electron scattering effects the intensity distribution and causes an extended fine structure. This EXAFS-like behaviour (Sect.2.6.3) of BIS and APS has been used among other EXAFS-like techniques [2.176] in extended X-ray bremsstrahlung isochromat fine structure (EXBIFS) and extended appeerence potential fine structure (EAPFS) techniques for structure analysis.

2.6 Absorption Spectra

2.6.1 General Features

Absorption of an X-ray photon by an atom in a material occurs when the photon energy $h\nu$ is comparable to or exceeds the binding energy E_i of one of the electrons in the atom. During the absorption process the electron is either ejected out of the atom with a certain kinetic energy or it makes a transition to a bound state above the filled shells of the atom (in the context of the one-electron approximation) which corresponds to an excited state of the atom. If excited states are neglected, absorption of photons with $h\nu < E_{1s}{}^*$ will expel all electrons except the 1s electrons. At $h\nu \geq F_{1s}{}^*$ there is a stepwise increase in the absorption coefficient which is called the K absorption edge (Fig.2.4). The dependence of the absorption-edge energy on the atomic number is given through Moseley's relation (Sect.2.1)

$$\sqrt{k/R} = Z - \sigma,$$ (2.52)

where wave number $k = 1/\lambda$ is measured in cm^{-1}, Z is the atomic number of absorbing atoms, R the Rydberg constant, and σ the screening constant.

The constant σ arises due to the screening of the 1s electron by all other electrons. Since the value of σ changes for the atom in chemical binding, one expects the energy values of the absorption edges in different compounds to change also. This effect, observed in many compounds, can amount to 10 eV or more.

Neglecting bound states, the following expression is obtained for the wavelength-dependent absorption coefficient τ in the non-relativistic approximation using hydrogenlike wave functions [2.5, 178-180]

$$\tau = C \, \frac{n^2}{1 + n^2} \, \frac{e^{4 + 2\alpha n}}{1 - e^{-2\pi n}} \, Z^4 \lambda^3$$

$$\tan\alpha = - \, \frac{2n}{n^2 + (\kappa/k)^2 - 1}$$ (2.53)

($n = Z/ka_0$, $k = mv/h$, $\kappa = 2\pi/\lambda$, a_0: Bohr radius, v: velocity of the ejected electron, C: constant).

Within a region of some 100 eV the velocity of the ejected electron is small, i.e., n is large, which simplifies (2.53), namely

$$\tau = C \, Z^{10/3} \, \lambda^{8/3} .$$ (2.54)

The absorption coefficient decreases towards higher energies above $h\nu = E_i$, which corresponds to the absorption edge. Figure 2.33 shows the

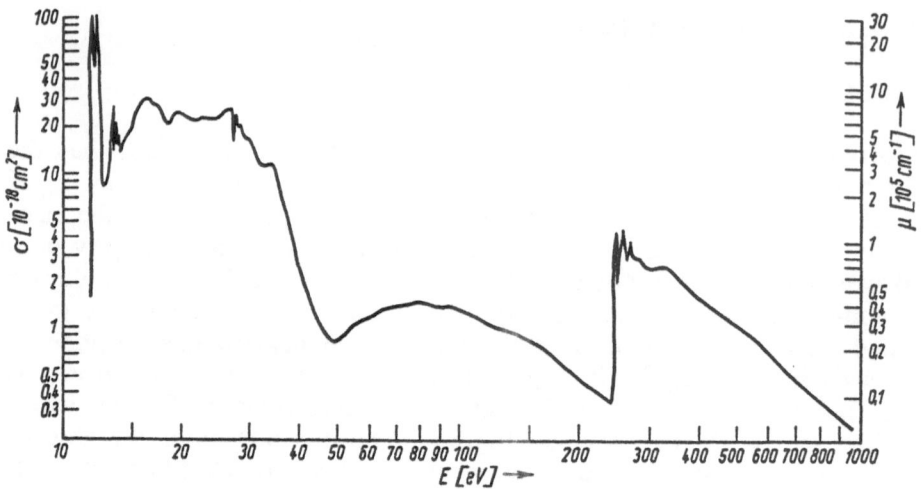

Fig.2.33. The absorption spectrum of argon [2.181]

absorption spectrum of Ar over an energy region that ranges from vacuum-UV to soft X-ray energies. The characteristic saw-tooth profile in the absorption curve is observable in the region of the $L_{II,III}$ edges whilst in the left part of the figure below 100 eV some peculiarities of the absorption of ultrasoft X rays are presented. Such effects are observed in transitions from elements in gaseous as well as in the solid state, are somewhat independent of bonding and can be generally interpreted using atomic models [2.182, 183].

Atomic effects are especially prominent in the fine structure of the spectrum from a solid in the region 10–20 eV above the ionization potential as in this region interactions *within* the excited atoms are important. This deviation of the shape from other X-ray absorption edges (in which the absorption coefficient decreases after the edge) is clearly seen for 3d and 4d transitions from heavier elements ($M_{IV,V}$ and $N_{IV,V}$ edges) (Figs. 2.34, 35). On the short-wavelength side of the discrete absorption lines (due to transitions of 3d and 4d electrons to bound states) the absorption coefficient rises and then starts to decrease after it has passed through a maximum around 10–30 eV on the high-energy side. This effect is independent of the state of aggregation and of binding, as is clear from a comparison of the spectra from gaseous and solid iodine and from the gases Xe and XeF_2 (Figs. 5.10, 15) and can be explained within the framework of free atoms [2.182, 183, 186] along the following lines.

In the one-electron model, the cross-section σ can be written in the dipole approximation as

$$\sigma_{n\ell}(\omega) = \frac{4}{3}\pi\alpha a_0 \left[N_{n\ell}(\epsilon - E_{n\ell})\frac{1}{2\ell+1}\right][\ell M^2_{\epsilon,\ell-1} + (\ell+1)M^2_{\epsilon,\ell+1}] . \tag{2.55}$$

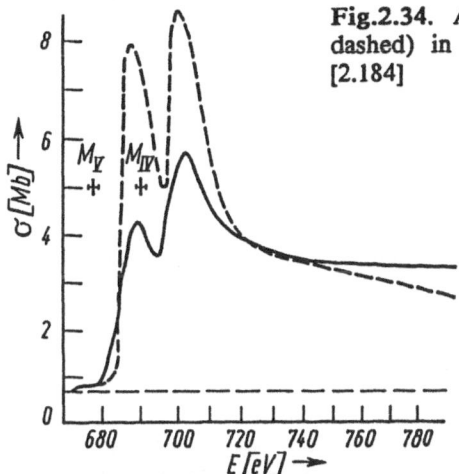

Fig.2.34. Absorption cross-section (theoretical curve dashed) in the region of the $M_{IV,V}$ edges of xenon [2.184]

Here α is the fine-structure constant, a_0 the Bohr radius, $N_{n\ell}$ gives the number of electrons in the subshell, $E_{n\ell}$ the binding energy, and ϵ the energy of the ejected electron.

The radial parts of the dipole-matrix elements

$$M_{\epsilon,\ell\pm1} = \int_0^\infty R_{n\ell}(r)r\, R_{\epsilon,\ell\pm1}(r)\, dr \qquad (2.56)$$

are formed from wave functions which obey an eigenvalue equation with the effective potential

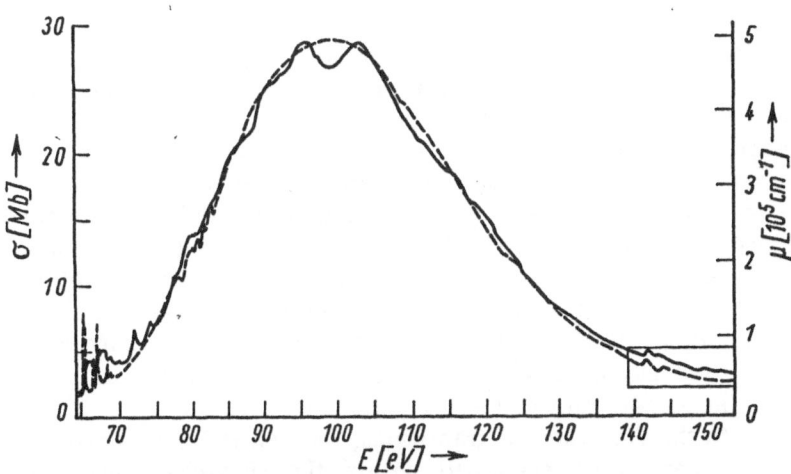

Fig.2.35. Absorption cross-section in the region of the $N_{IV,V}$ edges of xenon in gaseous (dashed) and solid state, $T = 45 \pm 5$ K (solid curve) [2.185]

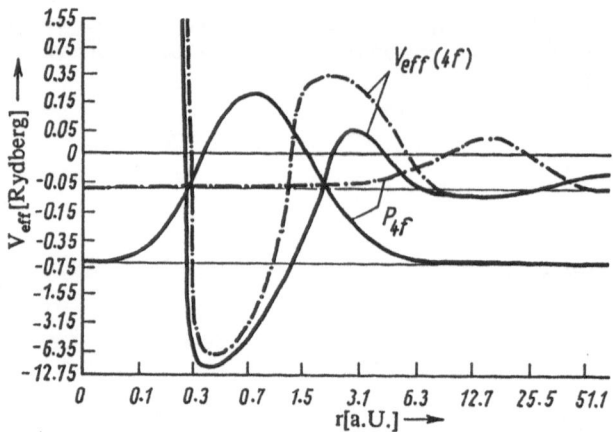

Fig.2.36. Effective potential V_{eff} (4f) and radial part of the wave function P_{4f} (r) for the electron configuration (Xe)6s4f of Ba (dashed-dotted) and (Xe)6s5d4f of La (solid) [2.187].

$$V_{eff} = V(r) + \frac{\ell(\ell+1)\hbar^2}{2mr^2} .$$

(2.57)

Thus absorption by electrons with higher angular-momentum quantum numbers is controlled by the second term in (2.57). This "centrifugal potential" involves a barrier which divides the atomic potential into an inner and outer well (Fig.2.36). States with a total energy below or close to the maximum of the barrier exist therefore as localized states within the barrier (inner-well states) and as diffusely distributed states outside of the barrier (outer-well states). Influence of the potential barrier on the spectra of atomic gases can be characterized by the two main cases below [2.188, 189].

For the first case, the broad maxima in the $M_{IV,V}$ and $N_{IV,V}$ spectra of Xe can be taken as an example. Here the initial state (3d or 4d vacancy) is localized within the potential barrier, whilst the possible final states at the onset of the absorption (energy eigenvalue ϵ of the continuum is zero) are localized outside the barrier, so that the wave functions at $\epsilon = 0$ practically do not overlap at all. The general increase of absorption cross-section above the threshold can then be explained by the delayed penetration of the barrier by the f waves which occurs only at some distance above threshold. At still higher energies the radial matrix element returns to zero and changes sign. Thus the "Cooper minimum" is produced in the 4d ionization channel [2.190]. If the f wave functions inside the barrier are characterized by their atomic analogs, then the above-mentioned cases can be described as 3d → "4f" and 4d → "4f" intershell and intrashell transitions. In contrast to the first kind of transitions, the delayed onset of the 4d → "4f" transitions cannot be explained within the Hartree-Slater model since the strong exchange interaction between the excited electron and the 4d vacancy is not considered.

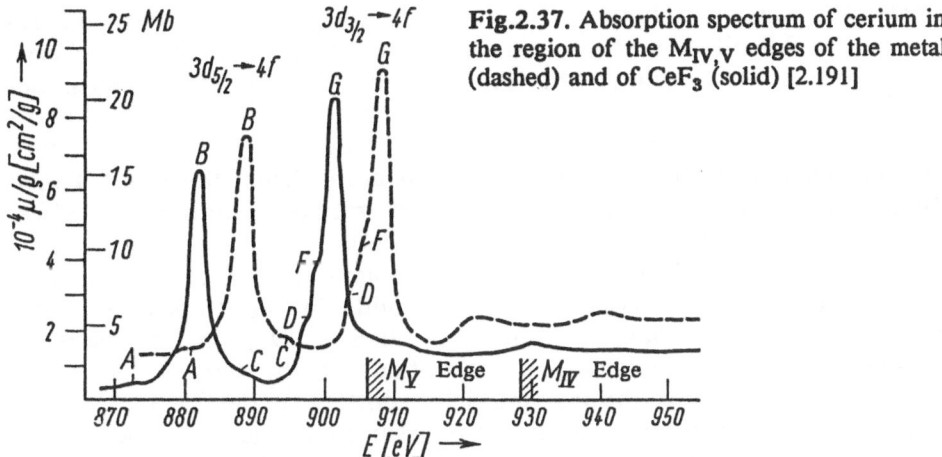

Fig.2.37. Absorption spectrum of cerium in the region of the $M_{IV,V}$ edges of the metal (dashed) and of CeF_3 (solid) [2.191]

A second case of transitions influenced by the potential barrier can be found from the $M_{IV,V}$ and $N_{IV,V}$ spectra of the rare earths (Figs. 2.37, 38). Even though the spectra have been obtained using thin solid films, the atomic models can still be used for interpretation since the states participating in the absorption process are localized within a barrier and are little influenced by the surroundings. In the spectra shown and in the similar absorption spectra from gaseous and solid cerium in the region of the 4d threshold [2.193], lines occur which can be assigned as transitions from 3d or 4d electrons to a final state with f symmetry inside the potential barrier (since the 4f subshell of a rare-earth atom is only partially occupied). In contrast with the $3d^{10}4f^n \rightarrow 3d^9 4f^{n+1}$ intershell transitions (which give lines only below the ionization limit of the 3d shell), the $4d^{10}4f^n \rightarrow 4d^9 4f^{n+1}$ intrashell transitions produce one or several intense maxima

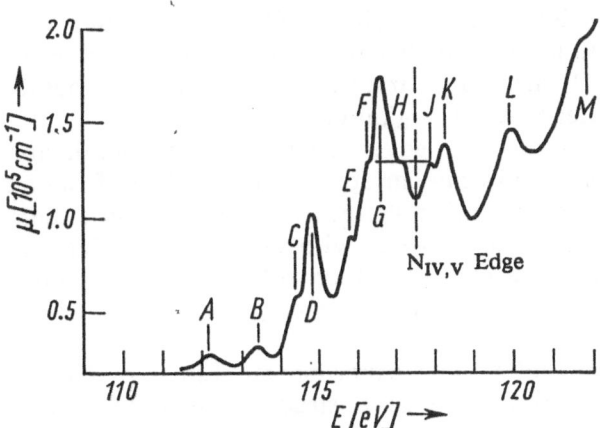

Fig.2.38. Absorption spectrum of neodymium in the region of the $N_{IV,V}$ edge [2.192]

above the ionization limit and a series of weak maxima below it. The reason for this difference is that the exchange interaction between 4f electrons and the 4d vacancy splits the levels of the $4d^9 4f^{n+1}$ configuration over a region more than 25 eV, so that individual terms are pushed above the ionization limit and give intense autoionization maxima. Atomic multiplet calculations with transition rates from perturbation theory on a relativistic Schrödinger equation gave reasonable results concerning the white-line structure for all rare-earth elements [2.194].

Atomic many-electron effects also explain the overall similarity of spectra from isolated atoms and from solids in the absorption region above the 3p threshold of transition metals [2.195]. In the threshold region, spectra are dominated by transitions of the type $3p^6 3d^n 4s^2 \rightarrow 3p^5 3d^{n+1} 4s^2$, whereby, through multiplet splitting of the $3p^5 3d^{n+1}$ configuration, a relatively large width of the absorption edge is produced.

Earlier calculations with different approximations (Hartree-Slater, Hartree-Fock-Slater, Hartree-Fock, random-phase) predict correctly many of the observed spectral features [2.196-202].

The centrifugal effects mentioned above as the most real among other sources of potential barrier are insufficient to explain all features of the near-edge region. A more successful framework for interpretation of this fine structure is provided by the theory of multiple scattering [2.203-205], which we will not treat in detail.

2.6.2 Near-Edge Structure of the Absorption Spectrum

In recording X-ray absorption spectra with instruments of high resolution, more or less pronounced oscillations are observed in many spectra in the immediate vicinity of the edge which extends some 100 eV on the short-wavelength side of the edge. In particular, large effects can be observed in X-ray absorption near-edge structure (XANES) changing the surrounding of an absorbing atom. Within this region of XANES (renamed from the older term Kossel structure), transitions to bound or quasi-bound states occur which extend over a region of about 20 eV.

Figures 2.39-42 exemplify the near-edge fine structure of atomic and molecular gases and also from solids. Similar spectra are available from an extended bibliography of core excitation spectra [2.206]. The prominent structures of the spectra from different gases with the same absorbing atom differ strongly. If studies are carried out on solids, some individual lines disappear completely while some small, strongly smeared maxima remain observable.

In recent years many theoretical and experimental studies of fundamental aspects of XANES and its use as a structural tool have been reported [2.207]. In the simplest model core-hole potential effects are ignored and the fine structure is interpreted in terms of a site and symmetry selected density of unoccupied states. Strong deviations, however,

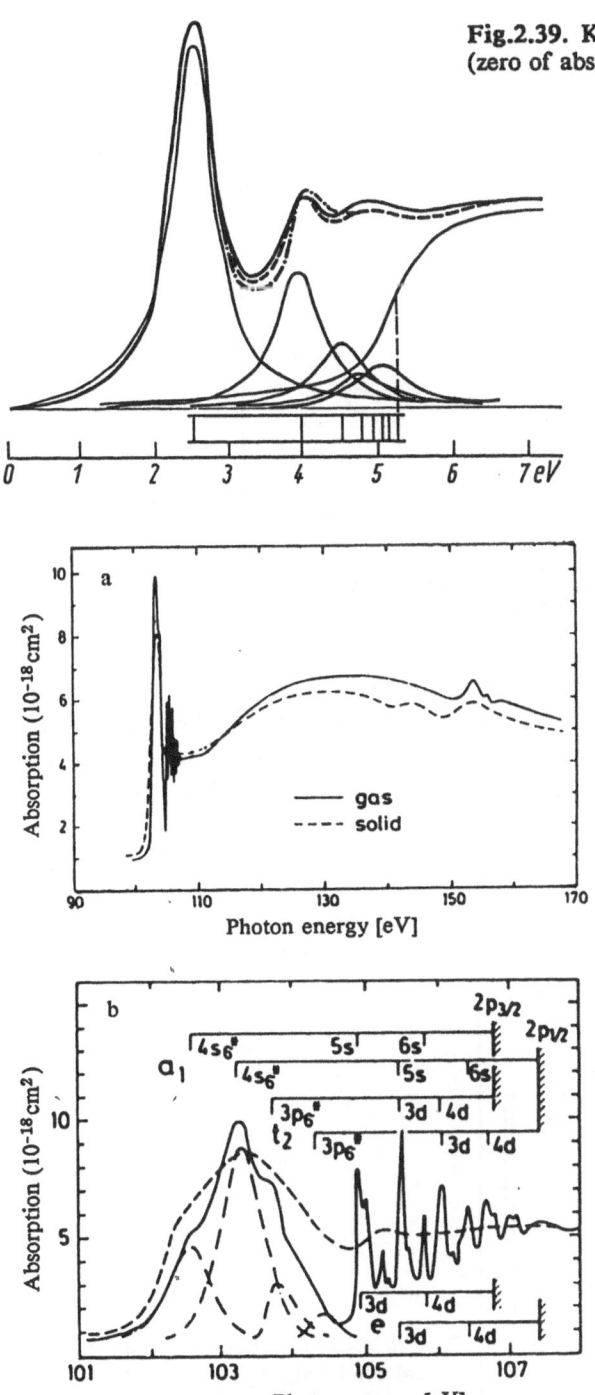

Fig.2.39. K absorption spectrum of argon (zero of abscissa arbitrarily chosen) [2.106]

Fig.2.40. Silicon L absorption spectra from molecular and solid SiH_4 (a) and a section in the region of the $L_{II,III}$ ionization limit (b). The absolute cross-section refers only to the gaseous sample [2.209]

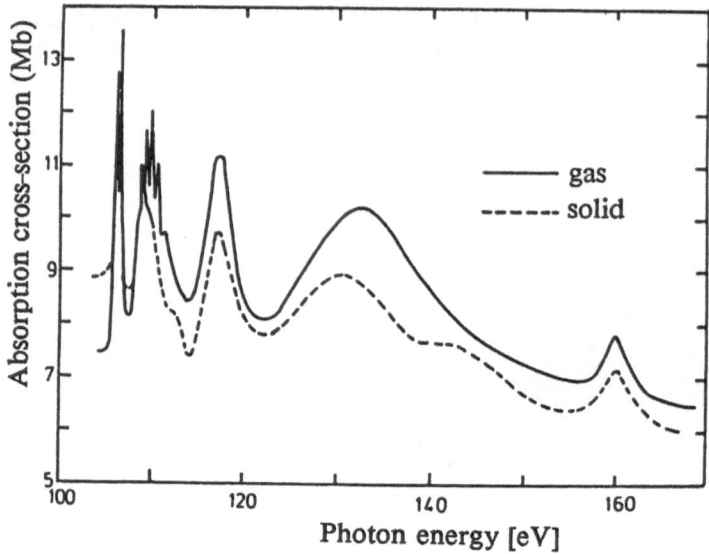

Fig.2.41. Silicon L absorption spectra of gaseous and solid SiF_4. The absolute cross-section refers only to the gaseous sample [2.209]

between the local density of states and the X-ray absorption features indicate clearly influences of core-hole potential [2.208]. Nevertheless the traditional view in the framework of the MO theory, the ligand-field theory and the band theory explain the essential details of the spectra and will be treated more in detail.

The production of the narrow absorption lines in the near-edge region of isolated atoms can be illustrated by the case of the K absorption spectrum of argon: When an X-ray photon ejects a K electron, the ion produced resembles the atom of the next element of the periodic table,

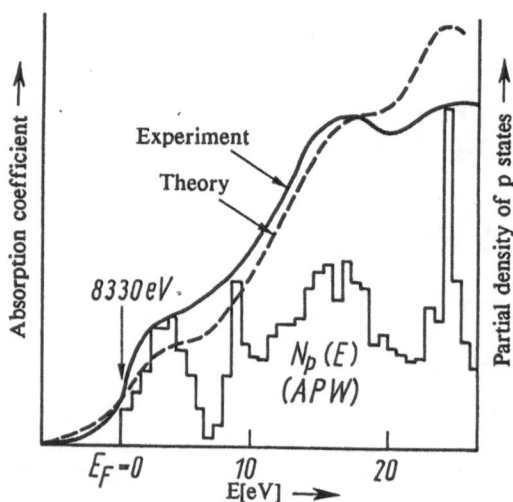

Fig.2.42. Comparison of experimental Ni K absorption spectrum with a curve calculated from the density of unoccupied p states [2.155]

potassium. Since the electron configuration of the argon atom agrees with the K^+ ion, the electron ejected from the K shell of the argon atom moves in a potential field resembling that of the K^+ ion [(Z+1) approximation]. Thus, absorption of X rays will produce excited states of the argon atom comparable to the excited states of the potassium atom. As the dipole selection rules are dominant, only empty p states are available for the transition of the 1s electron which for the argon atom starts with 4p. This transition is visible as the first line in the K absorption spectrum, and is followed by transitions of higher energies corresponding to 5p, 6p states, etc., which finally converge to a Rydberg series limit at ionization threshold. The distance between the first absorption line and the threshold of the continuum should correspond to the ionization potential of the 4p state of the potassium atom. Figure 2.39 shows the extremely good agreement with the corresponding values of potassium, which are deduced from optical values (shown as vertical lines below the components of the X-ray absorption spectrum). In the figure the individual components of the K absorption spectrum are shown. In comparison to the optical lines the large widths of the X-ray absorption lines are evident; except for the first line all the others merge. The energy values of the transitions are not sufficiently well resolved to analyze the X-ray absorption spectrum in detail. Instead a theoretical model which includes the relative intensities, the shape of the absorption lines and the shape of the threshold of the continuous absorption must be developed.

To calculate the above-mentioned parameters a series of equations has been obtained, which are based on the central-field approximation using the time-dependent perturbation theory developed by P.A.M. Dirac, and hydrogenic wave functions for the atom with different charges in the initial and final states. These equations for calculating the energies of the stationary states (more properly, quasi-stationary states), the limit of the continuous spectrum and the shape of the absorption spectrum will be summarized below [2.26].

The energy of the K absorption lines and the limit of the continuous spectrum can be obtained from

$$E_\infty - E_{n+a} = \frac{\eta^2}{(n+a)^2} R \quad (a = 0, 1, 2, ...) \tag{2.58}$$

(E_∞: energy of the continuous-absorption limit, E_{n+a}: energy of the absorption line with effective quantum number n+a, η: effective charge of the absorbing atom).

The distance between two successive absorption lines decreases with increasing a, i.e. by approaching the value E_∞ inversely proportional to $(n+a)^2$. Using the Rydberg formula (2.58) the energies of the Rydberg states are very accurately calculated, so that this gives a possibility to determine the binding energy of the core electrons from the series limit of the corresponding absorption lines.

The profile of the absorption line is given by the energy-dependent absorption coefficient $\tau(E)$ (the ordinate of the absorption spectrum)

$$\tau(E) = \gamma^2 \frac{\tau_{n+a}}{4(E - E_{n+a})^2 + \gamma^2/4} \tag{2.59}$$

(τ_{n+a}: absorption coefficient at the maximum of the line, E_{n+a}: energy of the maximum of the absorption line, γ: width of the absorption line). Widths of individual absorption lines equal the width of the core level.

The continuous spectrum has the shape of an arctan function

$$\tau(E) = \frac{\tau_\infty}{\pi} \arctan \frac{E - E_\infty}{\gamma/2} + \frac{\pi}{2} \tag{2.60}$$

(τ_∞: asymptotic limit of the absorption coefficient for E, γ: width of the core level).

The ratio of the integrated absorption coefficients in the discrete and in the continuous spectrum is given by

$$\frac{S_n}{\tau_\infty} = 2\eta^2 \frac{n^2 - 1}{n^5} \left[1 + \frac{6\eta^2}{n^2 Z^2} \right] \tag{2.61}$$

(S_n: integrated absorption coefficient, i.e., the area of the absorption line with quantum number n).

Equations (2.58-61) show that all the parameters of the spectra depend on three variables: the effective charge η, the quantum number n, and the nuclear charge Z. If the K absorption spectra of neon, argon, and krypton (Figs.2.39,43,44) are compared, then the decrease of the intensity of the absorption lines with increasing nuclear charge [according to (2.61)]

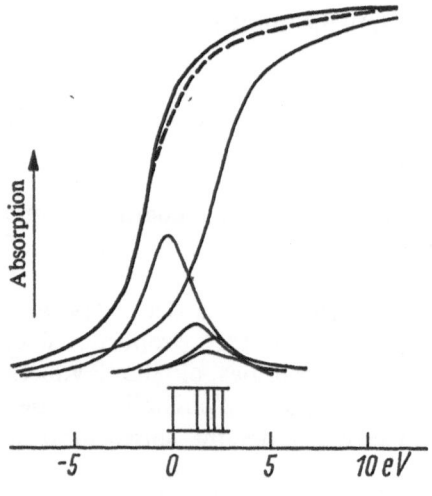

Fig.2.43. K absorption spectrum of krypton [2.210]

Fig.2.44. K absorption spectrum of neon [2.211]

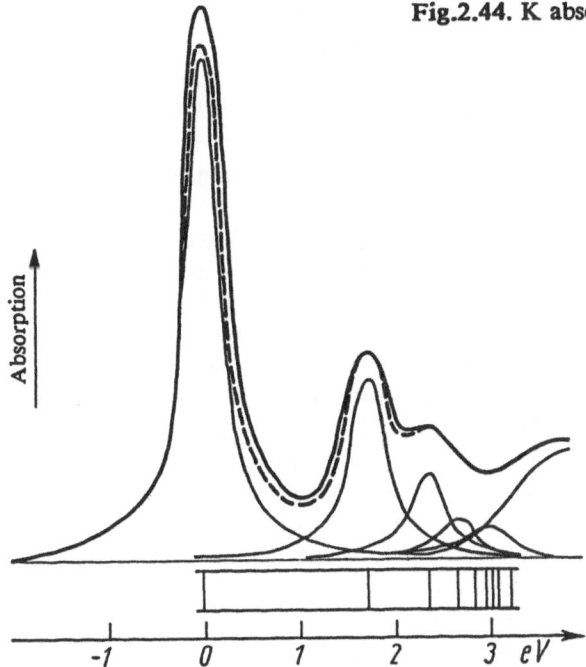

is obvious as is the simultaneously increasing width of the initial state. For krypton this has led to a complete smearing of the structure. If the values for the effective charge $\eta = 1$, the quantum number n (known from optical spectra of the alkali atoms) and also (known either from the spectrum or extrapolated from the Z^4 scaling) the values of the widths of the lines γ are included in the calculation of the absorption coefficient, besides which appreciable broadening due to Auger and instrumental effects must be considered, then the validity of the theory leading to (2.58-61) is demonstrated in an impressive way through the excellent agreement between theoretical and experimental curves. The dashed curves in Figs.2.39,43,44 are calculated accordingly and they agree except for small discrepancies with the experimental curves [2.26].

The good results given by the calculation of the experimental spectra encouraged some researchers to determine effective charges and the other variables from the spectra. Thus, by solving the system of equations (2.58-61) for different components of the spectrum using an iterative method, the variables can be determined [2.26]. The criterion for the validity of this approach is the agreement between the approximate and experimental spectra. The number of required iterations is determined from the characteristics of the structures in the absorption spectrum. Already in the first step good agreement is obtained between theory and experiment for neon, whereby the calculated values $\eta = 1$ and n = 2.12 agree well with those determined from optical measurements of the sodium atom (the agreement is also shown by the relative position of the absorption lines of the neon spectrum and the position of the absorption lines marked below in Fig.2.44). The value of the effective charge $\eta = 1$, obtained from the K absorption spectra for all noble gases, shows that the screening effect from the remaining electrons, caused by the change of the charge in the K shell after ejection of a 1s electron, is negligible, so that the assumption that a valence electron moves in a corresponding fixed central field is approximately valid.

In addition to the K absorption process in an isolated atom, corresponding processes take place in the L shells. The fine structure which occurs in the region of the $L_{II,III}$ edges (Fig.2.45) can be described in terms of transitions of 2p electrons into bound states with s and d symmetry. In the central-field approximation, the energy levels of the stationary states of the excited 2p electrons produce four Rydberg series, of which two with the same initial state (L_{II} or L_{III}, respectively) have a common limit which corresponds to the transition of the 2p electron into the continuum with a zero kinetic energy. Energy levels can be calculated from (2.58) with the same effective charges η but different quantum numbers n_s and n_d for the s and d series. Equations (2.58,59) for the K absorption spectra are valid also for these spectra and the ratio between the absorption coefficients for transitions of L_{III} electrons into the discrete, and the continuous spectrum can be calculated similarly to that for the K spectrum, for $\eta \ll Z$, viz.

$$\frac{S_n^s}{\tau_\infty^s} = \frac{2\eta^2}{n_s^3} \quad \text{and} \quad \frac{S_n^d}{\tau_\infty^d} = 2\eta^2 \frac{(n_d^2 - 1)(n_d^2 - 4)}{n_d^7} \tag{2.62}$$

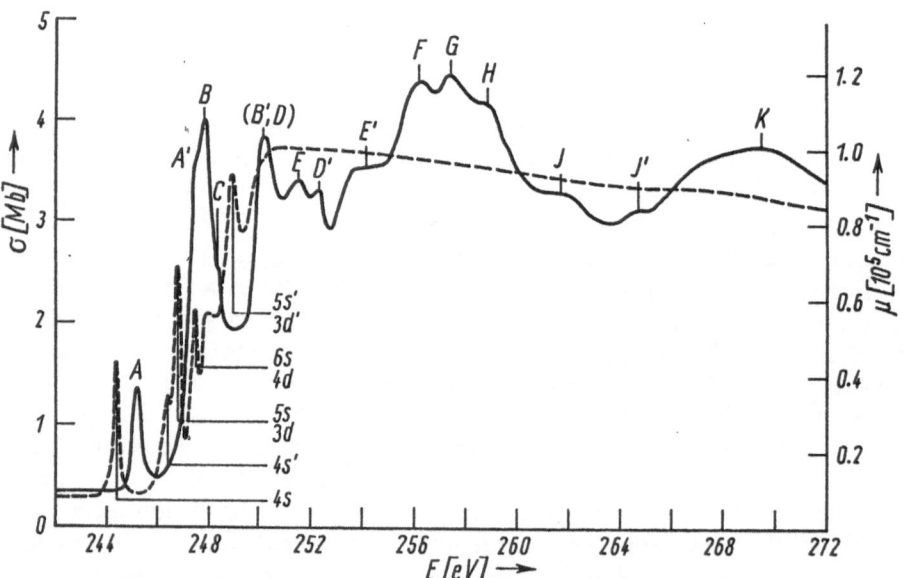

Fig.2.45. $L_{II,III}$ absorption spectrum of gaseous (dashed) and solid argon (solid) [2.181]

(S_n^s, S_n^d: integrated absorption coefficients of the absorption line for the transition $p \rightarrow ns$ and $p \rightarrow nd$; τ_∞^s, τ_∞^d: asymptotic values of the absorption coefficient of the continuous spectrum). Thus the L_{III} absorption edge of noble gases can be calculated in analogy to the K-edge case using optical data from alkali spectra (a prerequisite is that the ratio of the transition probabilities for the d and s states can be estimated, for example, by using hydrogenic wave functions [2.26]).

In contrast to the spectra from free atoms, the absorption spectra from molecular gases cannot be analyzed by using a Rydberg formula in many cases. The fine structure below the ionization limit of molecules can be described by one-electron transitions into unoccupied orbitals of Rydberg or valence character or a mixture of these. For simple molecules like some hydrogen compounds of elements from the third period of the periodic table there is, for example, the structure of SiH_4 shown in Fig.2.40. The absorption starts with transitions to antibonding orbitals which show spin-orbit splitting and are superimposed forming a band largely independent of the aggregation state. Sharp lines follow thereafter which are assigned to transitions into orbitals of Rydberg type, but they disappear in spectra from solid targets [2.209]. Comparison of absorption spectra of gaseous and solid targets can often be used to ascertain the mixing of valence and Rydberg states, and the local character of resonances (see below) [2.212].

In molecules, where the central atom is enclosed in a cage formed by a bonding partner with higher electronegativity and also for simple molecules such as N_2, CO and NH_3 [2.213-216], there are strong fluctuations in the region up to 20 to 30 eV above the ionization limit, in addition to the fine structure below the ionization threshold. These intensity variations are very similar for gaseous and solid samples, and have a tendency to become wider and weaker with increasing energy (Fig.2.44). These strong "shape resonances" close to the ionization threshold are caused by interactions of the excited electron with the surrounding atoms, which can be treated in the framework of multiple-scattering theory [2.203-205]. For explanation of the resonance states and the high intensity of the resulting absorption lines on one hand, and the occasional depression of Rydberg lines, on the other hand, earlier models for potential barriers [2.213, 217-220] turned out to be useful. As an example the potential for a unit with O_h symmetry is shown in Fig.2.46. If the electron is close to the central atom then the attraction of the partially screened nucleus is dominant (inner-well potential) whilst at larger distances from the central atom only the Coulomb attraction of the molecular ion is recognizable (outer-well potential). In between a barrier is produced from the electronegative ligands.

In other structural units, for example in planar molecules of symmetry D_{3h} (BF_3), the effective potential for a specific state is dependent on the direction of the corresponding orbital. Thus, the orbitals perpendicular to the BF_3 molecular plane are not affected so that no barrier is produced

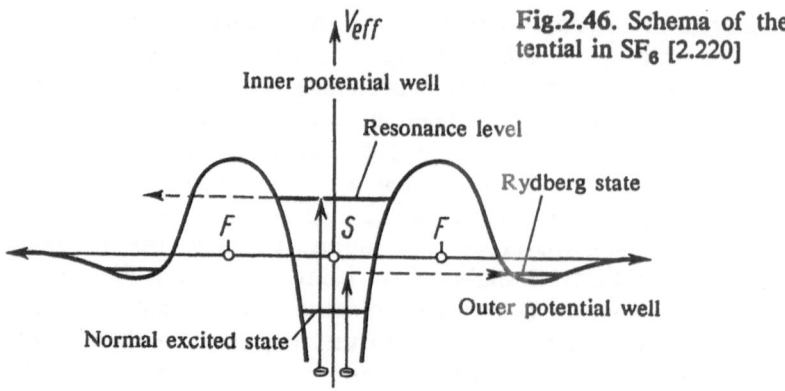

Fig.2.46. Schema of the effective potential in SF$_6$ [2.220]

Inner potential well

Resonance level

Rydberg state

F   *S*   *F*

Outer potential well

Normal excited state

corresponding to those excited states. In contrast, the effective potential for those states whose orbitals are oriented in the plane of the fluorine atoms contains a potential barrier.

Figure 2.46 shows the possible states in two different spatial regions separated by the barrier. Inner-well states do not form a Rydberg series. Localization of the excited electron within the molecule is effective in this case in limiting interactions with the environment outside the first coordination sphere. Thus the absorption spectra of solid Na_2SO_4, SiO_2 and SF_4 are similar to those from comparable molecular gases. Another consequence of the increased localization is the possibility that the localized excited electron may reemit producing corresponding lines in the K and L emission spectra of molecular gases (resonance fluorenscence, Chap.5).

The states of the outer potential well form Rydberg series. Since their wave functions overlap very little with those of the inner electrons due to the barrier, the intensities of the lines from the corresponding transitions are very weak.

First calculations of absorption spectra for compounds used, with some success, the model developed to calculate the spectra for free atoms [2.26, 221]. In spite of the fact that this work assumed a central field (which does not seem very appropriate), the predictions from this model showed surprisingly good agreement with experimental results. Obviously, the stationary states produced by X-ray absorption are highly excited, i.e., the electron ejected from an inner shell goes into an orbit outside the closed shells of the molecule and moves in a region far from the nucleus of the absorbing atom (a high-excitation approximation). In this way relatively small qualitative differences between the K absorption spectra from, for example, Ar and Cl, for a series of compounds can be explained (compare Figs.2.39,47).

In the central-field approximation, interaction of the nuclei and the remaining electrons of the molecule with the ejected electron is described by an effective field which does not explicitly contain the coordinates of the nuclei and electrons of the molecule. Determination of the excited

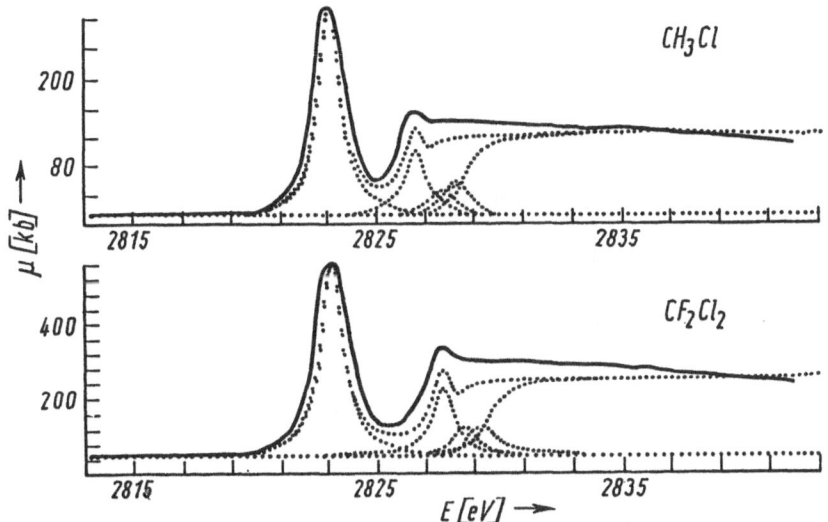

Fig.2.47. K absorption spectrum of chlorine in molecular gases. The solid curves show the experimental spectra; the dotted curves were calculated using a central-field model [2.222]

states of a molecule then follows from solving the problem of describing the motion of the excited electron in this effective potential.

Using a centro-symmetric potential such as

$$U(r) = \frac{\eta e^2}{r} - \frac{\alpha}{r^2} \qquad (2.63)$$

(η: effective charge of the remaining molecule, for r close to unity; α: potential of a dipole which is induced in the remaining molecule from the field of the excited electron), hydrogenic wave functions with effective charge η are obtained for the excited electron for large r. The corresponding part of the absorption spectrum yields a Rydberg series, as for free atoms.

The application of (2.58), explained already in conjunction with the absorption spectra of atoms to calculate the relative energy position of the components of a series, makes it possible to determine the binding energy of core electrons and thereby to investigate atomic charges using a simple electrostatic point-charge model (Sect.4.3).

Another possibility for the determination of effective charge is to apply (2.58) in conjunction with (2.59-61) for the shape and intensity of the absorption lines and to predict the edge using an optimum fit of a theoretical curve to the experimental spectrum (Fig.2.47). Since the effective charge η is influenced by the hole in the K shell of the absorbing atom and by the redistribution of the valence electrons in the atom upon formation of a chemical compound, the parameter

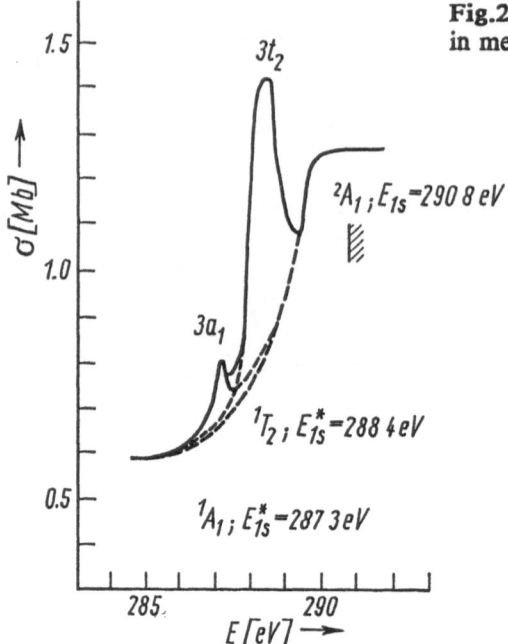

Fig.2.48. K absorption spectrum of carbon in methane, for explanation see text [2.224]

$$\eta' = \eta - 1 \qquad\qquad (2.64)$$

characterizes the distribution of the electron in the neutral molecule, neglecting the charge redistribution in the molecule upon formation of the inner vacancy. The above-described central-field model which regards any particular molecule in terms of the parameters η and α, leads in a number of cases to satisfactory agreement between theory and experiment [2.26]. By considering deviations of the real potential from centro-symmetry using perturbation theory, the splitting and broadening of absorption lines in fields of different symmetries can also be understood [2.26].

Even though the central-field model gives partly satisfactory results for the interpretation of absorption fine structure, many questions are unresolved in this picture. More accurate theoretical spectra can, of course, be deduced from ab-initio calculations of the total energy of the molecule in the ground state and in excited states with an inner vacancy, where the differences in energy give photon absorption energies. The interpretation of the C K absorption spectrum of methane in terms of a single configuration picture is demonstrated in Fig.2.48 [2.40,224]. The energies of the excited states $^1A_1$, $^1T_2$ and $^2A_1$ (ionized molecules) are also shown. Partial cancellation of errors caused by basis set, correlation and relativistic effects may be responsible for the excellent agreement.

If the K absorption spectrum of neon (Fig.2.44) is compared to that of methane (Ne and CH_4 contain the same number of electrons and occupy the analogous electron configurations $1s^2 2s^2 2p^6$ or $1a_1{}^2 2a_1{}^2 2t_2{}^6$ in the ground state), then another peculiarity of the absorption spectra from molecular gases is evident. While for Ne the first absorption line ($1s \rightarrow 3p$) is the most intense one, the absorption for methane starts with a less intense line which, following dipole selection rules, can be assigned to the forbidden transition $1a_1 \rightarrow 3a_1$. The reason for this divergence from dipole selection rules is the electron-vibrational coupling in the molecule. This leads to transition moments differing from zero even though the transition is forbidden in the case of unbroken symmetry. Although the simple single-configuration approach for small molecules in a few cases turned out to be useful, newer theories including configuration interaction appear more successful for XANES interpretation [2.209, 223, 225-227]. Nevertheless numerous studies with strong simplifications arrived at instructive results and will be treated therefore in detail.

The term values needed are often obtained from calculations of an ionic system with a (Z+1) core (equivalent ionic core-virtual orbital method) [2.228]. The term values for the core excited states, i.e., the energy difference between the excited states and the corresponding ionization limit, are here approximated by the negative values of the virtual orbital energies of the final orbital in the ionic system with the equivalent (Z+1) core. Good agreement with the experimental results is obtained for the term values of the argon atom with a 2p vacancy and of the molecules SiH_4 and SiF_4 with 2p vacancy from calculations for the systems K^+, PH_4^+ and PF_4^+ [2.209, 229]. The above-described method is similar to the calculation of an energy E_i^* of an absorption transition from the ground state [2.230-233]. The transition of an inner electron into an unoccupied orbital changes the energy of a molecule insofar as the one-electron Coulomb and exchange terms of the electron in the initial state disappear and instead the corresponding terms of the electrons in the final state arise. Thus the difference

$$E_i^* = \epsilon_f - \epsilon_i - J_{if} + K_{if} \pm K_{if} \qquad (2.65)$$

is obtained as the energy of the transition where ϵ_i and ϵ_f are the orbital energies of the electron in the initial and final orbitals, J_{if} and K_{if} denote the Coulomb and exchange-interaction terms corresponding to (2.8), while positive and negative signs distinguish the singlet and triplet states [2.36].

By neglecting the exchange interactions [2.232], it is possible to estimate the relative position of the absorption lines from the corrected orbital energies where the Coulomb interaction J_{if} of the excited electron with the inner electron is approximated by the interaction with a localized negative charge in the center of the atom [2.234]. Since in this approximation the change of the wave functions due to the adjustment to the inner vacancy is not considered, the theoretical results thus obtained are not accurate. An improvement can be obtained by considering the orbital relaxation using an optimized linear combination of single-electron wave functions in the ground state to describe the spin orbital of the excited state.

A good alternative to HF calculations of core-excited states, which is also applicable to larger molecules, is the rapidly converging multiple-scattering $X\alpha$ method which was developed firstly for discrete molecular states [2.75] and later was extended to cover continuum states, too [2.214]. Besides being used for discrete excited states [2.77, 131, 235], the method has been applied to discrete resonances in the continuum for simple molecules such as N_2 and CO [2.214] and also for highly symmetric molecules such as SF_6 [2.236-239].

Often extensive information can be obtained on the fine structure of absorption spectra of molecular gases using models and simple rules [2.228]. The applicability of the (Z+1) or equivalent-core model to describe the absorption spectra of atomic gases encouraged the interpretation of XUV core-excitation spectra of molecules with Z core through the well-known UV valence-excited spectrum of the (Z+1) molecule. Information on the otherwise difficult to investigate (Z+1) radicals from core-excited spectrum [2.216, 220, 240] was obtained by this method. If these assumptions are valid then there should be a strong similarity between the XUV spectrum of NO_2 and the UV spectrum of O_3. If geometrical and exchange corrections are included deviations are less than 0.1 eV [2.241]. Geometrical corrections are necessary for this case, since for Z and (Z+1) systems different bond lengths and angles occur and for the lighter elements also different core sizes appear [2.228]. Corrections for the exchange interaction occur because the exchange of the valence electrons with the closed shells of the (Z+1) core and with the open-shell core of the Z system often cannot be neglected.

If the XUV spectra of a Z system are compared with the UV spectra of the same molecule, then the term values are not transferable. Since Coulomb interaction between a diffuse Rydberg orbital (R) and a valence (V) or core orbital (C) is very similar, the term values depend on initial orbitals because of the exchange interaction. For core orbitals this is relatively small so that the following simplified relation is valid between the term values for the UV (V→R) and XUV (C→R) excitations because of the absence of the exchange term in the triplet excitation [2.209]:

$$T^1(V{\to}R') + 2k_{vr} = T^3(V{\to}R) \simeq T^{1,3}(C{\to}R) . \qquad (2.66)$$

Term values for core-excited Rydberg states are thus larger than those for valence-excited Rydberg singlet states. Using an analogous rule for excitations into empty valence orbitals explains the different position and valence Rydberg mixings of core- and valence-excited states.

The near-edge structure of the absorption spectra of solids with more or less pronounced band structure is shown in the experimental spectra of Figs. 2.42, 49, 50. A comparison with the partly available spectra from the corresponding monoatomic gases (Fig. 2.45) shows that prominent structures occur in the immediate vicinity of the onset of absorption in gases and also in solids with only small differences between the states, while

76

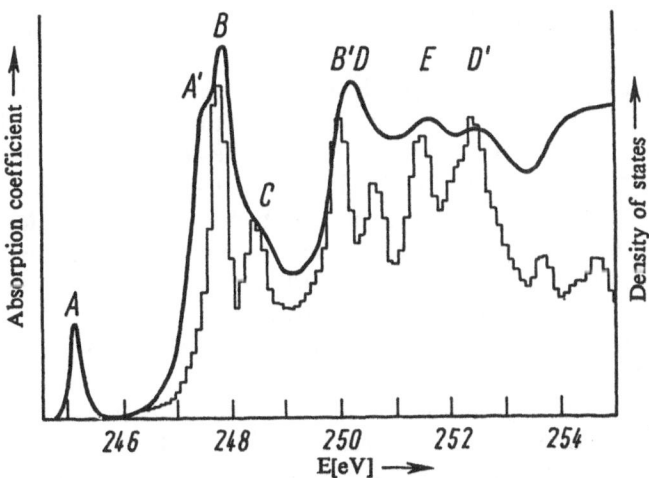

Fig.2.49. $L_{II, III}$ absorption spectrum of argon compared to a calculated density of states [2.242]

other gas-absorption lines in solid-target spectra either are strongly smeared out or visually completely disappear, and in other places of spectra structures of solids occur which have no counterpart in spectra of gases.

It is natural to explain structures in absorption spectra of solids on the basis of one-electron transitions from inner shells to the unoccupied valence band, whence the density of states should be visible in the fluctuations of the absorption coefficient. Interpretation of the relatively narrow lines near the onset of absorption is possible using certain excited states

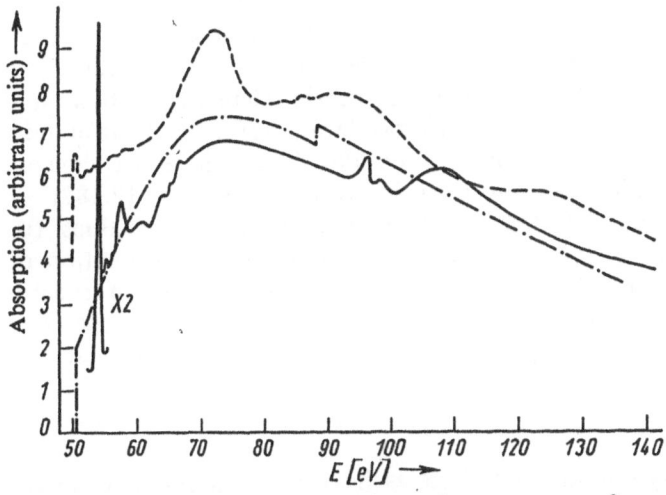

Fig.2.50. Experimental $L_{II, III}$ absorption spectrum of magnesium in metal (dashed) and $MgCl_2$ (solid) and theoretical curve for the free atom (dashed-dotted) [2.243]

(excitons) which play a prominent role in measurements in the optical region.

As discussed in Sect.2.3, bands are produced in k space upon aggregation of atoms to form a solid, due to splitting of the occupied and unoccupied electron states whose structure determines the density of states and thereby influences the intensity distribution of one-electron transitions considerably. The energy eigenvalues of the bands are distributed over the total region of the Brillouin zone and are no longer determined by the atomic quantum numbers but by the band indices and the wave vector k. Contrary to this, the large energy eigenvalues of core states in solids, whose wave functions overlap only a little, show only a weak k dependence. These bands are flat and atomic selection rules for transitions from these flat bands into the conduction band similar to transitions from the valence band to inner holes (Sect.2.6) can be applied. If the excited electron leaves a hole with p symmetry after absorption, then strictly speaking only transitions into some states in the conduction band are allowed whose wave functions contain components with s or d symmetry. The relation between absorption coefficient and photon energy is given by

$$\mu(E) \simeq \sum_j \int_{S(E)} \frac{P(E,k)}{|\nabla_k E_j|} \, dS \, , \qquad (2.67)$$

which is completely analogous to (2.30) for the intensity of the emission bands. Even if removal of the transition probability $P(E,k)$ from the integral due to its anisotropic character in momentum space [to permit the approximation $\mu(E) \simeq P(E)N(E)$] seems doubtful, the results still show that information can be obtained on the band structure and the partial density of states $N(E)$ from the absorption curves [2.155, 166, 242]. It must be emphasized, however, that many effects arise which cannot be predicted from the one-electron density-of-states model. Among these we have already alluded to energy-dependent matrix elements, intraatomic multiplet splittings and general electron-correlation effects, electron-hole interactions such as excitons (see below), besides which there are conspicuous edge anomalies which can also arise in absorption spectra [2.182].

Returning to the one-electron density-of-states picture, values for the width of the forbidden zone can be obtained from a combination of the intensity distributions of the emission bands and absorption spectra under the assumption that the absorption spectra in the region of the edge can be satisfactorily described by (2.67). In reality, the short-wavelength emission edge and the long-wavelength absorption edge coincide with the Fermi-edge positions for a series of metals, a fact known for a long time [2.3].

For some materials bound excited states have been found close to the bottom of the conduction band and have been marked with dashed lines

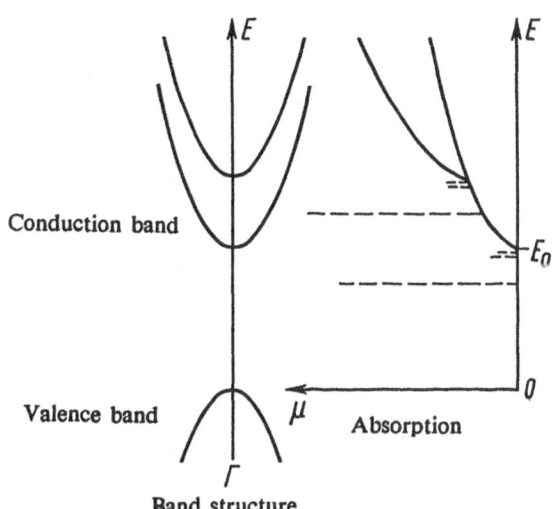

Conduction band

Valence band μ Absorption

Γ

Band structure

Fig.2.51. Schematic representation of the band structure and the absorption spectrum in the optical domain (exciton lines dashed) [2.244]

in Fig.2.51. Transitions into these excited states (so-called excitons) evidently also play a role in X-ray absorption spectra (Figs. 2.49, 50).

The utility of the X-ray exciton concept [2.245] depends on the extent to which the excited electron stays bound to the Coulomb field of the hole that it has left behind. Two limiting cases are recognizable. If the excited electron is in the same elementary cell as the residual hole, then one has the case of a strongly-bound or Frenkel exciton, whose excitation energy differs only slightly from that of a corresponding state of a free atom. In the other limiting case the electron is only loosely bound to the hole and moves around the hole as the electron moves around the proton in the hydrogen atom (Wannier model). The E_n of the excited states of the excitons form a Rydberg series:

$$E_n = - \frac{m^* e^4}{2\hbar^2 \epsilon^2 n^2} \quad n = 1,2,3... \tag{2.68}$$

and converge to a series limit which corresponds to the onset of transitions into the conduction band (m^*: reduced mass; $1/m^* = 1/m_e^* + 1/m_h^*$ with the effective electron mass m_e^* and the effective mass of the hole m_h^*, ϵ: static dielectric constant).

The applicability of the above-developed theoretical concepts will now be examined with the help of some examples. Figures 2.42,49 show that the density of states of the conduction band can satisfactorily explain most of the structures in those absorption spectra. For the Ar spectrum the question about the origin of the first maximum A remains. Due to the agreement of the energy and relative intensity of this line with those of the first gas-absorption line of argon (Fig.2.45), it is reasonably associated with the production of a Frenkel exciton. As the lower part of the conduction band is formed from wave functions with s symmetry, then the corresponding line lies below the onset of the conduction band (Fig.2.51). An analogous explanation can be given for the sharp absorption lines in the $L_{II,III}$ spectra of the magnesium halides (Fig.2.50). In the free Mg^{2+} ion somewhat larger differences arise between the $^1S_0 \rightarrow {}^1P_1$ and $^1S_0 \rightarrow {}^3P_1$ ($2p^6 \rightarrow 2p^5 3s$) transitions and between the given exciton lines. This can be caused by the smaller exchange interaction between electron and hole in the magnesium halides compared to the free Mg^{2+} ion.

However, as discussed, even if many peculiarities of the near-edge absorption spectra of solids can be explained using an atomic band and exciton models, several open questions still remain. Thus, for example, the ratio of the L_{III}/L_{II} near-edge intensities for transition metals deviates from the statistical ratio 2:1. This phenomenon has been shown to be sensitive to the charge state. Secondly, the lineshape reveals a decreased near-edge intensity with respect to the DOS, and thirdly, the spin-orbit splitting deviates from that observed in 2p core-line photoemission [2.247-251]. A good simulation of these spectral features is obtained taking into account the atomic-like splittings together with the solid-state DOS [2.252].

Similar intensity ratio effects have been observed in the $M_{IV,V}$ absorption spectra of rare-earth elements [2.194]. Excellent agreement between calculated dipol-allowed part of the $3d^9 4f^{n+1}$ multiplet and observed $3d \rightarrow 4f$ X-ray transitions qualifies XANES for giving information on oxidation state [2.253], mixed valence [2.254, 255] and relative populations of spin-orbit-split states [2.256] for rare-earth compounds.

Simulation of the spectra using many-body theory gives information also on the charge-transfer energies and d-d Coulomb interaction in the ground state [2.257]. In investigating the Cu K edge in the polarized absorption spectra of Cu(II) compounds [2.258] a splitting of the $1s \rightarrow np$ excitation turned out to arise from ligand to metal charge transfer (shake down) induced by the core hole [2.259]. Usefullness of XANES to determine magnetic structure of rare-earth materials [2.260] and electronic and geometric properties of adsorbed species [2.261] has been proved recently.

2.6.3 Extended Fine Structure in Absorption Spectra

The absorption coefficient of an atom which is surrounded in a molecule or solid by other atoms shows besides the near-edge structure a fine structure which extends over several hundred electronvolts above the edge and is called extended X-ray absorption fine structure (EXAFS). After the first EXAFS experiment with synchrotron radiation [2.262] and the development of theory [2.263-265] the application of the local-structure determination in complex or disordered systems is rapidly expanding [2.207, 266-272]. The main properties of this fine structure have been established over the last fifty years and show the following relations with the three-dimensional geometry of the environment of the absorbing a-tom:

1) The positions of extrema in the fine structure are, to a first approximation, determined by the geometry of the environment. Separation of maxima and minima decreases with increasing distance between absorbing and neighboring atoms.

2) As shown first by *Hanawalt* [2.273], the extended fine structure is temperature dependent. Distances between extrema and the amplitudes of the fine structure decrease with increasing temperature.

3) The extended fine structure is dependent on the symmetry of the core-hole states, so that, for example, K spectra resemble L_I spectra while L_{II} and L_{III} spectra are similar. The influence of the core state on the extended fine structure is mainly determined by the quantum number ℓ.

4) In anisotropic crystals studied with polarized X rays, the amplitude of the fine structure is dependent on the orientation of the crystal with respect to propagation and polarization vector of the radiation.

EXAFS is understood to arise via a diffraction process as the electron ejected from the core at the X-ray absorption is scattered by the neighboring atoms. The electron wave associated with the outgoing electron is thought of as superimposed on the backscattered waves forming the final state (Fig.2.52). The transition-matrix element for the absorption process is determined, inter alia, by the overlap of the wave function for the core

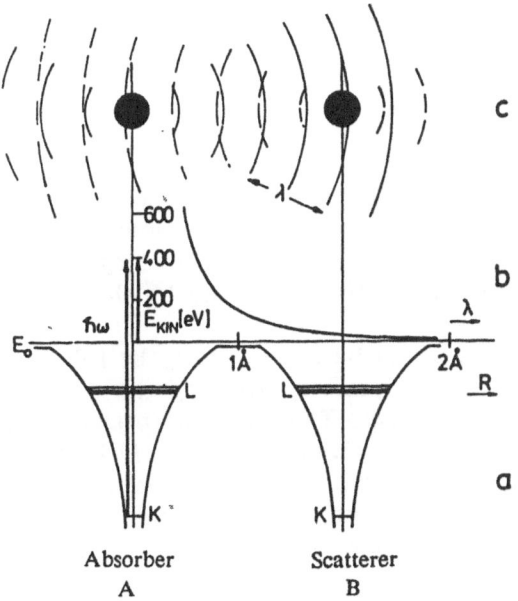

Fig.2.52a-c. Schematic description of the absorption process in the EXAFS domain [2.268]. a) Potential with core levels, $\hbar\omega$: photon energy; b) relation between kinetic energy and the wavelength of the photoelectron; c) emitted and scattered electron waves, the darkened circles represent the region of the 1s function

state and the final state and depends on the relative phase of the outgoing and backscattered waves which changes with the photon energy. We now proceed to look at this process systematically.

The absorption cross-section σ of an atom in the X-ray region can be separated into contributions from the individual atomic levels i (partial absorption cross-sections):

$$\sigma = \sum_i \sigma_i(n_i, \ell_i, j_i) . \tag{2.69}$$

Here (n_i, ℓ_i, j_i) is the set of quantum numbers which characterizes the level i, i.e., n_i is the main quantum number, ℓ_i the orbital momentum and j_i the total angular momentum quantum number. For sufficiently large energies of the photoelectron (wave vector large compared to the reciprocal vector in the chemical compound which contains the absorbing atom) $\sigma_i(k, n_i, \ell_i, j_i)$, the partial absorption cross-section, is proportional to

$$\rho(j_i) \, P(k, n_i, \ell_i) \, N(k) . \tag{2.70}$$

Here $\rho(j_i)$ is the degeneracy of a state j_i, $P(k_i, n_i, \ell_i)$ is the transition probability from the inner state i into a continuum state with wave vector k (neglecting the spin-orbit interaction) and $N(k)$ is the continuum density of states. The wave vector k is, with some limitations (see below), given by the dispersion relation

$$k = \sqrt{\frac{2m_e}{\hbar^2} (h\nu - E_0)} \tag{2.71}$$

and connected with the energy $E_{kin} = h\nu - E_0$ of the photoelectron (m_e: electron mass). The functions $P(k)$ and $N(k)$ are generally monotonic for an isolated atom so that, for this case, no extended fine structure is expected in single-vacancy transitions, neglecting scattering by the atomic shells themselves.

Next some older results on EXAFS will be mentioned to be followed by achievements reported in more recent years spurred by the growing interest in applications for structural analysis. The first interpretation of the extended fine structure in crystals was given by *Kronig* [2.274]. He considered the motion of a photoelectron in the crystal potential using the quasi-free-electron approximation. In this case, discontinuities occur in the dispersion relation where the Bragg relation gives for k

$$k^2 = (k - g)^2 \tag{2.72}$$

where g is a vector in the reciprocal lattice. These discontinuities occur as forbidden energy regions. Since an electron behaves as a free electron outside these discontinuities the function P(k) stays monotonic and N(k) goes to zero in the forbidden energy regions. Thereby inflection points in the function $\sigma_i(E)$ occur at the energies

$$E(g,\nu) = \hbar^2 \frac{g^2}{8m_e} \cos^2\nu \,, \tag{2.73}$$

where ν is the angle between g and k (the density of states goes through a maximum at the lower edge and a minimum at the upper edge). For polycrystalline substances and unpolarized radiation an average over all ν must be taken which gives

$$E(g) = \hbar^2 \frac{g^2}{8m_e} \,. \tag{2.74}$$

The number of extrema in the extended fine structure predicted by this relation is too large. Therefore, *Kronig* assigned to E(g) a diagrammatic representation whereby they are grouped into bands whose amplitude is proportional to the multiplicity of g, i.e., the number of vectors with the same magnitude. In this way, the number of expected undulations is dramatically reduced, leading to a more plausible but rather arbitrary picture.

One result of the early *Kronig* picture is the plausible explanation of the experimental tendency noted under Item 1 above, as long as one deals with a crystal with only one atomic species. It is not possible, however, to explain within this framework the dependence of position of extrema on the absorbing atom and especially the difference in spectra from two different atoms in a crystal.

A possible basis for this difference was suggested by *Hayasi* [2.275]. He considered the excitation of a specific atom as a disturbance of the lattice periodicity. *Hayasi* did not use plane waves to describe the photoelectron as in *Kronig's* method but rather spherical functions. The photoelectron would thus, according to *Hayasi*, undergo "reflection" by neighboring atoms in the crystal, remaining in the vicinity of the absorbing atom if its wave number satisfied the corresponding requirement for reflection. He called these states "quasi stationary" and explained the extended fine structure as caused by transitions to these states. The common basis for the pictures developed by *Kronig* and *Hayasi* is the assumption of complete lattice periodicity. Therefore they are usually referred to as long-range order pictures.

The temperature dependence of the fine structure can easily be understood from the long-range-order approach. The thermal lattice vibrations lead to a decrease of the periodic contribution to the lattice potential affecting the density of states in the forbidden zones or the width of the quasi-stationary states and a smearing of the fine structure occurs.

The temperature dependence of the amplitude of the fine structure can be described through the function $\exp[-2M(T,k^2)]$ [2.276] where 2M is the Debye-Waller factor

$$2M(T,k^2) = 3\hbar^2 \, \frac{k^2}{m_A \, k_B \, \Theta} \left[\frac{\Phi(x)}{x} + \frac{1}{4} \right] \tag{2.75}$$

with $x = \Theta/T$. Here h is Planck's constant, m_A the mass of the atoms in the crystal, k_B is Boltzmann's constant, Θ is the Debye temperature, T is the temperature, and $\phi(x)$ is the Debye function. The decrease of the distances between extrema in the extended fine structure with increasing temperature can evidently be explained by the decrease of the value of g following the increase of the spacing in the crystal lattice.

The effects given under Item 3 above appear comprehensible in the *Hayasi* model, as *Kronig* did not include the symmetry properties of the photoelectron. An energy dependence of the amplitude of the extended fine structure can be explained in terms of the long-range-order picture only via the energy dependence of the Debye-Waller factor.

For spectra from single crystals using polarized radiation, an angular dependence of the positions of the extrema is to be expected using the long-range-order picture as given by (2.73). This prediction gives rise to a contradiction with results from experiments, so necessitating alternative models.

Petersen and coworkers [2.277] developed an approach to the extended fine structure of molecules which differs from the earlier *Kronig* picture for crystals in that elastically scattered waves from the neighbors of the absorbing atom, calculated in the Born approximation, are added to the initial approximate wave functions for the final states (plane waves). In (2.70) the dependence of the cross-section on the environment of the absorbing atom is then changed to

$$P(E) \simeq \int |\langle \psi_f(k)|r|\psi_i \rangle|^2 \, d\tau \; . \tag{2.76}$$

Both the dispersion relation (2.71) and the density of states remain unchanged (ψ_i and ψ_f are the wave functions of the initial and final states of the photoelectron). For a free molecule with a statistical orientation of k, the contributions of the molecule to (2.76) have to be integrated over all k directions.

Petersen and coworkers [2.277] avoided this integration in the calculation of the Ge K absorption spectrum of $GeCl_4$ by averaging the effects of the Cl atoms to obtain a spherical potential. The full potential acting on the photoelectron is thus spherically symmetric so that only the changes of the r-dependent part of the wave function need be considered, thereby simplifying the calculation appreciably. This method was subsequently used by *Kostarev* [2.86] and *Kozlenkov* [2.278] for crystalline solids. The relation derived by *Kozlenkov* (generalized to arbitrary ℓ)

$$\sigma_i = \sigma_i{}^0 \left\{ 1 + \frac{2m_e}{\hbar^2 k} \int \overline{\Delta V}(r)\sin[2kr + 2\eta_\ell(k) - (\ell+1)\pi]\, dr \right\} \qquad (2.77)$$

is valid only for crystals with a highly symmetric lattice according to the approximation by *Petersen* et al.. *Izraileva* [2.30] could, however, show for K absorption that this equation is rigorous for polycrystalline substances. In (2.77) σ_i and $\sigma_i{}^0$ are the absorption cross-sections of the i-th shell of the absorbing atom in the crystalline compound and isolated, respectively; $\overline{\Delta V}(r)$ is the spherically averaged potential in the region of the absorbing atom at a distance r from its nucleus, and η the asymptotic scattering phase of the photoelectron in state $\ell = \ell_i + 1$. [The second allowed state following the ℓ-dipole selection rules with $\ell = \ell_i - 1$ in (2.77) can be neglected due to its small transition probability]. The dependence of the positions of the extrema of the fine structure on the absorbing atom and the symmetry of the inner state enters through $\eta_\ell(k)$.

Equation (2.77) demonstrates the important influence of geometry of the environment of the absorbing atom on the positions of the extrema. Let the potential $\overline{\Delta V}(r) < 0$ be expressed as a sum of separated coordination spheres s about the absorbing atom, i.e.

$$\overline{\Delta V}(r) = \Sigma\, V_s(r) . \qquad (2.78)$$

Since $V_s(r)$ at $r = r_s$ (radii of the s-th sphere) have rather sharp minima, then the contribution of the s-th sphere to the integral of (2.77) results in a maximum for those k given by

$$2kr_s + 2\eta_\ell(k) - (\ell+1)\pi = \frac{3}{2}\pi + 2\pi n , \quad n = 0,1,2... . \qquad (2.79)$$

The extended fine strucutre is thus a superposition of periodic contributions from the separate coordination spheres around the absorbing atom, yielding the total fine structure. Therefore, different spectra arise from different atoms in the same crystal since different atoms have generally different coordination. From (2.77), a decrease of the amplitude of the fine structure with increasing energy is predicted and also a dependence on the absorbing material [by way of $\Delta V(r)$]. The temperature dependence of the extended fine structure derived from the theory of *Kozlenkov* was calculated by *Shmidt* [2.279]. A particularly simple and transparent form is obtained from (2.77) when the following relation is used

$$\overline{\Delta V}(r) = -V_0 \sum_s \delta(r-r_s) , \qquad (2.80)$$

where $V_0 > 0$ is a constant, and $\delta(r-r_s)$ is the Dirac delta function. The periodically varying part of the absorption cross-section is then proportional to

$$\sum_s \frac{N_s}{kr_s^2} \sin[2kr_s + 2\eta_\ell(k) - (\ell+1)\pi] , \qquad (2.81)$$

where N_s is the number of atoms in the s-th coordination sphere. This expression gives a trigonometric series of the variable k. The frequencies are mainly determined from $2r_s$ and the amplitude from N_s/r_s^2. The periodicity of the extended fine structure is then mainly dependent on the fact that discretely separated r_s exist in molecules and crystals. If the r_s are distributed statistically (liquid metals, dense monoatomic gases), then the extended fine structure disappears almost completely.

Kostarev [2.86], *Kozlenkov* [2.278], and also *Shiraiwa* et al. [2.280] showed that the extended fine structure of X-ray absorption spectra from crystals could be fairly accurately described by including the potential from only nearest neighbors (about 3 to 7 coordination spheres) of the absorbing atoms. Therefore, this was called a short-range-order theory. The physical meaning of the short-range-order model was given by *Shiraiwa* et al. [2.280], who calculated the K spectrum of Cu using the *Kronig* method for molecules according to the following: Due to inelastic collisions with the neighbors of the absorbing atom, the lifetime of the excited electronic state is very short and during this time the electron can move only a few Å from the absorbing atom. The electron wave is damped due to penetration into the crystal so that its amplitude quickly decreases to zero. The damping constant of the electron wave was considered a constant by *Shiraiwa*. However, in reality it depends strongly on the energy of the photoelectron. From this arises the energy dependence of the amplitude of the extended fine structure in regions close to the edge [2.281].

The dependence of the extended fine structure of a single crystal on the orientation of a polarized X-ray beam was calculated for K absorption by *Kostarev* and *Weber* [2.282], and *Izraileva* [2.30] using the short-range-order approximation. In agreement with experiment, these calculations showed that the main influence of the orientation consists in only a change of the amplitude of the fine structure.

The basic validity of the short-range-order model was shown principally by experiments on substances which have different long-range order but the same short-range order, e.g., the crystalline and amorphous state of the same substance [2.276, 280, 283, 284] or crystalline form and aqueous solution of a complex salt [2.285, 286]. From these results it is clear that the extended fine structure is determined almost completely by the short-range order. To obtain information on the geometrical (and also electronic) structure of matter, the short-range-order picture is also more convenient than the long-range-order model since,

1) as shown above, only a few coordination spheres influence the spectrum so that the calculation of extrema from (2.79) is considerably simplified compared to (2.74);

2) in contrast to the long-range-order picture, the short-range-order approach, at least in principle, can be used to predict intensities of the fine structure so that it is possible to obtain information involved therein.

Until now the theoretical interpretation of EXAFS has been improved [2.263-265, 268]. These efforts lead, as will be apparent below, to descriptions which are almost indistinguishable from the results of the short-range-order model, as given above, especially see (2.77). The scattering phase, the scattering potential and the mean free path are included with full dependence on k and r; they are exactly deduced from a first-principles theory. According to this model the photoelectron wave created by the absorption process is scattered from the neighboring atoms of the absorbing atom whereby the amplitude and the phase of the back-scattered waves depend on the position, the kind of scattering atom, and the energy of the photoelectron. Through superposition of the scattered wave with the outgoing wave, the varying component of absorption probability and its energy dependence are derived. To calculate the oscillations of the absorption coefficient, a normalized expression is introduced,

$$\chi(k) = (\eta - \eta_0)/\eta_0 . \tag{2.82}$$

This eliminates the influence of the monotonic part η_0 of the absorption cross-section η and emphasized the dependence of the fine structure on the wave number of the photoelectron. If plane waves are used at the absorbing atom and the scattering atom then, for single scattering, the single contributions $\chi_s^{\ell,\ell'}(k)$ to the fine structure due to scattering of an electron at an atom in a coordination sphere s into a state ℓ' can be expressed by [2.268]

$$\Delta_s^{\ell,\ell'}(k) = k \, \chi_s^{\ell,\ell'}(k) = \int_0^\infty dr \, \frac{g_s(r)}{r^2} \, \mathrm{Im}\Lambda_s^{\ell,\ell'}(k,r)\exp(2ikr)$$

$$\Lambda_s^{\ell,\ell'}(k,r) = f_s(\pi,k)\exp(-2r/\lambda + i\delta_\ell + i\delta_{\ell'}) , \tag{2.83}$$

where $f_s(\pi,k)$ describes the complex backscattering amplitude. The phases δ_ℓ and $\delta_{\ell'}$ account for the influence of the potential of the absorber on the outgoing and scattered waves. The distribution of the atoms around the mean distance r_s is expressed through the correlation function $g_s(r)$ which is related to the number N_s of atoms in the coordination sphere s by

$$\int_0^\infty g_s(r)dr = N_s .$$
(2.84)

Damping of the electron wave by scattering processes and the limited lifetime of the core vacancy is taken into account using the mean free path λ.

If a summation is taken over all coordination spheres, then the EXAFS for K (or L_I) absorption spectra is given by

$$\chi(k) = -\sum_s \chi_s^{1,1} (k) .$$
(2.85)

For single crystals with noncubic symmetry the fine structure is dependent on the angle Θ_s between the polarization vector and the vector connecting the absorbing atom and the scattering atom s, according to

$$\chi(k) = -\sum_s 3\cos^2\Theta_s \; \chi^{1,1}(k) .$$
(2.86)

The EXAFS of the $L_{II,III}$ edges with transitions into s and d states is described by

$$\chi(k) = \sum_s \left[\tfrac{1}{2}(1 + 3\cos^2\Theta_s)|M_{21}|^2 \chi_s^{2,2} + \tfrac{1}{2}|M_{01}|^2 \chi_s^{0,0} \right.$$

$$\left. + (1 - 3\cos^2\Theta_s)M_{01}M_{21}\chi_s^{0,2} \right] \left[|M_{21}|^2 + \tfrac{1}{2}|M_{01}|^2 \right]^{-1/2} .$$
(2.87)

For core transitions the ratio of the radial dipole-matrix elements connecting the initial and final states with $\ell = 1$ (2p), $\ell = 0$ and $\ell = 2$ (s and d, respectively) is $(|M_{21}|/|M_{01}|) \simeq 5$ [2.265], so that the second term in this expression can be neglected. In isotropic substances the mixed term is eliminated.

Equations (2.83) or (2.86,87) are suitable, within the framework of one-electron pictures, to calculate the scattering amplitudes in a generally satisfactory way. Influence of relaxation and of many-electron processes on the absolute values of the amplitude and the contribution from multiple scattering [2.268] will not be discussed here.

Recent development of dedicated synchrotron-radiation facilities opened the potential of EXAFS for determination of the local geometry in numerous fields [2.207]. The surface-extended X-ray absorption fine structure (SEXAFS) is becoming the basis for a valuable technique for determination of local configuration of clean surfaces and surfaces with adsorbed atoms and molecules [2.207]. The high cost of synchrotron-radiation sources motivated, on the other hand, the development of various EXAFS-like techniques utilizing electron beams (Sect.2.5) [2.176].

3. Recording of Spectra

In this chapter the design of a high-resolution X-ray spectrometer (Sect. 3.1), the physical basis for the excitation, monochromatization and detection of the spectra (Sects.3.2-4), the main parameters of different spectrometer types (Sect.3.5), and some problems connected with the data evalution from the measured spectra (Sect.3.6) are described. Relationships for the excitation of the X-ray emission spectra with electrons and photons and also for the optimum sample thickness for measurements of the X-ray absorption spectra are given. The luminosity and the resolution of X-ray monochromators equipped with some types of single crystals and gratings are discussed. The different types of ionization counters are described. A detailed description of the dispersion and of the resolution of different crystal spectrometers (Johann-type, Cauchois-type, double-crystal-type) and of a grating spectrometer is given, and finally the influence of different experimental distortions on the shape of the spectra is discussed.

3.1 General Structure of a High-Resolution X-Ray Spectrometer

With respect to its principal, structural features a high-resolution X-ray spectrometer is similar to other spectrometers. Its components fall into three groups: radiation source, monochromator, and detector system, as described in detail in the following sections. Experimental arrangements are determined according to the kind of spectrum and recording technique (Sect.3.2). Available energy-dispersive techniques in X-ray spectroscopy using pulse-height analysis are not considered here since this approach does not yet give sufficient resolution to distinguish the fine structure in the spectra.

The particular arrangements necessary for a high-resolution X-ray spectrometer are determined by the wavelength region to be recorded:
- air or short-wavelength spectrometer, $\lambda < 0.2$ nm;
- vacuum or long-wavelength spectrometer, $\lambda = 0.2$-1.0 nm;
- high-vacuum or ultrasoft-wavelength spectrometer, $\lambda > 2$ nm;
and by the type of monochromator unit:

- single crystal as monochromator (flat crystal: spectrometer system, as introduced by W.H. and W.L. Bragg and W. Soller; focussing crystal: spectrometer system by H.H. Johann, Y. Cauchois, and T. Johansson)
- double-crystal spectrometer;
- grating spectrometer.

Commercially built high-resolution spectrometers are available in the USSR: the RSM500 with diffraction gratings for the ultrasoft wavelength region up to 50 nm and the SARF1 with curved crystals for $\lambda = 1-10$ nm. Otherwise most of the spectrometers used are home built and show large variations depending on their applications. The most important characteristics of an X-ray spectrometer are the sensitivity (throughput), dispersion, and resolving power. These parameters are suitable for comparison of characteristics of different instruments.

The throughput of an X-ray spectrometer is determined by the kind of excitation of the spectra (Sect. 3.2), the path length of the X rays, the reflection power of the dispersive element used (Sect. 3.3), and the efficiency of the detector system (Sect. 3.4). The angular dispersion given by the Bragg relation is

$$D_\psi = \frac{d\psi}{d\lambda} = \frac{n}{2d\cos\psi} , \qquad (3.1)$$

where n is the order of reflection, d the lattice constant, and ψ is the Bragg angle. It was common in the past to give the inverse linear dispersion $D_\ell' = d\lambda/d\ell$ in XU/mm, which can be calculated from the geometry of a spectrometer (Sect. 3.5).

The resolving power A of a high-resolution X-ray spectrometer is important for the characterization; it can be defined, in agreement with generally used notations in spectroscopy, as

$$A = \frac{\lambda}{\Delta\lambda} . \qquad (3.2)$$

Here $\Delta\lambda$ is the full width at half-maximum of the instrumental distortion curve, which depends on the diffraction at the monochromator (Sect. 3.3) and the geometrical configuration in the spectrometer. According to (3.2) the resolving power of an instrument increases with the wavelength of the recorded spectra, which is one of the reasons for the increasing importance of spectroscopy in the ultrasoft wavelength region. Substitution of $\Delta\lambda$ using the equation for the angular dispersion (3.1) gives

$$A = \frac{\lambda D_\psi}{\Delta\psi} \qquad (3.3)$$

from which it is obvious that the resolving power increases with increasing dispersion only if the width of the distortion curve does not increase simultaneously. An increase in the dispersion with a simultaneous increase of the width of the distortion curve - for example, through an increase of the spectrometer radius - will give only a possibility to increase experimental accuracy but not an increase in resolution.

Determination of A usually follows from determination of $\Delta\lambda$. Mostly A is obtained from the difference of the experimentally determined half-width of a spectral line $\Delta\lambda_{exp}$ and from the measurements of the natural width of the line $\Delta\lambda_{nat}$ using a double-crystal spectrometer

$$A = \frac{\lambda}{\Delta\lambda_{exp} - \Delta\lambda_{nat}} . \tag{3.4}$$

The assumption for this simple addition of linewidths requires lines which can be described by the general dispersion formula (Sect.3.6) [3.1]. Furthermore, $\Delta\lambda$ can also be determined from the sum of the geometrical line broadening of the spectrometer and the aberration of the monochromator [3.2], the resolution, half-widths, and intensity ratio of doublet lines [3.3] or through direct experimental measurement of the imaging properties of a spectrometer [3.4].

The best resolving power obtained with a crystal spectrometer is of the order of 10^4. The resolving power is often approximated by the half-width of the crystal diffraction curve or its reciprocal value in eV or 1/eV. Throughput, dispersion and resolution for different spectrometers are discussed in Sect.3.5.

3.2 Excitation of X-Ray Spectra

The first part of this section discusses excitation of X-ray emission spectra and the second part treats production of X-ray absorption spectra.

X-ray emission spectra are excited either by primary excitation, i.e., by energetic electrons, protons [3.5] or ions [3.6], or by secondary excitation or fluorescence excitation using X rays (this process might also be called "cold" excitation). The intensity of a characteristic X-ray line depends on the number, nature, and energy of the ionizing species of quanta j, on the angles of incidence and emergence (ϕ and ψ, respectively) and on the atomic number Z of the studied element. The intensity can be quantitatively expressed in number of photons emitted in a certain transition $i \rightarrow q$ of the q series $N_q{}^j$ compared to the number of the exciting particles or quanta j [3.7] (the energy of j must be larger than the ionization energy of the level q)

$$N_q{}^j = \frac{1}{4\pi} c \frac{N_A}{m_A} \omega_q \, p_{qi} \, \alpha_j \, \sigma_q{}^j(E_0) \int_0^\infty \frac{\Phi_j(x) e^{-xx}}{\sin\psi} \, dx , \tag{3.5}$$

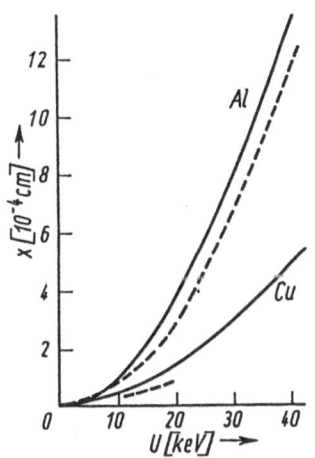

Fig.3.1. Penetration depth x of electrons in aluminum and copper as a function of the excitation potential U [3.8] (solid line: theory, dashed line: experiment)

where c is the concentration of atoms with atomic mass m_A, N_A is the Avogadro number, ω_q is the fluorescence yield, p_{qi} is the ratio of photons of the transition i → q divided by the total number of photons in the q series, α_j is the correction for the intensity gain of the studied X rays caused by the continuous and characteristic radiation of the sample, $\sigma_q^j(E_0)$ is the cross-section for the q level to be excited by particles j of energy E_0, $\Phi_j(x)$ is the ratio of the number of nonexcited atoms to the number of excited atoms as a function of layer thickness x, $\chi = \mu_m(\lambda_{qi}) \cdot$ cosecψ, $\mu_m(\lambda_{qi})$ is the mass absorption coefficient for photons with wavelength λ_{qi}.

Electron excitation has been investigated most often in connection with applications of X-ray microprobe analysis. Penetration depths required in estimating the intensity of the primary spectra are in good agreement with theoretical models (Fig.3.1) [3.8-11]

$$-\frac{\partial E_{kin}}{\partial x} = \frac{2\pi e^4 N_A Z}{E_{kin}} \ln\left(\frac{2E_{kin}}{E}\right) . \tag{3.6}$$

Here E_{kin} is the kinetic energy of the electrons and E is the mean excitation energy of the atom.

The following relation is valid for the penetration depth [nm] for electrons with an energy E = 1-10 keV [3.12]

$$x = \frac{25m_A}{\rho Z^{n/2}} E^n \tag{3.7}$$

where ρ is the density, and n = 1.2/(1-0.29logZ). This relation has been verified experimentally by measurements of the effective layer thickness d = xcotϕ [3.13].

The ionization function $\Phi_j(x)$ has a maximum as a result of electron backscattering [3.14,15]; the ionization cross-section for K levels is constant in the region of $3 < f < 6$ ($f = U/U_K$, U: acceleration voltage, U_K: excitation potential of the K level) [3.15].

The intensity of the characteristic radiation is related to the accelerating voltage according to

$$N_q^e \propto (U - U_K)^n ,$$ (3.8)

where n values between 0.8 and 1.7 have been reported; the most probable value is n = 1.6 [3.16]. The function $N_q^e = f(Z, \phi, U)$ is shown graphically in Fig.3.2. For smaller exit angles the signal-to-noise ratios are, however, larger [3.17].

The acceleration potential is optimized for K spectra if the higher value is chosen from U_1 and U_2 [3.7]

$$U_1 = \left[\frac{3 \cdot 10^5 \sin\phi}{\mu_m(\lambda_{qi})\sin\psi} \right]^{2/3} \quad [keV] ,$$ (3.9)

$$U_2 = 6U_K .$$

In the ultrasoft X-ray region, on the other hand, an optimum value of U near to $U \simeq 15U_q$ was recommended [3.18].

To investigate the X-ray spectrum of a substance using primary electron excitation, a sample is affixed to the anode of a demountable X-ray tube (Fig.3.3). Solid materials can be evaporated or a fine powder may be rubbed into the anode surfaces, whereas special X-ray tubes are needed to study gases [3.19-21].

Incandescent filaments are most frequently used as cathodes, made from a W wire of diameter 0.2-0.3 mm or a Pt wire coated with oxides (prepared from 55-60% $BaCO_3$, 30-40% $SrCO_3$ and 5-10% $CaCO_3$). The wire surface A [mm^2] and thereby the dimensions of the cathode needed to achieve a specified saturation emission current i_s follow from

$$A = 0.00166 \frac{i_s}{T^2} \exp(\phi/k_B T) ,$$ (3.10)

where ϕ is the work function, T is the temperature of the cathode, and k_B is the Boltzmann constant. The lifetime of the cathode is dependent on the vacuum; Pt oxide cathodes are preferred for use in poorer vacuum conditions. The size of the active target region can be controlled by using a focussing design. The anode needs to be both demountable and adjustable. Experience indicates that higher signal-to-noise ratios are obtained when the anode is operated at positive high voltage [3.22].

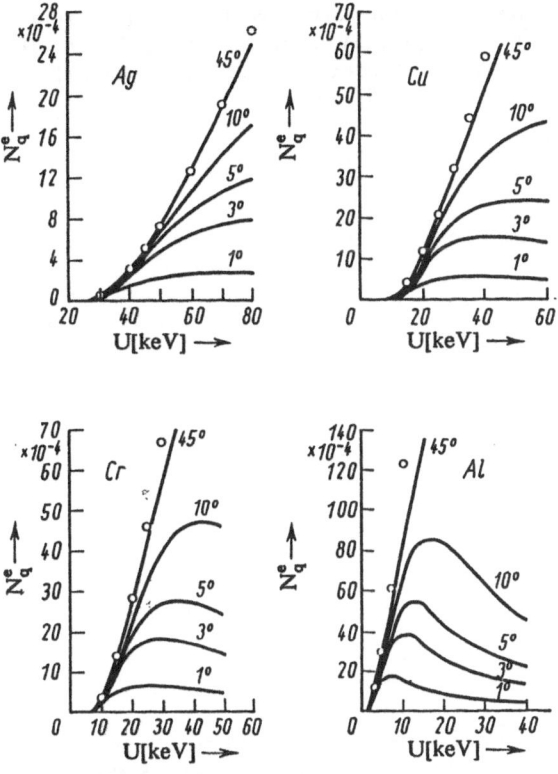

Fig.3.2. Plot of the function $N_q^e = f(Z, \phi, U)$. The points correspond to the functional dependence $(U-U_K)^{1.6}$

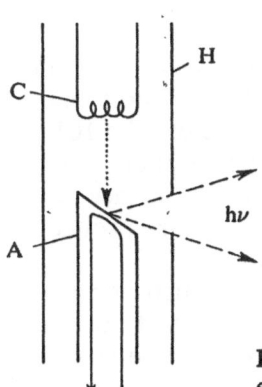

Fig.3.3. Principal components of an X-ray tube for primary excitation. (A: anode; C: cathode; H: X-ray housing; W: water cooling)

To avoid discharges in the X-ray tube a vacuum of $5 \cdot 10^{-2}$ Pa or better is required. For primary excitation, good thermal contact between the specimen and the anode cooling system is particularly critical since more than 96% of the incident electron energy appears as heat at least for

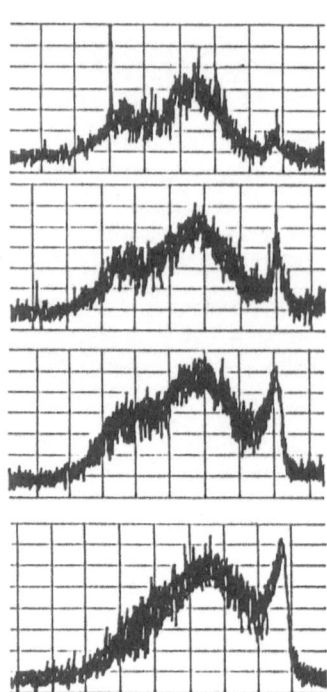

Fig.3.4. Temporal variations of the Al L band of Al_2O_3 using primary excitation (3kV, 2.2mA, 10^{-6} Pa). From top to bottom: 1 to 14 hours after start of measurements [3.25]

energies less than 1 MeV [3.23]. This is one of the main disadvantages of using primary excitation since intense heating of the sample on the anode can cause chemical changes which are hard to control. The number of substances exhibiting negligible response to electron bombardment is very small [3.24]. Thus, especially in the ultrasoft X-ray region, this is a source of considerable difficulty since the rapid decrease in fluorescence yield with decreasing Z requires primary excitation. Excitation yield using primary radiation for the elements of Z = 22 to 11 decreases proportional to $Z^{3.5}$ compared to Z^{12} using secondary radiation, i.e., a factor of $4 \cdot 10^3$ for the elements Ti to Na [3.7]. In addition, the effective penetration depth is smaller for longer wavelengths, and radiation damage in the surface layers is therefore very important [3.24]. Even using very weak excitation, chemical surface changes cannot be neglected (Fig.3.4).

Therefore such X-ray spectra should preferably be studied using secondary excitation. This, however, results in a rather low intensity in spectra of low Z elements because of the low fluorescence yield (due to the increased Auger yield for such elements). In addition to the advantages of secondary spectra noted above, high contrast is obtained compared to primarily excited spectra because of the absence of a bremsstrahlung continuum.

The effective penetration depth to be used in the integration according to (3.5) is determined by the absorption coefficient. For $\psi \simeq \phi \simeq 45°$ and assuming that the exciting radiation is of much shorter wavelength

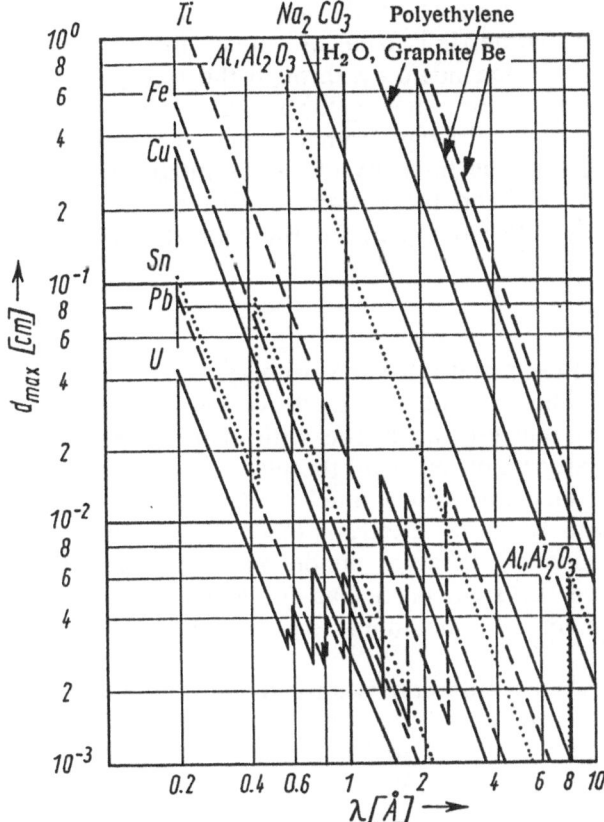

Fig.3.5. Maximum escape depth d_{max} of fluorescence radiation as a function of the wavelength [3.26]

than the fluorescence radiation, the maximum effective production depth d_{max} is given by [3.26]

$$d_{max} = \frac{4.9}{\mu_\ell} . \tag{3.11}$$

Here μ_ℓ is the linear absorption coefficient for the fluorescence radiation. From Fig.3.5 it is clear that the penetration depth of X rays exceeds that of electrons by orders of magnitude (Fig.3.1).

From (3.5) a relation can be obtained for the excitation by monochromatic X·rays [3.7]

$$N_q{}^x = \frac{1}{4\pi} \, c \, \frac{N_A}{m_A} \, \omega_q \, p_{qi} \, \sigma_q^x(E_0) \, \frac{\alpha_x \, b_x}{1 + \chi b_x \sin\psi} \left[1 + \eta_x \, \frac{\sin\psi}{\sin\phi}\right], \tag{3.12}$$

where $b_x = 1/\mu_m(\lambda)$, and η_x is the backscattering coefficient for the incoming photons. Excitation by radiation from an X-ray tube is, on the

97

other hand, less easy to describe because of the superposition of characteristic and bremsstrahlung radiation. It is true, of course, that the intensity of the excited radiation N_q^x depends on λ through the $\mu_m(\lambda)$ dependence (in the vicinity of an absorption edge $\mu_m(\lambda)$ has a maximum and a weak dependence on λ, whereas well away from the edge a variation according to λ^3 is valid [3.28]), but the intensity of the bremsstrahlung N_B^e depends on the excitation voltage and the atomic number according to a relation which differs considerably from (3.5) for the characteristic radiation

$$N_B^e \propto U^2 Z \; . \tag{3.13}$$

By a careful choice of the anode material of the primary radiation source, particularly favorable conditions can be obtained for producing strong fluorescence radiation. Either one uses an anode with characteristic radiation which is close to the absorption edge of the element to be studied, or one chooses as anode an element with a much higher atomic number and utilizes its intense bremsstrahlung spectrum [3.29]. This monograph contains a valuable comparison of different anode materials. Because of the influence of fluorescence yield (in particular for lower Z) and self-absorption (especially for higher Z) a maximum is observed in $N_q^x = f(Z)$ around Z = 50 [3.28].

The intensity of fluorescence radiation is strongly dependent on the distance between anode and secondary fluorescer. To minimize this distance *Alexander* and *Faessler* [3.30] designed a demountable X-ray tube in which the secondary fluorescer was inside the tube. To avoid difficulties accompanying this design (heating of the specimen, partly primary excitation, sample in high vacuum) several researchers [3.22, 31-33] have recently described high-intensity X-ray tubes with distances from anode to fluorescer of <10 mm and capable of currents of up to 800 mA. In particular, there has been considerable effort in the (ultra) soft X-ray region. *Henke* [3.34] and *Deslattes* [3.19] developed special demountable X-ray tubes (Fig.3.6) which can operate with various anodes at voltages in the range 5-20 kV and currents up to 330 mA. These sources produce secondary spectra from some tenths to more than 10 nm.

Very intense radiation in the ultrasoft X-ray region can be obtained from particle accelerators. Radiation from electron synchrotrons and mainly storage rings is being used now widely for X-ray spectroscopic investigations [3.35, 36]. Storage rings offer clear advantages over synchrotrons like higher photon flux, smaller electron beam dimensions, better geometrical and temporal stability, no dead time for the acceleration cycle, better vacuum conditions, and are therefore used exclusively in recent years. The intensity distribution of the radiation covering the full electromagnetic spectrum depends on electron energy, as shown in Fig.3.7. The position of the intensity distribution in the wavelength or photon-energy scale is defined by the characteristic wavelength λ_c (characteristic photon energy ϵ_c) given by

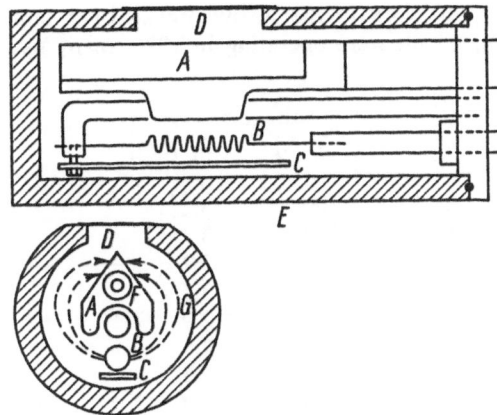

Fig.3.6. An X-ray tube developed by *Henke* [3.34] exciting fluorescence spectra in the ultrasoft X-ray region. (A: anode; B: cathode; C: focussing tube; D: window; E: X-ray tube housing; F: water cooling; G: electron orbits)

$$\lambda_c = \frac{0.56\,R}{E^3} = \frac{1.86}{BE^2} \quad [\text{nm}]\,,$$

$$\epsilon_c = \frac{2218E^3}{R} \quad [\text{eV}]$$

(3.14)

where R is the radius of curvature of the electron orbit in meters, E is the electron energy in GeV, and B is the field strength of the bending magnets in Tesla. The maximum of the photon flux per second at constant resolution and angle aperture lies at $2.3\lambda_c$. Usable radiation is present down to $\lambda \simeq 0.1\lambda_c$. The intensity of the radiation is directly proportional to the electron current. The number of emitted photons at $\lambda \gg \lambda_c$ per second within a mrad in the horizontal plane, within a resolution $\Delta\lambda/\lambda$ of 0.1% and at a beam current of 1 mA, is given by

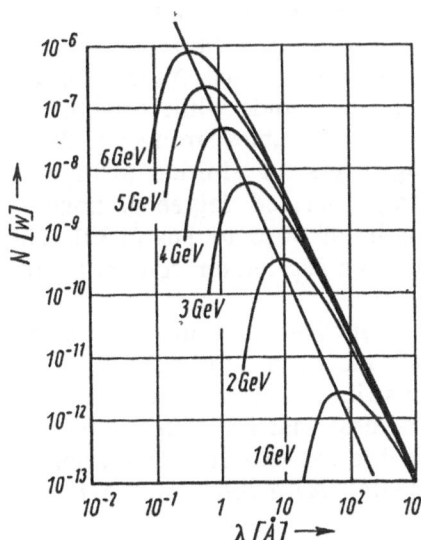

Fig.3.7. Irradiated power of synchrotron radiation as a function of wavelength and electron energy [3.37]

99

Table 3.1. Electron storage rings suitable for work in the X-ray region. (The table includes only storage rings with $\lambda_c \gtrsim 5nm$, which were operating in 1986; many new facilities are under construction) [3.40]

Location	E [GeV]	R [m]	I [mA]	λ_c [nm]
TERAS (Japan)	0.6	2.0	100	5.2
ACO (France)	0.55	1.11	100	3.7
EPA (Switzerland)	0.6	1.43		3.7
MAX (Sweden)	0.575	1.2	200	3.5
UVSOR (Japan)	0.75	2.2	500	2.9
NSLS I (USA)	0.75	1.9	1000	2.5
VEPP-2M (USSR)	0.67	1.22	100	2.3
BESSY (West Berlin)	0.8	1.78	500	1.9
ALADDIN (USA)	1.0	2.08	500	1.2
ADONE (Italy)	1.5	5.0	150	0.8
DCI (France)	1.8	3.82	400	0.4
SRS (UK)	2.0	5.55	500	0.4
VEPP-3 (USSR)	2.2	6.04	100	0.3
Photon Factory (Japan)	2.5	8.66	500	0.3
SPEAR (USA)	3	12.81	100	0.27
NSLS II (USA)	2.5	6.88	500	0.25
CESR (USA)	5.5	32	100	0.1
VEPP-4 (USSR)	6	35.7	10	0.1
DORIS II (FRG)	5.3	12.22	100	0.05

$$N(\lambda) = 4.34 \cdot 10^{10} \, R^{1/3} \, \lambda^{-1/3} \, , \tag{3.15}$$

where R is the bending radius in m, and λ the wavelength in nm. Since the dependence on R and λ is neglibible in this wavelength region, the photon flux is approximately proportional to the electron beam current.

The intensity obtained from an electron storage ring exceeds the bremsstrahlung intensity from the strongest conventional X-ray sources with rotating anodes by up to four orders of magnitude while for characteristic radiation in the ultrasoft region it is 10–100 times stronger [3.38].

Synchrotron radiation of a storage ring offers, in addition to its high intensity, further advantages: It has a strongly defined degree of polarization and is pulsed with typical pulse lengths of 50ps to 1ns [3.39]. The only disadvantage is that it is confined to large accelerator centers (Table 3.1).

Synchrotron radiation can be used to excite X-ray fluorescence spectra or to get absorption spectra. The former has been done only in a few cases [3.41,42] because the broad spectral distribution of radiation includes very intense UV radiation which damages most materials. In X-ray absorption spectroscopy synchrotron radiation has led to a real revolution in the fields of XANES [3.43] and EXAFS [3.44]. As a source for a broad,

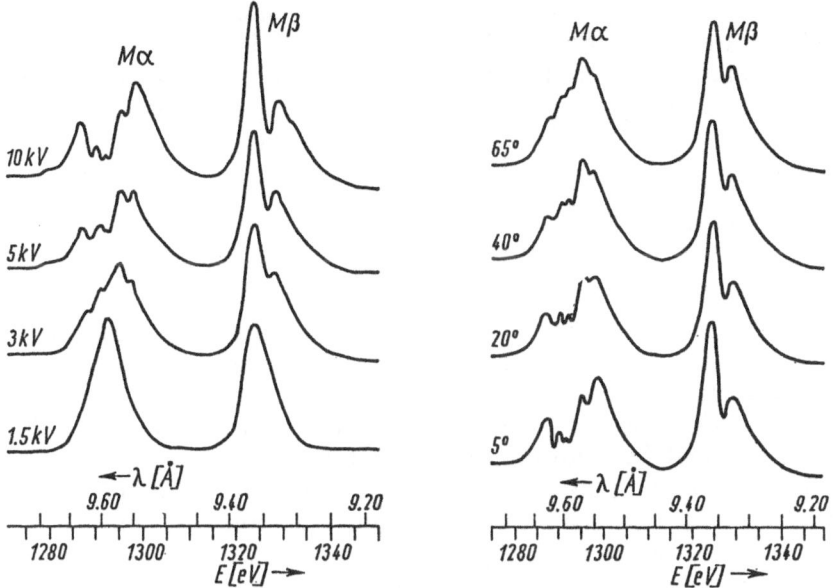

Fig.3.8. Influence of the excitation energy at a constant take-off angle of 30° (left), and of the take-off angle at a constant voltage of 4kV (right) on the shape of the M bands of dysprosium under primary excitation [3.48]

white X-ray continuum, synchrotron radiation is superior to all other sources in the soft X-ray and uniquely in the ultrasoft X-ray region.

In the construction of storage rings one distinguishes between two generations up to now - the first one, when high-energy physics was the main aim and radiation work had to be done in a symbiotic manner with all its disadvantages, and the second one in the design of the "dedicated sources" like BESSY, NSLS and the Photon Factory. Now a third generation is showing up, characterized by special magnetic structures like wigglers or undulators to get more intensity and smaller beam dimensions and to shift the λ_c values to the harder X-ray region [3.45-47].

It is important to minimize the effect of self-absorption in studies of the fine structure in emission spectra, i.e., to minimize the absorption of radiation upon exciting the sample. A particularly conspicuous example of the influence of self-absorption on a spectrum is shown in Fig.3.8. Minimum self-absorption is obtained when the radiation emerges normal to the target surface. Nevertheless, there remains an appreciable problem especially toward longer wavelengths where the problem is partially alleviated by using as low an excitation energy as is conveniently possible.

Absorption spectra are usually studied by recording changes in the intensity distribution of the continuous bremsstrahlung spectrum due to transmission through an absorber, mounted between the source and monochromator, if possible, to minimize the influence of fluorescence excitation. As indicated by (3.13), X-ray tubes with higher Z anodes are used to gain intensity. To avoid the superposition of higher orders of the bremsstrahlung, the accelerating voltage on the tube is chosen to be $U \le 2U_q$. The optimum absorber thickness d_{opt} for obtaining a high contrast in the absorption spectrum depends on whether the main edge of the absorption spectrum (3.16a) or its extended fine structure (3.16b) is to be studied [3.49]:

$$d_{opt} = 2.3 \frac{\log \mu_2 - \log \mu_1}{\mu_2 - \mu_1} \quad \text{or} \quad d'_{opt} = \frac{1}{\mu_2} . \tag{3.16}$$

Here μ_1 and μ_2 are the absorption coefficients on the long- and short-wavelength sides of the edge. Additional effects occur due to statistical fluctuations and matrix effects in the absorber.

Lukirskii [3.50] showed that even in the ultrasoft X-ray region absorption spectra could be studied in this way by using the bremsstrahlung spectrum of a W anode in an X-ray tube operated at 3.5 kV and 200 mA. Furthermore, discrete line spectra (for example, vacuum spark lines [3.51] and even certain X-ray lines [3.52]) can be used as sources of radiation needed to obtain information on absorption spectra.

In the short-wavelength region, fabrication of absorbers is fairly straightforward (by rolling of foils, electron sputtering or vacuum evaporation, loading of organic foils with pulverized materials, pressing with graphite, emulsions with zapon lacquer or collodium solutions or fabrication of appropriate cells for gaseous and liquid samples). However, in the ultrasoft X-ray region it is sometimes very hard to fabricate suitable

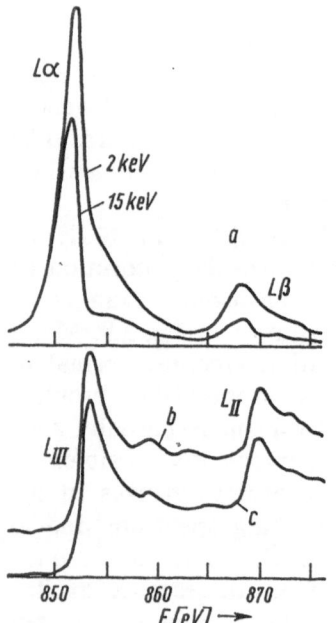

Fig.3.9a–c. Comparison of emission spectra with differing amounts of self-absorption (a). The absorption spectrum obtained using varying self-absorption is shown in (b) while the $L_{II,III}$ absorption spectrum of nickel obtained in a standard way [3.53] is given in (c)

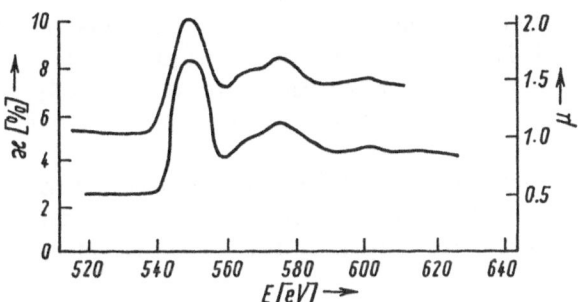

Fig.3.10. Comparison of the spectral dependence of the quantum yield of the external photoeffect (lower curve) with the O K absorption spectrum (top curve) of BeO [3.52]

absorbers (with $\mu \simeq 10^5$ to 10^6 cm^{-1}, $d'_{opt} \simeq 10$ to 10^2 nm using (3.16)). Especially in these cases other methods can be used which give information similar to that obtained in absorption spectra:

a) Recording two emission spectra with different amounts of self-absorption [3.53]. From a comparison of the spectral distribution of the two spectra the absorption spectrum can be deduced (Fig.3.9).

b) Recording the reflection spectrum (Sect.2.6). Especially by using synchrotron radiation, absorption spectra have thus been recorded in the ultrasoft X-ray region [3.54].

c) Recording the spectral dependence of the quantum yield of the external photoeffect $\kappa(\lambda)$. Following *Lukirskii* and colleagues [3.55-57], the relation $\kappa(\lambda) \simeq \mu(\lambda)$ (Fig.3.10) is valid especially in the vicinity of the absorption edge if surface effects on κ are neglected.

d) Recording the appearance-potential spectra of radiation transmitted by an absorber of the substance and measuring the differential change of the current on a photocathode on changing the voltage applied between photocathode and photoelectron collector. Even though the energy resolution does not match that of a dispersive spectrometer, prominent fine-structure details are clearly visible [3.58].

3.3 X-Ray Diffraction

Diffraction of X rays by single crystals or a grating is described by the Bragg relation. The properties of dispersive elements determine essentially the brightness and resolving power as also does the entire construction of the spectrometer (Sect.3.5).

Usually crystals are used for diffraction in the short-wavelength region ($\lambda < 2$nm), and gratings are employed in the longer-wavelength region ($\lambda > 2$nm). Recently, efforts have been made to produce crystals with high reflectivity in the wavelength region λ = 2-10 nm, particularly by growing layer-type crystals with large interplanar distances [3.59-63]. Additionally, holographic diffraction gratings have been introduced [3.64]. A quantitative comparison of properties of gratings and crystals in the overlapping wavelength region cannot be carried out on the basis of available data especially because of different geometrical factors of individual spectrometers. In general, a grating yields a higher intensity [3.65] while its resolving power is better than that of crystals only for longer wavelengths and especially for λ > 5nm [3.52,66].

Crystallographic perfection, adequate reflecting power, elasticity and chemical stability are all needed in crystals suitable for dispersive elements. Deviations from crystal perfection in the form of certain mosaic structures lead to higher reflecting power but to poorer resolving power. The reflecting power can be expressed as [3.67]

$$R = \frac{I_R}{I_0} = \frac{CN|F|R^*}{\sin2\phi} \quad , \tag{3.17}$$

where C is a constant which accounts for the effect of polarization, N is the number of lattice cells per cm^3, $|F|$ is the absolute value of the real part of the structure factor, ϕ is the Bragg angle, R^* is approximately $8(1-2.4|g|)/3$ for the dynamic case and $\pi|g|/4$ for the kinematic case (g $\simeq$ $-0.5(mc/e)^2(\mu/\lambda N|F|)$, μ is the linear absorption coefficient). In measurements using a double-crystal spectrometer, X-ray reflection properties are characterized by the percent or peak reflectivity P (ratio of the maximum intensity of the X rays reflected from the second crystal to the intensity incident on this crystal) and by the reflection coefficient or integrated reflectivity (R_0) (ratio of the integrated intensity of the beam reflected by the second crystal to the intensity reflected by the first crystal [3.1,68]).

According to diffraction theory for X rays a strictly monochromatic beam with infinitely narrow angular distribution is reflected as a divergent beam by the lattice planes. In addition, structural imperfections lead to a distribution of reflected rays around the exact Bragg angle ϕ_0 even when the incident beam is perfectly monochromatic. The ratio of the intensity distribution of this reflected, divergent beam $I(\phi-\phi_0)$ to the integrated intensity of the incoming beam is defined as the diffraction pattern. The combined diffraction pattern of two crystals in parallel position of a double-crystal spectrometer (Sect.3.5) gives the rocking curve of the crystal pair [3.1,68]; its full width at half-maximum, $\Delta\lambda_R$, is a measure of the physical resolving power of a crystal monochromator, which is calculated from the relation A = $\lambda/\Delta\lambda_R$.

Systematic investigations of reflection and resolving power, and their wavelength dependences have been performed only for a few crystals, some of which are exemplified in Figs.3.11-13. In Table 3.2 some results are collected from the literature, which, however, due to the different experimental situations, pertain only to an initial orientation [3.22].

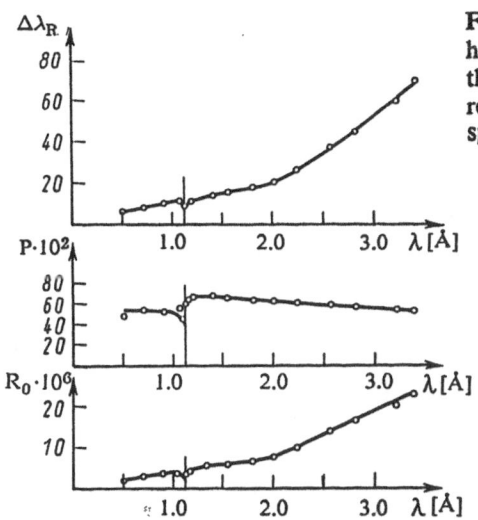

Fig.3.11. Wavelength dependence of the half-width of the rocking curve $\Delta\lambda_R$ (top), the peak reflectivity P (middle) and the reflection coefficient for a double-crystal spectrometer R_0 (below) for Ge (220) [3.69]

Because of the strong wavelength dependence of these parameters, the wavelength region is indicated. As evident from the table, resolving power and reflected intensity also depend on the order of reflection. Following (3.17) these changes in intensity should, to a first approximation, follow the changes in the structure factor. This has been experimentally verified in studies on mica up to high orders of reflection, as shown in Fig.3.14 [3.72].

Due to small penetration depth of soft X rays (Sect.3.2) the surface properties of the crystals used are important. Thus etching of optically polished crystals can sometimes considerably improve reflected intensity and resolving power [3.1, 22, 73, 74]; even an improvement in transmission

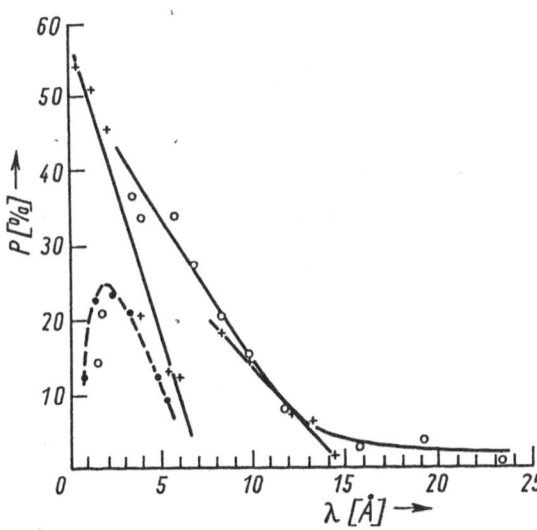

Fig.3.12. Wavelength dependence of the reflectivity P for KAP (circles), quartz (dots) and beryl (crosses) [3.70]

105

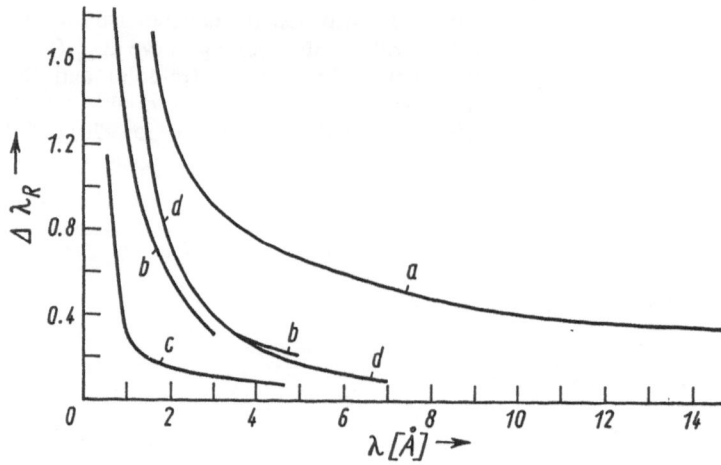

Fig.3.13. Wavelength dependence of the width of the rocking curve $\Delta\lambda_R$ (a) for beryl (10$\bar{1}$0); (b) calcite (100); (c) quartz (11$\bar{2}$0); and (d) quartz (10$\bar{1}$0) [3.71]

Table 3.2. Comparison of the intensity and the resolving power of different crystals for X-ray spectroscopy. (The data are to be considered a first orientation because they were collected from different sources where the experimental conditions were not comparable in all cases. The values in parentheses represent the wavelengths in Å, and the figures in front of the colons specify the reflection orders)

Crystal	d [Å]	Intensity (related to LiF=1)		Resolving power $\lambda/\Delta\lambda$
Topaz				
400	1.162			8900(0.7)
303	1.356	0.25		70000(0.7)
LiF				
200	2.013	1(1.0)	0.9(0.7)	
420	0.90	0.2(0.7)		clearly better than 200
422	0.823	0.1(0.7)		
Calcite				
211	3.036	0.3		11000(1.4)
				8900(0.7)
				2:70000(0.7)
Si				
111	3.135	0.5(2.3)		17...27000
220	3.838			(0.7...3)
Ge				
111	3.27	0.3	3:0.4	2300(1.5)
422	1.1			11000
Graphite				
002	3.358	5.5(1.5)	3(2.7)	
		2(6.1)		

Table 3.2 (cont.)

Quartz				
10$\bar{1}$0	4.255	0.1(0.07)0.5(6.9)		16000(1.4)
				2:48000(1.4)
				3:140000(1.4)
10$\bar{1}$1	3.343	0.3		
13$\bar{4}$0	1.180	0.2		
EDDT				
020	4.402	0.2(5.3)		3000(8.0)
ADP				
101	5.321	0.12(2.7...9,9)		830(10)
				1000(9.8)
Sorbitol-				
hexacetate				
SHA	7.00	0.06		
Gypsum				
020	7.590	0.12(14,6)		
Mica				
001	9.963	0.05(1.7...8.3)		400(10)
		3:0.35	4:0.1	1800(17.6)
		5:0.40	6:0.05	2700(19.4)
		8:0.15		2:4700(9.5)
TlAP	12.95	1.2(25.9)		3000(6.7)
RbAP	13.06	0.06(7.1)	0.15(13.6)	1500(21.4)
				1800(24.3)
KAP				
010	13.31	0.2(5.3)	0.1(18.3)	2...4000(10...23)
				670(10)
Clinochlore	14.196			1600(27.4)
OAO	93.8	0.2(93.8)		
Stearate	100	0.1		

properties has been obtained in this way [3.75]. Influence of uneven crystal surfaces has been discussed by *Bubakova* et al. [3.76].

In addition to their application as plane crystals in single-crystal Bragg monochromators and in double-crystal spectrometers, cylindrically or spherically curved crystals have also been used to improve the luminosity. The simplest method of obtaining a cylindrical surface is to place the crystal between two ground metal blocks having appropriate radii of curvature. This method, however, does not lead to satisfactory curvature for radii <1m and crystal widths of 20 to 30 mm [3.77]. In most cases the crystal bending for high-resolution spectroscopic application is made using four rods or rollers [3.78, 79]. In recent years crystal holders have been used which, in addition to the simultaneous adjustment of two rods, provide a possibility for fine adjustment of a single rod (deviations from parallelism of the rods <5 μm) [3.80, 81]. In this way very accurate bending can be achieved even for large crystals, which has made it possible to study very-low-intensity spectra.

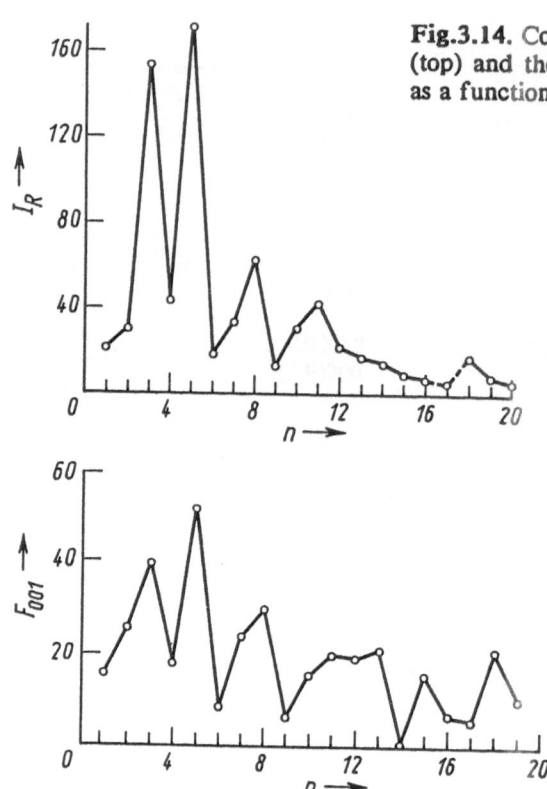

Fig.3.14. Comparison of the reflected intensity I_R (top) and the structure factor F (below) for mica as a function of the order of reflection n [3.72]

The progress of a bending operation is monitored using visible light, afterwards the exact focus is determined by X-ray methods [3.22]. Only particular regions of the bent crystal surface [3.77] need to be monitored to assure precise bending. If the bending is done according to a logarithmic spiral, which can be simply obtained by bending over four rollers and translating one pair of rollers, the focussing aberration caused by cylindrical bending can be minimized (Sect.3.5). Since even a minute change in the distance between the two rods changes the radius of curvature of the crystal appreciably, temperature stabilization of the entire crystal holder is rather valuable [3.31,33]. A temperature change dT entails a wavelength shift of dλ according to

$$d\lambda = -\alpha\lambda dT \ , \tag{3.18}$$

where $\alpha \simeq 10^{-5}$ K^{-1} is the thermal expansion coefficient. For λ = 0.1 to 1 nm typical values for the shifts are 10^{-1} to 10^{-2} pm/K.

The spherical bending of crystals, employed mostly in applications of isochromat spectroscopy but more recently also used to monochromatize X rays for ESCA studies, was developed by *Ulmer* and co-workers [3.82]. The crystal is, in this application, affixed to a corresponding spherical surface by adhesive.

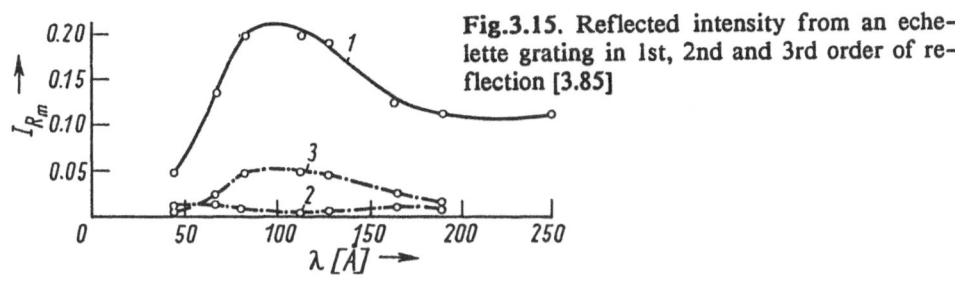

Fig.3.15. Reflected intensity from an echelette grating in 1st, 2nd and 3rd order of reflection [3.85]

Siegbahn [3.83] introduced ruled gratings as dispersive elements in X-ray spectroscopy. The angle of incidence must be very small since only under such conditions is "total reflection" needed for the efficiency realized. Later on, in addition to these gratings, with the area between the grating grooves planar while the whole grating is curved (Sect.3.5), blazed gratings (echelette gratings) have been introduced [3.51,84]. The advantage of such gratings with their sawtooth-like profile arises from the fact that they permit diffraction of maximum intensity into a chosen order of reflection, in contrast to grooved gratings (Fig.3.15), thereby increasing the efficiency and minimizing complications arising from the superposition of different orders in the spectra. Furthermore, to the plane and cylindrically curved gratings were added gratings with spherical and toroidal surfaces, holographic gratings, transmission gratings and zone plates [3.86,87].

Holographic gratings are made by recording interference fringes from laser light in a photosensitive medium. Therefore it is possible to make them on any surface which can be coated with photo-resist and exposed to interference fringes. The grooves of these kind of gratings have a sinusoidal shape, the efficiency is therefore spread out over a wider spectral range than for blazed gratings, but does not reach the maximum efficiency of a blazed grating. Holographic gratings can be made fully aberration-corrected, and the scattered light from them is lower than that from mechanically ruled gratings.

Transmission gratings, free-standing thin gold gratings on a random support structure, and zone plates as a special form of such a transmission grating are not widely used since they are fragile and not commercially available.

From Fig.3.16 one obtains the conditions for reflections as

$$n\lambda = PB - RA = \sigma(\cos\psi - \cos\phi) . \qquad (3.19)$$

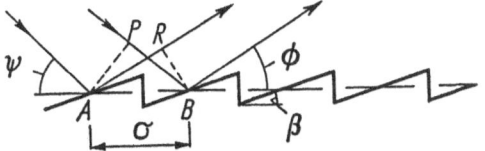

Fig.3.16. Reflection of X rays from a ruled grating and from an echelette grating. (For explanation, see text) [3.88]

109

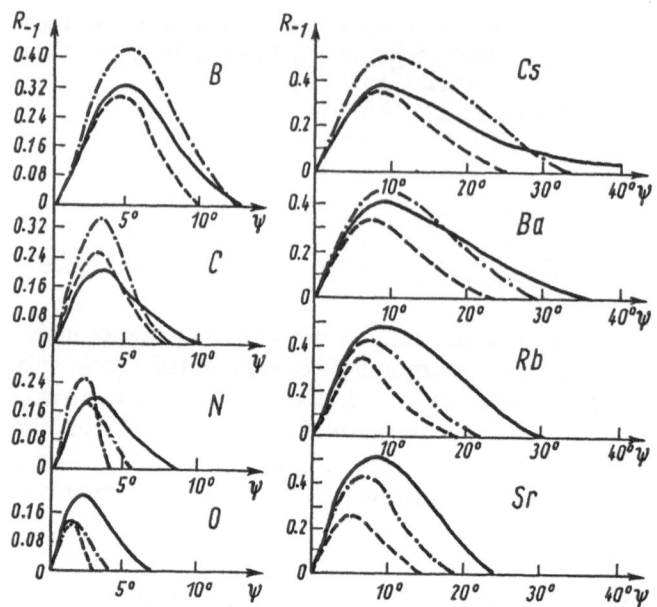

Fig.3.17 Calculated dependence of the maximum reflection coefficient R_{-1} in the 1st order on the angle of incidence ψ for different wavelengths (O: 2.36 nm; N: 3.14 nm; C: 4.44 nm; B: 6.7 nm; Sr: 10.865 nm; Rb: 12.866 nm; Ba: 16.46 nm; Cs: 19.03 nm) for an echelette grating with 600 lines/mm and different coatings (Au: solid line, Ti: dashed–dotted line, glass: dashed line) [3.85]

For a fixed angle of incidence ψ the angle β determines the critical angle for total reflection and thereby the limit of applicability of the grating. This critical wavelength limit is, however, not very sharp [3.88]. The reflected intensity is different for different orders as a result of different areas of the grating surface being effective in the reflection as determined by β. The maximum intensity is given by $\phi = \psi \pm 2\beta$, and for a particular order

$$R_n = f^2 r . \tag{3.20}$$

The factor f takes into account effects of different areas of the active grating surface caused by the shape of the grating. It is calculated from $f = 1 - \sin\beta / \sin(\psi + \beta)$. Application of (3.20) allows the determination of the maximum reflected intensity as a function of ψ for fixed r, the reflection coefficient of the grating material (Fig.3.17). The relation between β, ψ and λ is shown in Fig.3.18. It is clear that β can also be characterized using an optimal wavelength [3.85].

Experimental studies of the wavelength dependence of the reflected intensity agree generally with theoretical predicitons. Using such a method both weak and strong structures were determined in the region of the absorption edges for glass (Fig.3.19). For different metals such as Pt, Au, Ti, Ni, and Al different critical angles and an increased reflection inten-

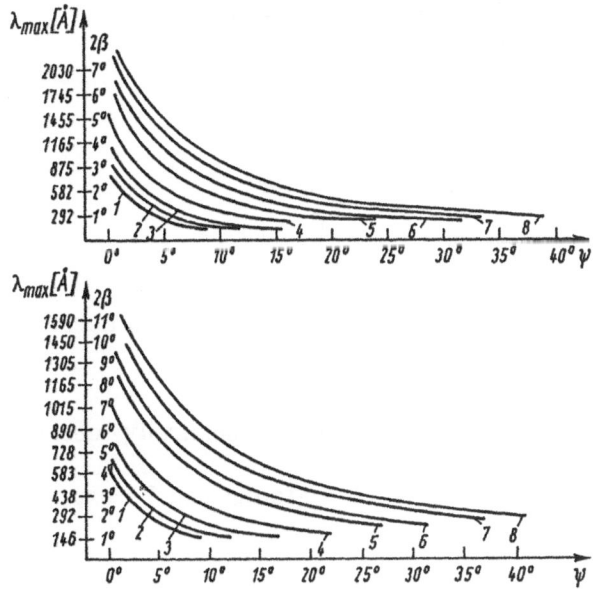

Fig.3.18. Variation of the wavelength yielding maximum intensity λ_{max} with the angles β and ψ for different radiation for an echelette grating with 600 (top) and 1200 (below) lines/mm. Curves *1* to *8* correspond to O to Cs as in Fig. 3.17 [3.85]

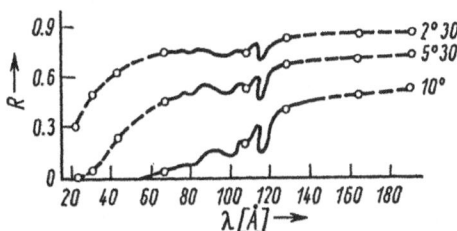

Fig.3.19. The spectral dependence of the reflectivity R of glass for different angles of incidence [3.85]

sity have been observed with much weaker structures (Fig.3.20) [3.85, 88, 89]. It is then possible to draw conclusions on the different reflected intensity distributions for glass gratings and gratings coated with a layer of evaporated metal. Furthermore, measurements of the reflection properties of polystyrene have shown the existence of a clear discontinuity in the reflectivity curve at around 4.5 nm. Therefore, a polystyrene-coated mirror is a suitable material to filter out the short-wavelength part of the spectrum [3.85] (Sect.3.5).

Reflecting mirrors of different geometry and from different materials are now widely used in synchrotron-radiation beamlines [3.86]. Surface roughness turns out to be the main factor influencing the reflecting and imaging properties of such a device; the best values reached by now are

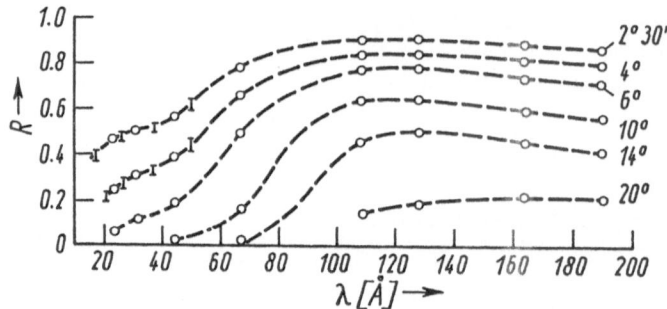

Fig.3.20. The spectral dependence of the reflectivity R of gold for different angles of incidence [3.85]

0.8-1.0 nm for bowl-feed polished glass and for chemically-mechanically polished electroless nickel.

The theoretical resolving power A of a grating can be expressed by [3.90,91]

$$A = \frac{\lambda}{\Delta\lambda} = N\cdot n = \frac{Bn}{\sigma} \ . \tag{3.21}$$

Here N is the total number of lines on the full grating width B. For fixed σ, n and λ a certain B is determined for giving optimal resolution (Sect. 3.5).

3.4 Recording Spectra

Spectra are recorded with either photographic plates, but nowadays mostly using various quantum detectors with electronic read-out systems. In this case some optimization is needed with regard to detector efficiency according to the wavelength region under study.

Figure 3.21 shows a block diagram of an electronic counting system. Ionization and scintillation counters, electron multipliers of various kinds, semiconductor detectors and lately also microchannel plates are included as belonging to so-called quantum detectors. The most significant properties of these detectors, namely efficiency and resolution, will briefly be discussed here. Efficiency or quantum yield of a detector is defined as the ratio of pulses obtained from the detector to the number of incoming photons. It depends firstly on the absorption of the radiation in the detector material, secondly on the particular physical processes leading to production of the electronic pulse after stopping an incoming photon (i.e., the relations between the intensity of the photopeak and the escape loss peak or the Compton scattering peak arising from fluorescence radiation

112

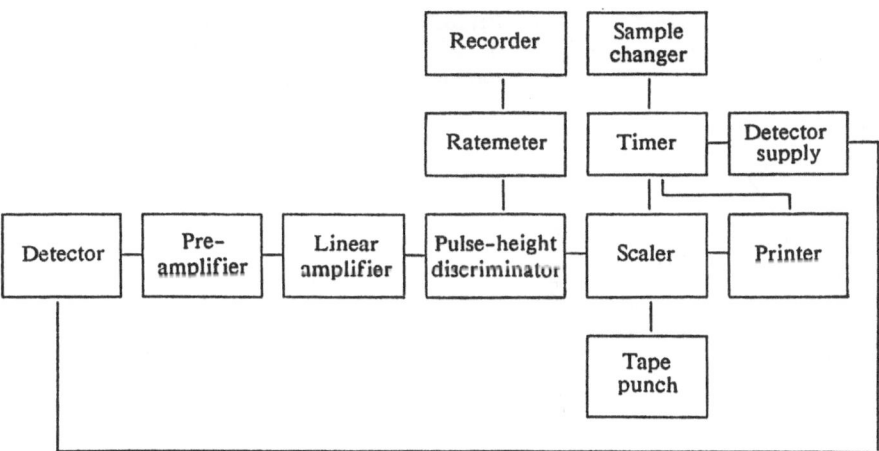

Fig.3.21. Block diagram of an electronic data acquisition system

leaving the detector [3.92]), and thirdly on the absorption and scattering in the detector determined by its particular design.

The resolution of a detector is given by the distribution of pulse amplitudes about their mean value, as shown in a graph of the number of counts as a function of pulse height. By disregarding the influence of the subsequent electronics [3.93], the resolution for the radiation energy E can be expressed as

$$\Delta E/E = C \cdot E^{-1/2} . \tag{3.22}$$

If further $C = \epsilon(F+f)^{1/2}$ (ϵ is the ionization energy, F is the Fano factor [3.94], f is the relative mean error of the gas amplification for ionization counters) and $\epsilon \propto 1/N$ (N is the number of elementary processes, which are excited by the energy E) [3.95], then

$$\Delta E/E \simeq N^{-1/2} . \tag{3.23}$$

Since the resolution of detectors in the spectral region of interest here is much less than that of the X-ray monochromator (Fig.3.22), this quantity plays only a minor role in high-resolution X-ray spectroscopy.

Among the various ionization counters available [3.22], the Geiger-Müller tubes and the proportional counters are most often used. In both cases ionization of gas molecules by the incoming radiation leads to a voltage impulse caused by the current pulse (an avalanche-amplified response to the initial ionization) in a resistance connected between the two electrodes of the counter tube, i.e., the housing and the inner electrode (for technical descriptions, see [3.22]). The two counters differ with regard to the high voltage applied, which leads to quite different output-voltage pulse characteristics. The applied voltage is sufficiently high in both cases for a secondary collisional ionization to occur by the collisions

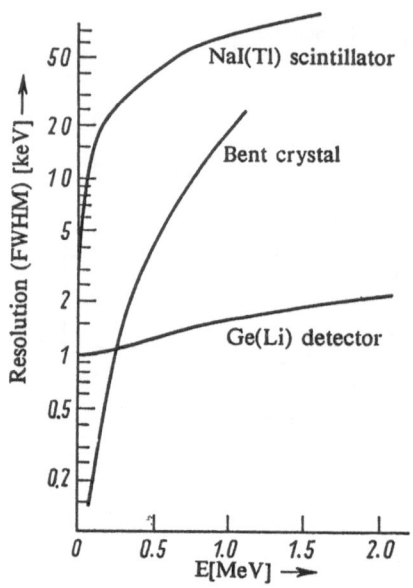

Fig.3.22. Comparison of the energy resolution of solid-state detectors with that of a crystal monochromator [3.95]

of electrons produced in the initial event with the atoms (gas amplification). For proportional counters the applied voltage leads to pulse heights of 10^{-2} to 10^{-4} V and the pulse height is proportional to the energy of the incoming radiation. The Geiger counter operates at still higher voltages so that upon even a single ionization event an intense collisional ionization cascade occurs, i.e., a self-sustained discharge. Then the proportionality between the incoming photon energy and the pulse height (pulse height is about 10 V for a 1 MΩ shunt) is lost. Another advantage of proportional counters over Geiger counters is the smaller dead time of the proportional counter, i.e., the time which is required to quench the collisional ionization. Proportionality is therefore assured at high count rates, too (up to about $4 \cdot 10^3$ counts/s) between the incoming number of photons and the number of pulses (for Geiger counters such proportionality is maintained only up to 500 pulses/s).

The efficiency of gas counters can be optimized with the proper choice of the type of gas and its pressure (Fig.3.23) and by proper geometry of the detector (minimize the absorption in the entrance window, minimize inhomogeneities in the electric field in the counter). Noble gases with an addition of a certain organic substance to decrease the dead time are usually used. Most frequently employed is a gas mixture of 90% Ar and 10% CH_4. Mica and Be are often used as window material in the shorter wavelength region (Fig.3.24), while at longer wavelengths various organic foils (Fig.3.25) [3.34,96] as well as thin Be foils were utilized. For spectroscopy in the ultrasoft X-ray region windows of, e.g., nitrocellulose (Fig.3.26) [3.97] either free-standing or supported by a fine metal mesh [3.34,98] and counter gas of Ar/C_2H_5OH, CH_2O or 75% CH_4/25% Ar have been used (Fig.3.27) [3.18,34,97-99]. Since foils are not vacuum

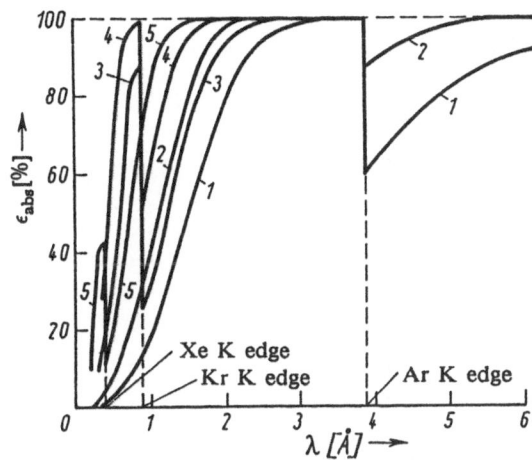

Fig.3.23. Wavelength dependence of the fractional absorption in a 10 cm gas path. (Curve *1*: Ar at 30 kPa, Curve *2*: Ar at 70 kPa, Curve *3*: Kr at 30 kPa, Curve *4*: Kr at 70 kPa, Curve *5*: Xe at 20 kPa) [3.22]

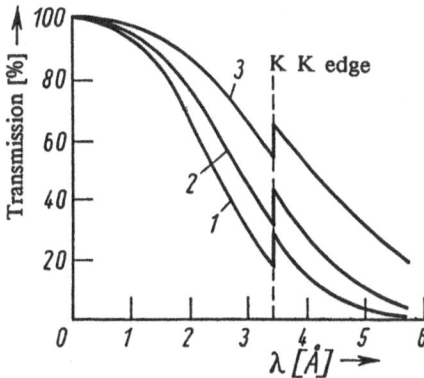

Fig.3.24. Transmission of mica for several thicknesses (Curve *1*: 15 μm, Curve *2*: 10 μm, Curve *3*: 5 μm) [3.22]

tight, flow counters with a continuous flow of gas have to be used [3.28, 100, 101].

For studies of harder X rays, scintillation counters are particularly suitable [3.102]. The X rays generate an electron in a particular crystal - most frequently used is NaI doped with Tl, the excited atoms deexcite with emission of photons in the visible or UV region. These light flashes are recorded using a secondary electron multiplier with a light-sensitive cathode. The advantage of a scintillation counter is the very small dead time which allows linearity between the number of quanta and pulses up to count rates of $2 \cdot 10^4$ counts/s and its very high quantum efficiency (Fig.3.28). The disadvantage stems from the sensitivity to light, which requires metal foils as entrance windows, and from the noise caused by the secondary-electron multiplier. These two factors limit the applicability of scintillation counters to the wavelength region between 0.03 and 0.3 nm. A comparison of efficiency between proportional and scintillation counters as a function of λ is given in Fig.3.28.

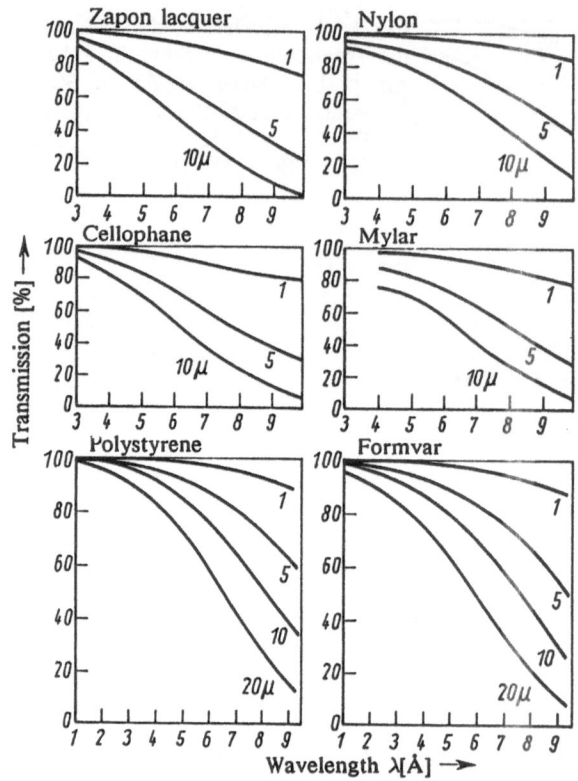

Fig.3.25. Transmission of organic foils of various thicknesses [3.96]

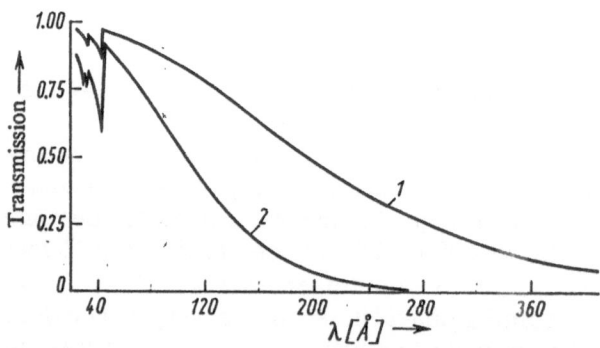

Fig.3.26. Transmission of nitrocellulose films (Curve *1*: 30nm, Curve *2*: 100nm) [3.52]

Using substances with not too small work functions as photocathodes, secondary electron multipliers can detect softer X rays directly. Studies of the wavelength dependence of the quantum yield of the photocathode and the influence of the incidence angle of X rays for different materials were undertaken, in particular, by *Lukirskii* and his colleagues [3.103],

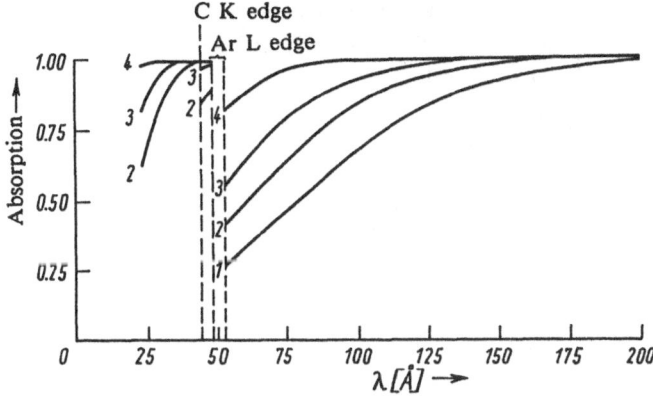

Fig.3.27. Absorption of various argon-ethanol mixtures (Curve *1*: 0.5kPa Ar, 1kPa C₂H₅OH; Curve *2*: 1 and 1.6kPa; Curve *3*: 2 and 2kPa; Curve *4*: 7 and 1.5kPa) [3.99]

and values for the quantum yield κ larger than 80% were achieved (Fig.3.29). Electron multipliers can be constructed with many different geometries [3.104] and optimized for particular experiments.

The photoconductivity which arises in semiconductors irradiated with X rays is utilized in semiconductor detectors which, in particular, operate like diodes. Due to the technical difficulty of making the simple pn junction thick enough to absorb the X rays, Li-drifted Ge or Si diodes have preferentially been used. Lithium is situated in interstitial crystal plane positions compensating for the acceptors. It thus constitutes a layer with very high resistivity. Figure 3.30 shows the efficiency of such detectors as a function of wavelength. The obvious drop in efficiency towards the long-wavelength region is caused by absorption in the detector window. The drop on the short-wavelength side is caused by decreased absorption of X rays. The increased resolution of semiconductor detectors versus

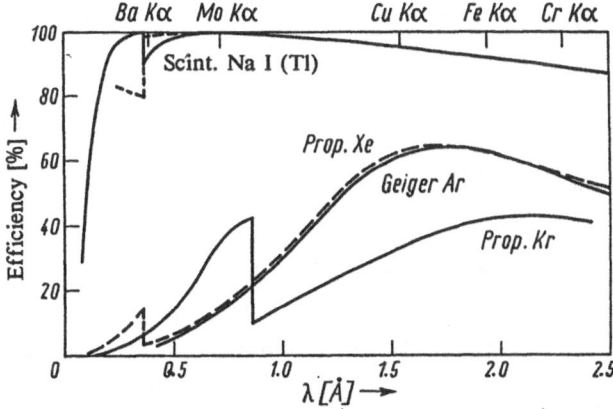

Fig.3.28. Efficiencies of several common detectors [3.29]

117

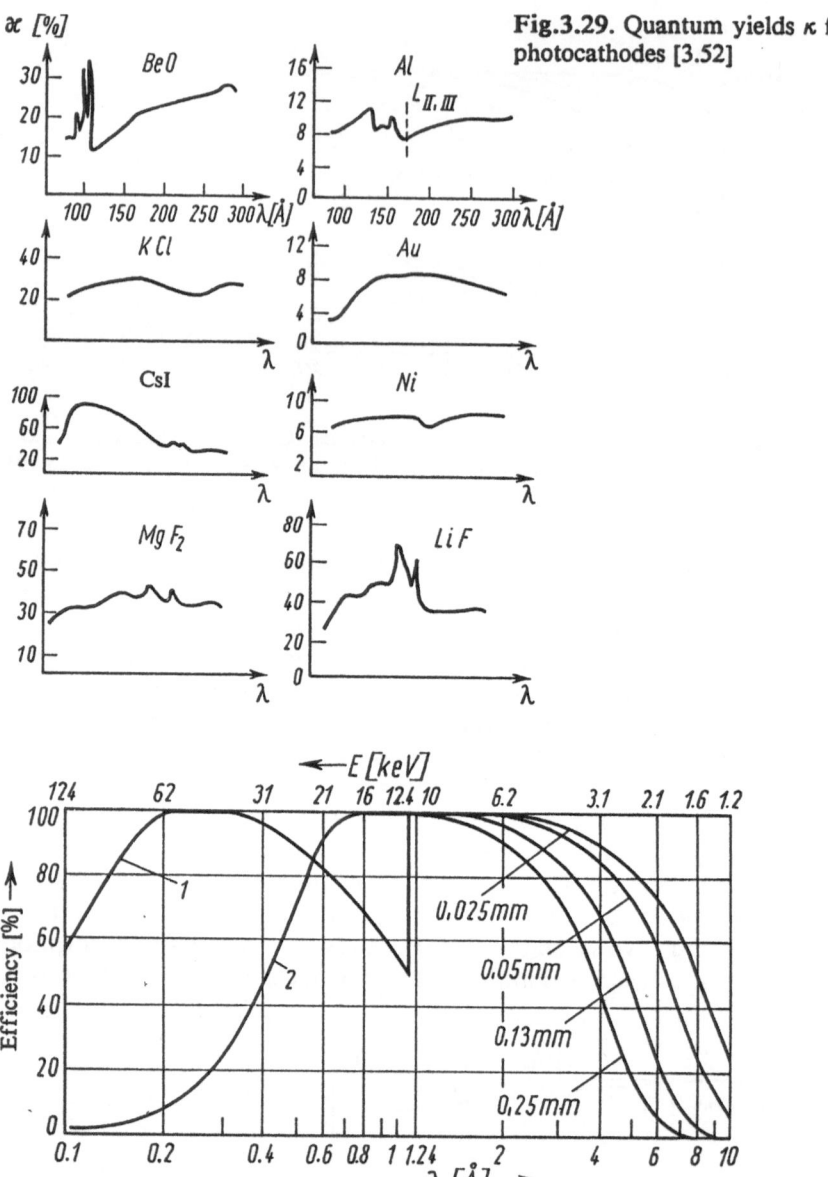

Fig.3.29. Quantum yields κ for various photocathodes [3.52]

Fig.3.30. Efficiencies of Ge(Li) (Curve *1*) and Si(Li) detectors (Curve *2*) with several different Be window thicknesses [3.93]

proportional and scintillation counters (Fig.3.22) can be understood from (3.23) as arising from the smaller energy required for the ionization (for semiconductor counters: $\epsilon \simeq 3$eV, proportional counters: $\epsilon = 30$eV, NaI (Tl): $\epsilon \simeq 300$eV). As a consequence of the very small pulse heights of ca. 10^{-3} V coupled with the simultaneous requirements to be able to distinguish differences in pulses of a magnitude of 10^{-5} V, extremely-high-performance preamplifiers are needed [3.93, 95].

A considerable reduction in recording time for spectra can be expected by application of multichannel plates (schemes with many detectors having extremely small openings in a plane) [3.104, 105] or by the electronic detection of charged insulated layers caused by photoemission [3.106] and other position-sensitive detectors [3.107]. With these detectors the advantage of photographic plates in terms of simultaneous recording of a larger spectral region is combined with the advantages of electronic registration by both multichannel [3.108] and video systems.

For a single-channel detector, the profile can either be obtained by continuously scanning with a counting rate meter (whose inertia, however, will introduce distortions [3.22]), or using point-by-point recordings with preset time or preset number of events. To minimize additional distortions, the entrance slit of the detector should not exceed 1/5 of the width of the recorded line or the sum of $\Delta\lambda_R$ (Sect.3.3) and $\Delta\lambda_{nat}$ (Sect.3.6). To avoid effects due to instabilities of the read-out electronics, counter voltage, voltage on the secondary-electron multiplier, and amplifier gain may be chosen so that they have only small influence on the count rate. Therefore, a large plateau should exist for the count rate with respect to variation in the instrumental parameters. With the exception of Geiger counters, the pulse height increases with the energy of the incoming radiation. Using this property of quantum detectors, introduction of energy discriminators or single channel analyzers, which have an upper and lower cutoff for the pulse height, can considerably improve signal-to-noise ratios. Furthermore, in principle it is possible to construct an energy dispersive spectroscopy in this manner with multichannel recording without using angular dispersive monochromators; this, however, cannot as yet be applied to high-resolution X-ray spectroscopy.

In particular cases it is advantageous to use a combination of point-by-point recording and change of samples to avoid experimentally determined wavelength errors and fluctuations in electronics [3.109, 110]. In this case several samples are studied one after the other for each position of the spectrum. This allows the utilization of different count-rate strategies by varying the number of samples, the order of changing and the counting time, and will also avoid the effect of instabilities of the excitation source [3.81]. The latter is particularly important for demountable X-ray tubes and can also be achieved by using a monitor, i.e., with the help of a counter which is in a fixed position in the spectrometer and seems to normalize the event rate in the principal detector [3.22, 111].

Optimization of recording conditions using quantum detectors is determined by the kind of detectors (dead time), the experimental conditions (stability of the electronics, ratio of line to background) and the counting statistics. In radiation measurements the counting statistics follow a Poisson distribution. Since, in practice, completely constant parameters are not achieved, a Gaussian distribution is superposed on the Poisson distribution. The standard deviation of a measurement at a specific point in the spectrum is given by

$$\sigma = \sqrt{N} \tag{3.24}$$

and the relative standard deviation can be expressed by

$$\sigma_{rel} = 1/\sqrt{N} . \tag{3.25}$$

In order to study drift in the electronics or the excitation source simple tests according to (3.24) are sufficient. Very high count rates are needed to detect the same deviations of all measured variations, i.e., to distinguish an instrumental Gaussian distribution from the Poisson distribution in an actual measurement. Following (3.25) the relative standard deviation of the Poisson distribution decreases with increasing N and therefore the influence of the instrumental standard deviation increases. If increasing deviations from a pure Poisson distribution can be detected, a practical limit of the measuring method is reached [3.112]. For a specified background N_B there is an optimum time t which should be used to measure the intensity when it lies m times above the background N_B to obtain a relative error of k%. The time can be expressed as

$$t = \frac{4(m + 2)}{9m^2 k^2} \frac{10^4}{N_B} . \tag{3.26}$$

The optimal arrangement minimizes the relative errors in the difference between intensity and background [3.29].

3.5 Description of Some Spectrometers

In this section the basic properties of the more important spectrometers will be described briefly and typical examples of each kind will be presented. The division into different types is according to the form taken by the monochromator (Sect.3.1).

Figure 3.31 shows schematically a spectrometer with a flat crystal (Bragg geometry). A spectral line occurs as an image of the entrance slit. The spectral range observable is determined by the finite extension of the radiation source and the angular aperture of the reflecting crystal. The reciprocal linear dispersion can be written as

$$D_\ell' = \frac{\lambda}{2R} \cot\phi , \tag{3.27}$$

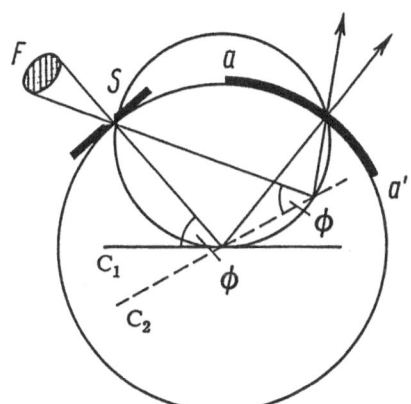

Fig.3.31. Schematic representation of the spectroscopic arrangement due to Bragg (F: focus of X rays; S: slit; C_1, C_2: different crystal positions; ϕ: glancing angle; a-a' recording film) [3.22]

where R is the spectrometer radius. The luminosity, neglecting its obvious dependence on the radiation source (Sect.3.2), is determined by the slit width and the vertical angular aperture. The resolving power, however, is largely determined by the horizontal angular aperture while the vertical angular aperture introduces an asymmetry in the intensity distribution [3.113].

By oscillating the crystal it is possible to focus the X rays (this is not quite self-evident); the luminosity is, however, weak since only a small part of the crystal can be used due to the geometry. Introduction of Soller slits allows one to use the full area of the crystal which considerably improves the luminosity. However, this decreases the resolution, which in this case is limited by the opening aperture between two parallel plates in the Soller slits. Bragg geometry is seldom used in high-resolution applications due to its low efficiency in comparison with focussing instruments and the limited resolution of the Soller slit versions; moreover, in the focussing case each position in the spectrum comes from a different point of the source. *Fischer* and *Baun* [3.62] used a vacuum spectrometer (10^2 to $4 \cdot 10^{-4}$ Pa) built according to Soller's method to record moderately high-resolution X-ray spectra using various flat crystals as dispersive elements (a wavelength region up to 10nm was covered). The radius of curvature of the spectrometer was 30 cm. The two collimators in front of and behind the crystal were 11.3 cm long, the plate separations were 0.05 and 0.012 cm. Furthermore, Ar/CH_4 was used at reduced pressure as counting gas for the proportional counter.

A newer application of a Soller-type spectrometer was suggested by *Romand* and co-workers [3.114]. They used a gas-discharge tube as source of low-energy electrons with energies from 0.5 to 5 keV for the excitation of primary spectra. The X rays for the wavelength range 0.3-10 nm are dispersed by flat crystals and pseudocrystals, respectively. Regardless of the drawbacks of primary excitation (Sect.3.2), this approach is attractive since electrons of such low energies probe only a thin surface layer because of the low penetration depth.

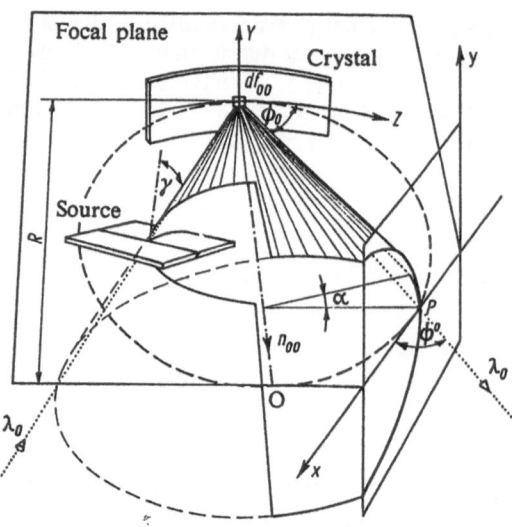

Fig.3.32. Geometry of the Johann focussing (R: crystal radius of curvature; O: origin of curvature; df_{00}: surface element of the crystal; n_{00}: surface normal; λ_0: reflected wavelength; ϕ_0: Bragg angle) [3.33]

Two orders of magnitude improvement of the luminosity compared to a flat crystal spectrometer without a loss of resolution can be achieved by using a bent crystal. A detailed description of focussing spectrometers, with special attention to the Soviet ones, has been given in [3.22]. Only the most frequently used techniques based on horizontal focussing (introduced by H.H. Johann, Y. Cauchois, and T. Johansson) are briefly discussed here. *Johann* [3.115] was the first to build such an instrument in which a spectral line was imaged using horizontal focussing by a bent crystal (Fig.3.32). A slit is not required in front of the radiation source in this geometry. The imaging properties of a Johann spectrometer were thoroughly studied by *Läuger* [3.33]. Assuming that the crystal has a precisely cylindrical form, the illumination is uniform, and that the radiation density is independent of the position on the radiating surface, the image formed in the focal plane is a conical section with a radius of curvature r = Lcos ϕ_0, L being the distance between df_{00} and P (Fig.3.32). The total intensity I is given as a product of an intensity factor W and the convolution I^* of the window function and the true intensitiy distribution

$$I(x) = W_\lambda I^*(x,\gamma) .\tag{3.28}$$

This relation gives firstly the possibility to correct the spectra with respect to intensity ratios and profile distortions, and secondly it is a useful expression for luminosity and resolving power. The intensity factor W (and thereby the luminosity) depends on geometrical factors and the glancing angle for a specific lattice plane and order of reflection as follows

$$W_\lambda \simeq \frac{HBhb}{L^2} f(\phi) , \qquad (3.29)$$

where $f(\phi) = \sin^4\phi$ for $\phi > \pi/4$ and $f(\phi) = \sin^2\phi\cos^2\phi$ for $\phi < \pi/4$ (H and B are the total height and width of the crystal, h and b are the height and width of the detector slit). The resolving power is determined by the window function F which depends on instrumental parameters as

$$F(x) = F_1(x)*F_2(x)*F_3(x)*F_4(x) . \qquad (3.30)$$

The widths contributed by these factors are expressed as

$$\Delta x_1 = (H + h)^2 \frac{\tan\phi}{8L\sin\phi_0} ,$$

$$\Delta x_2 = -B^2 \cos\phi \frac{\sin\phi}{8L\sin\phi_0} , \qquad (3.31)$$

$$\Delta x_3 = L\frac{\Delta\phi}{\sin\phi_0} ,$$

$$\Delta x_4 = \frac{b}{\sin\phi_0} ,$$

where $\Delta\phi = \Delta\lambda_R/C$, C is a number between 1 and 1.32, and $\Delta\lambda_R$ the width of the rocking curve.

The following conclusions can thus be drawn:

a) Line position and shape are both invariant with respect to changes in the position of the radiation source.

b) Neither luminosity nor resolving power depends on the distance between the source and the crystal.

c) For the minimum source height H_m (projection in y direction) the following relation can be written

$$H_m \geq H + 2Qh/L , \qquad (3.32)$$

where Q is the distance between crystal and center of radiation source.

d) From the relation for Δx_1 it follows that broadening due to the crystal height and the detector slit height are not independent of each other; the relation used earlier [3.22] is wrong.

e) The relation for Δx_2 is identical to the relation employed earlier to express geometrical line broadening [3.22].

A schematic representation of the vacuum spectrometer of the Johann type built by *Läuger* [3.33] is shown in Fig.3.33. It is equipped with a high-intensity X-ray tube (maximum 12kV, 800mA), which facilitates excitation of fluorescence spectra outside of the X-ray tube (the distance between anode and sample is 8 to 14 mm). The samples are brought into the spectrometer by a vacuum manipulator. Spectra are recorded with a flow-proportional counter (with a window of $2\mu m$ mylar, gas is Ar/CH_4)

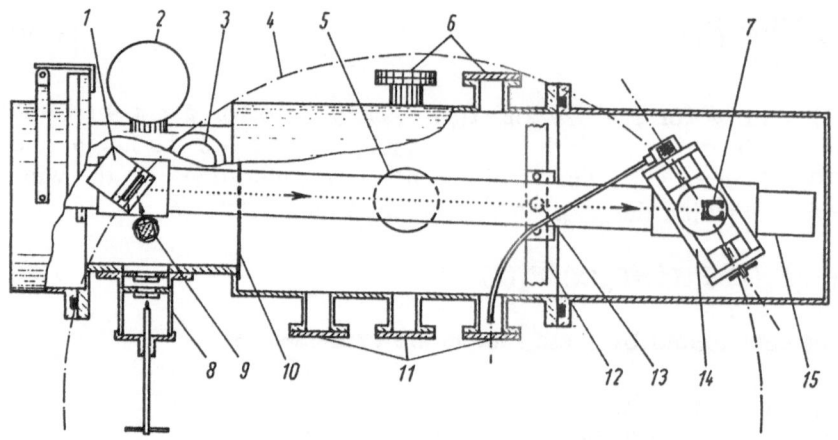

Fig.3.33. X-ray spectrometer with Johann focussing according to *Läuger* [3.33] (*1:* crystal holder; *2:* evacuation valve of the X-ray tube; *3:* view port for observation of the X-ray window; *4:* focal circle; *5:* pump support; *6, 11, 12:* electrical and mechanical feedthroughs; *7:* detector; *8:* vacuum lock; *9:* secondary emitter; *10:* aperture to screen the primary radiation; *13:* axis for rotation of *15*; *14:* detector carriage; *15:* motion rail for *14*)

moveable along tangents to the focal circle. Other newer Johann spectrometers have been built by *Blokhin* [3.22], *Heintz* [3.31], *Gilberg* [3.20], *Leonhardt* [3.81], *Ehlert* and *Mattson* [3.116], and *Davidson* [3.60]. With the last two instruments spectra were studied in the ultrasoft X-ray region using organic crystals.

A detailed analysis of (3.31) shows that especially Δx_2 increases rapidly with smaller glancing angles (with $L = R\sin\phi$ and $\phi = \phi_0$ one obtains $\Delta x_2 \propto \cot\phi$). For short-wavelength X rays the effective crystal width must therefore be severely limited which according to (3.29) leads to a decrease in luminosity. These difficulties can be overcome by using a transmission geometry introduced by *Cauchois* [3.117] (Fig.3.34). Geometrical analysis gives a line broadening [3.22] caused by the opening angle Θ of the crystal (with $R\Theta = B$) and the thickness K of the crystal

$$\Delta x' = \frac{1}{2}\left[\frac{R\Theta^2}{4} + K\right]\tan\phi \tag{3.33}$$

and including the finite height H of the crystal

$$\Delta x'' = \frac{H^2}{8R}\tan\phi . \tag{3.34}$$

According to calculations by *Läuger* [3.33] the line broadening $\Delta x''$ is at least questionable. From the ϕ dependence it follows that the Cauchois geometry should not be used for large glancing angles.

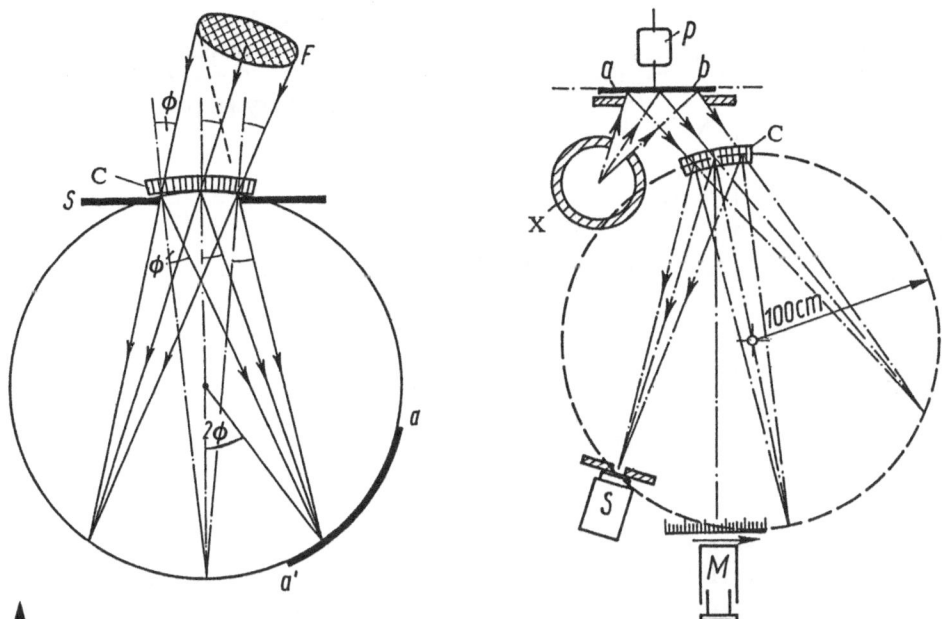

Fig.3.34. Schematic representation of the focussing Cauchois spectrometer. For explanation, see Fig.3.31

Fig.3.35. X-ray spectrometer with Cauchois focussing, as described by *Sumbaev* [3.110]. (X: X-ray tube; P: sample changer; a and b: different samples; C: crystal; M: monitor; S: secondary-electron multiplier)

The spectrometer described by *Sumbaev* [3.110] and shown in Fig. 3.35 has a Rowland circle radius of 1 m. Recording is done by a stepwise movement of crystal and detector; these motions are measured by microscopes. For a careful determination of line shifts a standard sample is brought to the source position at each point in the spectrum alternating with the substance under study. A modernized Cauchois spectrometer with radioactive isotopes as primary light source was built by *Jewell* [3.118].

For latest developments of spectrometers, it is characteristic that they are designed with special regard to non-conventional X-ray sources such as synchrotron radiation [3.86] or plasma sources [3.119]. In plasma diagnostics, e.g. from tokamak sources, the characteristic line emission for highly ionized impurity elements is used to study the plasma conditions. Therefore, the spectral range of interest must be recorded in a short time interval, with medium to high resolution to resolve the lines in the multi-line spectra, and with high maximum count rates. In most use for this purpose is the Johann geometry with large-area crystals and fast, large-area, position-sensitive X-ray detectors without geometrical scanning [3.119].

Another experimental possibility for achieving horizontal focussing was suggested by *Johansson* [3.120]. A thin quartz plate was ground to a

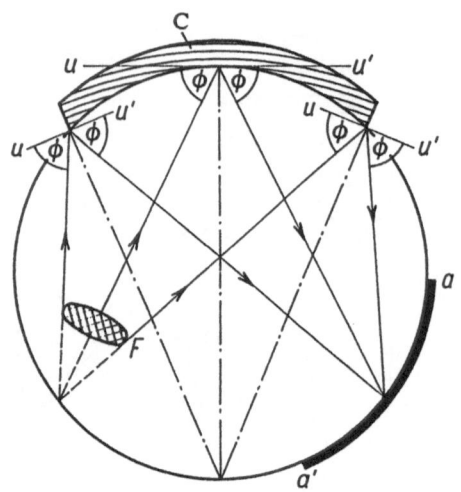

Fig.3.36. Schematic representation of Johansson focussing geometry. For explanation, see Fig.3.31. Tangents indicated by u–u′ lie along the crystal planes

cylindrical shape with radius R and thereafter bent on a cylindrical surface of $r = R/2$ (Fig.3.36), thus eliminating the line broadening and improving the resolving power compared to a spectrometer in the Johann or Cauchois geometry. However, the Johansson type has found fewer applications, in part because of the technical difficulties of grinding and bending with adequate precision. A comparison of focussing properties with Johansson or Johann geometry was presented by *Cermak* [3.121]. From calculations of the window function for each of the monochromators described one sees that the difference in performance between the two approaches becomes smaller with increasing Bragg angle.

A double-crystal spectrometer uses two flat crystals and the radiation is reflected off the two crystals in series. Two different positions of the crystals are possible (Fig.3.37). The angular dispersion for such an arrangement can be written as

$$D_\phi = D_{\phi_1} + D_{\phi_2} = \frac{1}{2d} \left[\frac{n_1}{\cos\phi_1} + \frac{n_2}{\cos\phi_2} \right]. \tag{3.35}$$

It follows that for the same order of reflection from two equal crystals the parallel position has $D_\phi = 0$ and the antiparallel one $D_\phi = n/d\cos\phi$. Although there is no vertical slit in the spectrometer, a double-crystal spectrometer reflects purely monochromatic radiation for each position of the crystals. The first crystal works to a certain extent as a collimator with an angular aperture corresponding to the width of its rocking curve (Sect. 3.3). Recording of a spectrum is done most often by rotating only the second crystal (although coordinated scanning of both crystals and the detector has been used). In scans where only the second crystal rotates, some beam walking occurs placing certain requirements on the uniformity of the crystals and detector efficiency depending on the size of the assumed

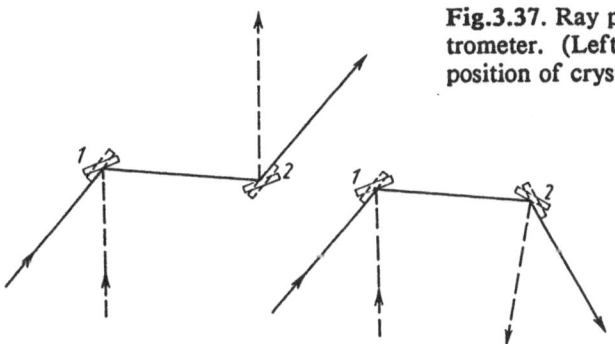

Fig.3.37. Ray paths in a double-crystal spectrometer. (Left: parallel; right: antiparallel position of crystals *1* and *2*)

angular region [3.49]. A limitation in the horizontal plane is not necessary but the vertical divergence affects both the luminosity - the intensity increases quadratically with vertical divergence - and the resolving power, the latter decreasing with increasing divergence (lower monochromaticity on the short-wavelength side and a shift of the intensity maximum) [3.122]. The geometrical resolving power of a double-crystal spectrometer can be written as

$$A = \frac{\lambda}{\Delta\lambda} = \frac{4}{\Omega_m^2} , \tag{3.36}$$

where $\Omega_m = (h_1 + h_2)/2L$ (h_1 and h_2 are the heights of two horizontal slits, L is the distance between the two slits) [3.69]. The big advantage of a double-crystal spectrometer lies in the fact that by recording the rocking curves in the (n,-n) position, the spectrum recorded in the (n,n) position can be corrected for the physical resolving power of the crystal [3.69, 123]. Correction is required to obtain intrinsic line profiles (Sect.3.6).

Blokhin and *Nikiforov* [3.124] described a double-crystal spectrometer in which the radiation from the X-ray source impinges on the first crystal after vertical divergence limitation by a 16 cm long collimator (plate separation: 0.08 cm). The construction of the crystal holder follows the design given by *Brogren* [3.78]. By mounting the crystal holder on a plate in vacuum an adjustment is possible after evacuation. Detector and X-ray tube are rotatable through ±120° and the first crystal through ±60°. The angular orientation of the second crystal could be determined with an accuracy of 2". Other double-crystal spectrometers have been built by *Drahokoupil* and *Fingerland* [3.125], *Azaroff* [3.126], *Schnopper* [3.127], *Deslattes* [3.128], *Chopra* [3.70], *Brytov* [3.129], and *Bhalla* [3.130] (the last with application particularly to absorption measurements in the region of 0.5 to 7 nm); the last four instruments were built as vacuum spectrometers.

A new design of double-crystal spectrometers has come up with synchrotron radiation use since there is the demand for fixed entrance and exit beams [3.131, 132]. This is reached by mounting two Bragg reflecting

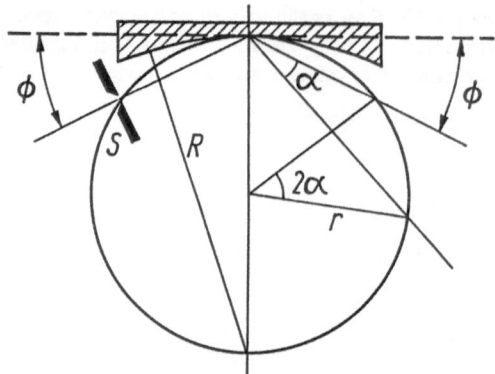

Fig.3.38. Scheme of the geometry for spectroscopy using a concave grating. α angle between directly reflected and diffracted beam

crystals separately in parallel position constrained by a linkage system. The wavelength scanning is done by angle scanning both crystals while one of them is moved on a linear trajectory. Such systems are now widely applied. The monochromatized X ray is used for different kinds of experiments with XANES [3.43] and EXAFS [3.44] at front places.

For studies in the ultrasoft X-ray region only grating spectrometers with concave gratings are important (Fig.3.38); spectrometers with a plane grating have too small luminosity apart from the spectrometers for synchrotron radiation where this drawback is corrected by focussing mirrors [3.132, 133].

From the grating equation (3.19) for recording a spectrum tangentially to the Rowland circle (Sect.3.3), the following relation can be deduced for the inverse linear dispersion

$$D_\ell' = \frac{\sigma}{nR} \sin\phi \ . \tag{3.37}$$

By recording perpendicularly to the reflected beam (Fig.3.38), a considerable increase in luminosity can be obtained particularly for shorter wavelengths; the luminosity is connected with the dispersion by an inverse relation

$$\frac{I_\perp}{I_\parallel} = \frac{\cos\alpha_0}{\sin(\phi+\alpha_0)} \ , \tag{3.38}$$

where $\alpha_0 = \phi-\psi$. Beyond the focal point line broadening occurs which is proportional to the distance from the focal point [3.134]. Since for most cases $\alpha < 5°$ the following relations for the focus and beyond the focus point are valid

$$D_\ell' \simeq \sigma/R \quad \text{at focus} \quad \text{and}$$

$$D\ell' \simeq \frac{\sigma}{R} \frac{\sin(\phi+\alpha_\lambda)}{\sin(\phi+\alpha_0)} \quad \text{beyond focus.} \tag{3.39}$$

According to *Wiech* [3.25] the resolving power of a grating spectrometer, recording perpendicular to the reflected radiation, can be expressed as

$$A = \frac{n\lambda R}{\sigma} \frac{1}{S_1 + 0.2S_2}, \tag{3.40}$$

where R is the diameter of the Rowland circle, S_1 and S_2 are the widths of the entrance and exit slits, respectively. The relation given by *Fischer* [3.135] and *Holliday* [3.18] is valid when an exit slit placed perpendicular to the Rowland circle is used. To obtain optimum luminosity and resolution the following usable width B of a concave grating is necessary

$$B_{opt} = 2.36 \frac{4\lambda R^3}{\pi} \frac{1}{\cos\psi\cot\psi + \cos\phi\cot\phi}. \tag{3.41}$$

The optimum recording slit width S_{2opt} and the optimum resolving power A_{opt} can be expressed as [3.90]

$$S_{2opt} = \frac{\lambda R}{B_{opt}}, \tag{3.42}$$

$$A_{opt} = 0.92 B_{opt}\, n/\sigma. \tag{3.43}$$

The resolution for $S_2 \neq S_{2opt}$ given by these relations

$$A = 0.92 \frac{n\lambda R}{\sigma S_2} \tag{3.44}$$

is nearly identical to (3.40).

Figures 3.39,40 show schematic diagrams of instruments suitable for the wavelength region 1–50 nm, using as an example the RSM500 suggested by *Lukirskii* [3.136] and the grating spectrometer by *Wiech* [3.137] which operates over a range of $\lambda = 4$–60 nm. The recording in the RSM500 is done by a simultaneous movement of grating and detector permitting a measuring apparatus of moderate size. The vacuum in the spectrometer and the excitation source is $5\cdot10^{-3}$ and $5\cdot10^{-4}$ Pa, respectively. To minimize radiation of higher orders the instrument is equipped with a reflector located behind the excitation source (the limit of the reflected spectral range can be changed by choice of an appropriate reflector coating and by changing the geometrical arrangement). The spectrometer can be operated with a grating with either 2 or 6 m radius of curvature and with a flow-proportional counter or a secondary-electron multiplier as detector.

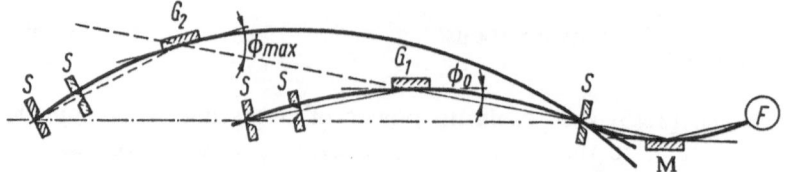

Fig.3.39. Ray paths in RSM500. (F: X-ray focus; M: mirror; S: slit; G_1, G_2: grating in different positions of reflection; ϕ: angle of incidence) [3.52]

In contrast to other designs the spectrometer by *Wiech* [3.137] has the X-ray tube mounted inside the actual instrument. As a consequence a vacuum of 10^{-6} Pa can be obtained with an entrance slit connected to a differential pumping stage between the spectrometer with 10^{-3} to 10^{-4} Pa and the X-ray tube with 10^{-6} Pa. The dispersive element is a 2 m replica echelette grating with 600 lines/mm. The detector is an open Bendix electron multiplier with W cathode and spectra are recorded by slit motion along the Rowland circle.

Other instruments have been built by, e.g., *Holliday* [3.18], *Lukirskii* [3.138], and *Andermann* [3.139]. A new spectrometer type was designed and built by *Sawada* [3.140] in which neither the detector nor the grating is moveable. *Jaegle* [3.141] designed a double-grating spectrometer. Spectrometers using both primary and secondary excitation especially for the

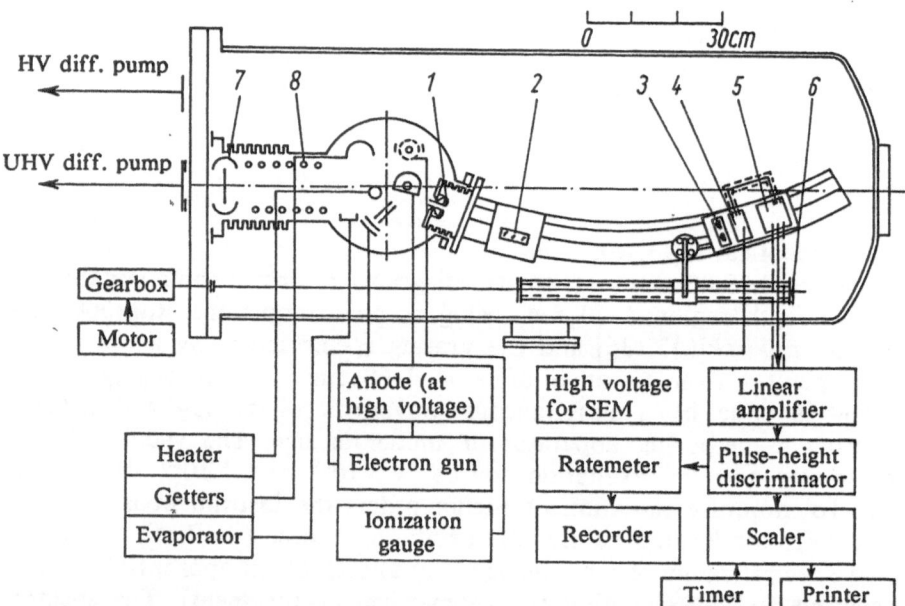

Fig.3.40. A grating spectrometer according to *Wiech* [3.137]. (1: entrance slit; 2: grating; 3: exit slit; 4, 5: secondary-electron multiplier; 6: drive spindle; 7, 8: cold traps)

130

study of gases were constructed by *Siegbahn* and *Nordling* [3.105, 142], and by *Gilberg* [3.143].

A newer type for monochromatization of synchrotron radiation is the toroidal grating monochromator [3.86, 132, 144]. It is strikingly simple: The wavelength scan is possible via one single rotation of the grating, while fulfilling approximately the condition for focussing at the exit slit; the electron beam of a storage ring, due to its small size, can be directly used as the entrance slit. This design is characterized by high luminosity at medium resolution.

3.6 Analysis of Measurement Results

Analysis of emission and absorption spectra to determine the true spectral shape i = f(E) can be separated into establishment of energy positions and determination of relative intensities in the spectrum. Here it is assumed that there are no instrumental errors (Sect. 3.4). In particular, the intensity distribution of the spectrum can be considerably distorted by different instrumental effects, making the evaluation of the true shape of the spectrum very difficult.

To determine the energy position in the spectrum the relation between the position of the detector and the wavelength or the energy must be accurately determined either using calibration wavelengths (e.g., from [3.145]) or by a very accurate angle measurement in a spectrometer fitted with appropriate devices. (It is important to notice, however, that different features of the lines have been utilized to determine the wavelength [3.146, 147].) In this way an experimental dispersion value is obtained, which should agree with the dispersion calculated from the geometrical configuration of the spectrometer used (Sect. 3.5). Using linear or more complex interpolation energy positions can be deduced for the lines under study. Due to imperfect geometrical arrangements in actual spectrometers, deviations from expected dispersion arise, which can be reduced by using a sufficient number of additional reference lines. The calibration curve can be obtained as a polynomial by adopting an approximation to the measured values.

The true intensity distribution of the spectrum is influenced by a variety of experimental factors:

a) A principal problem in determining the true intensity profile arises from the finite resolving power of the spectrometer, the so-called spectral window or window function. This is dependent on the finite physical resolving power of the dispersive element and the geometrical resolving power of the complete spectrometer. An exact determination of the former is possible either through recording the rocking curve with a double-crystal spectrometer or by the dynamical theory, when the resolving power can be calculated. The latter can be estimated from the equations given in Sect.3.5. The distortion of the spectrum by the spectrometer window function can be expressed as an integral equation of the following form

$$I(\lambda) = \int_{-\infty}^{+\infty} i(\lambda_0) F(\lambda - \lambda_0) d\lambda_0 . \tag{3.45}$$

Here $I(\lambda)$ is the measured spectrum, $i(\lambda_0)$ the undisturbed, true spectrum, and $F(\lambda-\lambda_0)$ the window function. The correction using an unfolding procedure requires knowledge of the function $F(\lambda-\lambda_0)$ and is mathematically a very complicated procedure even using symmetric instrumental functions (e.g., Lorentz or Gauss profiles).

The determination of the linewidths is possible only with the assumption that both the true spectrum and the distortion function have a Lorentz distribution. Then the experimental half-width $\Delta\lambda$ is given by

$$\Delta\lambda = \Delta\lambda_0 + \Delta\lambda_i , \tag{3.46}$$

where $\Delta\lambda_0$ is the linewidth of the true spectrum, and $\Delta\lambda_i$ is the instrumental linewidth. If a Gaussian distribution is assumed for the two functions, it follows that

$$\Delta\lambda^2 = \Delta\lambda_0{}^2 + \Delta\lambda_i{}^2 . \tag{3.47}$$

Investigations by *Nelson* [3.148] using a Cauchois spectrometer recording a γ line with very small half-width showed that the instrumental function for this spectrometer could be described by a Gaussian distribution. The experimental lines are then approximated by a superposition of this Gaussian distribution and a Lorentz distribution which describes the natural line shape.

Correction of the full $I(\lambda)$ spectrum is usually done by using Gauss or Lorentz distributions for the $F(\lambda-\lambda_0)$ function, whose width is determined either directly by measuring the rocking curve and calculating the geometrical broadening, or by comparing the width of an emission line measured by the actual instrument with the width determined by a double-crystal spectrometer, or as a sum of the atomic levels [3.49, 149] participating in the transition to give the width $\Delta\lambda_0$ of the emission line. Different unfolding methods have been developed for high-resolution X-ray spectroscopy, in particular by *Nikiforov* [3.150], *Sachenko* [3.22] and *Porteus* [3.151].

b) Due to X-ray scattering and bremsstrahlung obtained by primary excitation, a wavelength-dependent background is usually present. In most cases a linear shape can be assumed and it is consequently possible to subtract the background from the measured spectrum in a straightforward way.

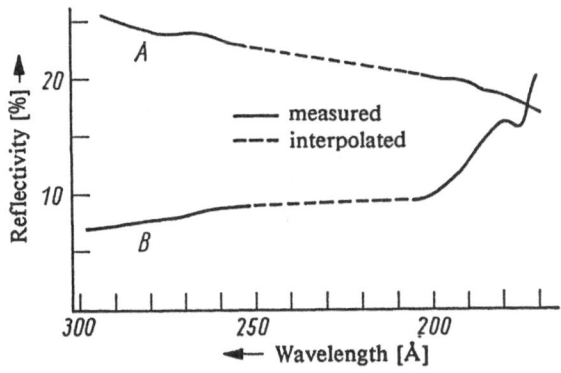

Fig.3.41. Wavelength dependence of the reflectivity of two gratings (R = 1999.5 mm, 600 lines/mm) A and B. Grating A: blaze angle 3°31′ with evaporated gold; grating B: blaze angle 1°31′ with evaporated Al [3.25]

c) The spectra are furthermore distorted by the influence of self-absorption (Sect.3.2). Even if experimental conditions are chosen so that this effect is minimized, it may still be necessary to introduce corrections. By measuring the effective layer thickness for the X-ray emission, the influence of self-absorption can be quantitatively calculated for known $\mu(\lambda)$ and thereby correcting the spectrum [3.28,89].

d) As was discussed in Sect.3.3, the reflectivity of a dispersive element, in particular that of a grating, is wavelength dependent (Fig. 3.41) which can lead to an evident distortion of the spectrum. This effect can be determinated if one studies the reflectivity for individual monochromatic lines [3.85] or if one places a grating to be examined behind the exit slit of a spectrometer and measures the intensity distribution at the exit slit and behind the second grating. Correction can then be applied using a straight line as an approximation to the experimentally determined reflectivity function or another simple function.

e) As a result of the energy dependence of the quantum yield and nonlinearity between quanta and the number of pulses, distortions arise in the spectra due to the detector. The quantum yield can be calculated (Sect.3.4) or determined experimentally through a comparison with a detector with known efficiency (or by performing an absolute measurement of the X-ray intensity, for example, using calorimetric measurements). The nonlinearity of detectors is determined from the dead time τ. If τ is known (from the determination of pulse length using an oscilloscope [3.22]) the true count rate N_0 can be derived from the experimental count rate N through

$$N_0 = \frac{N}{1 - N\tau} \,. \tag{3.48}$$

However, this relation is valid only for relative errors $(N_0-N)/N < 10\%$. A direct measurement of the calibration curve between N and N_0 can be performed by measuring the radiation absorbed in a stack of foils or by measuring the variation in N with varying current in the X-ray tube [3. 22].

f) When recordings covering large parts of the Rowland circle are performed, it is important to consider that a fixed slit width at the detector leads to an effective width Δx which varies over the region studied. This necessitates a correction to the measured intensity N_x to obtain comparable intensities $N(\nu)$ at various points as follows

$$N(\nu) = \frac{d\lambda}{d\nu} \frac{dx}{d\lambda} \frac{N_x}{\Delta x} . \tag{3.49}$$

The coefficients needed can be obtained from the calibration curve and by using the relation $c = \lambda\nu$; Δx can be deduced from the geometry of the spectrometer [3.25]. Wavelengths and energy widths follow with the help of the relation

$$E\lambda = 1.2378 \quad [\text{keV·nm}] . \tag{3.50}$$

All these corrections – in particular that discussed first – are facilitated by the use of a numerical display of the results via multichannel analyzers. This is also very useful for deconvoluting superposed spectra, for example, to distinguish satellite lines, and to resolve multiplets. In addition, the reliability of most results is considerably improved [3.148].

4. Determination of Bonding Parameters and Effective Atomic Charges

The relations between the parameters of the chemical bond (for example, oxidation number, ionicity, atomic charge, coordination number, atomic distance) and the X-ray emission and absorption spectra are discussed. In Sect.4.1 different empirical explanations for the chemical shifts of the X-ray emission lines and the X-ray absorption edges and also for the shift and shape of the X-ray emission bands are given. The determination of the geometrical structure from the extended absorption fine structure (EXAFS) is discussed in Sect.4.2. In Sect.4.3 the definition of atomic charges in molecules and solids and their determination by means of the chemical shifts of the X-ray emission lines and absorption edges are described.

4.1 Oxidation Stages and Bonding Ionicity

The chemical state of an atom emitting or absorbing X rays influences both the position of lines and edges of the spectrum as well as its shape and intensity. The concept "chemical state" can be defined by the following factors:
- valence of the atom,
- kind of neighboring atoms, and
- geometry of the surrounding region of the atom.

All three factors can influence the X-ray spectrum. The direction of the phenomenon is, however, often different while its size is of the same magnitude, so that it is difficult to infer which effects from the observed sum of effects are from which individual parameter; this is possible only when two of the parameters can be kept constant.

X-ray emission and absorption spectra in the near-edge region, where the excitation remains largely confined to the region around the excited atom, is almost exclusively dependent on the nearest neighbors of the excited atom participating in the chemical bonding. It is questionable to describe the complicated interaction of the neighboring atoms with the excited atom using numerical quantities; nevertheless certain correlations emerge when observed spectra are arranged according to such parameters as electronegativity and bonding ionicity.

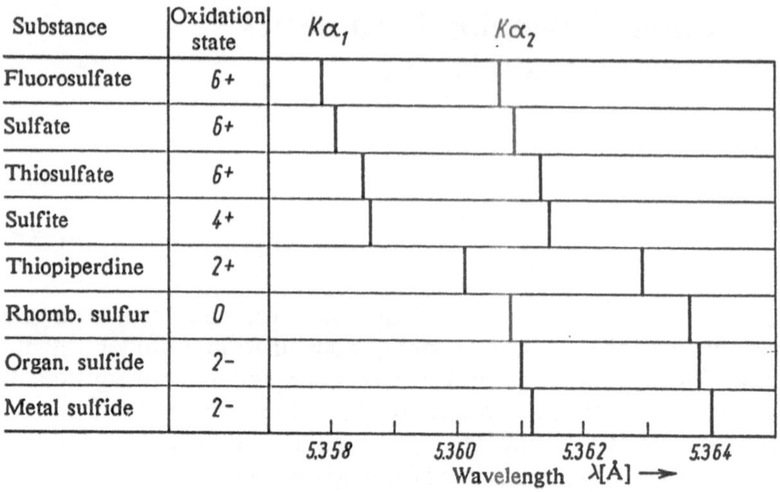

Substance	Oxidation state	$K\alpha_1$	$K\alpha_2$
Fluorosulfate	6+		
Sulfate	6+		
Thiosulfate	6+		
Sulfite	4+		
Thiopiperdine	2+		
Rhomb. sulfur	0		
Organ. sulfide	2−		
Metal sulfide	2−		

5.358   5.360   5.362   5.364

Wavelength λ[Å] →

Fig.4.1. Sulfur $K\alpha$ shifts for compounds with the indicated sulfur valencies [4.1]

Towards the end of the 1920s and beginning of the 1930s *Faessler* [4.1] found shifts for the $K\alpha$ lines and *Stelling* [4.2] for the K absorption edges of the 3rd period elements amounting to several eV in compounds compared to the pure element as a result of change in valence. Thus, a positive oxidation state causes a short-wavelength shift and a negative oxidation state a long-wavelength shift (Fig.4.1). *Johnson* [4.3] found a quadratic relation between the shift δE of $K\alpha$ for oxygen and fluorine compounds of the third-period elements with valence w, i.e.,

$$\delta E \propto w^2 . \tag{4.1}$$

Kunzl [4.4] studied shifts of the K absorption edge for transition elements of the 3rd and 4th periods and of the L_{III} edges of the main-group elements of the 5th period in oxygen compounds. He found a linear dependence on the valence within the same period.

It was thus hoped that (possibly unique) valence values could be obtained using X-ray spectroscopic measurements. However, this hope was frustrated for the transition elements, where bonding involving removal of d and f electrons [4.5-9] contributes to the shift of the $K\alpha$ doublet in the direction opposite to that associated with removal of s and p electrons, so that even long-wavelength shifts can arise (Sect. 4.3.2).

Furthermore, the well-known correlation between valence and edge shift for elements of the 3rd period has limited validity for heavier elements [4.10] and especially for transition elements. In fact, its validity seems confined to the case of ionic bonding, i.e., only when the initially unoccupied states have a simple structure. For covalent bonding and when d electrons participate, the edge structure is mostly so complicated that already the concept of "the position of the edge" is questionable [4.11].

Consequently, an alternative approach attempts to infer the ionicity of the bonding from the measured shifts instead of from valence only, the latter being merely a concept associated with chemical stoichiometry with little bearing on actual bonding. It will be seen that the spectral shift and its magnitude, caused by electrons removed or added through bonding, is better described by and related to the concept of ionicity.

Accordingly, *Ovsyannikova* et al. [4.12] suggested the following relation between edge shift δE, valence w, bonding ionicity f_i (as defined by L. Pauling) and the coordination number N of the absorbing atom

$$\delta E \propto w - (1 - f_i)N \ . \tag{4.2}$$

In addition, *Dey* and *Agarwal* [4.10] found an empirical relation for the L_{III} edge of Hg, Tl, Pb, and Bi in binary compounds

$$\delta E - E_{s \to p} \propto w^2 f_i \ , \tag{4.3}$$

where $E_{s \to p}$ is the promotion energy of an s electron to a p electron.

The strong influence of the type of bonding between the excited atom and its neighbors can be observed for predominantly covalent bonding where the valence of the excited atom plays only a minor role; for example, the shape and position of the K absorption spectra of Fe in $[Fe(CN)_6]^{4-}$ and in $[Fe(CN)_6]^{3-}$ [4.12] or from Co in $[Co(NO_2)_6]^{4-}$ and in $[Co(NO_2)_6]^{3-}$ [4.13] are practically identical.

Relatively strong correlations between energy shifts and bonding ionicity have been observed for the elements of the main groups [4.14-17]. Figure 4.2 shows the shifts of the Cl $L_{II,III}$ emission bands from some metal chlorides and their dependence on the electronegativity of each metal. Following L. Pauling, the electronegativity determines the bonding ionicity through

$$f_i = 1 - \frac{w}{N} \exp[-0.25(\chi_A - \chi_B)^2] \ , \tag{4.4}$$

where $\chi_{A(B)}$ is the electronegativity of the bonding partner. Here, the Cl $L_{II,III}$ bands shift towards shorter wavelengths corresponding to increasing filling of the Cl valence band and band-edge shifting towards higher

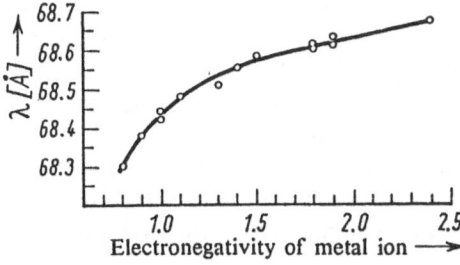

Fig.4.2. Correlation between position of Cl $L_{II, III}$ emission band and electronegativity of the bonding partner [4.16]

energies with decreasing bonding ionicity (i.e., decreasing electronegativity of the bonding partner), in contrast to the case of the $K\alpha$ line. An analogous result was obtained by *Domashevskaya* and *Ugai* [4.14] in the study of the Al $K\beta_1$ band for the series Al, AlSb, AlP, AlN, Al_2O_3, i.e., for continuously increasing bonding ionicity.

In addition to the energy positions in the X-ray spectra, the intensity distribution is also dependent on the kind of bonding in which the excited atom participates. This is most noticeable for transitions in which outer states are involved, i.e., for emission bands and absorption edges. *Drahokoupil* [4.19] showed that in the series Ge, GaAs, ZnSe, the intensity of the $K\beta_{2,5}$ band of the "cation" decreases as a function of increasing ionicity of the bonding as a result of loss of valence electrons on the "cation". Similarly, it can be seen from absorption spectra that the intensity of the first absorption line decreases with increasing covalency of the bonding. This presumably reflects loss of atomic character by the first unoccupied states with a corresponding decrease in transition probability [4.20]. The effect of individual ligands of the excited atom on the intensity of the spectra is mostly additive [4.11, 20-22].

Covalent bonding leads to complicated structure in the emission bands and absorption edges. According to *Srivastava* and *Nigam* [4.23] the following relation should be valid between the width of the absorption edge ΔE of an atom (energy difference between edge and first absorption maximum) and the quantity

$$\Delta\chi = \sum_i (\chi_A - \chi_i) \, , \quad \sqrt{\Delta E \Delta\chi} = C \, , \tag{4.5}$$

where i is the ligand of A, χ_i the corresponding electronegativity, and C is a constant for each element.

Considerable spectral shifts and intensity changes have been measured for satellites as a function of valence of the excited atom and the ionicity of the bonding. *Malkovska* [4.24] found significant differences between covalent and ionic bondings for iron compounds by studying the position of the $K\beta'$ satellite. Predominantly covalent bonding gave negative shifts while ionic bonding led to positive energy shifts. The wavelength separation between $K\beta_1$ and $K\beta'$ gives nearly constant values for Ca of 3.7 XU for ionic and 2.4 XU for covalent bondings. *Domashevskaya* and *Ugai* [4.14] observed steadily increasing positive shifts for the Al $K\alpha_{3,4}$ satellites with increasing ionicity for the above series. *Chun* and *Hendel* [4.25] determined the separation of the O $K\alpha$ satellite from the parent line $K\alpha_{1,2}$ for several metal oxides, for which a decrease of separation was observed with increasing electronegativity of the metal.

A strong influence of chemical bonding of the excited atom has been observed on the relative intensity of the satellites; in particular, for the lighter elements (2nd and 3rd periods) the valence electrons considerably affect the decay probability for doubly ionized states [4.26]. *Demekhin* and *Sachenko* [4.26] estimated the following limiting values for the ratio $\kappa = I(K\alpha', \alpha_4)/I(K\alpha_3)$ for elements of the third period (Mg, Al, Si):

κ = 0.7 for purely metallic bonding ,
κ = 1.2 for purely ionic bonding .

In comparison, the following values have been obtained for

Mg	MgO	Al	Al_2O_3	Si	SiC	SiO_2
κ = 0.54	1.27	0.60	1.48	0.69	0.81	1.10 .

Similar values of the intensity ratio $I(K\alpha_4)/I(K\alpha_3)$ have been obtained for the series of Al compounds discussed above [4.14]. Relative satellite intensities are particularly suitable for characterizing bonding conditions especially for the lighter elements.

The dependence of X-ray spectra on bonding has been discussed in a large number of papers to clarify questions of a chemical nature. Generally, a comparison is made between the spectrum of the element of interest in unknown compounds and in compounds with well-known structure. In this way, for example, *Blokhin* [4.27] determined the valence of chromium in a spinel from the Cr $K\beta$ spectrum. *Boehm* et al. [4.28] deduced the valence of cobalt in vitamin B_{12} from the Co K edge shift. *Vainshtein* and co-workers [4.29] studied the Mn K absorption edges of Co-Mn spinels ($CoMn_2O_4$ and $MnCo_2O_4$) in comparison with those of the oxides MnO, Mn_3O_4, Mn_2O_3, MnO_2 (Fig. 4.3). Thus, a significant shift of the edge was observed with increasing values towards higher energies, and an additive superposition of the spectra is present for Mn in multivalent Mn_3O_4. Analogous conclusions can be drawn from the presence of Mn^{2+} and Mn^{4+} in the spinels studied.

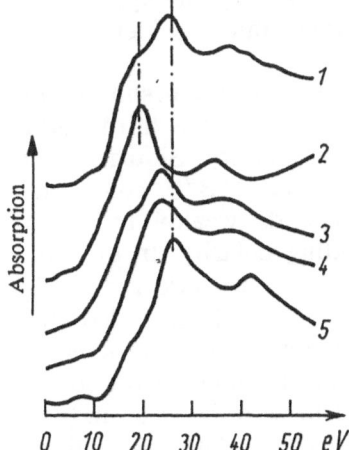

Fig.4.3. Shifts of Mn K absorption edge as seen in $CoMn_2O_4$ (Curve *1*), MnO (Curve *2*), Mn_3O_4 (Curve *3*), Mn_2O_3 (Curve *4*) and MnO_2 (Curve *5*) [4.29]

Similarly, *Kirichok* [4.30] found increasing shifts of the Fe K edge with an increasing ratio of Fe^{3+}/Fe^{2+} in the (Ni,Zn)-Fe spinels. X-ray spectroscopic evidence of two different states of valence in a compound was also given by, for example, *Srivastava* and *Nigam* [4.23] for Cu(I,II) in a complex with thiomalonic acid from observing the Cu K absorption spectrum, and by *Vainshtein* and co-workers [4.31] for Pr(III,IV) in Pr_6O_{11} using the Pr L_{III} edge and Sm(II,III) in SmB_6. *Meisel* and *Köstler* [4.32] proposed a method for determining the ratio of Ni/NiO in carrier catalysts based on the presence of two valence states causing asymmetry of the Ni $K\alpha_1$ line.

Gusatinskii and *Nemnonov* [4.33] showed that the In L_{III} absorption spectrum from "In_2O" condensed from the gaseous phase is identical to a corresponding mixture of In and In_2O_3, i.e., that this oxide disproportionates upon condensation. On the other hand, *Krämer* [4.34] gave direct proof of the existence of Si suboxides from X-ray spectroscopic measurements of shifts of the Si $K\alpha$ line.

Vainshtein and co-workers [4.31,35-37] studied the L_{III} absorption edges of rare-earth metals in borides, silicides, phosphides, germanides, hydrides, and intermetallic compounds in comparison with their oxides. They found evidence, in the case of d states, of a strong similarity in binding type between the above-mentioned compounds and their trivalent oxides, except for Eu and Yb, which can be understood in terms of the particular electronic configurations of these two elements. *Meisel* and co-workers [4.8] determined the valence of titanium in $LaTiO_3$ and $EuTiO_3$ to be three and four, respectively, as follows from the long-wavelength shift of the $K\alpha$ doublet caused by increasing d-electron transfer.

The relation found between X-ray spectra and bonding ionicity for a constant valence has been used to clarify chemical bonding in some cases: *Blokhin* and co-workers [4.38] developed an incremental model for the contributions of different atoms bonded to sulfur to the total S $K\alpha$ shift and used this model to differentiate between various possible structures for organic sulfur compounds. For example, the structures of methylene blue, sulfur-containing chelate complexes and rhodanites could be identified in this way.

Faessler [4.39] found an increase in the long-wavelength shift of the $K\alpha$ line of the B component in $A^N B^{8-N}$ compounds of the third period with decreasing N, i.e., with increasing bonding polarity. Similarly, SiC shows mainly covalent bonding whilst $A^{II}B^{VI}$ and $A^I B^{VII}$ compounds show a large long-wavelength shift, which depends weakly on the A partner, if at all, for compounds with high bonding polarity. In this context it is worth noticing, in contrast to earlier models, that also for $A^{III}B^V$ compounds a moderate ionicity was found. *Faessler* [4.39] concluded from results for alkali polysulfides that the S $K\alpha$ shifts decrease with increasing chain length. The negative charge is spread over the total chain so that already the measured shift for trisulfides is close to zero. *Ugai* et al. [4.40] used the Zn $K\alpha$ shifts in ZnSb, Zn_3Sb_2, and Zn_4Sb_3, related through the formal valency, as a measure of the bonding ionicity of these compounds. The values can be compared with the widths of the forbidden gaps of the substances studied.

4.2 Structural Analysis

4.2.1 Analysis Using Emission and Near-Edge Absorption Spectra

Studying the influence of geometry of the surroundings of the excited atom on its X-ray spectra, it becomes meaningful to distinguish between the emission spectrum and the near-edge part of the absorption spectrum (in which the electron remains localized in the vicinity of the excited atom) on one hand, and the extended part of the absorption spectrum (which is mainly influenced by the geometry of the neighboring atoms) (Sect.2.6.3) on the other hand. For the first case, geometrical factors influence the X-ray spectra only to the extent that the electron density of the excited atom is influenced as well. This is evident for the spectra of stereoisomeric compounds. *Stelling* [4.2] obtained already in 1928 a separation of 4.5 XU in the K edge position of black and white phosphorus. An analogous result was achieved by *Fichter* [4.41] with respect to the position of the P $K\alpha$ doublet. Furthermore, *Stelling* observed a shift of the position of the Cl K edge in cis- and trans-isomeric complexes $[Co(NH_3)_4Cl_2]^+$. The effect of particular neighboring atoms on the electron distribution of the excited atom (and thereby on the emission and near-edge absorption spectra) depends not only on their number and distances from the excited atom, but also on the geometrical structure. In *Stelling* (op.cit.), it was shown that the Cl K edge shifts towards longer wavelengths in the series LiCl, KCl, RbCl (NaCl lattice) with increasing ionic distances. Similar results have been obtained for the S K edge for the series MgS, CaS, BaS, MnS (NaCl lattice); BeS, ZnS, CdS (ZnS lattice) and FeS, CoS, NiS (NiAs lattice). Due to increased ionic distances the inclusion of H_2O molecules in the lattice of metal chlorides causes a long-wavelength Cl K edge shift.

Aoyama and co-workers [4.42] found, for the same lattice type and the same ionic configuration, a linear relationship between the short-wavelength shifts of Cl K and S K edges in chlorides and sulfides (compared to the free ion), and the inverse ionic distances. Analogously, *Kirichok* [4.30] obtained for the Fe K edge in the previously discussed (Ni, Zn)Fe$_2$O$_4$ ferrites an increasing short-wavelength shift with increasing lattice constant. On the assumption that the coordination number is constant, similar correlations between bond lengths and band shifts were obtained by *White* and colleagues [4.43,44] for Al $K\beta$ and Si $K\beta$ emission spectra of alumosilicates and by *Freund* and *Hanich* [4.45] for Mg $K\beta$

spectrum from Mg silicates. Already from *Stelling's* results [4.2] it was evident that the coordination number had an influence of the same order of magnitude as ionic distances. Thus, the Cl K edge of KCl and CsCl has the same position in spite of different ionic distances since the larger coordination number in the CsCl lattice causes a short-wavelength shift of the Cl K edge which compensates the long-wavelength shift caused by larger ionic distance.

Ovsyannikova [4.46] observed different positions of the metal K edge for CoS and NiS depending on whether the sulfide crystallized in millerite or NiAs type lattice (coordination number 5 or 6, respectively). Analogously, she found identical edge positions for ZnS as sphalerite or wurtzite (equal coordination number 4). *Day* [4.47] observed different Al $K\alpha$ shifts depending on whether Al was present in six- or fourfold or mixed coordinations. Using this result it was possible for him to assign coordination numbers not otherwise known to Al in several compounds ($Li_2O \cdot Al_2O_3$, $Al_2(SO_4)_3 \cdot 18H_2O$, $Al(NO_3)_3 \cdot 9H_2O$). Furthermore, *Shuvaev* et al. [4.48] determined the coordination number of Al in glass and ceramic using the $K\alpha$ shifts while *Iseki* and *Tagai* [4.49] did the same for Ca aluminates. *Suwa* and coworkers [4.50] showed that for Mg alumosilicates, the Al $K\alpha$ and Mg $K\alpha$ shifts depend on the ratio of six- to fourfold coordinated Al and of eight- to sixfold coordinated Mg, respectively, and can be used to determine these ratios. *Läuger* [4.51] showed that the Al $K\alpha$ line is broadened and composed of two components in mixed coordinations compared to that in a single coordination, each being the same as the two mixed ones. Based on different coordinations causing different shifts of the Cu K edge in CuO and $CuAl_2O_4$ (spinel), it was possible for *Wolberg* and *Roth* [4.52] to explain the formation of this spinel as a surface phase of CuO on Al_2O_3 carriers. This could not be observed with X-ray diffraction in earlier studies.

The shape of the X-ray spectrum is more sensitive to the geometry of the environment just as is its position. Using the fine structure of the absorption edge, the environment of an absorbing atom can be relatively easily identified. *Keeling* [4.53] studied the Co K absorption edge of CoO on SiO_2 and Al_2O_3 carriers. He compared the spectra of diluted CoO with those of CoO, Co_3O_4 and $CoAl_2O_4$ and found (besides the formation of Co_3O_4 and $CoAl_2O_4$ for low CoO concentrations) the formation of a surface layer consisting of a so-called δ phase with octahedrally coordinated Co.

The near-edge fine structure of the X-ray absorption spectrum is determined by the structure of the outer orbitals of the absorbing atom, which depends on the coordination number of this atom. Thus, for example, tetrahedrally coordinated ions with d^3s hybridization (V^{5+}, Cr^{6+}, Mn^{7+}) show an intense preabsorption (as a result of the 4p orbitals which do not participate in the bonding) caused by transitions into these orbitals. Using this feature, *Lytle* [4.54] could distinguish between $CrVO_4$ and $VCrO_4$, and assign the structure to the latter compound. On the contrary,

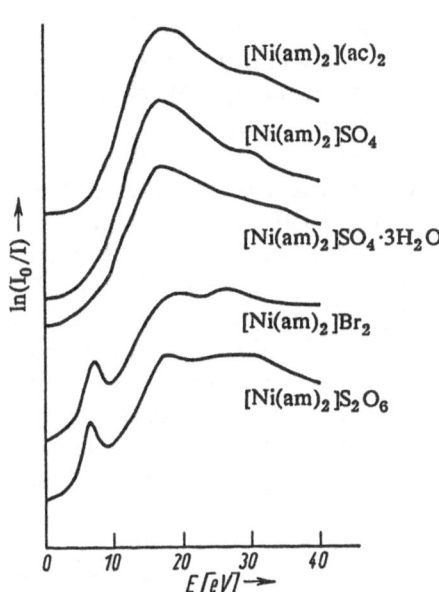

[Ni(am)₂](ac)₂

[Ni(am)₂]SO₄

[Ni(am)₂]SO₄·3H₂O

[Ni(am)₂]Br₂

[Ni(am)₂]S₂O₆

Fig.4.4. Nickel K absorption spectra from Ni complexes having tetrahedral stucture (without preabsorption, upper part of figure) and planar structure (with preabsorption, lower part) [4.11] (am: aminine, ac: acetate)

tetrahedrally coordinated ions with sp^3 hybridization show no preabsorption, except for fourfold coordinated planar complexes with dsp^2 hybridization, since for this case there is also a p orbital participating in the bonding [4.11, 55-57]. This result has been used by several investigators [4.13, 58-60] to determine the structure of fourfold coordinated transition-metal complexes. For example, *Böke* [4.11] was thus able to demonstrate the existence of two different forms of Ni aminine complexes (Fig.4.4).

Further discussion of local geometry effects, as seen in X-ray emission bands and near-edge absorption spectra, are given in Chaps.5 and 6 because of the close association of these with electronic structure.

4.2.2 Analysis Using EXAFS

Recently, it has been shown that the extended fine structure of an X-ray absorption edge (EXAFS) has a significant role in structure determination. Following initial studies of this type [4.61, 62], interest in EXAFS increased rapidly with the development of intense sources, especially synchrotron radiation sources, together with improved data analysis and parallel developments regarding the theoretical framework including electron scattering. In this context structural analysis of disordered systems is particularly important since no radial distribution information could be obtained using methods available earlier, such as X-ray, electron, and neutron diffraction.

Considering the objectives of this book and the abundance of published EXAFS literature, this chapter gives only a general overview of experimental methods of EXAFS recordings and data analysis. An extensive compilation of EXAFS studies carried out by transmission technique may be found in several sections of the Appendix, along with details concerning objectives of these studies. More extensive and detailed reviews can be found in a series of articles [4.63-68] which emphasizes results from organometallic compounds and biochemical systems [4.63,67,69], amorphous systems and glasses [4.64], catalysts [4.70] and ionic conductors [4.71]. Other studies have concentrated on spectra from highly dilute systems, ions in solution [4.72], inorganic complex compounds [4.63,73,74], thin films and surfaces [4.75], salt solutions [4.76], and liquid metals [4.77].

Modulation of the absorption coefficient in the EXAFS region, caused by superposition of the ejected photoelectron wave and waves scattered from neighboring atoms (Sect.2.6.3) can be investigated via the primary process, i.e., through transmission and reflection experiments or by means of secondary processes, for example recording of X-ray fluorescence or Auger electron emission.

Transmission experiments give the simplest configuration for EXAFS studies. With careful preparation of optimum homogeneous samples, structural information concerning all materials which have a sufficiently large contribution from inner-shell absorption μ_s to the total absorption $\mu = \mu_b + \mu_s$ (μ_b: background absorption) can be obtained.

Reflection experiments are particularly advantageous for those cases where geometrical structures of surfaces or different depths in the sample are to be investigated. This can be easily understood since the reflectivity below the critical angle for total reflection is determined by the imaginary part β of the index of refraction $n = 1-\delta-i\beta$; i.e., it is determined by the absorption coefficient and can be calculated with different angle-dependent penetration depths of the X rays [4.75].

X-ray fluorescence experiments measure the dependence of the emitted characteristic X rays on the incoming photon energies [4.78] (X rays are emitted when a vacancy created by photoabsorption, is filled). This technique benefits from the strong reduction of the background signal due to depression of the signal of the matrix atoms and is particularly suitable for studying samples present in small concentrations, thin films and adsorbants on solids.

Auger-electron yield spectra give information on EXAFS from the alternate secondary processes in an analogous way to X-ray fluorescence [4.79,80]. As a consequence of the smaller mean-free path of the Auger electron, only a few atomic layers on the surface contribute to the fine structure, causing considerable experimental difficulties due to small yields but giving very useful sensitivity for surface studies.

Electron-yield spectra combining all the secondary processes, carry information from depths of 10-20 nm and can be recorded with less difficulties than Auger electron spectra (due to the lower yields in the latter case) [4.75,81-83]. Electron-yield spectra are particularly useful for studying the surroundings of light atoms for edge energies below 1 keV.

Analysis of EXAFS spectra is done using Fourier transformation techniques [4.62, 84, 85] combined with curve-fitting analysis [4.86, 87]. It is assumed that multiple scattering can be neglected, the amplitudes for each type of backscattering atom and the phase functions for each pair A-B (A: absorbing atom, B: backscattering atom) can be separated. The latter implies that the phase functions do not depend on chemical binding. The calculation of structural parameters proceeds through the following steps: The modulation χ of the total absorption coefficient μ is extracted as a function of photon energy E assuming that the unknown monotonous absorption coefficient μ_0 can be approximated by a low-order polynomial. By transforming E into the wave vector k of the photoelectron using (2.71), Eq.(2.82) is obtained. Using the relation between $\chi(k)$ and the scattering amplitude of the individually observed atoms, the Debye-Waller factor, the phase shift of the photoelectron and the distance of the scattering atom to the absorbing atom (2.83), the structural parameters are obtained by a curve-fitting technique with amplitude and phase as parameters. Using Fourier transformation, a series of Gaussian peaks is obtained, which correspond to individual coordination spheres and whose maxima, after shifts of 0.02 to 0.04 nm, are related to the interatomic distances which are definable with an accuracy better than 1% (Figs. 4.5, 6). The widths of the peaks arise from a disturbance of the crystal symmetry caused by motion and crystal imperfections (disregarding limited instrumental and theoretical resolving power of the finite transformation process). The peak heights are proportional to the coordination numbers. The Fourier-transformed spectrum thus represents a radial-distribution curve of the absorbing atom. Individual contributions to the function $\chi(k)$ can be separated from an inverse Fourier transformation. In this way, backscattering amplitudes, coordination numbers, and Debye-Waller factors can be extracted.

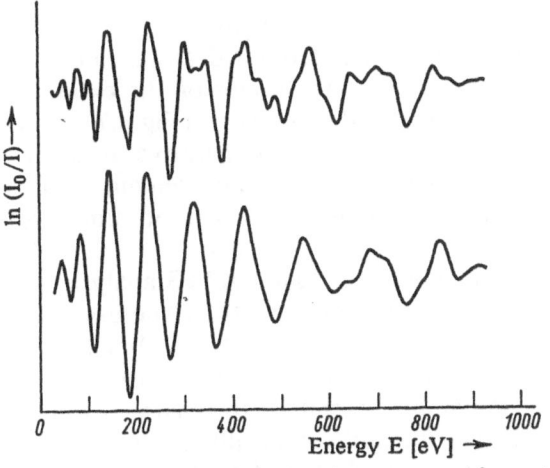

Fig.4.5. Smoothed K absorption spectra of crystalline (upper curve) and amorphous germanium (lower curve) [4.88]

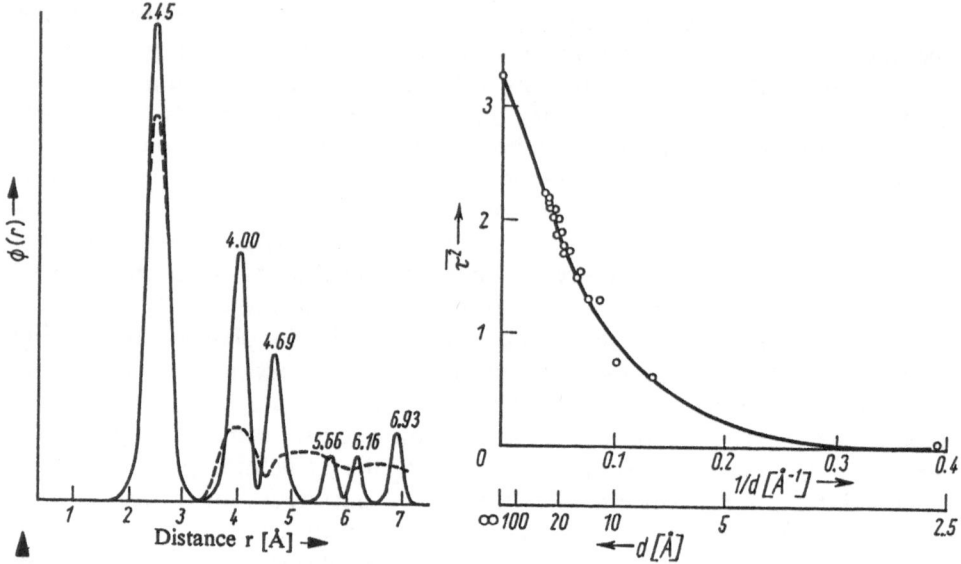

Fig.4.6. Fourier-transformed spectra from Fig.4.5. (Solid curve: crystalline Ge, dashed curve: amorphous Ge) [4.88]

Fig.4.7. Dependence of the mean square fluctuation, $\overline{r^2}$ of the extended fine structure of the K absorption edge of Cu on the crystallite size d [4.89]

Just as amorphous materials can be considered as crystals with a large number of displacements of the atoms in the crystal lattice, so can small crystallites with large specific surface area (high-area materials) be represented as crystals with a large number of lattice vacancies, i.e., for atoms on or near the surface, the coordination number is considerably decreased in the crystallite compound compared to that of the bulk atom. For sufficiently small crystallites (<10nm) a measurable decrease of the amplitude of the extended fine structure is to be expected following (2.81). This effect can be theoretically understood if one starts from a definite model of the crystallites. Using the relation between the amplitudes of the extended fine structure and crystallite sizes obtained theoretically (Fig.4.7), it is possible to determine mean crystallite sizes from comparative measurements on micro- and macrocrystallite samples of the same element. In this way *Keilacker* and *Meisel* [4.89] obtained average sizes of platinum crystallites spread onto charcoal.

With the results available today it is evident that EXAFS analysis is a very useful addition to older and well-known methods for structural analysis. However, it is worth noting that a complete three-dimensional structural analysis is not feasible and angular-dependent information is possible to obtain only when studying single crystals with polarized X rays [4.90-92]. Further development of this method, especially improving focussing monochromators and X-ray tubes and extending the wavelength region to lower energies, will greatly extend standard methods for structural analysis. In this regard, particular attention to the development of intense, nanosecond soft X-ray sources from laser-produced plasmas [4.93] should prove useful.

4.3 Effective Atomic Charges

4.3.1 Concepts, Interpretation and Determination of Effective Atomic Charges

The properties of a substance are largely determined by the spatial and energetical distribution of all of its electrons. The electronic structure of the valence electrons is especially interesting since it changes characteristically upon formation of chemical bonds. To give an accurate description of chemical bonding it is necessary to determine the total electron density which, however, is possible only with a very considerable experimental effort. The redistribution of valence electrons can, however, be classified using the limiting cases of chemical bonds, which can be understood merely as a description of the spatial electronic structure. Using such a simplification, which necessarily introduces a loss of information, the characterization of the bonding is possible using only a single parameter (e.g., ionicity, covalency, atomic charge). This can be very useful when relations are found between these parameters and specific properties of the substances.

By the effective atomic charge, one understands the charge an atom has when it is incorporated in the corresponding compound and which can adequately describe the properties of the compound. From this, a

Table 4.1. Atomic charges for $A^{III}B^{V}$ compounds obtained with X-ray diffraction [4.96]; q: determination from constant volumes; q_{HF}, q_{TF} : determination from structural amplitudes (with Hartree-Fock and Thomas-Fermi wave functions, respectively)

	q	q_{HF}	q_{TF}
AlP		0.80	0.32
AlAs	1.00	0.60	0.40
AlSb	0.67	0.63	0.45
GaP		0.75	0.62
GaAs	0.80	0.51	0.36
GaSb	0.58	0.59	0.43
InP		0.70	0.58
InAs	0.49	0.50	0.35
InSb	0.38	0.27	0.18

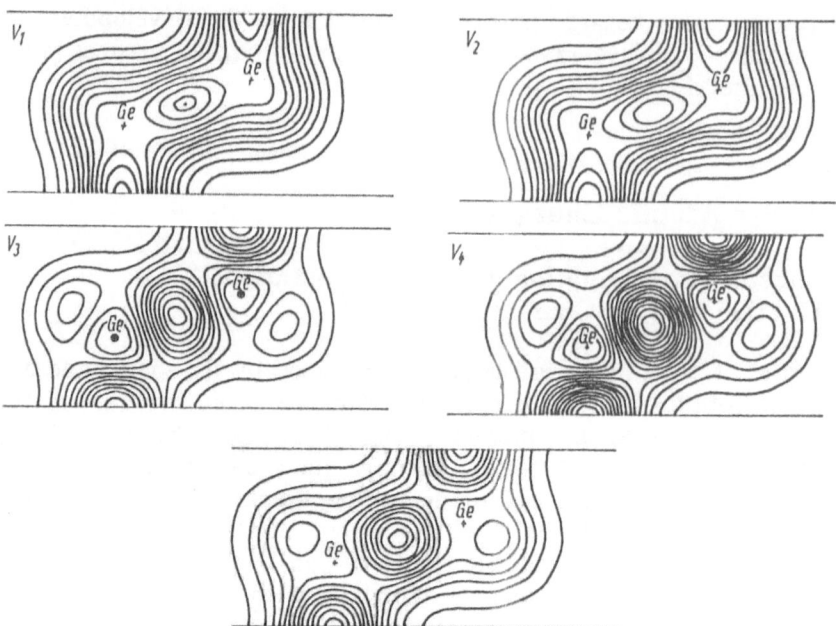

Fig.4.8. Calculated electron density maps for the indicated valence bands of Ge. (Top and middle: valence bands v_1 to v_4, bottom: sum of these bands) [4.100]

considerable difficulty emerges. The atomic charge, if it is understood as the difference between the electron densities in the compound and in the neutral atom, is not constant in space. Therefore, it is necessary to choose criteria for dividing the valence electrons in the discussed compounds and their assignment to individual atoms [4.94]. Since the individual methods are sensitive to different regions of spatial electron densities and thereby lead to different criteria for the assignment of electrons to the atoms, these different experimental methods do not always give comparable values for the atomic charges.

Determination of valence-electron densities from X-ray scattering has so far been pursued only for a few substances. From such investigations the determination of atomic charge is possible either from determining the atomic volume or from structure amplitudes [4.95-97] (Table 4.1). Lately, especially theoretical electron-density distributions of simple solid-state compounds have been published for covalent bondings, which show an increase of the electron density between the bonding partners (bonding charge) and furthermore give relations between spatial and energetic electronic structure [4.98-100]. It is thus possible to determine which part of a special wave function participates in the chemical bond (Fig.4.8).

If the wave functions are expanded in an LCAO representation, the atomic charges q_A can be calculated from a population analysis [4.101–103]

$$q_A = n_A - \sum_i n_A(i) \, . \tag{4.6}$$

Here n_A is the number of valence electrons of the neutral atom A. The number of electrons $n_A(i)$ of atom A in orbital i is given by

$$n_A(i) = N(i)[C_A^2(i) + C_A(i)C_B(i)S_{AB}] \, , \tag{4.7}$$

where C_A, C_B are eigenvectors in the space spanned by the molecular orbitals, S_{AB} is the overlap integral, and $N(i)$ is the total number of electrons in the molecular orbital i. The influence of choice of the basis set and the validity of the definition of charge must be considered.

One of the oldest methods is to determine the ionicity f_i of bonding from the electronegativity χ [4.104] (4.4), where χ is determined, for example, following L. Pauling from thermodynamical values or following R.S. Mullikan from ionization energies and electron affinities. A direct relation exists between the effective charge through $q = f_i w$ (w: charge assuming a pure ionic bonding). However, it must be recalled that the electronegativity is not a well-defined quantity from a quantum-mechanical standpoint [4.105, 106].

For solids a simple relation has recently been found between the ionicity f_i and the width of the energy gap E_g between bonding and antibonding states [4.107, 108]. Assuming that E_g arises from the existence of a symmetric and antisymmetric part of the potential (in a homopolar and heteropolar division), according to $E_g^2 = E_h^2 + C^2$, the following relation is valid

$$f_i = C^2/E_g^2 \, . \tag{4.8}$$

The E_g and E_h are obtained experimentally from the static electronic dielectric constant of the compound and the corresponding elements, respectively. Beyond that, f_i can be deduced from the structure of the valence band alone according to

$$f_i = \sin^2\left[\frac{\pi}{2} - \frac{E_{AS}}{E_{VB}}\right] , \tag{4.9}$$

where E_{AS} is the width of the antisymmetric energy gap which corresponds to the distance between the two lowest energy bands, E_{VB} is the full width of the valence band.

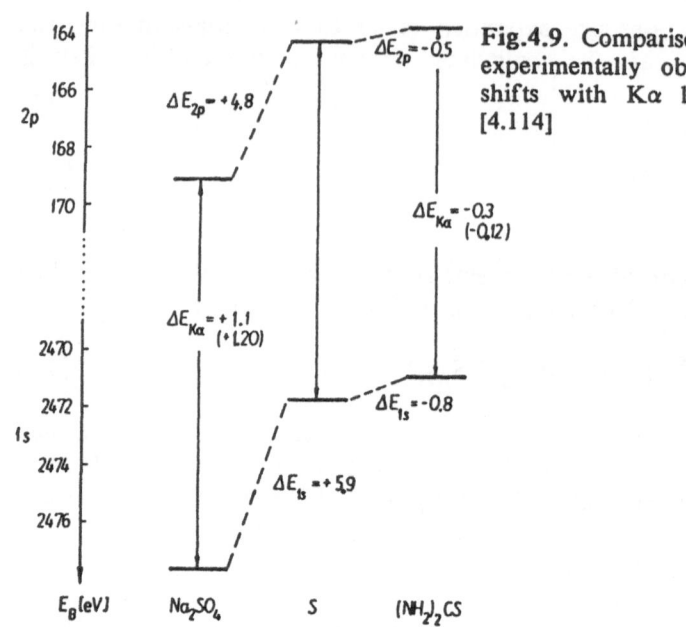

Fig.4.9. Comparison of the differences of experimentally obtained 1s and 2p line shifts with Kα line shifts (in brackets) [4.114]

Of the different spectroscopic methods those which depend on changes in mainly atomic properties should be particularly useful for determining atomic charges. Besides measurements on the nuclei itself (Mössbauer spectroscopy [4.109] and NMR spectroscopy [4.110]), measurements of the chemical shifts of the inner shells (photoelectron line shifts [4.111,112] and X-ray line shifts [4.94,113]) are particularly useful. Determination of atomic charges from chemical shifts of both photoelectron and X-ray spectra depend of the assumption of a point-charge model (Sect.4.3.2). Limitations of this model are obvious for X-ray transitions including levels with small binding energies, and are less severe for the ESCA method due to the larger shift values measured. The advantage of ESCA, namely larger chemical shifts, is partly counteracted by the generally lower experimental accuracy. Clearly, differences between ESCA shift values correspond to the chemical shifts of X-ray lines (Fig.4.9), both methods give, in principle, the same result [4.114]. For solids the interpretation of ESCA shifts must also include changes in work functions [4.111, 112,115,116]. Numerous studies exist giving relations between ESCA shifts and atomic charges also obtained by other means [4.111,112, 117–119].

4.3.2 Determination from X-Ray Emission Spectra

The justification for determining effective atomic charges from the shifts of inner-electron levels is based on the fact that the position of these levels is directly related to the electrostatic potential [4.118, 120, 121]. Assuming the validity of Koopmans' theorem (2.12), one obtains the chemical shift for each level by (2.19). The integrals (one- and many-center integrals of the one- and two-electron Coulomb and exchange interactions) can then be simplified in the following way [4.117, 122]:

a) Among the one-electron integrals H only the one-center terms H_{AA} are considered exactly. For the many-center integrals use is made of a point-charge approximation, viz.,

$$\sum_{B \neq A} Z_B{}^{core}/R_{AB} \; .$$

b) Among the two-electron integrals G the many-center exchange integrals are neglected.

c) For the many-center Coulomb integrals also a point-charge approximation

$$\sum_{B \neq A} Q_B/R_{AB}$$

is used, where Q_B is the total valence-electron density on atom B. The one-electron energy of the orbital i is then given by

$$\epsilon_i = H_{AA} + G_{AA} + V_A \qquad \text{with} \tag{4.10}$$

$$V_A = \sum_{B \neq A} \frac{Q_B - Z_B{}^{core}}{R_{AB}} \; .$$

From the remaining one-center integrals all one-electron integrals cancel in the calculation of the chemical shifts according to (2.19) and all two-electron integrals cancel as well, except those with n valence electrons

$$\delta E_i = \epsilon_i(2) - \epsilon_i(1) = \sum_n G_{AA} - \sum_n G'_{AA} + (V_A - V_A') \; . \tag{4.11}$$

If one assigns a magnitude k to the remaining two-electron interaction G_{AA} between the inner electrons and the valence electrons, a value q to the difference in the valence-electron density, and V to the difference in potential, the following relation can be deduced which was obtained empirically [4.111, 112]

$$\delta E_i = k_i q_A + V \; . \tag{4.12}$$

151

For the many-center potential it can be assumed that all levels have the same Madelung energy

$$V = \sum_{B \neq A} \frac{q_B}{R_{AB}} .$$ (4.13)

Since q and V depend only on the valence-electron densities, the chemical shift has been described as arising from the changed potential for the inner electrons brought about by the valence electrons:

$$\delta E_i = \langle i | \Delta V_n | i \rangle .$$ (4.14)

In this connection there are possible, naturally other approaches to ΔV_n [4.118, 123].

From this discussion we have obtained a free-ion model which can be applied to describe the differences of level shifts measured by X-ray spectroscopy

$$\delta E_{ij} = \delta E_j - \delta E_i = (k_j - k_i)q = k_{ij}q .$$ (4.15)

It is, of course, assumed that both levels participating in the X-ray transition do not experience different relaxation, many-electron and correlation effects. Using (4.15) even though the magnitude of k_{ij} is not known, from experimental values of the shifts one can make qualitative statements concerning charge transfer and the direction of charge transfer in a series of compounds - at least for substances without participation of d or f electrons (see below) - and the sign of the charge.

Quantitative estimates of the parameters k_{ij} are obtained empirically and semitheoretically or from quantum-mechanical calculations. By neglecting the exchange interaction and approximating the Coulomb interaction by Slater functions, the $K\alpha$ line shift can be expressed as [4.124]

$$\delta E_{K\alpha} = a \, E_{K\alpha} \sum_q \sqrt{I} ,$$ (4.16)

where a is an empirically determined parameter, and I is the ionization energy.

The curve thus obtained for $\delta E = f(q)$ agrees satisfactorily with the relations obtained from quantum-mechanical calculations (Fig.4.10). *Sumbaev* and co-workers [4.9] obtained empirical values of k_{ij} for elements of higher atomic numbers from the shifts δE_{ij} and the ionicity $f_i{}^P$ according to (4.4) using experimental determinations of the dependence of chemical shifts and using the ionicity determined following L. Pauling according to

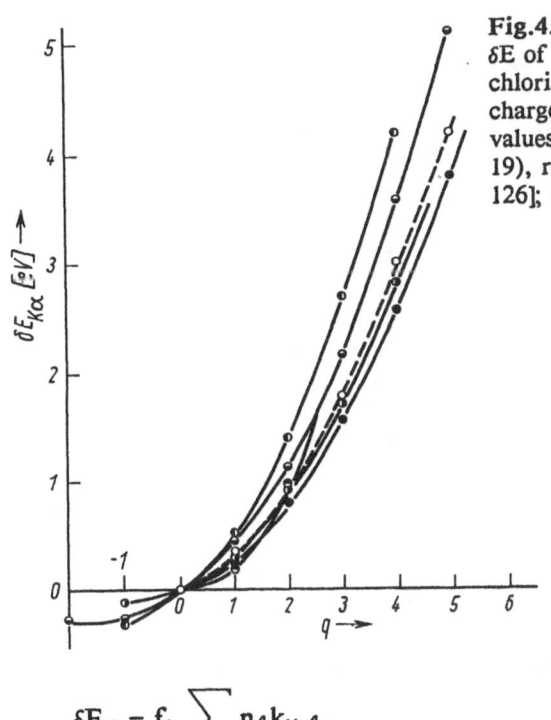

Fig.4.10. Variation of the chemical shifts δE of $K\alpha$ lines of sulfur (solid curves) and chlorine (dashed curve) with effective charges q: ○, ● from (2.19) [4.113] with values from [4.125]; ◑, ◐ from (2.18 and 19), respectively [4.111]; ⊖ from (2.19) [4.126]; ⊜ from (4.16) [4.124]

$$\delta E_{ij} = f_i \sum_\ell n_\ell k_{ij,\ell} , \qquad (4.17)$$

with ℓ being the angular momentum quantum number, and n_ℓ the number of electrons with quantum number ℓ. One can also determine k_{ij} from the change in electron density in spherical volumes with the covalent radius whose magnitude is determined so that the charges q obtained agree satisfactorily with those derived from dipole moments [4.127].

Quantum-mechanical values can be obtained from calculations of screening (2.21) as

$$k_{ij} = \sigma_{ik} - \sigma_{jk} , \qquad (4.18)$$

where the screening constants σ_{ij} are obtained from calculations of free atoms [(2.20), $k_{ij} \stackrel{\wedge}{=} \delta E_{K\alpha}$], of atoms and ions (2.19) and by including inner vacancies for the initial and final states following (2.18). The corresponding k values for the last two cases can be estimated from a calibration curve of $\delta E_{ij} = f(q)$ (Fig.4.10), where q is the charge for the free ion. Since for both light and heavy elements such shifts δE_{ij} exist [4.125, 128, 129], this method is generally applicable. By comparing results from different methods of calculation, the following conclusions may be drawn:

1) Calculation of $\delta E_{ij} = f(q)$ using different quantum-mechanical methods gives surprisingly good agreement (Fig.4.10) [4.113].

2) The dependence is not linear, but rather it can be described by a function of the form $\delta E_{ij} = aq^b$. For the elements of the third period a

least-square fit gave the values a = 0.2 and b = 1.6 [4.130]. A more accurate derivation of the dependence, (4.12), also shows a deviation from a linear relation.

3) Calculations following (2.18) gave higher values of the shifts for the same q value compared to the results using (2.19). With calculations for Z+1 this difference can be partly accounted for [4.113] (Fig.4.10). This corresponds to the model recently called the equivalent-core method for the interpretation of ESCA shifts [4.131, 132].

4) To a first approximation, loss of s and p valence electrons leads to an equal short-wavelength $K\alpha$ shift while the loss of d and f valence electrons of transition elements or of rare earths gives a long-wavelength $K\alpha$ shift [4.7, 94, 113, 127, 133]. Small differences have, however, been observed for the loss of s and p electrons [4.122, 134]. For elements with atomic numbers $Z \geq 36$ such results have been obtained both from relativistic Hartree-Fock calculations [4.128] and from experimental investigations [4.112, 135]. The difference in k_{ij} values for s and p electrons, on the one hand, and for d and f electrons, on the other, can be understood in terms of the larger amplitudes near the nucleus for d and f wave functions; their removal causes a larger change in shape for the inner wave functions. This means that for loss of s and p electrons a perturbation calculation of first order is sufficient, while for d and f electrons a perturbation calculation of second order is required [4.136].

For the elements of the third period, quantitative values can be obtained for the atomic charges from X-ray emission lines either directly according to (4.15) or by using a calibration curve for Z+1. The charges obtained in this way [4.94] are in qualitative agreement with general chemical understanding of a series of compounds, although quantitative values based on a homopolar bonding partition, e.g., in the alkali halides, are lower than the formal oxidation numbers. For many compounds with an otherwise unspecified charge distribution, quantitative values appear to have been obtained in this way. One can compare charges determined from LCAO-MO calculations [4.137-139] and reach conclusions concerning approximations used in the calculations.

Another advantage of the method besides simple interpretation of experimental data is the possibility it gives to study each of the bonding partners individually. Reasons for different effective atomic charges of bonding partners in binary compounds are, besides the uncertainty in determining quantitative values, the influence of effects related to the changes due to the inclusion of the free ions in a solid and the asymmetric distribution of the valence-electron density (presence of bonding charges) [4.96].

For the 3d transition elements the charge is composed of two parts according to expected differences of d, s, and p electrons. One may write

$$q = a + b \, ,$$

$$\delta E_{K\alpha} = a k_{1s2p}^{s,p} + b k_{1s2p}^{d} \, . \tag{4.19}$$

154

Table 4.2. Atomic charges determined from Kα shifts using $q = (\delta E_{K\alpha}/a)^{1/b}$ with a, determined from different quantum-mechanical calculations and b = 2 (if no other value was mentioned) [4.130]

	Missing p electron Method of calculation			Missing s,p electrons Method of calculation			(Z+1) appro:
$\delta E_{K\alpha}$(Ga) ±0.007eV	Hartree a=0.003	Clementi a=0.039	Watson a=0.101	Hartree a=0.056	Clementi a=0.067	Watson a=0.153	a=0.13 b=1.5
GaP	0.028	0.92	0.85	0.51	0.71	0.65	0.39	0.36
GaAs	0.030	0.95	0.88	0.53	0.73	0.67	0.40	0.38
GaSb	0.018	0.72	0.66	0.40	0.55	0.51	0.32	0.27
Ga_2O_3	0.203	2.51	2.31	1.39	1.93	1.78	1.07	1.37
$Ga(NO_3)_3$	0.250	2.75	2.53	1.52	2.12	1.95	1.17	1.58
$Ga_2(SO_4)_3$	0.269	2.86	2.63	1.58	2.20	2.02	1.21	1.62
$\delta E_{K\alpha}$(As) ±0.01eV	a=0.039	a=0.083		a=0.018	a=0.100		a=0.14 b=1.7
GaAs	0.10	-1.61	-1.11		-1.46	-1.00		-0.80
InAs	0.06	-1.26	-0.86		-1.14	-0.76		-0.53
As_2S_3	0,04	+0.96	+0.67		+0.88	+0.60		+0.48
K_3AsO_4	0.22	+2.36	+1.62		+2.15	+1.46		+1.01
As_2O_3	0.16	+2.04	+1.39		+1.84	+1.26		+1.30
NaH_2AsO_4	0.22	+2.36	+1.62		+2.13	+1.46		+1.30

Thus, two measurements are required to determine q. An example of this was given from a combination of (4.19) with Mössbauer shifts [4.94] or X-ray Kβ line shifts [4.113] using

$$\delta E_{K\beta_{1,3}} = a k_{1s3p}^{s,p} + b k_{1s3p}^d \;. \tag{4.20}$$

However, interpretation of the $K\beta_{1,3}$ line shifts introduce certain difficulties [4.113], which evidently are related to the limit of applicability of the free-ion model (see below). Due to the multiplet structure [4.94] the shifts of the centers of gravity must be used for determining line shifts instead of using the line maxima. Charges and electron population determined in this way agree partly very well with theoretically determined values using the MO method [4.113, 134, 140].

For the main-group elements with $30 \leq Z \leq 36$ good calculations are available only for free atoms and for singly ionized ions. This means that to determine q, either a linear relation or an analogous dependence as for the elements of the third period (see above) must be assumed. Effective atomic charges for Ga and As compounds are given in Table 4.2 [4.141].

For the elements $32 \leq Z \leq 74$ numerous investigations have been made especially of the Kα shifts [4.9, 135]. A survey of experimental data is given in Fig.4.11. From these data k_{1s2p} values have been deduced semi-theoretically (for the loss of 4sp electrons +0.09; 5sp: +0.08; 4d: -0.2; 5d:

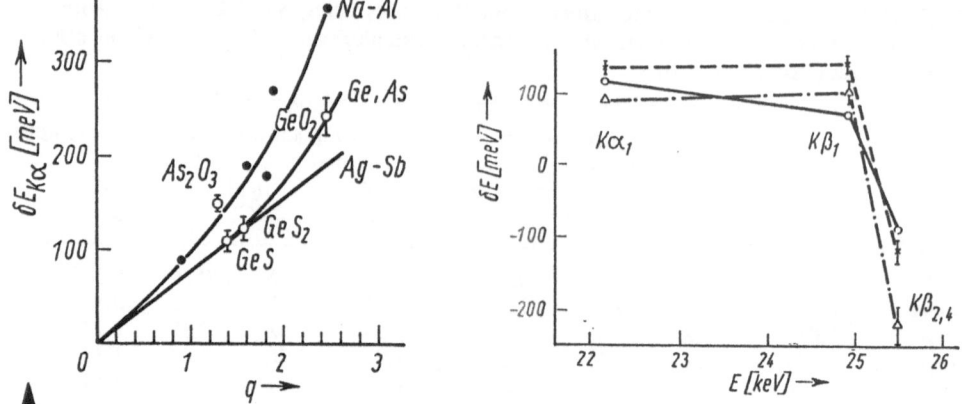

Fig.4.11 Dependence of chemical shift of Kα lines, $\delta E_{K\alpha}$, on the charge q for elements of the 3rd, 4th, and 5th periods [4.9]

Fig.4.12. Comparison of theoretical (solid curve) and experimental (AgCl-Ag: dashed, Ag_2S-Ag: dashed-dotted) X-ray line shifts [4.135]

-0.11; 4f: -0.57). To determine directly the atomic charges for compounds with d and f electrons, in analogy to the case for 3d elements, it is necessary to record, for example, a second X-ray line. Comparative studies have been done between experimentally measured shifts and those determined with (2.19) from relativistic calculations for free atoms and ions of heavy elements for the $K\alpha_1$, $K\beta_1$, and $K\beta_{2,4}$ lines. An example of this is shown in Fig.4.12.

Only a few studies of the L line shifts exist. For the transition elements these studies show, both from calculations [4.128] and experiment [4.142-144], a result which is contrary to the case for the Kα line: Short-wavelength shifts occur due to loss of d electrons and long-wavelength shifts due to loss of s electrons. Contrary to this, measurements of uranium compounds (Fig.4.13) can be interpreted in agreement with wavelength shifts for Kα lines for elements with Z < 36.

Interpretation of chemical shifts for X-ray transitions involving levels with smaller binding energies using the point-charge model discussed

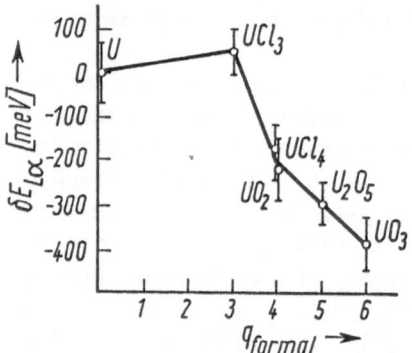

Fig.4.13. Chemical shifts of U Lα lines [4.145]

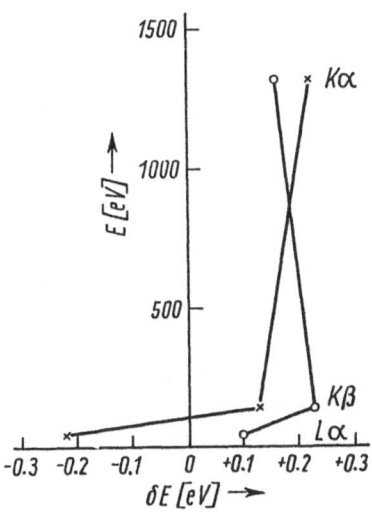

Fig.4.14. Chemical shifts δE of As_2O_3-As (circles) and $AsO_4{}^{3-}$-As (crosses) [4.114]. E is the energy of the final state of the X-ray transition

above introduces difficulties which obviously show the limits of this model. Thus, deviations from this model are found even in interpretation of $K\alpha$ shifts of elements from the third period [4.146]; these deviations are still clearer for the $K\beta_{1,3}$ lines of the 3d elements [4.113] and the $K\beta_{2,4}$ lines of the elements from the fifth period [4. 135]. This is apparent, for example, in the chemical shifts of the $L\alpha$ lines of As and Sb (Fig.4.14) [4.114, 147-149]. The values obtained cannot be interpreted by the one-electron model since the shifts of the $K\beta_5$ lines in As_2O_3 and $AsO_4{}^{3-}$ correspond to the sum of the $K\alpha_1$ and $L\alpha_1$ shifts [4.130]. The explanation must be found in the inability of the point-charge model to explain chemical shifts of levels with lower binding energies. Following [4.151] we can, however, obtain a condition for validity of the point-charge model

$$J - \frac{K}{2} \neq \langle \phi_v | 1/r_A | \phi_v \rangle , \qquad (4.21)$$

where J is the Coulomb integral, K is the exchange integral, ϕ_v is the wave function of the valence electrons, and r_A is the atomic radius. Calculations

Table 4.3. Test of the point charge model according to (4.21). The shorthand $<..>$ stands for $\langle \phi_v | 1/r_A | \phi_v \rangle$

ϕ_v		$<..>$	J − K/2					
			1s	2s	2p	3s	3p	3d
Si	3p	0.4779	0.4739	0.4448	0.4451	0.3242		
Ge	4p	0.4455	0.4438	0.4348	0.4181	0.4358	0.4169	
As	4p	0.5100	0.5076	0.4955	0.4733	0.4711	0.4968	0.4720

using Clementi functions (Table 4.3) support this estimate of the limit of validity, as do the experimental data.

Determination of atomic charges is possible not only from wavelength shifts of the main lines but also from chemical shifts of satellite lines and from intensity changes in the valence emission band. Using calculations for free ions for the $K\alpha$ satellites produced by multiple ionization, a relation can be derived between the charge and the chemical shift similar to that for the main lines. Thus, atomic charges can be determined. Such calculations have been done for Mg, Al, Si, P [4.151], and S [4.152] (Table 4.4 and Fig.2.20). The 3p population distribution for sulfur was derived from the intensity distribution of the S $K\beta$ band in organic compounds. The agreement of these values with those derived from chemical shifts of the $K\alpha$ line using the results of semitheoretical MO calculations provides a generally satisfactory physical basis for charges and electron populations derived from X-ray spectroscopic studies (Table 4.5).

Table 4.4. Theoretical and experimental shifts of $K\alpha$ lines and the resulting atomic charges [4.151]

	Calculated shifts [eV]			Experimental shifts relative to the metal [eV]			Effective atomic charge		
	$Mg{\rightarrow}Mg^{2+}$	$Al{\rightarrow}Al^{3+}$	$Si{\rightarrow}Si^{4+}$	MgO	Al_2O_3	SiO_2	MgO	Al_2O_3	SiO_2
$\alpha_{1,2}$	0.86	1.37	3.23	0.29	0.40	0.58	0.67	1.21	0.99
α''	1.41	2.06	4.78	-	-	-	-	-	-
α'	1.45	2.01	4.19	0.75	-	1.22	1.03	-	1.60
α_3	1.44	2.00	4.18	0.60	0.74	1.09	0.83	0.95	1.47
α_3'	1.41	1.93	4.60	-	-	-	-	-	-
α_4	1.49	2.05	4.72	0.55	0.67	0.99	0.74	0.88	1.00

Table 4.5. Comparison of charges and electron populations obtained from $K\alpha$ shifts and $K\beta$ intensities with those obtained from MO calculations (EHT and CNDO methods) [4.153]

	Charge			Atomic net population			
	$q_{K\alpha}$	q_{EHT}	q_{CNDO}	$\left(\dfrac{3p}{3s}\right)_{EHT}$	$(3s+3p)_{K\alpha}$	$\left[\dfrac{3p_{compound}}{3p_{sulfur}}\right]_{K\alpha}$	$\left[\dfrac{3p_{compound}}{3p_{sulfur}}\right]_{K\beta}$
Thiourea	-0.35	-0.9	-0.46	3.10	6.35	0.93	0.92
Sulfur	0	-	-	-	6.00	1.00	1.00
Dimethyl sulfone	+1.46	+2.85	+0.36	2.37	4.54	0.62	0.67
Sulphanilamide	+1.54	+3.07	-	2.31	4.46	0.63	0.64

Table 4.6. Comparison of ionization energies (in eV) obtained from X-ray absorption spectra and ESCA measurements [4.154]

	Level	XAS	ESCA
BCl_3	1s	199	199.8
CO	1s	293.3	296.2
N_2	1s	409.5	409.9
O_2	1s	543.5	543.1
SF_6	1s	696	694.6
$SiCl_4$	2p	110	110.2
SF_6	2p	180.6	180.4
BCl_3	2p	207	207.0

4.3.3 Determination from X-Ray Absorption Spectra

Effective atomic charges can be obtained by X-ray absorption spectra from the position of the main edge and from the near-edge fine structure. For the chemical shift of an absorption edge δE, which corresponds to ionization of an inner level, there should be a relation analogous to that for shifts of inner levels. Therefore, the position of the main edge should agree with the corresponding ESCA value, and chemical shift of the absorption edge could be described using a point-charge model. The position of the main edge has been extracted experimentally for certain absorption edges in the ultra-soft X-ray region using synchrotron radiation (Table 4.6). Difficulties arise, however, because the position of the main edge, due to the presence of Rydberg states and transitions into virtual orbitals of molecules and exciton levels (Sect.2.6.2), is often very difficult to locate accurately from the observed spectra. Therefore, no accurate comparisons exist up till now between δE values and charge. Semiquantitative efforts to correlate δE and charge or ionicity have been discussed in Sect.4.1. Deviations from an exact quantitative relation similar to (4.12) can be understood as discussed above.

A method developed mainly by *Barinskii* and *Nefedov* [4.94] to determine effective atomic charges is based on the description of Rydberg states in the near-edge fine structure in molecules using a central-field approximation (Sect.2.6.2). Applying an iterative procedure, in which both the parameters n (the effective quantum number) and η (the effective charge) of the absorbing atom are changed, (2.58-61) must be solved to obtain best

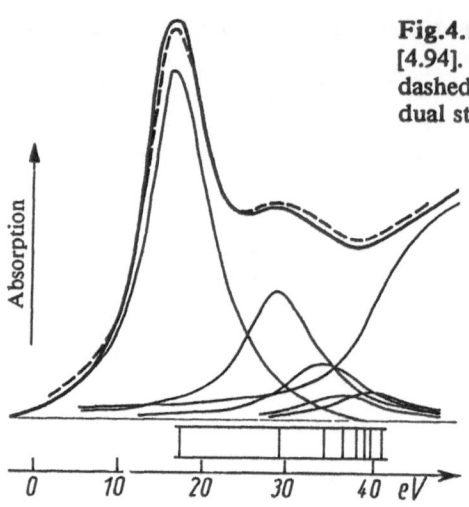

Fig.4.15. K absorption spectrum of Zn^{2+} ions [4.94]. Heavy, solid line: experimental curve, dashed: curve obtained from addition of individual states from a Rydberg series

agreement with the experimental spectra (Fig.4.15). Since both the effective atomic charge q and the vacancy of the inner shell are included in the parameter η, one obtains

$$q = \eta - 1 . \tag{4.22}$$

Numerous charges obtained with this method can be found in [4.94]. For a comparison of q values due to other methods, it is necessary to consider the physical significance of the charge determined in this way. The effective volume, in which the charge q is localized, is not identical with the volume defined by other methods for changes in the electron density $Q(r)$. The charge η used in the central-field approximation is given by

$$\eta(r) \simeq Z - \int_0^r Q(r)dr . \tag{4.23}$$

Thus, e.g., the q value decreases with increasing formal charge (oxidation stage) or remains constant. In comparison, the charges determined from the shifts of the X-ray diagram lines increase with increasing oxidation stage. Ions in aqueous solutions have very high effective charges which approximately correspond to formal charges but the inner spheres of complex compounds show only small atomic charges. Additional information on atomic charges can be obtained from absorption spectra by studying absorption lines split due to the Stark effect [4.94].

5. Investigation of the Valence Electron Structure of Molecules and Isolated Groups in Crystals

In this chapter the connection between energy and intensity of X-ray e-mission and absorption features, and the electronic structure of molecules and metal-ligand clusters is treated in more detail. It is suggested that the reliability of conclusions can be increased by a combined study of X-ray spectra, photoelectron spectra, electron yield spectra, and electron energy loss spectra. The method for evaluation of energy levels of occupied and empty electronic states, and their symmetry is explained together with a discussion of the (Z+1) analogy and the application of polarized X radiation (Sects.5.1,2). Experimental X-ray emission and absorption spectra of the molecular gases N_2, I_2, HCl, H_2O, SO_2, XeF_2, C_2H_2, CH_3Cl, and SF_6 (Sect.5.3), of some sulfur containing organic compounds (Sect.5.4), of oxyanions of the elements of the 2nd and 3rd period, and complex compounds of d elements are represented and discussed including theoretical results and photoelectron spectra (Sect.5.5).

5.1 Initial Remarks

In Chap.2 we discussed the connection between the X-ray spectra of an atom in a molecule or crystal and its electron distribution. From this treatment it is evident that information can be obtained on the valence electron structure, and thereby also chemical binding, from the relative positions and intensities of structural features in X-ray emission and absorption spectra.

The spectra of interest here, in contrast to inner-shell transitions (such as Kα) in heavy elements, are subject to various kinds of splitting and are richly structured. Thus analysis of these spectra must have as its goal a detailed understanding of the observd features so that an adequate picture emerges of the "pure" emission band arising from transitions between singly ionized states. In the course of such an analysis, some steps are relatively easy while others are not. For example, one can with modest effort (Sect.3.6) correct the observed spectra for distortions due to the apparatus; although even in this step it is necessary to be aware of the possibility of additional maxima appearing due to anomalous scattering in the crystal. After this, one is ready to try identifying those features in the corrected quantitative intensity profile arising from transitions between multiply ionized and excited states, i.e., the satellites. This is a particularly difficult exercise when the satellites occur close to or, as is frequently the case, under components of the main band.

Observable satellites can be identified from comparison of intensity distributions in the photoelectron spectra in the region corresponding to the valence electrons. For some cases it is possible to determine the satellites from calculations of their transition energies (Sect.2.4). Larger difficulties are encountered in quantitative estimation of satellite intensities in the X-ray emission bands of molecules when the molecular system cannot be adequately approximated by an atomic model. Whilst the identification of observable satellites is important for the determination of the energy levels of a molecule, uncertainty in satellite intensity interferes with the possibility of obtaining information on the shape of the wave function from observed data.

The problems discussed above suggest that combined study of electronic structure using X-ray spectra, photoelectron spectra and yield spectra is desirable. Even though these methods can yield analogous information on electronic structure, peculiarities emerge which influence the nature and reliability of conclusions obtainable by the various methods.

X-ray emission and absorption spectra (XS) yield energy differences between ionized states (XES) and between excited states with one inner vacancy and a neutral initial state (XAS). It is possible to determine the arrangement of valence levels and energies of excited states with accuracies of ±0.05 eV (or better). Since all one-electron transitions discussed here follow dipole selection rules, one can obtain information about symmetries of the states involved and the shapes of their wave functions. In polyatomic molecules, e.g., in complexes with extended organic ligands, it is possible to obtain information on the electronic states in the vicinity of a certain atom due to the localization of the core electron which is involved. Emission spectra can be studied from gases, liquids and solids, and fluorescence excitation can be used in the ultrasoft wavelength region (<10nm). Absorption spectra in the ultrasoft wavelength region can be observed in transmission only when extremely thin absorbers ($\simeq 10^2$ nm) can be fabricated. X-ray spectra are relatively insensitive to surface impurities and give results whose resolution is limited by experimental resolution and the natural width of the inner level(s) involved. Thus, while it is possible to resolve vibrational states separated by ca. 0.2 eV in the Kα emission band of molecular nitrogen (Fig.5.8), individual components of the S Kβ emission band in sulfates have a natural half-width of 1.6 eV. The K emission bands of the central atoms of 3d complexes show components which are barely resolved, being separated with less than 1-2 eV (Fig. 5.47), and in the region of the 4d complexes structure in the L emission band is obtained for only a few cases. Similar conclusions can be reached regarding achievable resolution in X-ray absorption spectra.

Ultraviolet photoelectron spectra (UPS) are excited with generally monochromatic lines having photon energies above 20 eV (He I: $h\nu = 21.2$ eV; He II: $h\nu = 40.8$ eV; Ne I: $h\nu = 16.3$ eV; Ne II: $h\nu = 26.9$ eV, etc.) or with monochromatized synchrotron radiation. It is possible, using the relation $E_i = h\nu - E_{kin}$, to determine ionization energies E_i of the valence orbitals of molecular gases [5.1-4]. Since the relative intensity of a particular maximum (corresponding to a valence orbital) contains, in general, contributions from all the atoms, information on detailed shape of the wave function is harder to extract from UPS than it is from X-ray spectra. Qualitative conclusions concerning localization and bonding can, however, be obtained from analysis of the vibrational structure. The resolution of the spectra is such that individual states from gaseous molecules are well resolved and energy differences as small as 0.01 eV can be measured. This is a result of the small width of the exciting lines and a negligible instrumental broadening. In contrast to X-ray spectroscopy, thin layers near the surfaces of solids ($\simeq 1$ nm) can be studied due to the small mean free path of electrons and the method is thus very sensitive to surface contaminations and provides an excellent method to study them. At the present time both gases and solid surfaces are frequently studied.

X-ray photoelectron spectra (XPS) are excited with characteristic X-ray lines (Zr Mς: $h\nu = 151.4$ eV; Mg Kα: $h\nu = 1253.6$ eV; Al Kα: $h\nu = 1486.6$ eV, etc.) or with monochromatized synchrotron radiation of variable energy, and give for molecular gases, using the same relation as for the UPS method, absolute ionization energies for valence and core orbitals with an accuracy of 0.1 eV [5.5-7]. In studies of solids there are experimental problems such as charging effects and some problems of a more fundamental kind, such as definition of the Fermi level in insulators. These interfere to some extent with the determination of absolute binding energies. Accurate quantitative results concerning wave functions of the valence states can be obtained only with some difficulty from relative intensities using effective cross-sections. Resolution in XPS is poorer than that obtained in UPS due to the larger width (about 1 eV) of the exciting line (Fig.5.8), so that vibrational structure of individual photoelectron bands cannot be observed. A considerable improvement in the resolution can be obtained using a monochromator ($\simeq 0.2$ eV). The sensitivity of the method to surface contaminations is smaller compared to UPS due to the longer mean free path of the electrons, so that studies of solids proceed without great difficulty and, with increased experimental effort, even liquids can be studied.

Yield spectra from external photoeffect in the extreme UV and ultrasoft X-ray region resemble corresponding absorption spectra. This similarity follows from the fact that the variation in detected electron yield (either total yield or partial yield within a suitably chosen energy window) with incident photon energy is proportional to the absorption coefficient. This proportionality arises from the fact that the escape depth of photoelectrons is generally much smaller than the penetration depth of the incident radiation, assuming that external yield electrons arise, for the most part, from multiple secondary processes [5.8-10]. In contrast to absorption spectroscopy the yield technique does not require fabrication of thin (fragile) absorption foils. Using specially designed experimental geometries unoccupied surface states can also be studied in this way. The information content and resolution of yield spectra are similar to those of absorption spectra.

Electron energy loss spectra can be used to study energies, widths and vibrational structures of core-excited states which arise in transitions of core electrons to unoccupied valence or Rydberg orbitals of molecules. The cross sections for inelastic small-angle scattering of monoenergetic electrons with 1-10 keV energies are directly related to those for photon absorption. Therefore electron energy loss spectra are restricted to dipole transitions and lead to information on excited states analogous to that obtained from X-ray absorption spectra. Excitation due to inelastic electron scattering has the advantage of higher energy resolution (50-100 meV) for excitation energies above 200 eV and permits simple energy calibration [5.11-13].

5.2 Information Content of X-Ray Emission Bands and X-Ray Absorption Spectra

5.2.1 Determination of Energy Levels

If a transition occurs in a molecule from a state with a vacancy in a core shell i to a state with a vacancy in a valence shell j, the energy difference

$$h\nu = E_{ij} = E_i - E_j \tag{5.1}$$

is emitted in radiation if one assumes that the states involved are stationary. Since the dipole selection rules, in certain cases, allow many possibilities for transitions from the initial states i in the valence manifold, correspondingly many maxima appear in the X-ray emission band whose separations are determined by energy differences of the final states, that is to say from the ionization potentials of the corresponding valence shells:

$$E_{ij} - E_{ik} = E_k - E_j . \tag{5.2}$$

According to dipole selection rules, in a K emission band all valence levels appear whose wave functions contain p-symmetry components with respect to the emitting atom in the molecule. Valence states with s and d components can be studied using L emission bands (initial state with 2p vacancy). The relative position of certain valence states can be estimated

164

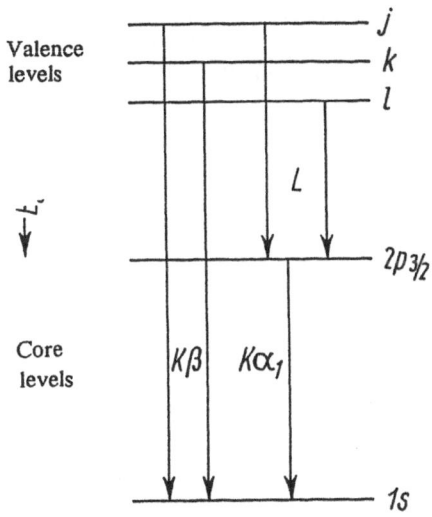

Valence
levels

Core
levels

j

k

l

L

$2p\,{}^{3}\!/_{2}$

$K\beta$ $K\alpha_1$

$1s$

Fig.5.1. Simplified energy level diagram for a molecule with K and L emission transitions.

with both K and L spectra, using the energy difference between the initial states and the $K\alpha_{1,2}$ doublet (Fig.5.1)

$$E_{1s,j} - (E_{2p,\ell} + h\nu_{K\alpha}) = E_\ell - E_j \; . \tag{5.3}$$

If the emission bands from different atoms in a molecule are studied with different initial states, the relative position of all valence levels can be obtained if the experimental resolution is sufficient to separate individual components. If this is not the case, the emission band can be resolved into the individual components using numerical methods where either group theoretical consideration or calculations of the MO structure assume that the number, relative position and degeneracies of the levels are predictable in a first approximation [5.14-16].

Calculation of ionization and binding energies from photon energies of the band components requires, following either the energy of a valence level (from optical measurement or UPS data) or a core level (from XAS or XPS data). Since narrow X-ray bands and binding energies of the valence electrons of molecular gases can be determined with an accuracy better than ±0.05 eV, the binding energies of core electrons can be determined relatively accurately with this method. Otherwise the above-mentioned difficulties in recording UPS from solids can be bypassed if the valence levels can be calculated from the X-ray and the X-ray photoelectron spectra.

Even if the final state of a molecule after emission of an X-ray band is the same as that produced by photoionization, differences are observed in the XES and UPS in certain cases for gases, which can be understood in terms of different vibrational excitation (Fig.5.8). If a 1s orbital is considered as nonbonding the molecular geometry (bond lengths and bond angles) is not appreciably changed by 1s ionization, and the K state, with

165

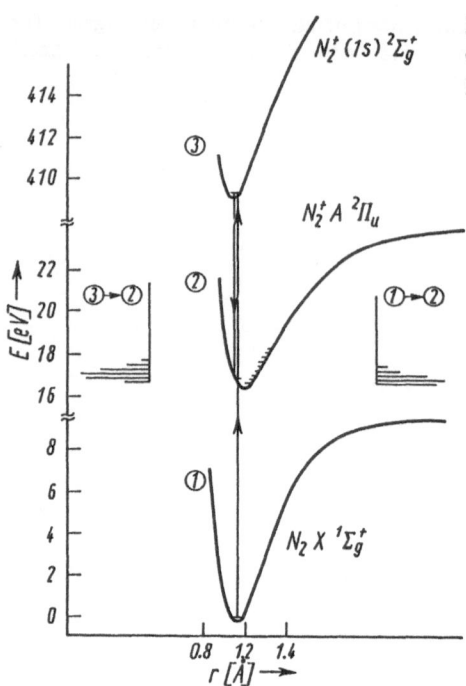

Fig.5.2. Potential curve and transitions for different states of the N_2 molecule (for explanation, see text) [5.17]

respect to molecular geometry, would be regarded as equivalent to the ground state of the neutral molecule, then the vibrational excitation in the ionization and emission processes should be similar. Differences are, in fact, observed, explained in Fig.5.2. It shows potential curves for the ground state $X^1\Sigma_g^+$ of the neutral N_2 molecule, for the state $^2\Sigma_g^+$ of N_2^+ with a 1s vacancy and for the state A $^2\Pi_u$ of the ion with a vacancy in the π_u orbital. Since the calculation of the Franck-Condon factors requires knowledge of the equilibrium distances for the nuclei of the molecule with a 1s vacancy which are difficult to obtain, the effect of 1s ionization is approximated by an increase of the nuclear charge with one unit, so that for example $N_2^+(1s^{-1})$ and $CO^+(1s^{-1})$ can be described by NO^+. In this approximation, as shown in Fig.5.2, the minimum of the potential curve of N_2^+ $(1s^{-1})$ is shifted a distance of 3.6 pm towards smaller bond lengths compared to the neutral molecule so that 1s ionization necessarily involves vibrational excitation. Thus it follows that the maximum of the X-ray emission band $(^2\Sigma_g^+ \rightarrow A^2\Pi_u)$ corresponds to a transition 0→2 (left level diagram) and the photoelectron spectrum starts with two almost equally intense lines (right level diagram) and the maximum can be assigned to the transition 0→1. Thus, differences which appear occasionally between ionization potentials determined by XES and UPS can be traced back in a similar way to different vibrational excitations in the two processes. Good agreement between XES and UPS energies is obtained after correct assignment of the vibrational structures of CO [5.17]: $E_{3\sigma}^{0\rightarrow0} =$

14.01 eV, $E_{CK\alpha}^{0 \to 0}(3\sigma) = 282.27$ eV; $E_{CK\alpha}(3\sigma)+E_{3\sigma} = E_{1s} = 296.28$ eV; E_{1s} (XPS) = 296.1 eV. This leads to the conclusion that relaxation effects, which reflect different time scales for adjustment to the changed potential due to ionization and emission processes, do not exceed the magnitude of the experimental errors.

Structure in the near-edge region of the absorption spectrum from molecular gases, as discussed in Sect.2.6.2, is associated with transitions of core electrons into virtual unoccupied molecular orbitals. Among these, one must distinguish between MO's with mostly antibonding character formed from the valence orbitals of the atoms, and Rydberg orbitals with nonbonding character. Since the separation between the Rydberg and the valence orbitals is considerable, a description not only of X-ray transitions from the inner orbitals but also of optical Rydberg spectra of molecules in the framework of the one-electron model is adequate for transitions in Rydberg orbitals to be comparable to the optical spectra of the analogous (Z+1) molecule, as in case for free atoms. Following vacancy creation in the core created by an X-ray excitation, screening of the nuclear charge is decreased by one unit. Thus the Rydberg lines in, for example, the C K absorption spectrum of CO agree in relative line positions with the optical spectrum of NO.

Since the Rydberg spectra of molecules are complicated due to their lower symmetries compared with atoms, identification of energy levels and thereby the extrapolation of a series is more difficult than for atomic gases. Comparison with optical spectra of (Z+1) analogous molecules considerably simplifies the assignment process. In this way (2.58) gives a series limit E_∞ (or binding energy E_i) for core electrons participating in the absorption process from the measured photon energies E_n of the absorption lines. A simple calculation shows that this binding energy E_i can be related to the energies of analogous Rydberg lines $E_n(Z)$ and $E_n'(Z+1)$ in the X-ray and optical spectrum through the ionization energy E_1 of (Z+1) analogous molecules as

$$E_i \simeq E_n(Z) - E_n'(Z+1) + E_1(Z+1) . \tag{5.4}$$

The (Z+1) analogy opens a possibility to determine ionization potentials and excitation energies of radicals. Figure 5.3 shows schematic potential curves for the CO molecule in the ground state, for the ion CO^+ with a vacancy in the O 1s shell, for the excited molecule CO^* with an O 1s vacancy and for the corresponding (Z+1) analogous CF radical. The shift of the $CO^*(\pi^*)$ potential curve is caused by occupation of the antibonding π^* MO. If one considers the vibrational excitation caused by a vertical transition and the energy difference between the $^2\Pi$ ground state of CF and the $^1\Pi$ state of CO^* using the exchange integral K_{O1s,π^*}, then the ionization potential E_1 of CF is given by the difference between $E_{1s}(O)$ in CO and the energy of the maximum in the O K absorption spectrum of CO which

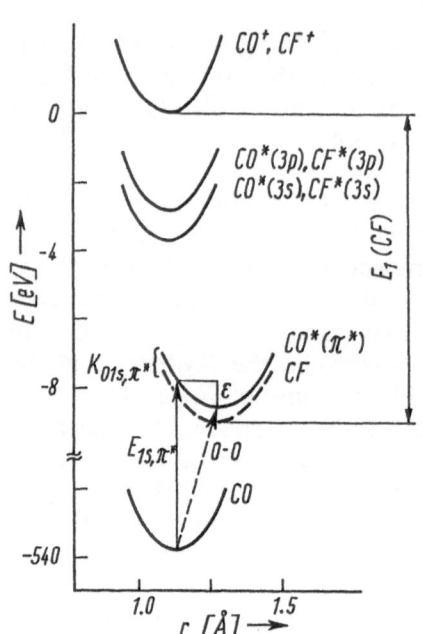

Fig.5.3. Schematic potential curves for core excitation of the CO molecule and optical excitation of the CF radical [5.18]

arises from a transition from the O 1s shell into the antibonding virtual π^* MO.

As for emission spectra, electron transitions from a mainly nonbonding core into a nonbonding Rydberg orbital are not (or only slightly) influenced by vibrational excitation since the potential curves are practically unshifted (Fig.5.3). In contrast to this the absorption maxima resulting from transitions into antibonding MO are broadened on account of increased bond lengths in the final state of the molecule.

5.2.2 Determination of the Symmetry of the Levels

Different approaches are possible to identify the energy levels [5.19-21]. First of all, one applies dipole selection rules. In the octahedral group AX_6 the p orbitals of the atom A are transformed according to the irreducible representation t_{1u}, the s and d orbitals according to the representations a_{1g}, t_{2g} and e_g. Accordingly, maxima of the K spectra correspond to levels of t_{1u} symmetry while the levels a_{1g}, t_{2g}, and e_g are observed in L spectra (Fig.5.20 below). In the tetrahedral group AX_4, the p orbitals are transformed to a_1 and the d orbitals to t_2 and e. Corresponding to maxima in the K spectrum of the atom A are levels with symmetry t_2 and the structures in the L spectra can be assigned to the symmetries a_1, t_2 and e (Fig.5.32 below).

For other symmetry groups one can obtain selection rules for X-ray transitions from character tables of irreducible representations (Table 5.1).

Table 5.1. Active levels for K and L transitions from dipole selection rules in molecules with different symmetries [5.20]

Molecule	Symmetry	Active levels					
		AK	AL	XK	XL	YK	YL
AX_6	O_h	t_{1u}	$a_{1g}(s)$, t_{2g}, e_g	$a_{1g}, e_g, t_u,$ t_{1g}, t_{2g}, t_{2u}	a_{1g}, e_g, t_{1u}		
AX_4Y_2	D_{4h}	a_{2u}, e_u	$a_{1g}(s,d)$, e_g, b_{1g}, b_{2g}	a_{1g}, b_{1g}, e_u, a_{2u}, b_{2u}, e_g, a_{2g}, b_{2g}	a_{1g}, b_{1g}, e_u	a_{1g}, a_{2u}, e_u, e_g	a_{1g}, a_{2u}
AX_4	D_{4h}	see AX_4 in AX_4Y_2					
AX_2Y_2	D_{2h}	b_{1u}, b_{2u}, b_{3u}	$a_{1g}(s,d)$, b_{2g}, b_{1g}, b_{3g}	a_{1g}, b_{3u}, b_{1g}, b_{1u}, b_{2g}, b_{2u}	a_{1g}, b_{3u}	a_{1g}, b_{2u}, b_{1g}, b_{1u}, b_{3g}, b_{3u}	a_{1g}, b_{2u}
AX_4	T_d	t_2	$a_1(s), e, t_2$	a_1, t_2, e, t_1	a_1, t_2		
AX_3Y	C_{3v}	a_1, e	$a_1(s,d), e$	a_1, e, a_2	a_1, e	$a_1 e$	a_1
AX_2Y_2	C_{2v}	a_1, b_1, b_2	$a_1(s,d)$, a_2, b_1, b_2	a_1, b_1, a_2, b_2	a_1, b_1	a_1, b_2, a_2, b_1	a_1, b_2
AX_3	C_{3v}	see AX_3 in AX_3Y					
AX_2	C_{2v}	see AX_2 in AX_2Y_2					
AX_3	D_{3h}	e', a_2''	$a_1'(s,d)$, e', e''	e', a_1', e'', a_2'', a_2'	e', a_1'		

Given the selection rules for transitions from a core state to valence states of various symmetries, it is possible to identify corresponding components of the spectrum by analysis of relative intensities. Since the intensity is proportional to the degree of degeneracy (multiplicity), it follows, for example, that the integrated intensity of a σ component should be smaller than that of a π component for a linear molecule (Fig. 5.8 below). For assignment of levels having equal multiplicity, relative intensities can be estimated from the composition of the corresponding orbitals. Thus, the most intense maximum of the O Kα spectrum of H_2O (Fig.5.12 below) should correspond to a transition into level b, since it is formed exclusively from a 2p electron pair of oxygen, while the levels a_1 and b_2 also contain H 1s and O 2s components which do not contribute to the transition probability.

Similar considerations permit an assignment for the short-wavelength maximum $K\beta_x$ in the S $K\beta$ emission spectrum of H_2S. In this molecule the $3p_x$ electron pair is completely localized on sulfur so that there is a large amplitude of the wave function in the region of the K shell and accordingly high intensity is expected for the X-ray transition corresponding to this band component.

Information on the symmetry of the valence orbital can also be deduced from the intensity differences between corresponding band components in the spectra of different compounds. If, for example, a comparison is made between the S $K\beta$ emission band of H_2S and that of diphenyl sulfide (Fig.5.4) a strong decrease is observed in the relative intensity of the short-wavelength maximum $K\beta_x$. In the case of bonding with the phenyl ring, which is capable of accepting π electrons, an appreciable delocalization of the S $3p\pi$ electron can be expected, so that, on the one hand, the intensity decrease can be understood and, on the other hand, assignment of the maximum to a b_1 and a $3p\pi$ level can be confirmed [5.22].

The high resolution available in the ultrasoft X-ray region gives another possibility for identifying energy levels by means of vibrational

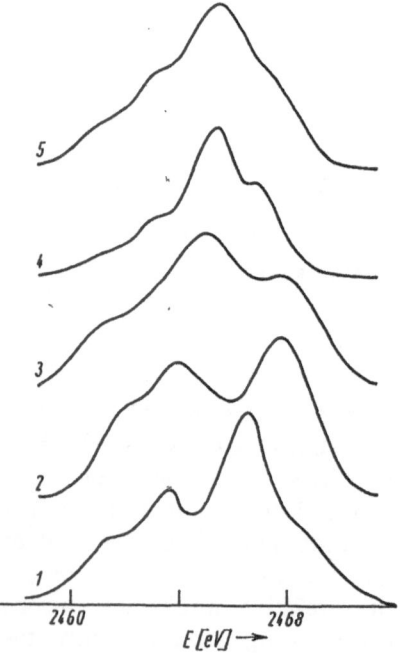

Fig.5.4. S $K\beta$ emission band of H_2S (Curve 1), S_8 (Curve 2), $(C_6H_5)_2S$ (Curve 3), Co[S-C(C_6H_5)=CH-CH=NCH$_3$]$_2$ (Curve 4), and Ni[S-C(C_6H_5)=CH-CH=NCH$_3$]$_2$ (Curve 5) [5.22]

5

4

3

2

1

2460   2468

E [eV] →

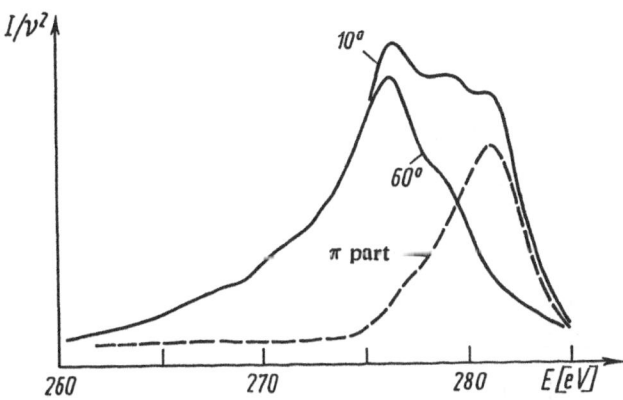

Fig.5.5. C Kα emission band of graphite and its dependence on the (take-off) angle [5.23]

structures observable in this spectral region. As discussed in Sect.5.2.1, a linear molecule should show little or no vibrational structure in a transition from a core level into a valence level formed by ionization of a nonbonding orbital due to the expected small change in the molecular geometry. In contrast, ionization of a bonding orbital causes an increase in the equilibrium distances and thus strong vibrational excitation. In nonplanar tri- or poly-atomic molecules it is necessary to consider vibrational excitation due to changes in bond angles arising from ionization of valence orbitals.

A further possibility for determining symmetries of the levels, which seems feasible in spite of some unsolved questions, is from the study of the polarization of X-ray emission spectra from anisotropic single-crystal samples [5.23-31]. The method can be exemplified with the C Kα band of graphite (Fig.5.5) as follows. From an analysis of the transition moments it is possible to deduce that the radiation from a (1s → π) transition is polarized along an axis perpendicular to the cleavage plane. Since propagation of radiation parallel to the polarization vector is impossible, emission from the π band can be suppressed by a special arrangement of the graphite sample and analyzing crystal so that only the σ band is recorded. Figure 5.5 shows the C K emission band of a graphite single crystal, recorded using radiation at 10° with respect to the cleavage surface (σ and π components), and 60° (π component suppressed). Investigation of the angular dependence of the anisotropic characteristic radiation emitted by single crystals permits not only determining the symmetry of energy levels but also separating a band into components without any theoretical modeling. The method has also been applied to electronic structure investigation of graphite intercalation compounds [5.32-35].

Results from MO calculations can also be helpful for the interpretation of X-ray spectra. The sequence of states given by the calculations is often sufficient to identify observed maxima. In still more detailed inves-

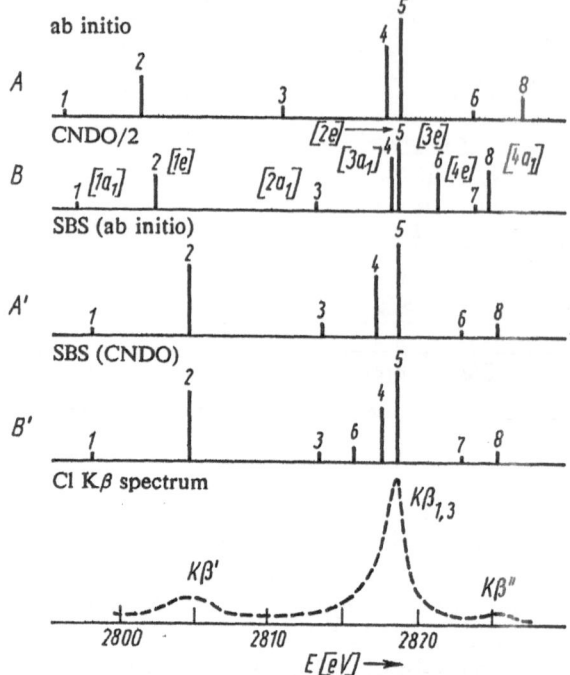

Fig.5.6. Components of the Cl Kβ emission band of KClO$_3$ (A′,B′) and theoretical initial values (A,B) for the spectrum analysis (eigenvalues and eigenvectors) [5.14]

tigations experimental line shapes were compared with theoretical curves constructed using coefficients of the wave functions and level energies (Sect.5.2.3). When calculations of adequate accuracy are available this procedure leads to convincing results and gives a possibility to check the reliabilty of calculations (Figs.5.6, 19, 34 below).

5.2.3 Determination of MO LCAO Coefficients

From the model developed in Sect.2.3 on the relationship between relative intensities of the components of an X-ray emission band and the transition probability, relative and absolute values of the MO LCAO coefficients can be derived from the spectral profiles provided certain conditions are fulfilled, discussed next.

From (2.6, 37) it follows that the intensity I_i of a component in the K emission band of the atom A in a molecule A$_x$B$_y$ can be written as follows, neglecting the energy-dependent factors

$$I_1 \propto |\langle \psi_i|r|\phi_{1s}(A)\rangle|^2 = \left| \langle \sum_{Zn\ell} C_{iZn\ell}\phi_{n\ell}(Z)|r|\phi_{1s}(A)\rangle \right|^2 . \tag{5.5}$$

172

Since the corresponding one-center terms

$$\langle\phi_{ns}(A)|r|\phi_{1s}(A)\rangle \quad \text{and} \quad \langle\phi_{nd}(A)|r|\phi_{1s}(A)\rangle$$

are equal to zero according to dipole selection rules, the intensity I_i can be expressed as [5.36-38]

$$I_i \propto C_{iAnp}^2 |\langle\phi_{np}(A)|r|\phi_{1s}(A)\rangle|^2 \qquad (5.6)$$

for the case when the integral $\langle C_{iBn\ell}\phi_{n\ell}(B)|r|\phi_{1s}(A)\rangle$ vanishes or differs only slightly from zero. This assumption is justified for compounds from the 2nd and 3rd periods. It is, however, necessary to include cross-transition terms as predicted by DV Xα calculation for the L emission bands [5.39]. An analogous relation can be derived for the other components of the K emission band so that the relative intensities of individual maxima in the intensity distribution, as given by (2.40), can give the relative values for the MO LCAO coefficients as long as one assumes that the integrals $\langle\phi_{np}(A)|r|\phi_{1s}(A)\rangle$ have the same value for all the transition moments. This entails identical shapes for the atomic orbitals ϕ_{np} in all the molecular orbitals, an assumption which is not always valid (Sect.2.3).

Absolute values can be obtained when the sum of the squares of the coefficients can be determined or when the magnitude of one coefficient can be determined. The latter occurs, for example, for wave functions of symmetries t_{2u} and t_{1g} (octahedral) and t_1 (tetrahedral), which are formed exclusively from p states of the ligands (f states of the central atom are thereby not included). The absolute value of a coefficient can be calculated for some cases [5.38] from the overlap integral in a simple way. Quadratic sums of the coefficients can be determined according to the following possibilities [5.19].

Using the assumption that the positive charge of the central atom in the oxyanion of an element of the 3rd period can be explained by a displacement of only the 3p electrons, then $\Sigma_i C_{i3p}^2$ can be uniquely determined from the X-ray spectroscopic determination of effective atomic charges (Sect.4.3). Alternatively, absolute coefficients can be obtained from the ratios of the intensities of all the components in the X-ray emission band to the intensity of an inner X-ray line, assuming $\Sigma_i C_i^2$ is known for a compound [5.40-42]. For KCl, for example, the occupation of the 3p orbital of Cl can be assumed to be 6. The ratio of line intensities of Kβ and Kα decreases from this definite value in KCl as the 3p occupational number decreases for compounds with more positive charge on the Cl atom.

The spectral deconvolution often required for the determination of the LCAO coefficients leads to satisfactory results when height, position and also width of individual components are taken as adjustable parameters [5.14]. The reliability of this method (which starts with a-priori determined eigenvectors and eigenvalues due to the large number of variables) is evident from Fig.5.6, which illustrates the results of a spectral deconvolution (lines 3 and 4) of the experimental spectrum (line 5) shown as "stick" diagrams together with the different sets of initial estimates from two different approximations (lines 1 and 2). For a large number of compounds of elements of the 2nd and 3rd periods (see [5.14, 19, 38] and the following sections) good agreement is to be found between theoretically and experimentally determined LCAO coefficients.

Determination of MO LCAO coefficients from the model under discussion requires that term $\langle C_{n\ell}\phi_{n\ell}(B)|r|\phi_{1s}(A)\rangle$ be negligible compared with $\langle C_{np}\phi_{np}(A)|r|\phi_{1s}(A)\rangle$, so that the contribution of the multicenter term in (5.5) is relatively small. This is not correct, however, when, for example, the intensity of the $K\beta_5$ emission band (4p→1s) of the 3d complexes is derived using (5.5). Since the 4p orbital is not occupied in the ground state of the free atom, the contribution of 4p states to the MO in the complex is small, i.e. $C_{4p}(A) \ll C_{n\ell}(B)$, and the multicenter term in (5.5) is more important. The requirement of a complete calculation of the matrix elements, which has been done only for a few cases so far [5.43, 44], can be circumvented by calculating the relative intensities of individual components of the $K\beta_5$ emission band from the squares of the amplitudes of the MO functions of the ligands (for polyatomic ligands) in the region of the 1s orbital of the central atom using various simplifying assumptions [5.45]. The results thus obtained show that the energy difference between the maxima of the $K\beta_5$ emission band agrees satisfactorily with the energy level differences of isolated ligands, for which ab-initio calculations can be performed without great difficulty (Fig.5.43 below).

Two examples will serve to demonstrate the results obtainable from this model concerning chemical bond. For compounds with an element A from the second part of the 3rd period and a nonmetal B, the bond is dominated by A 3p valence electrons. Their distribution in the MO can be surmised from the relative intensities of the components of the $K\beta$ emission band. In most cases, besides a main maximum due to electron transitions from the σ bonding orbital into the K vacancy, these show adjacent maxima on both the long-wavelength and short-wavelength sides, which are produced by transitions from the MO with mostly B ns symmetry or from π binding orbitals. As indicated by the relative intensities of the three components in the series AO^{n-}-$AO_2{}^{n-}$-$AO_3{}^{n-}$-$AO_4{}^{n-}$ [5.46] (Sect. 5.5), the A 3p electrons take part in the π bonding only for anions of low symmetry, while the corresponding $K\beta$ components are missing or are very weak for the tetrahedral anion $AO_4{}^{n-}$. On the other hand, the portion of 3p orbitals of atom A in the wave function formed mainly from O 2s AO's increases with increasing positive charge on the central atom A. In contrast, for highly symmetric molecules with a central atom from the 2nd period (e.g., CF_4, CCl_4), a feature is observed on the high-energy side of the $2p\sigma$ maximum of the $K\alpha$ emission band which can be identified as a $2p\pi$ component [5.38]. The greater tendency of elements of

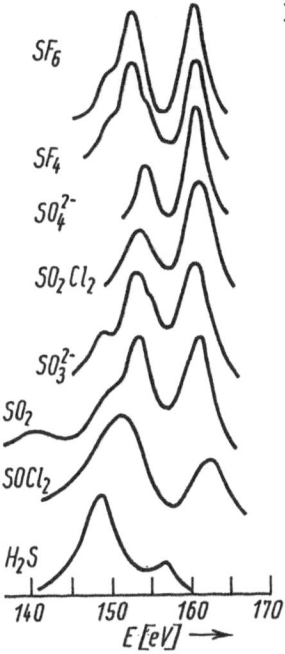

Fig.5.7. S $L_{II,III}$ emission bands of several compounds [5.47]

SF_6

SF_4

SO_4^{2-}

SO_2Cl_2

SO_3^{2-}

SO_2

$SOCl_2$

H_2S

140   150   160   170

$E\ [eV]\ \longrightarrow$

the 2nd period to form $p\pi$ bands is evident and thus the formation of double bonds is observed (double-bonding rule) in contrast to the polymer-forming tendency of compounds with elements of the 3rd period.

A second example can be derived from studies of the structure of L emission bands of compounds with an element of the 3rd period. Following the theoretical model, relative intensities of components of the L emission band give information on the s and d parts in the MO of the molecule. If the experimental spectra, for example, from different sulfur compounds (Fig.5.7), are compared it is evident that a short-wavelength maximum occurs above 160 eV when sulfur carries a positive charge in the molecule. From analysis of other emission bands of the molecule and from electron spectroscopic data (Sect.5.3.2) it may be concluded that this maximum is dominated by contributions from S 3d orbitals in the MO from which an interaction of O $2p\pi$-S $3d\pi$ is verified [5.37]. If the S L emission band from Fig.5.7 is resolved into its components, a linear relationship between the $K\alpha$ shift and the relative intensity of the 3d and 3s components is evident (Fig.5.29 below). Conclusions regarding π-acceptor and σ-donor interactions between ligand and central atom in complexes can be obtained by comparing the relative intensities of individual components (belonging to the same valence state) of the emission band of the free ligand with those of the ligand in the complex and of the central atom (Sect.5.5.2).

Table 5.2. Collection of experimental and theoretical studies in fine structures of X-ray emission band and absorption spectra (core–excited spectra) of molecular gases

Elements	Reference	Oxides	Reference
N_2	5.17,60,66–77	CO	5.12,38,64,69,74, 77,82,86–88
O_2	5.75,77–82	CO_2	5.64,70,79,81,87,89–92
Cl_2	5.48,49,53,71,83,84	N_2O,NO,NO_2	5.17,51,66,72,74,93–95
I_2	5.85	SO_2	5.47,51,71,96–103

Hydrogen compounds		Halogen compounds	
HF	5.51,106,107	$COCl_2$	5.48
HCl	5.48,49,53,59,83, 99,101,108	XeF_n	5.131–133
H_2O	5.50,76,109–114	NF_3	5.75,134
H_2S	5.56,59,97,99,102, 103,108,110,116–120	BF_3,BCl_3	5.134–137
NH_3	5.13,51,74,76,89, 121	$CH_{4-n}Cl_n$, $CF_{4-n}Cl_n$	5.11,38,64,91,139–144,147
PH_3	5.59,108,120,122,123		
CH_4	5.12,61,64,87,91, 124	$CH_{4-n}F_n$	5.12,13,64,81,124,140,148
SiH_4	5.59,108,120,123, 125–128	SiF_4	5.103,123,127,128,137,154
		$SiCl_4$	5.171–173
		SF_4	5.170
		SF_6	5.47,74,90,97,103,117,134, 141,157–168

Hydrocarbons		Miscellaneous	
C_2H_2	5.12,38,64,69,87,152	CS_2	5.51,74,89,98,129
C_2H_4	5.12,38,64,87,151	COS	5.51,71,89
C_6H_6	5.87,150	CH_3OH	5.114
C_nH_{2n+2}	5.38,64,87		

5.3 Molecular Gases

The X-ray spectra from gases containing elements of the 2nd and 3rd periods occur in the soft and ultrasoft X-ray region, so their registration entailed appreciable experimental difficulties even as late as the sixties. Due to development of more intense X-ray sources [5.48–55] use of synchrotron radiation as an X-ray source for spectroscopy [5.56–60] and significant improvements in gas cells [5.48, 49, 53, 61, 62] with differential pumping technique [5.3, 63, 65] obstacles have been largely overcome. As a result, recording of K and L emission bands and also absorption spectra of gases in the wavelength range between 0.5 and 10 nm is relatively straightforward and both electron and photon excitations can be used (Table 5.2). Ionization potentials of simple molecules are almost completely tabulated from results obtained mostly from UPS [5.4] and XPS studies [5.7] (for further references, see [5.20, 21]) so that together with semiempirical or ab-initio calculations, for most cases studied, it has been possible to achieve relatively convincing results for the electronic structure of molecules. In the following sections, selected examples are given for diatomic and polyatomic gases of problems regarding interpretation of X-ray spectra and the conclusions which can be drawn from their study.

5.3.1 Diatomic Molecules

N_2: For the nitrogen molecule with valence electron configuration $\sigma_g^2(2s)$ $\sigma_u^2(2s)\ \pi_u^4(2p)\ \sigma_g^2(2p)$ in the ground state, assignments for the components of the N $K\alpha$ emission band are given in Fig.5.8. With a N 1s binding energy $E_{1s} = 409.5$ eV [5.18], energy values for the components of the vertical ionization potential are obtained as follows: $E_{\sigma_u} = 18.63$ eV (18.75eV); $E_{\pi_u} = 17.03$ eV (16.91eV); $E_{\sigma_g} = 15.6$ eV (15.37eV). The agreement with UPS data [5.4], which are given in parentheses, is very

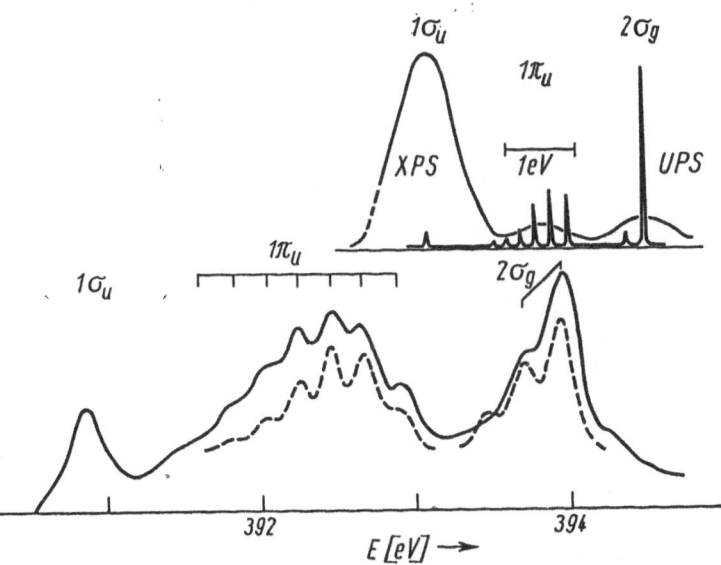

Fig.5.8. Experimental (bottom, solid curve) and (dashed) theoretical $K\alpha$ emission band [5.17] together with XPS and UPS curves in the region of the valence electrons [5.4, 7] from molecular nitrogen

Table 5.3. Interpretation of N K absorption spectrum (XAS) of molecular N_2 [5.75] and Rydberg spectrum (UV) of NO [5.174] (all values in eV)

Peak	XAS Photon energy	Transition	UV Photon energy
A	401.0	$\sigma_u(1s)\rightarrow\pi_g^*(2p)$	
B	406.1	$\sigma_u(1s)\rightarrow 3s$	5.48
C	407.2	$\sigma_u(1s)\rightarrow 3p\pi$	6.49
D	407.4	$\sigma_u(1s)\rightarrow 3p\sigma$	6.61
E	408.5	$\sigma_u(1s)\rightarrow 4s$	7.55
F	408.7	$\sigma_u(1s)\rightarrow 3d$	7.78

good. The different relative intensities of the vibrational structures (UPS and XES) can be understood in terms of the smaller bond lengths in the $N_2^+(1s^{-1})$ ion relative to the N_2 molecule (Fig.5.2). The vibrational profile of the emission band can be well predicted by various ab–initio calculations if relaxation following the 1s ionization is included [5.72, 74].

The assignment given in Table 5.3 of the maxima in the N K absorption spectrum shown in Fig.5.9 results in good agreement with energies of the Rydberg lines of the optical spectrum of the analogous (Z+1) NO molecule. Using an ionization potential of $E_\pi(NO) = 9.27$ eV, (5.4) gives a binding energy of $E_{1s}(N_2) = 410$ eV. If one extrapolates using the series formula (2.58) and the components B and E (effective quantum number n = 1.82; $\eta \simeq 1$), one finds that $E_{1s}(N_2) = 410.2$ eV [5.75]. Even though this value is in very close agreement with the XPS value of 409.9

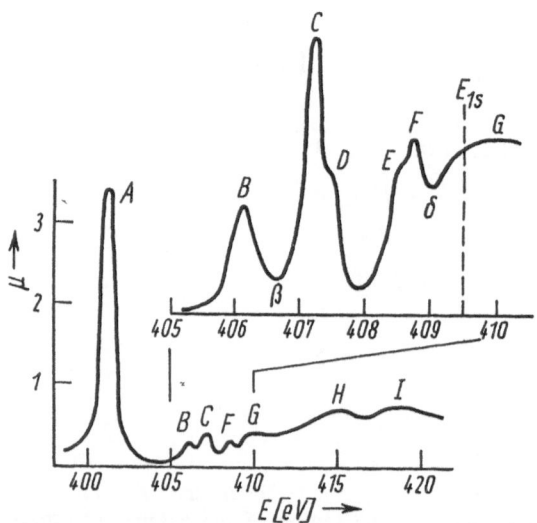

Fig.5.9. The K absorption spectrum of molecular nitrogen [5.75]

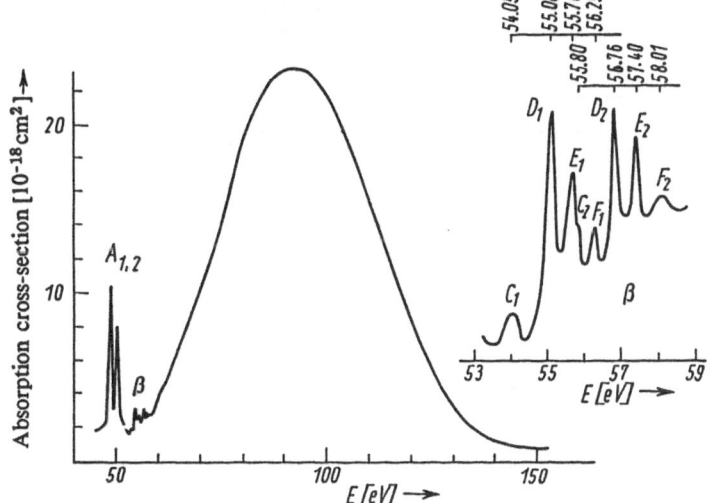

Fig.5.10 $N_{IV, V}$ absorption spectrum of molecular iodine [5.85]. To the right is an enlargement of the β region.

eV [5.7], still better agreement is obtained between energies of the valence levels derived from XES and UPS if the extrapolated value of $E_{1s}(N_2) = 409.5$ eV is taken from the N K absorption spectrum recorded using synchrotron radiation (see above).

Oscillator strengths and transition energies derived from energy loss spectra [5.70] and XAS experiments using synchrotron radiation [5.66] can be reproduced in the region of the Rydberg lines between B and F (Fig. 5.9) as well as the shape resonances A and I using ab initio HF calculations [5.69,71]. Similar results are also obtained from a multiple-scattering calculation [5.67]. The structure H can be attributed to a multiple-excitation process.

I_2: The $N_{IV, V}$ absorption spectrum of gaseous iodine (Fig.5.10) shows a relatively complex structure due to spin-orbit splitting. Several absorption lines caused by transitions due to antibonding MO and Rydberg orbitals are observed on the long-wavelength side of an intense continuum with a maximum at around 93 eV. The maximum can be understood as arising from a delayed onset of $4d \rightarrow \epsilon f$ transitions with increasing energy (Sect.2.6.1). Table 5.4 gives a summary of the energies and assignments of the observed maxima. Since the ligand-field splitting of the 4d atomic orbitals can be neglected for the I_2 molecule, a 4d spin-orbit splitting of 1.68 eV can be deduced from the separation of the maxima A_1 and A_2. The small difference between the experimental intensity ratio $A_1/A_2 = 1.4\pm0.1$ and the theoretical value of 1.5 support the interpretation in terms of a transition into the lowest antibonding MO and can be used as an argument for a small two-electron interaction and consequently for the applicability of a one-electron model. The shift of the potential curve of

Table 5.4. Interpretation of $N_{IV,V}$ absorption spectrum and energy levels of molecular I_2 [5.85]; XUV: determination from X-ray spectrum; UV: from optical spectrum [5.175] (all values in eV)

Peak	Photon energy	Transition	Level energy XUV	UV
A_1; A_2	49.27; 50.93	$4d_{3/2},5/2 \rightarrow \sigma_u(5p)$	8.5-9.0	8.0
C_1; C_2	54.05; 55.80	$4d_{3/2},5/2 \rightarrow 6s$	3.15	3.05
D_1; D_2	55.08; 56.76	$4d_{3/2},5/2 \rightarrow 6p,5d$	2.2	2.1-2.3
E_1; E_2	55.70; 57.40	$4d_{3/2},5/2 \rightarrow 4f$	1.55	1.3-1.6
F_1; F_2	56.29; 58.01	$4d_{3/2},5/2 \rightarrow 7p,...$	0.95	

I_2^* ($4d^{-1}\,\sigma_u^*$) towards larger bond lengths compared to the neutral molecule causes increased width and a shift of the absorption maximum A (as compared to the adiabatic transition) of about 0.75 eV [5.85]. This introduces an uncertainty to the $\sigma_u(5p)$ level energy (Table 5.4). From the ionization energy of the I_2 molecule, $E_{\pi_g} = 9.4$ eV ($^2\Pi_{g3/2}$) and 10.0 eV ($^2\Pi_{g1/2}$), and the excitation energy of 1.5 eV yields a binding energy of 8...8.5 eV for the σ_u orbital (see the last column of Table 5.4); thus binding energies of $E_{4d3/2} = 58.5$ eV and $E_{4d5/2} = 57.0$ eV are obtained from the energies of the maxima A_1 and A_2 with an uncertainty of ±1 eV.

Due to the splitting of the 4d levels (1.68eV) individual series overlap in pairs in the region of Rydberg transitions, making identification difficult. From the components between D and F, which are likely to belong to one series, series limits can be obtained using the above-described method so that the level energies given in the 4th column of Table 5.4 and the binding energies ($E_{4d3/2} = 58.95$eV and $E_{4d5/2} = 57.25$ eV) can be obtained with an accuracy of 0.2 eV. Agreement between these data and the levels deduced from optical spectra supports the interpretation for both cases. The Rydberg series observable in the optical spectrum can be calculated using a one-electron model (deviations are less than 0.1eV). Use of this model for the X-ray absorption spectrum is justified since the interaction between the excited electrons with the partly filled valence shell is stronger than that with the 4d vacancy.

HCl: Interpretation of the X-ray emission and absorption spectra of the HCl molecule with the valence-electron configuration $\sigma^2(3s)\sigma^2(3p)\cdot\pi^4(3p)$ in the ground state is shown in Fig.5.11. The identification of the emission maxima (Table 5.5) gives an energy difference between the corresponding valence states of 3.9 eV in agreement with calculations (3.4eV [5.99], 3.9eV [5.101], 4.2eV [5.108]) and with the difference of the UPS

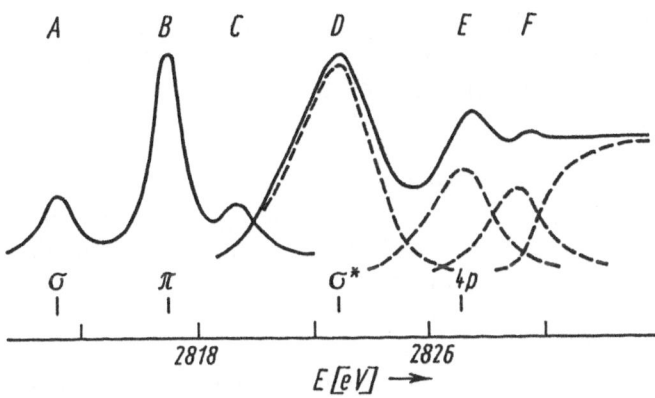

Fig.5.11. Cl Kβ emission band (left) and K absorption spectrum (right) of HCl [5.48, 49, 53]

Table 5.5. Interpretation of Cl K emission and absorption spectrum and energy levels of molecular HCl (all values in eV)

Peak	Photon energy [5.48, 49]	Transition	Level energy UPS [5.4]	Photon energy (theory) [5.99]	[5.101]	[5.101][a]
A	2813.1	$1s^{-1} \to \sigma^{-1}(3p)$ $^2\Sigma^+$	16.6	2806.8	2805.5	2836.0
B	2817.0	$1s^{-1} \to \pi^{-1}(3p)$ $^2\Pi_{3/2}$	12.74	2810.2	2809.6	2839.9
		$^2\Pi_{1/2}$	12.82			

| | | | Level energy [5.99] | | | |
			XAS	theory		
D		$1s \to \sigma^*(3p)$	6.15	6.0		
E		$1s \to 4p$	2.15	2.25		
F		$1s \to "5p"$	0.7	1.05		

[a]) These values have been calculated using Koopmans' theorem

ionization potentials (3.8 eV [5.4]). As a consequence of transitions in molecules with double vacancies, satellite structures arise on the short-wavelength side of the maximum which accounts for the third maximum in the region of 2820 eV. For the energy differences between the main line $1s^{-1} \to \pi^{-1}(3p)$ and transitions like $1s^{-1}\sigma^{-1}(3p) \to \sigma^{-1}(3p)\pi^{-1}(3p)$ and $1s^{-1}\pi^{-1}(3p) \to \pi^{-2}(3p)$, SCF calculations with open shells [5.101] give values of 2.76 and 2.53 eV, respectively. Therefore it can be concluded that the assignment given is correct.

The predicted satellite transitions associated with the $1s^{-1} \to \sigma^{-1}(3p)$ line A should occur at a distance of 3.5–4.5 eV on the short-wavelength side of the main line and are obscured by the B component. Since the potential curve of the molecular ion HCl^+ $[\sigma^{-1}(3p)]^2\Sigma^+$ is shifted towards larger bond lengths as a result of ionization of the bonding $\sigma(3p)$ orbital, the corresponding component A in the $K\beta$ emission spectrum is broader than B and is shifted towards lower photon energies compared to an adiabatic transition. Calculations of the relative intensities of the $K\beta$ and $L_{II,III}$ emission bands with a one-center model including a vacancy show that the relative intensity changes by about 20% compared to a calculation using ground-state wave functions of the molecule. Therefore, systematic error assignments of the spectrum should not arise due to neglect of the effect of the vacancy [5.108].

Decomposition of the K absorption spectrum into the components D, E, and F and also into the arctan function of the continuous absorption curve makes it possible to determine the energy levels given in Table 5.5. In this context it must be considered, however, that the line F is superimposed on the barely observable higher members of the Rydberg series with $n \geq 5$. The energy levels given in the last column of Table 5.5 are calculated from the excitation energies of the states $1s^{-1}np$ ($n = 4, 5$) and for the ion HCl^+ ($1s^{-1}$). Comparison with experimental values shows that in the K absorption spectrum of HCl a Rydberg series starts with the second line. For interpretation of the L absorption spectrum, see Fig.5.14 below.

5.3.2 Triatomic Molecules

H_2O (C_{2v}): The electronic configuration of the nonlinear H_2O molecule is $a_1^2(1s)a_1^2(2s)b_2^2(2p)a_1^2(2p)b_1^2(2p)$, $^1A_1$. Therefore, we expect three components in the O $K\alpha$ emission band (Fig.5.12): transitions into non-bonding b_1 orbital with exclusively O 2p character (this can be inferred from the observed high intensity and the small width of this maximum) and into σ-bonding orbitals a_1 and b_2. The energy differences between the emission maxima b_1 and a_1 or b_2 (2.0 and 5.4 eV) agree with the differences (2.12 and 5.89 eV) between the vertical ionization potentials (E_{b_1} = 12.62eV, E_{a_1} = 14.74eV, E_{b_2} = 18.51eV [5.4]). A long-wavelength shift of the b_2 component causes the deviation of the second value, which arises from a shift of the potential curve of H_2O^+ ($1s^{-1}$) ion as compared to the neutral molecule as for the N_2 molecule (Fig.5.2).

H_2S (C_{2v}): In contrast to the above example, L spectra (as well as K spectra) can be studied for the H_2S molecule. It has the valence-electron configuration $a_1^2(3s)b_2^2(3p)a_1^2(3s,3p)b_1^2(3p)$, so that information can be obtained on the level $a_1(3s)$ and the S 3s and 3d components of the MO. The relative positions of the K and L spectra of sulfur can be obtained from the energies of the S $K\alpha$ doublet of H_2S (2307.5eV [5.116]), as shown in Fig.5.13. The most intense maximum, A, of the $K\beta$ emission band arises from transition into the level b_1 (Table 5.6) so that the energy

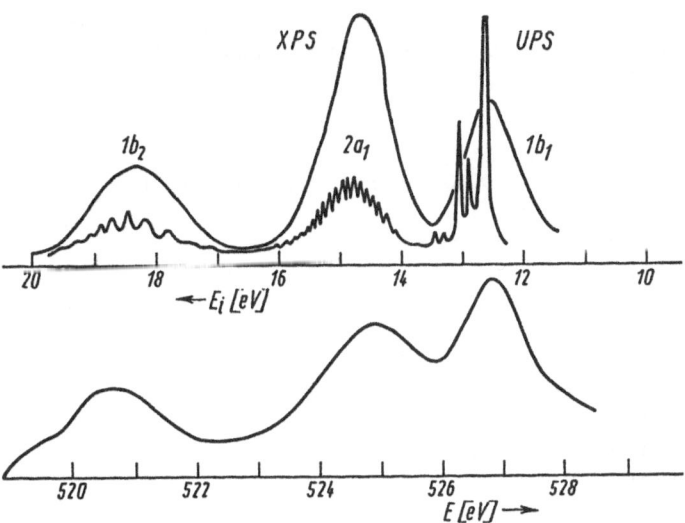

Fig.5.12. O $K\alpha$ emission band (bottom) [5.111], XPS and UPS profiles (top) [5.4,7] of the valence electron region of H_2O

levels given in the fifth column of Table 5.6 can be calculated from the vertical ionization potential E_{b_1} = 10.47 eV and the peak energies B to E. The binding energy $E_{1s} = E_{1s,b_1} + E_{b_1}$ = 2477.5 eV agrees with the value obtained from absorption spectra [5.18]. From the experimental relative intensities and the MO coefficients calculated using (2.40), it follows that the MO a_1(3s) and b_1(3p, lone-pair electrons) practically do not participate in the bonding. Since the relative intensity of the maximum C is smaller than that of B, it follows that the fraction of the H 1s AO in orbital b_2 is larger than in orbital a_1(3s, 3p). From the structure of the S L emission spectrum it follows that there is only a weak participation of the S 3d in bonding.

In agreement with calculations, the absorption main peak can be explained as caused by transitions into the unoccupied MO b_2^*(3p) and a_1^*(3s, 3p) [$E_{b_2^*}$ (1s^{-1}) = 4.5eV, $E_{1s}-h\nu_{A'}$ = 5.2eV], while structure on the short-wavelength side corresponds to transitions into Rydberg states ns(a_1) and np(a_1, b_2, b_1) (Figs.5.13, 14).

$SO_2(C_{2v})$: In the nonlinear SO_2 molecule with valence-electron configuration $1a_1^2$(O 2s)$1b_2^2$(O 2s)$2a_1^2$(S 3s)$2b_2^2$(S 3p)$3a_1^2$(S 3p)$3b_2^2$(O 2p, S 3d)$1a_2^2$(O 2p, S 3d)$4a_1^2$(O 2p) the O $K\alpha$ emission band occurs as an extra source of information regarding electronic structure of the molecule in addition to that already provided by the emission bands of sulfur. Due to the large number of valence electrons compared to the cases of H_2S and H_2O, the level structure is complicated so that it is impossible to determine the valence states without a decomposition of the spectrum into individual components.

Table 5.6. Interpretation of S, K, and L emission spectra, energy levels (values in eV), relative intensities of the components in the emission spectrum I(exp) and MO coefficients $C_{n\ell}$ for H_2S molecule [5.116, 119]

Peak	Photon energy	Transition	Level energy UPS [5.176]	XES	I(exp)	C_{3p} exp. [5.177]	theor.	C_{3s} exp.	theor. [5.177]
A	2467.0	$1s^{-1} \to b_1^{-1}(3p)$	10.47	10.5	1	1	1	0	0
B	2464.3	$1s^{-1} \to a_1^{-1}(3s, 3p)$	13.33	13.2	0.6	0.77	0.75		
D	156.9	$2p^{-1} \to a_1^{-1}(3s, 3p)$		13.1	0.3			0.55	0.52
C	2462.5	$1s^{-1} \to b_2^{-1}(3p)$	15.47	15.0	0.4	0.63	0.61		
E	148.7	$2p^{-1} \to a_1^{-1}(3s)$	22.2	21.3	1	0	0	1	1
A'	2472.3	$1s \to b_2^*(3p)$		5.2					
B'	2475.3	$1s \to np \ (n \geq 4)$							

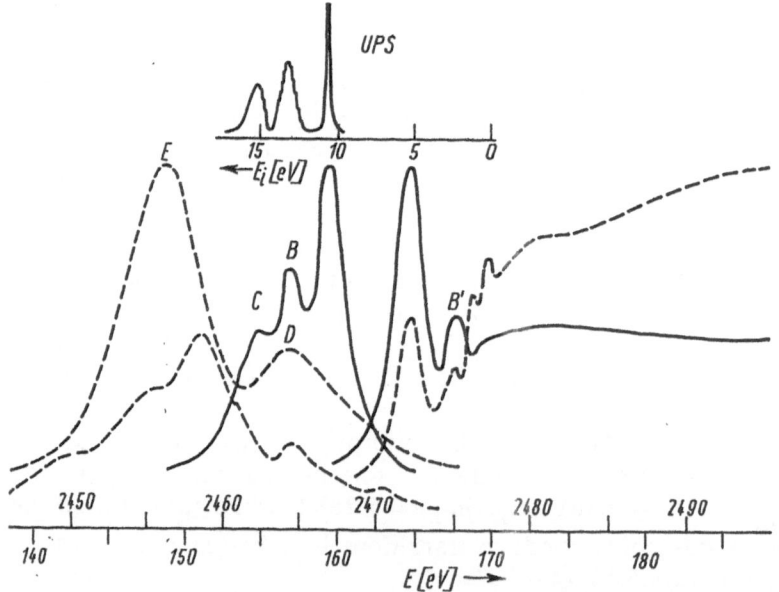

Fig.5.13. X-ray emission (left) and absorption (right) spectra [5.97, 103, 116, 119] and UPS curve [5.4] of H_2S. The K spectra are shown as solid curves while the L spectra are dashed (L emission results from [5.97, 116] are shown)

Figure 5.15 shows the spectra using the $K\alpha$ energy of $h\nu = 2307.9$ eV [5.100] and the binding energies $E_{S2p} = 174.8$ eV and $E_{O1s} = 539.6$ eV [5.7] on a uniform energy scale which includes the energies of the final state of the transition and the binding energy of the valence electrons. In the top part of the figure, estimates of the energy positions of the levels

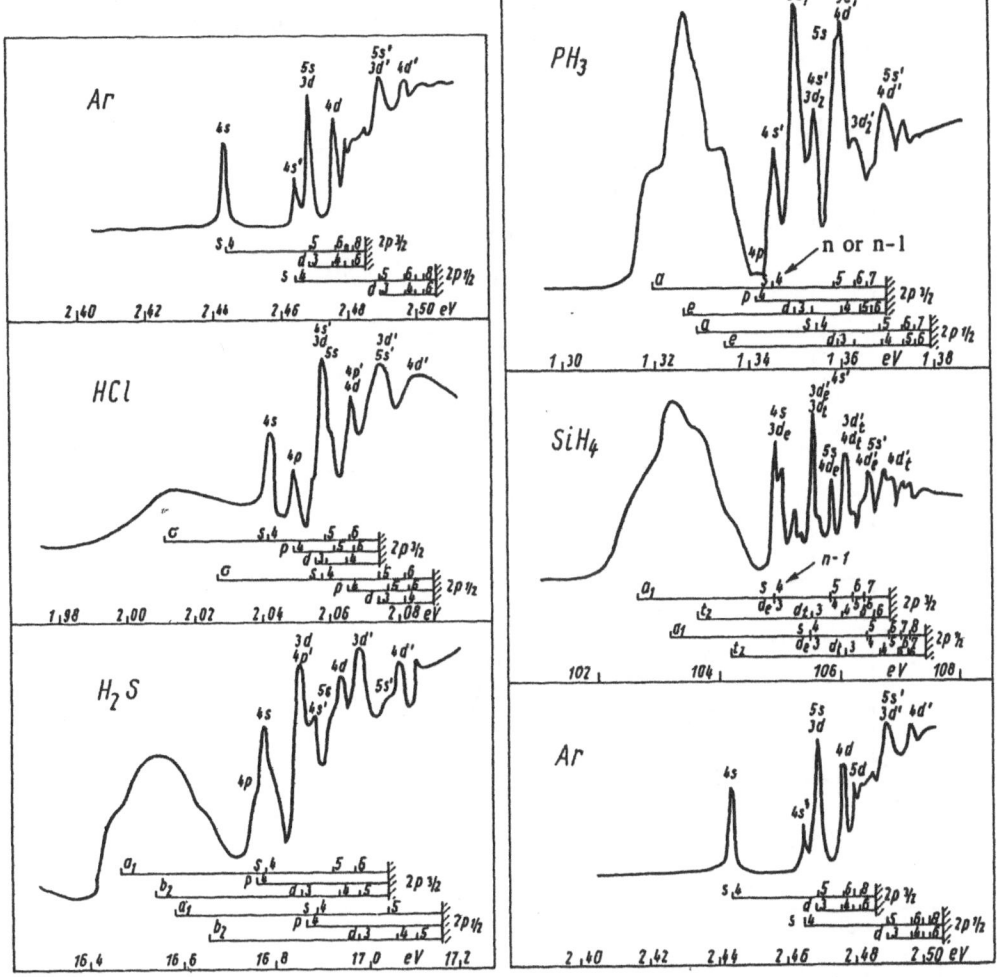

Fig.5.14. Interpretation of the $L_{II,III}$ absorption spectra of Ar, HCl, H_2S, PH_3, and SiH_4 [5.59, 120]

are given for some cases where uncertainties in the measurement and in the decomposition of the spectrum lead to ambiguities due to overlapping lines. The maxima labeled A and B should be disregarded since they arise, respectively, from anomalous scattering in the analyzing crystal (B) and a radiative Auger transition (A). Remaining features can then be associated with the valence states using group theoretical arguments (Table 5.1) and energies of the atomic levels of S and O except for the levels $4a_1, 1b_1$ and $3a_1$. Relative intensities and the transition energies agree with ab-initio calculations [5.71]. The value for $E_{4a1} = 12.0$ eV determined from X-ray spectra and from the O 1s and S 2p binding energies is in good agreement with the vertical ionization potential (12.5eV [5.4]). Existence and obser-

185

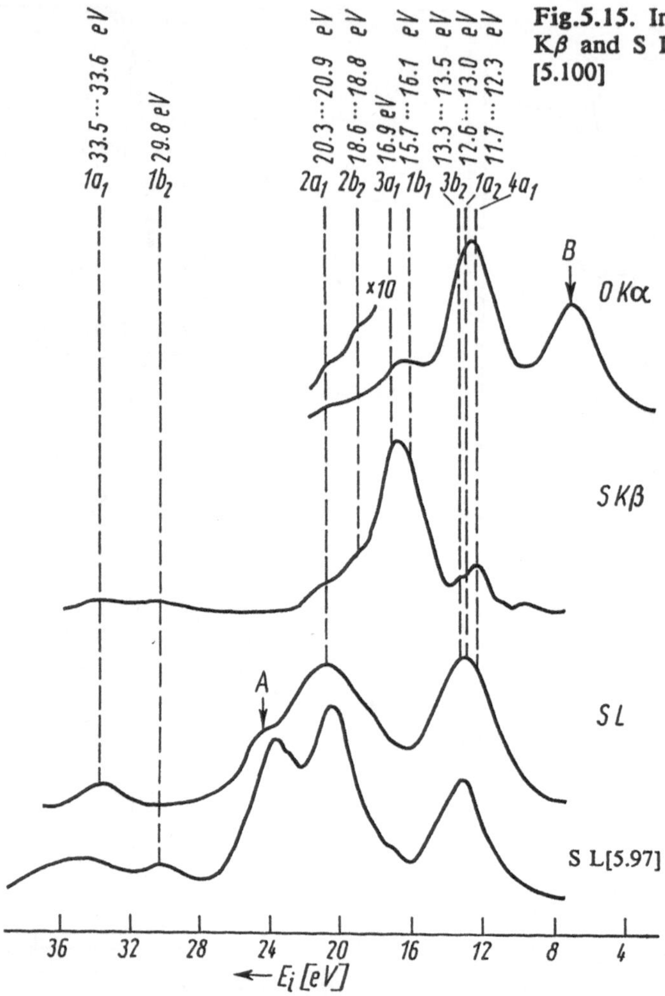

1a1 33.5···33.6 eV
1b2 29.8 eV
2a1 20.3···20.9 eV
2b2 18.6···18.8 eV
3a1 16.9 eV
1b1 15.7···16.1 eV
3b2 13.3···13.5 eV
1a2 12.6···13.0 eV
4a1 11.7···12.3 eV
x10
B
O Kα
S Kβ
A
S L
S L[5.97]
36
32
28
24
20
16
12
8
4
← Ei [eV]

Fig.5.15. Interpretation of O Kα, S Kβ and S L emission bands of SO_2 [5.100]

ved strong intensity of the $1a_2$ maximum in the L emission spectrum confirm the strong S 3dπ-O 2pπ bonding, as predicted by calculations [5.178]. This result agrees with the pronounced vibrational structure observed in the photoelectron spectrum [5.4] in which the second band shows strong vibrational excitation upon ionization of the $1a_1$ MO nonbonding only in absence of S 3d AO and indicates a change in bond length upon ejection of an electron.

For analysis of the absorption spectrum of SO_2, considerations analogous to those made in the previous example are useful [5.71]. A binding energy of E_{S2p} = 175 eV was determined from the S L absorption spectrum which is in good agreement with the XPS value [5.18].

XeF_2 ($D_{\infty h}$): In the absorption spectra of XeF_2 and the fluorides XeF_4 and XeF_6 five regions A to E can be distinguished between 50 eV

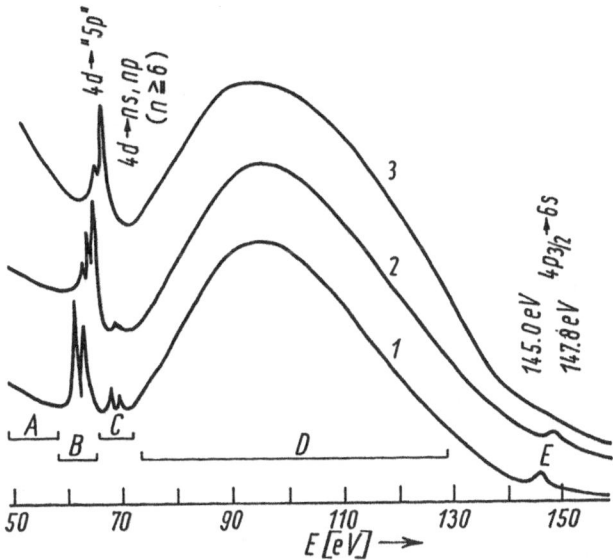

Fig.5.16. Xe $N_{IV,V}$ absorption spectra of XeF_2 (Curve *1*), XeF_4 (Curve *2*), and XeF_6 (Curve *3*) [5.131, 132]

and 150 eV (Fig.5.16) [5.131, 132]. On the long-wavelength side there is a continuum which decreases with increasing energy, which is missing in atomic Xe but becomes increasingly stronger with increasing number of fluorine atoms in the molecules, due to larger probability of F $2p$ excitations. Between 60 and 70 eV, a few intense lines follow, which arise only in the compounds of the noble gases and therefore can be understood as transitions of Xe $4d$ electrons into the lowest unoccupied MO. The successive maxima in the region around 70 eV, caused by transitions into Rydberg states of the Xe atom (partly split due to the F ligand field), can be observed only for XeF_2 and XeF_4, since the potential barrier introduced by the fluorine atom (Sect.2.6.2) effectively suppresses the overlap between the core and the Rydberg orbitals of XeF_6. After the Rydberg lines an intense continuum follows with a maximum at 95 eV caused by delayed onset of $4d \rightarrow \epsilon f$ transitions like that for I_2 and Xe (Figs.5.10 and 2.33). The short-wavelength tail shows structure around 145 eV, which can be ascribed to the $N_{II,III}$ absorption edges.

From absorption spectra one can extract information concerning core and the excited states [5.131]. Following mixing of the occupied $5p$ AO of the free Xe atom with the antibonding unoccupied $2p\sigma$ MO of free F_2 molecule, a lowest unoccupied $7\sigma_u$ MO is formed in XeF_2 with mostly Xe $5p$ character. Transitions from the nonbonding $4d$ orbital into the antibonding $5p(7\sigma_u)$ MO are broadened due to vibrational excitations. Although an upper limit for the natural $4d$ level width of 0.2–0.25 eV can be inferred from the half-widths of individual Rydberg components (mainly due to Auger width), the $4d \rightarrow$ "5p" absorption line has a half-

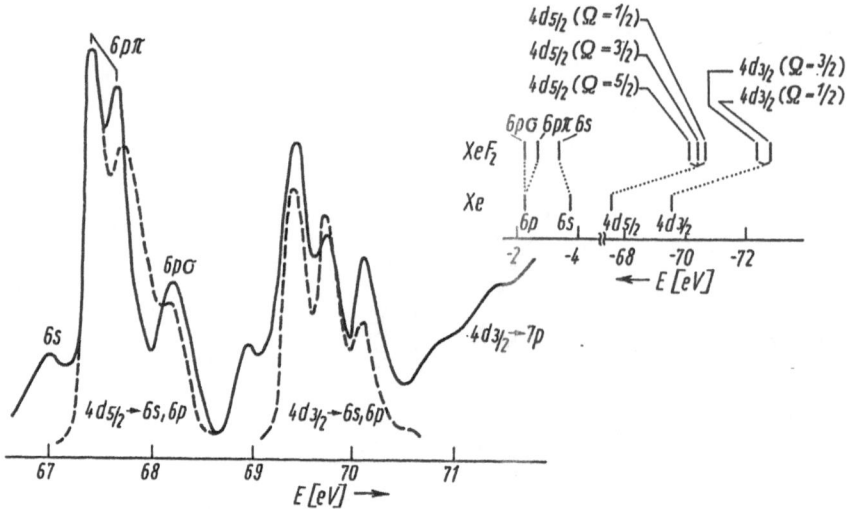

Fig.5.17. Interpretation of the 4d Rydberg spectrum (theoretical curve is dashed) of XeF$_2$ [5.131] and one-electron energy levels for Xe and XeF$_2$ [5.18]

width of about 0.9 eV. The interval between the components of 1.1 eV corresponds to the spin-orbit splitting of the 4d levels of the free Xe atom. In contrast with the theoretical intensity ratio of 1.5 (calculated from the multiplicities of the d$_{5/2}$ and d$_{3/2}$ levels a ratio of 1.2 is observed. This discrepancy arises from exchange interaction and ligand-field effects. The different widths of the d$_{5/2}$ and d$_{3/2}$ components in the 4d → "5p" transition can be understood in terms of different ligand-field splitting of the 4d levels. This result can be derived from the separations of the Rydberg transitions 4d$_{5/2}(\Omega=5/2)$ → 6pπ, 4d$_{5/2}(3/2;1/2)$ → 6pπ (0.24eV), 4d$_{3/2}(3/2)$ → 6pπ, and 4d$_{3/2}(1/2)$ → 6pπ (0.32eV) (Fig.5.17).

Such a ligand-field splitting of the fivefold degenerate d levels (which is appreciable in complexes with 3d metals) was detected in heavy elements for the first time in xenon fluorides. Since overlap and delocalization effects play only a minor role for d-core orbitals, the observed effects can be well represented by an electrostatic ligand-field model [5.131]. From the separation of the 4d$_{3/2}(1/2)$ → 6pπ and 4d$_{3/2}(1/2)$ → 6pσ components, a ligand-field splitting of 0.35 eV can be deduced for the 6p level (spin-orbit splitting < 0.05eV) and the energy level diagram as shown in Fig.5.17 is obtained for XeF$_2$. The calculated theoretical absorption curve for 4d → 6p transitions (dashed in Fig.5.17), based on the above interpretation using a one-electron model, agrees well with the experimental spectrum. In the energy region 69.5 to 70.5 eV 4d$_{5/2}$ → 7p components overlap, so that the rather high intensity there can be easily understood.

From the absence of a structure in the region E of the 4p transitions, the following conclusions can be drawn concerning the geometrical structure of the molecule [5.132]. Transitions into 5d and higher Rydberg orbitals are suppressed in connection with excitation of Xe 4p core electrons as a result of the above-described potential barrier. Since dipole selection rules prevent radiative transitions of 4p electrons into the unoccupied Xe 5p valence level, a structure can be expected only when the Xe 5s shell is partly occupied. There appear to be two possibilities for the structure of XeF_6. Firstly, it might be an octahedral molecule, albeit distorted by unharmonic vibration, with an electron configuration of $5s^2$ at the central atom. Alternatively, it might be better described as a mixture of electronic/geometric isomers (octahedral singlet plus distorted triplet), in which case the valence electron configuration would be $5s5p$ ^3P. The absence of absorption structure in the region of the 4p transitions implies that the first structure variant is the more probable [5.133].

5.3.3 Polyatomic Molecules

C_2H_2 ($C_{\infty v}$): Interpretation of the C $K\alpha$ emission band of C_2H_2 with the valence-electron configuration $\sigma_g^2\sigma_u^2\sigma_g^2(C2p)\pi_u^4(C2p)$ $^1\Sigma_g^+$ is shown in Fig.5.18. In agreement with semiempirical and ab-initio calculations [5.179, 180] which attribute 2p symmetry to all valence orbitals (Table 5.7), the main maximum and the long-wavelength structure can be explained by dipole transitions of the type $1s \rightarrow$ "2p". If one starts out from the experimental ionization potential of the molecule and calculates the emission band using the theoretical 2p coefficients, as given in (5.6), good agreement is obtained between the calculated spectrum and experiment. Comparison between variously estimated theoretical 2p coefficients (Table 5.7) with values obtained by decomposing the $K\alpha$ band into individual components shows the usefulness of the above-described model for intensity calculations. From the energy of the principal maximum and the

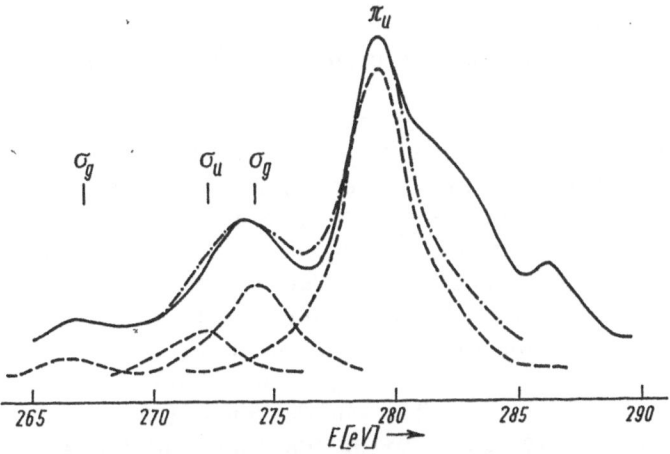

Fig.5.18. Experimental C $K\alpha$ emission band of C_2H_2 (solid line), theoretical curve (dash-dot) and expected components (dashed) [5.38]

Table 5.7. Ionization energies (in eV, 2nd column UPS data, 3rd column values deduced from relative position of the components of the $K\alpha$ band) and experimental C_{2p} coefficients (4th column) compared to theoretical values (5th and 6th columns) for C_2H_2

Level	Ionization energy [5.4]	[5.38]	[5.38]	C_{2p} [5.180]	[5.179]
π_u	11.4	11.4	0.61	0.61	0.61
σ_g	16.4	16.4	0.46	0.46	0.47
σ_u	18.4	18.4	0.27	0.29	0.31
σ_g		24.0	0.19	0.18	0.19

corresponding ionization energy, an energy value of $E_{C\,1s} = 290.4$ eV [5.38] is obtained. Better agreement is expected with the XPS value of 291.1 eV± 0.4 eV when the X-ray spectrum can be obtained with higher resolution, enabling assignment of transitions to different vibrational states [5.152].

The short-wavelength structure above 280 eV which appears in similar form for many C K emission bands (e.g., in CH_4, C_2H_4, CO, CO_2) and other spectra can be explained in terms of several processes. In addition to the main process under discussion, interfering signals appear in the same region both from transitions between multiple-ionized states of the molecule and from resonance fluorescence transitions (which are expected to coincide with absorption maxima) [5.61]. Energies and relative intensities of 1s core excitation (Rydberg and resonance transitions) calculated using ab-initio MO calculations with large basis sets agree well with energy loss spectra [5.69].

CH_3Cl (C_{3v}): The 14 valence electrons of this molecule are divided between the levels $1a_1$, $2a_1$, $1e$, $3a_1$, and $2e$, for which information on the relative positions is available from the C K, Cl K, and Cl L emission bands shown in Fig.5.19. The spectra have been placed on a common energy scale using UPS ionization energies [5.182] $E_{2e} = 11.3$ eV (to determine the main maximum of the Cl $K\beta$ band), $E_{1e} = 15.4$ eV (maximum of the C K band) and the energy of the $K\alpha_{1,2}$ doublet, $h\nu = 2621.3$ eV [5.183] (to determine the position of the Cl L emission band). The synthesized spectrum gives the energies of the valence states participating in the transition. One arrives at a similar result [5.38] using the series limit of the Rydberg components in the absorption spectrum of chlorine to determine the Cl K and L emission bands in the binding energy scale together with the C K emission band using the XPS binding energy $E_{C\,1s} = 292.4$ eV [5.184]. In this case the energy levels of the valence states can be determined without the UPS data. The observed deviations (for E_{1e} they amount to more than 1 eV) are caused by an uncertainty of the wavelength of the C $K\alpha$ band and the application of the adiabatic ionization

190

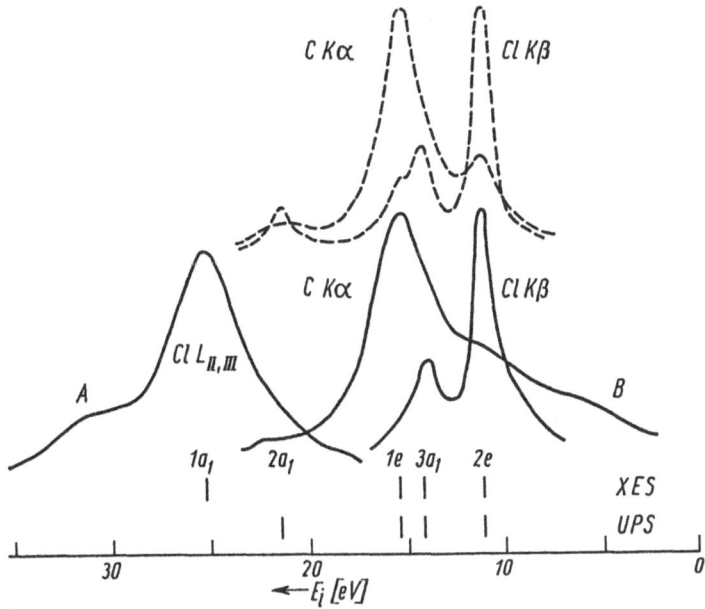

Fig.5.19. Experimental (solid line) and theoretical (dashed line) X-ray spectra [5.48,63,64,143] from CH_3Cl. Energy levels [5.182] are indicated below

potential. If the UPS and XES energy levels, drawn in the lower part of Fig.5.19, are compared, a good agreement is observed for the independently determined values for $3a_1$. The binding energy for the lowest valence level not given from the photoelectron spectrum, is 25.4 eV.

The relative intensities among the individual spectral curves give information on the distribution of the Cl 3p electrons (Cl $K\beta$), the C 2p electrons (C $K\alpha$) and the Cl 3s electrons (Cl L) into MO. Using Table 5.1 the following predictions can be made in agreement with the results from the corresponding spectra from chlorinated derivatives of methane $CH_{4-n}Cl_n$: The lowest occupied valence orbital ($1a_1$) is formed by 3s electrons of chlorine which do not participate appreciably in the bonding. The MO's with energies between 14 and 24 eV form σ bonds while the MO's with ionization energies below 14 eV are mainly formed by Cl 3p AO which separate into π-bonding and nonbonding orbitals. In agreement with theoretical results there is a short-wavelength component of the C $K\alpha$ emission band which coincides with the Cl $K\beta$ maximum and suggests an appreciable $Cl3p\pi - C2p\pi$ interaction. In contrast to the case of the highly symmetric oxyanions of elements from the 3rd period, this interaction also arises for the molecules CF_4 and CCl_4 (Sect.5.2.3), as indicated by the corresponding maximum in the C K emission spectrum. Some conclusions could be drawn about the $2a_1$ level on the basis of the Cl $K\beta$ emission spectrum. A unique identification of the relatively close-lying levels 1e and 3a is possible from the relative positions of the maxima of C $K\alpha$ and Cl $K\beta$. The structures marked as A and B are identified as satel-

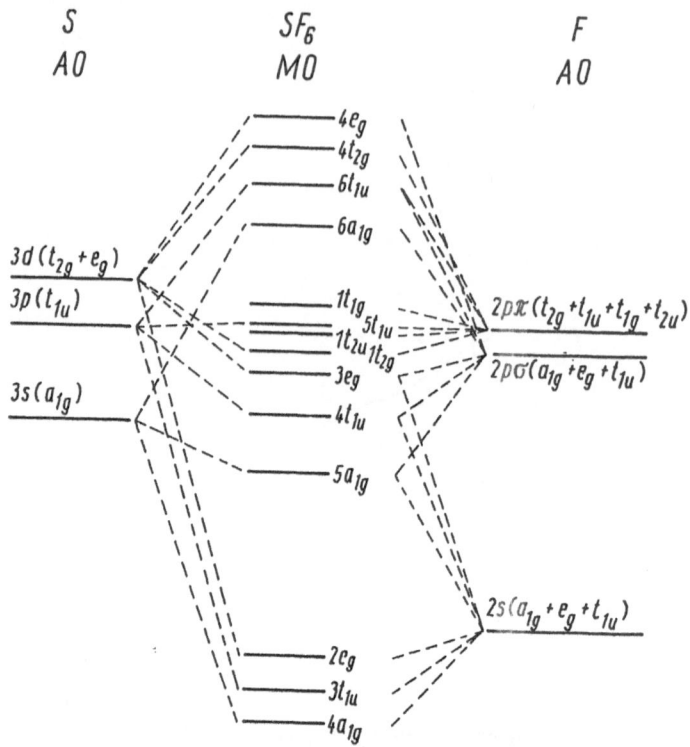

Fig.5.20. AO-MO energy level diagram of the SF_6 molecule

lites produced by radiative Auger transitions and transitions in multiply ionized molecules and by resonance transitions (Sect.2.4).

SF_6 (O_h): The 48 valence electrons of the molecule SF_6 are distributed according to octahedral symmetry into 10 levels, viz., a_{1g}, t_{1u}, e_g, t_{2g}, t_{1g}, and t_{2u} (Fig.5.20). The "F 2s" states (with small admixtures of S 3s and 3p), formed mainly of F 2s AO's, are found in the lower part of the circa 30 eV broad region of valence levels while the essentially nonbonding F 2p states are found in the upper part. The MO's, consisting of s, p, and d AO's of sulfur and p AO's of fluorine, which are responsible for the σ and π bonding, are found in between. If the energies of the valence levels are calculated from the photon energies of the maxima in the S and F emission bands and from energies of the initial states: $E_{1s}(S) = 2490\,eV$; $E_{1s}(F) = 694.6\,eV$; $E_{2p_{1/2}}(S) = 181.7\,eV$; $E_{2p_{3/2}}(S) = 180.4\,eV$ [5.7, 162], the following conclusions are obtained for the series of the valence states [5.162]: The selection rule allows filling of a $1a_{1g}$(S 1s) vacancy by transitions from t_{1u} valence orbitals. Consequently, the three maxima of the S $K\beta$ spectrum determine the energies of the t_{1u} levels ($3t_{1u} = 41.5\,eV$, $4t_{1u} = 22.9\,eV$, $5t_{1u} = 17.5\,eV$). Agreement with UPS and XPS curves is very good as can be seen in Fig.5.21. The energies of the two "F 2s" levels (close to $3t_{1u}$) were determined from XPS to be $E_{4a_{1g}} = 44.2\,eV$ and E_{2e_g}

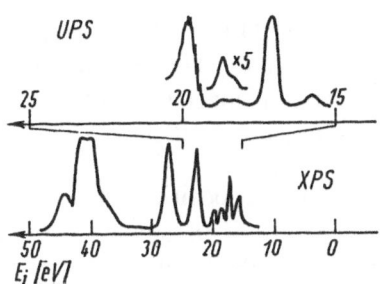

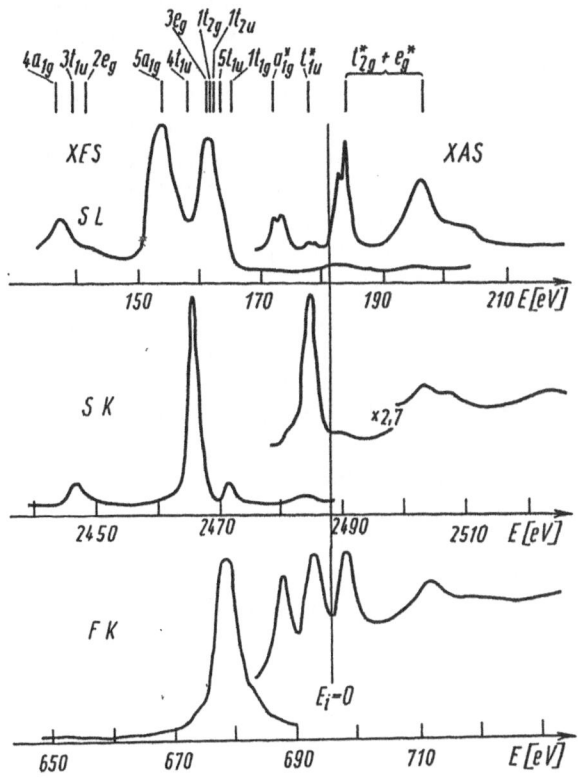

Fig.5.21. X-ray emission bands and absorption edges for SF_6 [5.162, 168]. Photoelectron spectra [5.1] are shown at the top

= 39.3 eV. According to the selection rules, the S L emission band is produced by transitions $a_{1g}(3s)$, $e_g(3d)$, $t_{2g}(3d) \rightarrow 2t_{1u}(2p)$, so that the binding energy of the $5a_{1g}$ level can be estimated as 27.0 eV from the maximum at 154 eV (Fig.5.21) in the L spectrum in agreement with the photoelectron spectrum. The value given was obtained from XPS since the accuracy of the wavelength measurement in the region of the S L emission band is lower. If the possible occurence of satellites is negotiated, the second intense maximum at 162 eV in the L spectrum suggests appreciable participation of the S 3d orbitals in the bonding and allows estimation of the position of the $3e_g$ and $1t_{2g}$ levels. More accurate values for these energies can be obtained from better resolved UPS curves which also give values for the remaining "F 2p" levels in the upper edge ($E_{1t_{1g}} = 15.7eV$; $E_{1t_{2u}} = 18.4eV$; $E_{1t_{2g}} = 18.7eV$; $E_{3e_g} = 19.9eV$). The best agreement between theoretical intensities and the L band spectrum recorded with considerably higher resolution than the one shown in Fig.5.21 is obtained when the calculation is performed including d-orbital population and the many-center terms in (5.5) [5.50]. The relative position of the F K emission spectrum agrees with the given distribution of levels. From the various ab-initio and nonempirical calculations available, no consistent picture emerges of the valence levels, particularly within the approximately 5 eV region spanned by the F 2p orbitals [5.141, 159, 161, 186, 187].

By procedures similar to those described for SF_6, one can interpret the analogous X-ray and photoelectron spectra of the isoelectronic anions AlF_6^{3-} and SiF_6^{2-}, thereby also confirming the results obtained for the molecule SF_6 and allowing the following general conclusions (Fig.5.22) to be drawn [5.26, 188]. The tendency of the $3t_{1u}$ and $5a_{1g}$ levels formed by F 2s and S 3s states to come closer together while the $5a_{1g}$ and $5t_{1u}$ levels (the latter determined by F 2p states) move farther apart, can be explained by the increasing binding energy of the 3s valence electrons of the central atom with increasing mass number. The decreasing ionic charge and increasing binding energies of the 3s and 3p valence electrons of the central atom with increasing atomic number follows simply from the increase of binding energy for all valence electrons. Finally, the decreasing ionicity of the bond in going from AlF_6^{3-} to SF_6 leads to increased splitting of both "F 2s" and the σ and π levels which, in the limiting case of pure ionic bonding, should form only two levels in the anion F^- (2s and 2p).

In contrast to the spectra for molecular gases discussed in the previous sections, the absorption spectra of fluorine and sulfur in SF_6 are dominated by a few strong maxima below and above each ionization threshold. This peculiar structure can be understood by considering the

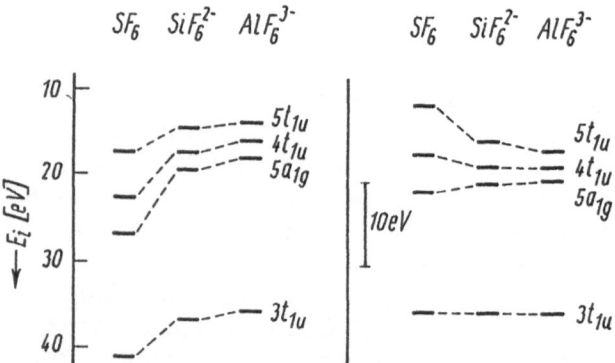

Fig.5.22. Experimental (left) and theoretical (right) energy level diagram for isoelectronic compounds XF_6^{n-} [5.187, 188]. Theoretical levels are positioned relative to $3t_{1u}$

effect of the potential barrier formed by the fluorine ions (Sect.2.6.2). A number of theoretical [5.160, 166, 167] and experimental [5.158] studies have helped elucidate the physical origin of such structure. The presence of a barrier in the effective molecular potential, when its effects are properly understood, explains the missing absorption edge, the high absorption cross-section in the region of the maxima and the strong suppression of the Rydberg transitions [5.160, 164]. Maxima in the K spectra are associated with odd parity levels (t_{1u}), while maxima in the L spectra correspond to even parity levels (a_{1g}, t_{2g}, e_g) according to dipole selection rules. Detection of Rydberg transitions of the type 2p $\rightarrow$ ns (n$\geq$4) and 2p $\rightarrow$ nd (n$\geq$3) in the region between 177 eV and the edge has been possible from absorption spectra obtained using synchrotron radiation with an experimental resolution of 0.08 eV [5.83]. Some very faint structures in the K and L absorption spectra which coincide with the strong lines of the complementary spectrum can be explained as vibrationally stimulated transitions [5.160]. The splitting of the L absorption maxima (1.1 eV) corresponds to the spin doublet splitting of the 2p levels. The relative strengths of the experimentally determined absorption cross-sections as compared to the statistical weights of the L_{II} and L_{III} levels can be explained by interaction of the inner vacancy with the excited electrons [5.103, 160]. The structure in the L absorption spectrum above 200 eV has not been satisfactorily explained by theoretical efforts to date [5.166]. However, a two-electron transition - a 3p electron of sulfur is excited simultaneously with a 2p electron - could possibly lead to an increase of the absorption cross-section in this region. The probability for such a process could amount to 20% [5.189, 190]. In the S K and S L emission spectra weak structures have been observed in the energy region of the corresponding absorption maxima which arise following resonance transitions (inverse of the absorption process), supporting identification of some of the absorption maxima.

5.4 Organic Compounds

Recording of high-resolution emission bands from elements of the 2nd period using fluorescence excitation was, for many years, associated with severe experimental difficulties. On the other hand, interest in organic compounds, especially those with heteroatoms, such as Si, P, S and Cl, has increased steadily. Therefore, such compounds have mainly been investigated up to now. Even though the X-ray spectra of these substances usually were recorded using solid targets, it is possible to use MO models in interpreting them since, to a first approximation, interaction between molecules in the crystal can be neglected compared to the interaction between the atoms in the molecule.

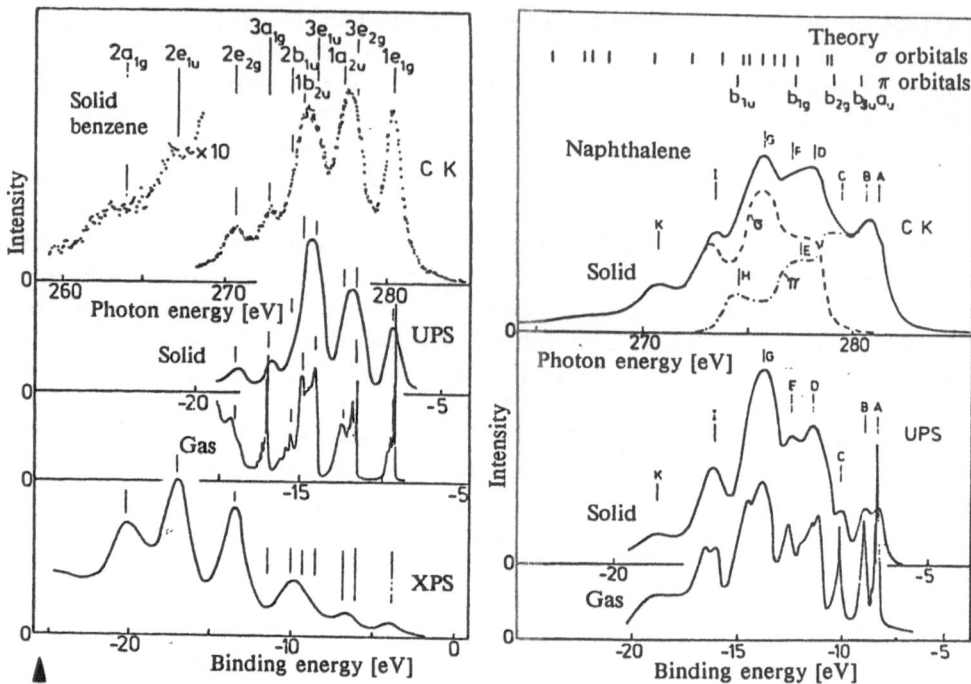

Fig.5.23. C K emission spectrum for solid benzene along with XPS and UPS curves for both solid and gas phases [5.27]

Fig.5.24. C K emission spectrum of polycrystalline naphthalene decomposed into π and σ bands and UPS curves from the solid and gaseous compound and theoretical results [5.27]

Several difficulties are entailed in obtaining results on the electronic structure of organic molecules from the fine structure in the emission bands of the atoms. Since one is often concerned with complex poly-atomic molecules, the one-electron energy levels in the region of the valence states are numerous, hardly separated from each other, and their wave functions are rather complicated in structure. Consequently, different components of the emission band are strongly superimposed, much more than for the similar molecular gases discussed in previous sections.

Some progress has occurred in the last ten years due to successful use of synchrotron radiation [5.27] and intense X-ray tubes [5.50] to obtain under fluorescence excitation the C K emission bands of hydrocarbons with an improved resolution compared to previous measurements [5.17, 64, 197]. These studies have enabled more detailed analyses of the valence states. Figures 5.23, 24 show the results for solid benzene and naphthalene together with the photoelectron spectra [5.27]. Neglecting the broadening of the UPS curve in the solid target compared to that from the gas phase, the structural details agree in both cases so that the C K emission band of the sample in the solid state can aid interpretation of the MO structure.

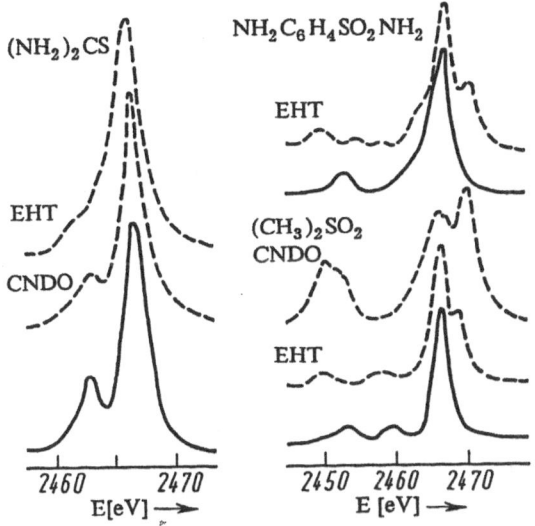

Fig.5.25. Theoretical (dashed) and experimental (solid) S Kβ emission bands from organic compounds [5.41]

The relative intensity of the C Kα components (C 2p $\rightarrow$ 1s) agrees nicely with the theoretical C 2p components.

For naphthalene the π and σ components were separated using spectra obtained with different take-off angles from the single crystal [5.192]. All observed five π states can be proven to be in good agreement with theoretical results in the π part of the C K emission band which also verifies the relatively high ionization energy for the lowest π orbital.

Decomposition of the spectrum of the emission bands from organic compounds is complicated because, due to the required large number of basis functions, only semiempirical MO calculations for molecules of organic compounds have been performed up till now. That the reliability of these calculations is limited is suggested by the fact that they do not seem to account satisfactorily for observed X-ray emission band structures. This is particularly evident from Fig.5.25. The S Kβ emission bands shown, calculated using semiempirical methods (EHT and CNDO/2), reproduce details of the structure of experimental curves correctly in a few cases but do not support general applicability of one or the other theoretical procedure. Thus, the S Kβ spectrum for thiourea calculated using CNDO methods agrees much better with the experimental spectrum than that calculated using EHT methods, while for sulfone compounds the situation is reversed.

However, in spite of these difficulties it is possible to obtain quantitative results if a mathematical analysis is performed using values from the experimental intensity distribution to obtain initial estimates [5.14, 38]. As shown already in Fig.5.6, reproducible results can be obtained in this way when the accuracy of calculated initial parameters is moderate or

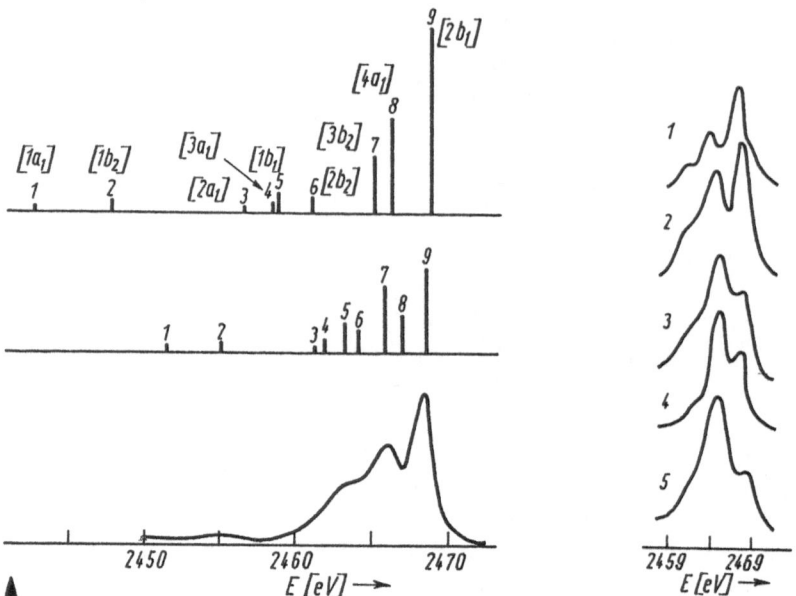

Fig.5.26. Experimental profile of S Kβ emission band of dibenzyl sulfide (bottom) and its components (middle). Top part shows the theoretical initial values (eigenvalues and eigenvectors) for the suggested analysis [5.14]

Fig.5.27. S Kβ emission band of thioether and heterocyclic sulfur compounds [5.22, 84, 99] - H_2S (Curve *1*), $(CH_3)_2S$ (Curve *2*), $(C_6H_5)_2S$ (Curve *3*), methyl thiophene (Curve *4*), and benzothiophene (Curve *5*)

poor. Figure 5.26 shows the components of the emission band determined from the experimental S Kβ spectrum of dibenzyl sulfide and from the theoretical eigenvalues and eigenvectors (CNDO/2), which can serve to calculate the energy levels and the LCAO coefficients using (5.2, 6).

Qualitative predictions on specific characteristics of chemical binding can be obtained without a profile analysis requiring a full theoretical calculation simply by comparing the different intensity distributions of the emission bands in different compounds. A few examples will be discussed in some detail. The S Kβ emission bands of thioethers and H_2S (symmetry C_{2v}) are very similar (Fig.5.27). On the long-wavelength side of an intense maximum A, which occurs as a result of transitions into a final state of b_1 symmetry (nonbonding, S $3p\pi$ AO in agreement with calculation results), there are two more or less resolved structures which can be identified with the levels a_1 and b_2 (bonding, S 3p AO). The relatively smaller intensity of the maximum A in compounds which contain a π bonding system (diphenyl sulfide, thiophene) verifies within the framework of the intensity model used (5.6) decreasing participation of the S $3p\pi$ AO (decreasing values for C_{3p} in the topmost occupied states) and as a result increased delocalization in the π system. The decrease in electron density at the sulfur atom accompanying delocalization confirmed by

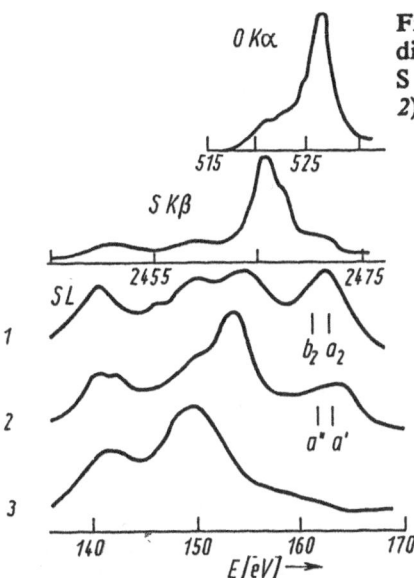

Fig.5.28. O K, S Kβ, and S L emission bands of dimethyl sulfone (top and Curve *1*) together with S L emission bands of dimethyl sulfoxide (Curve *2*) and alkyl sulfide (Curve *3*) [5.47]

simple Hückel calculations, can thus be determined from the relative intensity of the short-wavelength maximum [5.22, 84, 99, 193, 194]. Analysis of the Si Kβ and P Kβ emission bands in organic compounds have led to similar results [5.195, 196].

In the S Kβ spectrum (Fig.5.25) of thiourea (C_{2v}), theory [5.41, 89] indicates that two components, namely S 3p transitions from π bonding (b_1) and nonbonding (b_2) MO, merge to form a single large maximum. On its long-wavelength side there is a third component at a distance of about 3 eV which is produced by transitions from a σ bonding MO (a_1). Since this MO is composed of both S 3s and 3p AO's, a prominence is seen in the S L emission spectrum which coincides with the weak maximum in the Kβ spectrum (a_1) when the two spectra are superimposed using a common energy scale. Neither in the L emission band (transition s, d → p) of thiourea nor in that of alkali sulfide R_2S is it possible to find evidence of S 3d electron participation in the bonding. However, the S L emission bands of sulfoxide and sulfone (Fig.5.28) have components in the short-wavelength region which occur in the region of the O Kα emission band, evident when the spectra are superimposed on a consistent energy scale. In the absence of S 3d AO the MO's (levels b_2 and a_2) participating in the formation of the most intense components of the O Kα band have nonbonding character and are formed from O 2p AO. The maximum observed in the S L spectrum above 160 eV supports the assumption of a formation of a S 3d π-O 2pπ bond, which is also the case for SO_2 and SF_6. A detailed analysis of the P Kβ and some of the P L emission bands for a series of phosphinic oxides, phosphinates and phosphonates gave similar results [5.195].

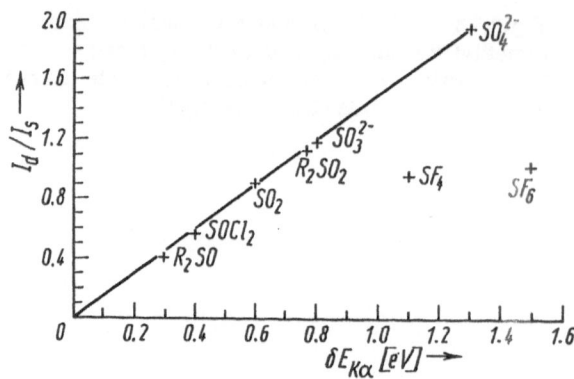

Fig.5.29. Dependence of the relative S 3d population determined from the intensity ratio I_d/I_s of the S L emission band on the Kα shift [5.47]

The relative S 3d population in compounds is evidently dependent on the positive atomic charge of sulfur. If the ratio of the integrated intensities I_d/I_s of the 3d and 3s components of the L spectrum is plotted against the measured Kα shifts (these are related to the atomic charge), a linear relation is obtained from which it is possible to determine the 3d population in sulfur compounds with oxygen and chlorine (Fig.5.29).

5.5 Coordination Compounds

5.5.1 Oxyanions of Elements of the 2nd and 3rd Periods

Results on the electronic structure of the anions $XO_4{}^{n-}$ and $XO_3{}^{n-}$ of elements of the third period have been obtained from a large number of theoretical calculations [5.43, 186, 193, 197-203], X-ray spectroscopic measurements [5.14, 47, 97, 163, 200, 204-209] and electron spectroscopy determination with UV [5.44] and X-ray excitation [5.43, 188, 210-212] supplementing one another. In the last ten years emission bands for elements from the second period in oxyanions $XO_3{}^{n-}$ (X = B, C, N) were recorded using synchrotron radiation and powerful X-ray tubes [5.136, 213]. Furthermore, O Kα spectra were recorded with excellent resolution in a series of oxyanions of the elements from the second to fifth periods [5.213].

The relations between XS, XPS, UPS, and theoretical results and the problem of interpreting X-ray spectra are exemplified in the collected results shown in Figs.5.30, 31 which were obtained for calcite and alkali salts with the anion $SO_4{}^{2-}$. The X-ray K emission bands of carbon and oxygen from calcite single crystals are shown together with the photoelectron spectrum and theoretical results in Fig.5.30 [5.28]. Due to the polarization of the radiation, both spectra show clear angular dependence. The observed anisotropy of emission makes it possible to distinguish between the

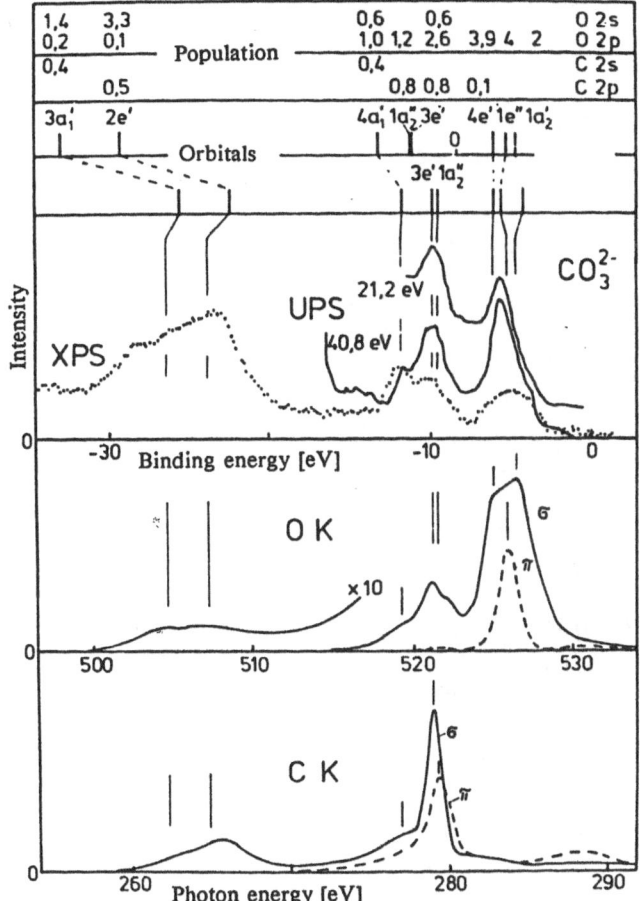

1,4	3,3			0,6	0,6				O 2s
0,2	0,1	Population		1,0	1,2	2,6	3,9	4 2	O 2p
0,4				0,4					C 2s
	0,5				0,8	0,8	0,1		C 2p

$3a_1'$ $2e'$ — Orbitals — $4a_1'$ $1a_2''$ $3e'$ $4e'$ $1e''$ $1a_2'$

$3e'$ $1a_2''$

Fig.5.30. π and σ parts of the O $K\alpha$ and C $K\alpha$ bands of crystalline CaCO$_3$, together with UPS and XPS curves, and theoretical results [5.28]

contributions from the σ and π valence electrons to individual components of the emission band. Only the states $1a_2''$ and $1e''$ of the eight occupied valence states of symmetries a_1', a_2', a_2'', e' and e'' have π character. According to dipole selection rules, electron transitions into O 1s vacancies are allowed from all orbitals, but the C 1s vacancy can be filled only from transitions from the e' and a_2'' orbitals (see population in Fig.5.30). Radiation emitted in the $\pi \rightarrow$ 1s transition is polarized perpendicular to the plane of the CO$_3^{2-}$ ions while in the $\sigma \rightarrow$ 1s transition parallel polarization occurs. If the radiation is recorded perpendicular to the plane of the ions only the σ component is observed while the π component can be obtained from the difference between spectra recorded in the parallel mode and the σ curve. The result shows that the π orbital $1a_2''$ has a smaller ionization potential than the σ orbital $3e'$. Consequently, it is pos-

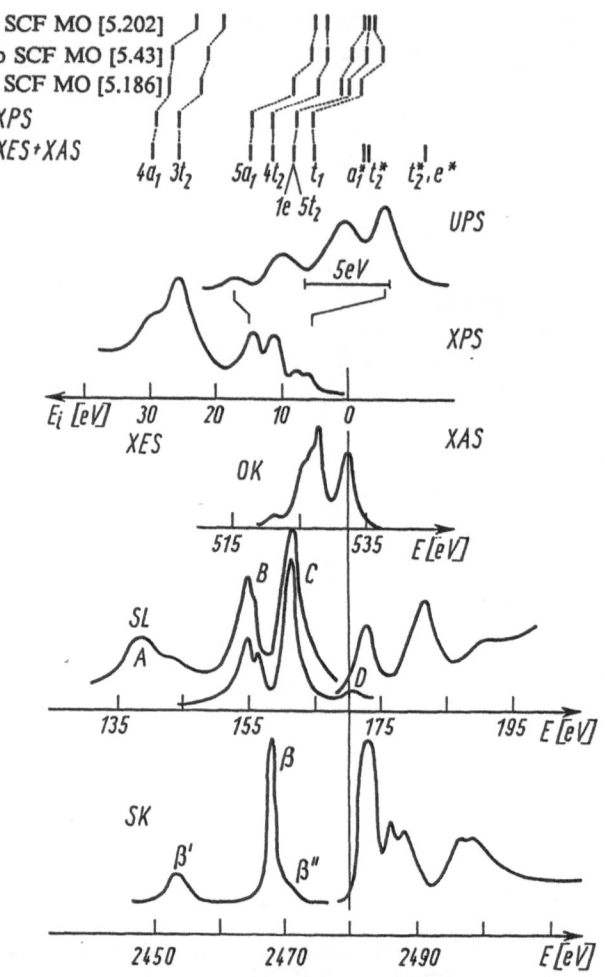

Fig.5.31. X-ray emission bands [5.47,97,163,205-207] and absorption edges [5.163, 209] together with photoelectron spectra (XPS [5.43,210,212] and UPS [5.44]) of sulfates (L emission spectra according to several authors are also shown). The energy levels calculated for the sulfate ion are also given

sible to observe a splitting of only 0.4 eV even though the experimental resolution is 0.7 eV [5.28].

While the S Kβ and S L emission bands from sulfates can easily be put on a consistent energy scale (Fig.5.31) using the S Kα₁ photon energy (hν = 2309eV) [5.183,215], the relative position of the O Kα emission band, however, can be obtained only from the O 1s and S 2p energies. The scale of the binding energy thus determined gives not only an independent determination of the valence levels for all the X-ray spectra but also allows a comparison with the XPS curves in the region of the valence electrons.

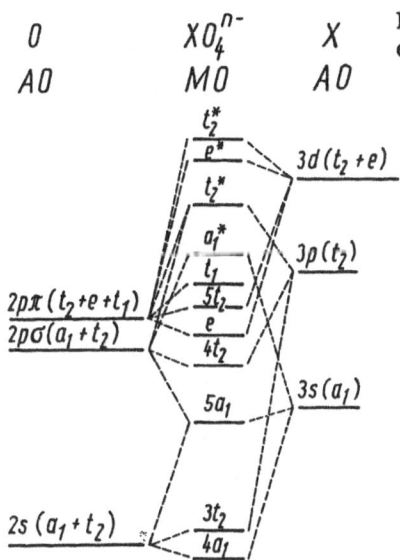

Fig.5.32. Energy level diagram of tetrahedral oxyanions of elements from the 3rd period

Due to charging effects in solid insulators, an absolute measurement of the binding energy requires a careful calibration of the energy scale [5.6] (the C 1s line from carbon impurity compounds is often used with $E_{1s}(C) = 285eV$ as a reference line). Consistent results can be obtained only when the XPS curves from core and valence levels can be recorded under the same conditions using the same reference line. The surface contamination is a particularly difficult problem since H_2O or OH^- groups easily influence the O 1s line to be measured. Particularly questionable is the determination of absolute binding energies of insulators as Na_2SO_4 from UPS curves.

Considering the difficulties discussed above, the spectra in Fig.5.31 were combined using a binding energy $E_{1s}(O) = 532.3$ eV [5.188] and $E_{2p}(S) = 168.9$ eV [5.216]. The UPS curve can, in general, give only relative positions of the levels. Using an energy level diagram based on group theoretical considerations, for the tetrahedral structural unit SO_4^{2-} (Fig. 5.32), one obtains the following results concerning interpretation of the spectra: The maxima of the S $K\beta$ band correspond to transitions to valence states $3t_2$ (O 2s, S 3p, S 3d), $4t_2$ (O 2p, S 3p), and $5t_2$ (O 2p, S 3d). The majority of S 3p electrons occupies the σ bonding $4t_2$ orbital in agreement with the observed high intensity of the associated $K\beta$ maximum, while the $3t_2$ level is mainly formed by O 2s AO. Transition into the $5t_2$ level, which occurs with much smaller transition probability since the S 3p component is absent from this MO and S 3d components do not contribute to the transition moment, is explained by two-center terms such as $\langle \phi_{2s}(O)|r|\phi_{1s}(S) \rangle$ [5.44]. The interpretation given above explains the changing distance between the long-wavelength satellite $K\beta'$ and the main component in the series SiC - Si_3N_4 - SiO_2 - SiF_6^{2-}: This distance increases with increasing binding energy E_{2s} from C to F [5.19, 217, 218].

Maxima in the L emission bands arise from transitions into the valence states $4a_1$ (O 2s, S 3s), $3t_2$, 5a (O 2s, O 2p, S 3s), 1e (O 2p, S 3d), and $5t_2$ according to dipole selection rules. The structure D in the region of the absorption maximum can be explained by a resonance transition (inverse of the absorption process, or re-emission) [5.163]. The spin-doublet character of the $L_{II,III}$ levels can be seen in the splitting of the maximum B. A considerable S 3d electron population related to the formation of a $3d\pi$-$2p\pi$ bonding between S and O can be inferred from the maximum C in the region of the orbitals 1e and $5t_2$ nonbonding in the absence of 3d participation.

The nonbonding MO, $1t_1$, composed almost entirely of O 2p AO, can be attributed to the main maximum in the O $K\alpha$ emission spectrum. It constitutes the highest occupied MO in the SO_4^{2-} ion as can be seen from the relative positions of the spectra shown in Fig.5.31. As in the case of SO_2, the O K maximum seen at about 533 eV due to anomalous scattering in the analyzing crystal should be discounted in an interpretation. The disturbing maximum disappears when the spectrum is recorded using a grating spectrometer [5.213].

The following energies are obtained from the K and L emission bands [5.47, 163, 206] for the valence levels in sulfates (the XPS values in parenthesis are the average values from [5.210, 211]):

$$E_{4a_1} = 30.4 \text{ eV} (29.9\text{eV}), \qquad E_{3t_2} = 24.2 \text{ eV} (25.8\text{eV}),$$
$$E_{5a_1} = 14.2 \text{ eV} (14.6\text{eV}), \qquad E_{4t_2} = 10.4 \text{ eV} (11.8\text{eV}),$$
$$E_{1e,5t_2} = 8.0 \text{ eV} (8.2\text{eV}), \qquad E_{t_1} = 5.3 \text{ eV} (6.2\text{eV}).$$

If the short-wavelength maximum of the L emission band is resolved into individual components, a value of about 1 eV is given for the energy difference between the levels 1e and $5t_2$ [5.47, 97]. The level energies are higher by 5-6 eV when absolute binding energies comparable to those of the free molecules [5.188] are given.

Differences between X-ray spectroscopic values and XPS values arise from several causes. For example, uncertainties in wavelength values and unresolved splitting of the $L_{II,III}$ levels both in the L emission spectra and in the S 2p electron spectrum can contribute to such differences. Moreover, one must include relaxation effects since the initial states for X-ray emission (ion with a K vacancy) and photoionization (ion in ground state) are different.

The experimentally determined sequence of valence levels agrees with theoretical results as is clear from the comparison given in the top part of Fig.5.31. Calculations of relative intensities for the components in the X-ray S $K\beta$ emission band using (5.6) with wave functions obtained from different approximations are improved (Fig.5.33) when taken from first-principle calculations. The very good agreement of the EHT curve in Fig.5.33 with the experimental band may be fortuitous since the same

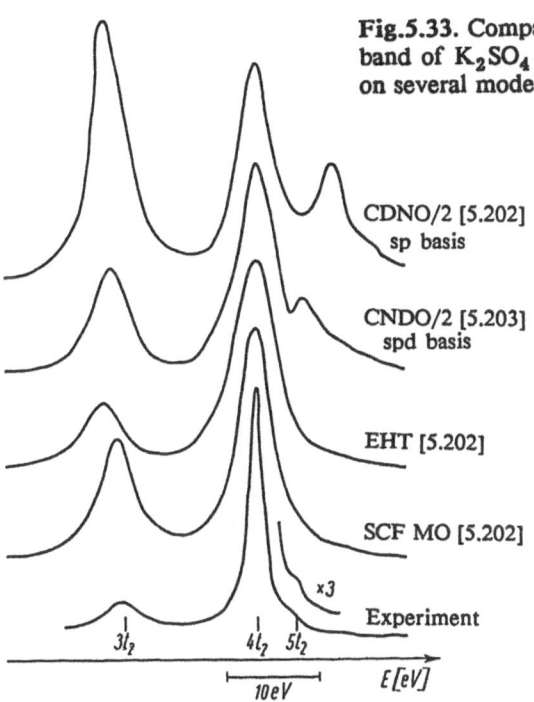

Fig.5.33. Comparison of experimental S Kβ emission band of K_2SO_4 [5.205] with theoretical profile based on several models

CDNO/2 [5.202]
sp basis

CNDO/2 [5.203]
spd basis

EHT [5.202]

SCF MO [5.202]

Experiment

approach, for example, for the ClO_3^- ion, gave a much less satisfactory theoretical spectrum (Fig.5.34). Agreement between experimental and theoretical spectra can be considerably improved both for the K bands and the L emission bands if a basis is used to account for the important 3d population (Fig.5.35).

For Kβ emission band calculations from (5.5), neglecting energy-dependent terms, theoretical relative intensities are directly estimated by the squares of the eigenvectors. The eigenvectors can thus be folded with Lorentz functions (taken to have the same width for all components) to give a theoretical spectrum although there is a possibility of different broadening of the components amounting to several tenths eV, as a result of splitting of the t_2 levels due to symmetry reduction arising from molecular vibrations [5.200]. Satisfactory agreement has also been obtained between experimental intensities for the O K emission band and the O 2p components of the MO for a series of oxyanions [5.213]. This method does not give satisfactory results for ultrasoft X-ray emission bands. Obviously, the L emission bands contain components with rather different widths occurring in pairs according to the well-known spin-doublet splitting of the 2p levels evident from mathematical decomposition of the emission band into paired Lorentz curves [5.97].

The ratio of integrated intensities (areas) obtained in this way agrees well with theoretical intensities obtained using (5.6) from ab-initio calculations (Fig.5.36). A relation between the peaks and integrated intensities

205

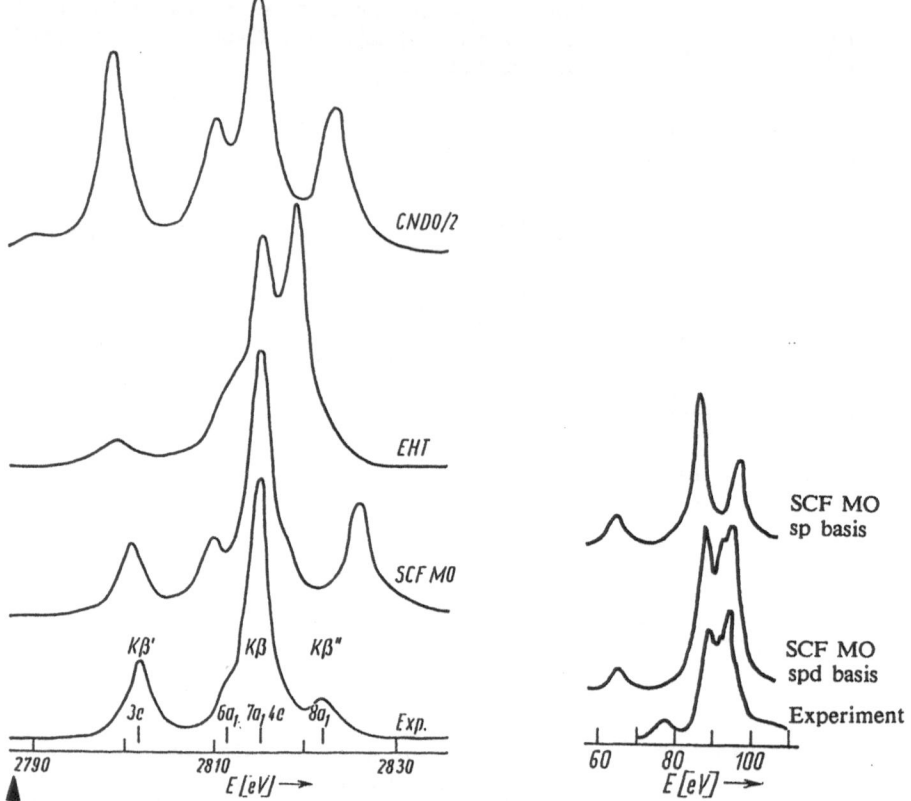

Fig.5.34. Comparison of experimental Cl $K\beta$ emission band of $KClO_3$ with theoretical profiles [5.202]

Fig.5.35. Comparison of experimental Si $L_{II,III}$ emission band of silicate with theoretical profiles [5.198]

of the detailed band structures exists only in specific regions of the spectrum (compare the S L emission band in Fig.5.31 with the experimental integrated intensities in Fig.5.36). DV $X\alpha$ calculations for XO_4^{n-}, XO_3^{n-} and XO_2^{n-} (X = P, S, Cl) evidenced that cross-transition terms have to be taken into account to improve agreement between theoretical and experimental L emission bands [5.39].

From investigations of the X-ray and photoelectron spectra of the oxyanions XO_4^{n-} and XO_3^{n-} (X = Si, P, Cl) the following general conclusion can be obtained regarding electron structure [5.188], which is also valid for octahedral fluorine complexes such as XF_6^{n-} (Sect.5.3.3). With increasing atomic number and thereby decreasing negative ionic charge, energies of the valence levels increase. In the series ClO_4^- - SO_4^{2-} - PO_4^{3-} - SiO_4^{4-} the distance between the valence levels $4a_1$ and $3t_2$ decreases, as also occurs for $5a_1$, $4t_2$, e, $5t_2$ and t_1 (Fig.5.37), since on the one hand, an increasing ionic character of the bond in this series of the

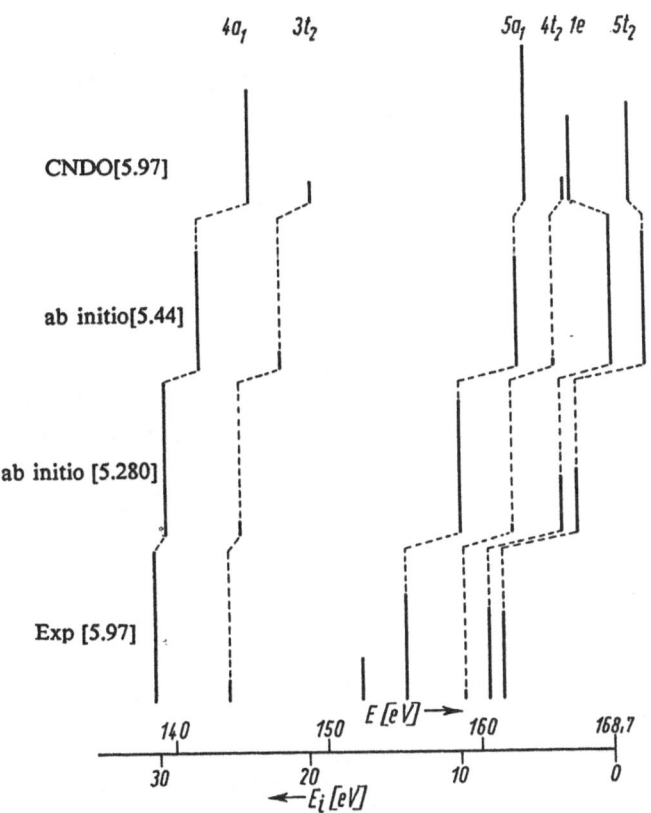

Fig.5.36. Comparison of experimental integral intensities of the components of the S L emission band of K_2SO_4 with several theoretical intensities

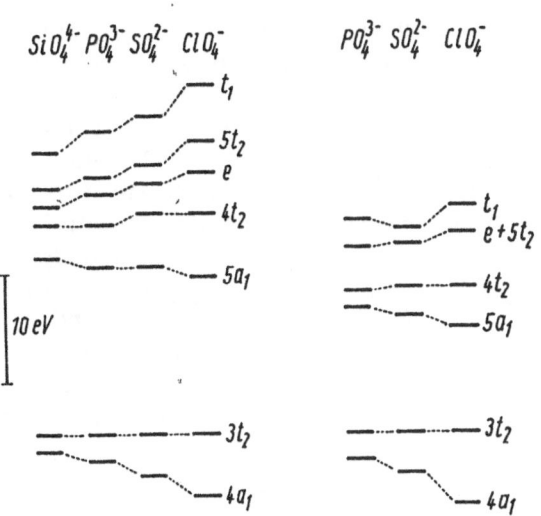

Fig.5.37. Theoretical (left) and experimental (right) energy level diagrams for isoelectronic compounds XO_4^{n-} [5.188, 199]

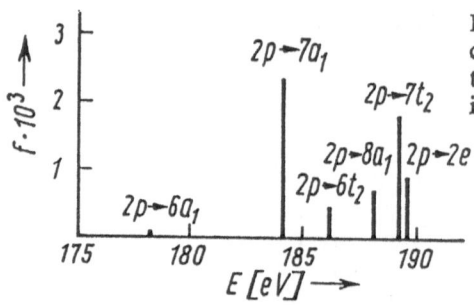

Fig.5.38. Diagram based on calculated oscillator strengths f for $L_{II,III}$ absorption transitions in the $SO_4{}^{2-}$ ion (theoretical ionization potential $E_{2p} = 175 eV$) [5.197]

limiting state of the O^{2-} ion with only valence levels (2s and 2p) is approached; on the other hand, with increasing atomic number, the increasing difference between the 3s and 3p energies of the central atom causes the σ levels $5a_1$ (X 3s) and $4t_2$ (X 3p) to separate farther and farther. As a consequence of the strongly increasing 3s binding energies from silicon to chlorine, the levels $5a_1$ (X 3s) and $3t_2$ (O 2s) approach each other.

For explanation of the K and L absorption spectra of sulfur in sulfates calculations for the free $SO_4{}^{2-}$ ion are available. Results for transition energies and relative intensities (Fig.5.38) from a CI calculation including singly excited configurations and considering, at least in part, orbital relaxation during the transition can be brought into agreement with experimental data only with difficulty. While the theoretical ionization energy E_{2p} of 175 eV for the free ion, taking into account the work function of the crystal (5-6eV), corresponds nicely to the experimental value (see above), the relative energies for the eleven transitions into the virtual orbitals $6a_1$, $7a_1$, $6t_2$, $8a_1$, $7t_2$, and 2e differ appreciably from experimental values (the transition is described as a single-electron transition from the degenerate inner orbital $2t_2$ into one of the individual virtual orbitals by approximate reduction of the CI to a single configuration). If the $2p$-$6a_1$ components which have rather small oscillator strengths are neglected (the corresponding first A_1 state can be described by a diffuse S 4s orbital with a very large mean radius, lying outside the potential barrier (Sect.2.6.2)), then the observed absorption maxima at 181.6 and 191.5 eV can be ascribed to the transitions $2p \rightarrow 7a_1$ (the corresponding second A_1 state can be described by an antibonding MO with mainly S 3s and O 2s character) and $2p \rightarrow 7t_2$,e (the corresponding t_2 and e states are represented by relatively diffuse MO with a large mixture of S 3d states) (compare Figs.5.31 and 38) [5.197]. As already shown for the anions $XF_6{}^{n-}$ [5.125], the absorption structures of the anions $XO_4{}^{n-}$, especially in the region of the ionization limit, can be treated using multiple-scattering theory, as done for shape resonances in molecules.

5.5.2 Complex Compounds of 3d Elements

The existence of unfilled 3d subshells in the transition metals and the participation of the 4p AO (unoccupied in free atoms) in chemical bonding, led to certain peculiarities of the electronic structure of complex compounds which must be considered in analyzing their X-ray and photoelectron spectra.

As discussed already in Sect.2.2, several ionized states arise upon ionization of systems with open shells through coupling of spin and orbital moments. This multiplet splitting causes asymmetry in the $K\alpha$ doublet of the transition metal compounds and splitting in the $K\beta$ doublet. Since the splitting of the different ionic terms, which can also be observed in the 3s and 3p photoelectron spectra, can amount to several electron volts, a multiplet structure of the X-ray emission band is to be expected.

Another peculiarity arises from the different relaxation energies for np valence levels of the ligand atoms and of the 3d levels of the transition metal atoms. The differences amount to some electron volts [5.5,20] so that the ordering of energy levels obtained by applying Koopmans' theorem to SCF MO calculations of the ground state does not agree with that given by experimental data and that following from ΔSCF MO calculations [5.219-227].

A third peculiarity follows from the small contribution of the metal 4p AO to the MO in the complex [5.228]. The profile of the $K\beta_5$ emission band of the central atom in the complex shows that the energy levels in the complexes as given by the MO of the ligands are only slightly influenced by interaction with the 4p states and thus they coincide approximately with the energy levels of the free ligand. From this result arises the possibility of estimating theoretically $K\beta_5$ intensities using accurate ab-initio calculations for the ligand molecule alone while avoiding a mathematical analysis of the full complex [5.45].

Some experimental peculiarities related to the recording of the X-ray spectra in the region of the valence electrons arise due to the relatively high atomic number of the transition metal atoms in the complex. Due to the small 4p $\rightarrow$ 1s transition probability, the $K\beta_5$ emission band present on the short-wavelength tail of the strong $K\beta_{1,3}$ lines has a very small intensity. Furthermore, due to the low fluorescence yield for excitation of the $L\alpha$ emission band (3d $\rightarrow$ 2p) of the transition metals in the wavelength region between 10 and 30 Å, recording of this band has also been difficult. Thus experimental results for the $K\beta_5$ and L emission bands using fluorescence excitation began only to appear in the late sixties [5.112, 229-236].

Information about the electronic structure of 3d complexes obtained from X-ray and photoelectron spectra, based on the peculiarities discussed above, will be exemplified in the following paragraphs.

Information on the multiplet splitting expected to arise upon ionization of valence shells in systems with unpaired electrons can be obtained from

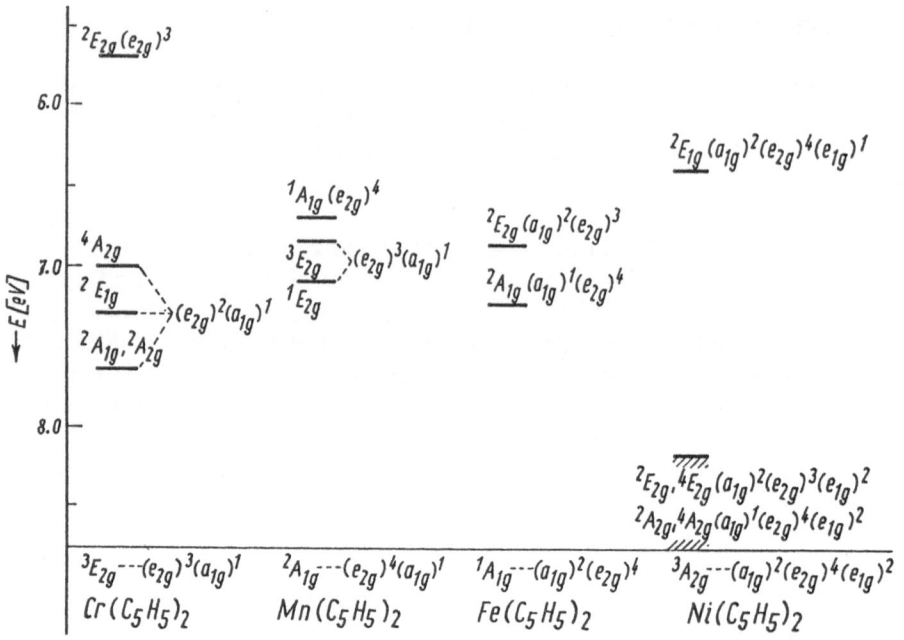

Fig. 5.39. Energy level diagram for molecules of certain metallo-organic compounds after ionization of the a_{1g} and e_{2g} MO [5.238]

photoelectron spectra of paramagnetic coordination compounds. In the HeI (584 Å) produced photoelectron spectrum of paramagnetic chromocene with an electron configuration $e_{2g}^3 a_{1g}^1$ in the ground state $^3E_{2g}$, individual components can be observed which arise from transitions into the ionic states $^2A_{1g}$, $^2A_{2g}$, $^2E_{1g}$ and $^4A_{2g}$ of the configuration $e_{2g}^2 a_{1g}^1$ (Figs. 5.39, 40) [5.237, 238]. The separation between the lowest and highest terms a-mounts to only about 0.6 eV which makes it almost impossible to observe the corresponding components in the X-ray spectra. Estimates of the full width of the valence levels of ferrocene can still be obtained from an ana-lysis of the Fe $K\beta_5$ (4p → 1s), Fe $L\alpha$ (3d → $2p_{3/2}$), and the C $K\alpha$ (2p → 1s) emission bands if the individual components, derivable from the fine structure of the spectra, are presented on a common energy scale, using XPS energies of the core levels (Fig. 5.40). Following dipole selection rules the maximum of the $L\alpha$ band is identified with the a_{1g} and e_{2g} levels which, in agreement with theoretical results [5.222, 239], occur in the re-gion of the high-energy components of the UPS curve. Individual compo-nents of the C $K\alpha$ and Fe $K\beta_5$ bands give the ionization energies of the σ and π orbitals of the pentadienyl ring. Several long-wavelength structures in the Fe $L\alpha$ emission band show that there is a transfer of C 2p electron density into the d orbitals of the central atom and that not only the top e_g state but also more strongly bound states of symmetries e_{1g} and e_{2g} contain 3d components [5.232]. The observed coincidence of components of the C

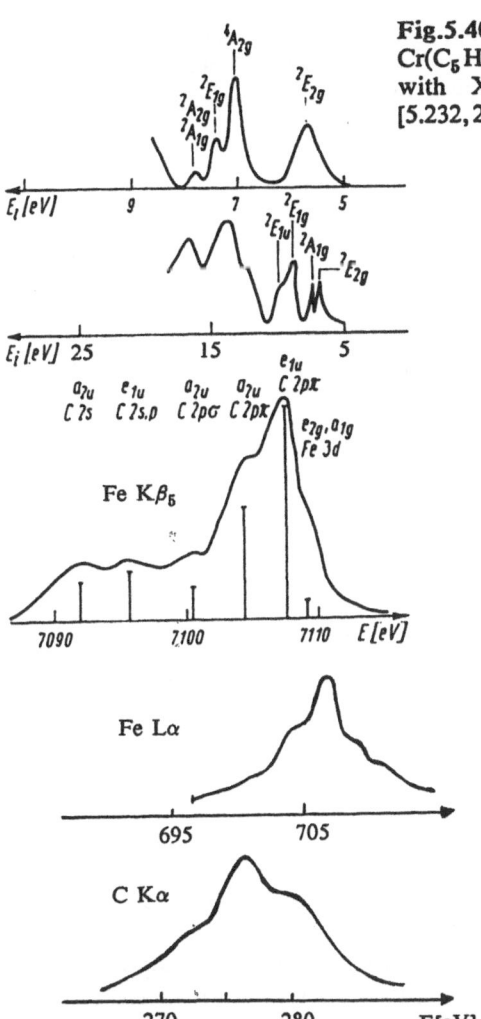

Fig.5.40. Interpretation of UPS curves of $Cr(C_5H_5)_2$ (top) and $Fe(C_5H_5)_2$ (below) [5.238] with X-ray emission bands from ferrocene [5.232, 233]

$K\alpha$ band with the maximum of the $L\alpha$ band points towards a delocalization of the 3d electrons of the metal and a shift towards the ligand.

Ionization of the d^5 system $[Fe(CN)_6]^{3-}$ corresponding to the $t_{2g}^5 \rightarrow t_{2g}^4$ transition leads to the ionic terms $^3T_{1g}$, $^1A_{1g}$, ^1E_g, and $^1T_{2g}$ whose splitting can be obtained from a calculation of the t_{2g}^2 system [5.240]; in this way one obtains a maximum energy difference of 3.5 eV [5.210]. The separation of the corresponding components in the XPS curves of $Li_3[Fe(CN)_6]$ [5.212] and $K_3[Fe(CN)_6]$ [5.210] is not visible both because of inadequate resolution and overlap of different valence electron regions. Even so, the example shows that the different terms in a configuration can be sufficiently separated to appear as resolved components in the spectra. The common method applied to systems with closed shells, i.e., to associate each component in the spectrum (disregarding satellites) with a

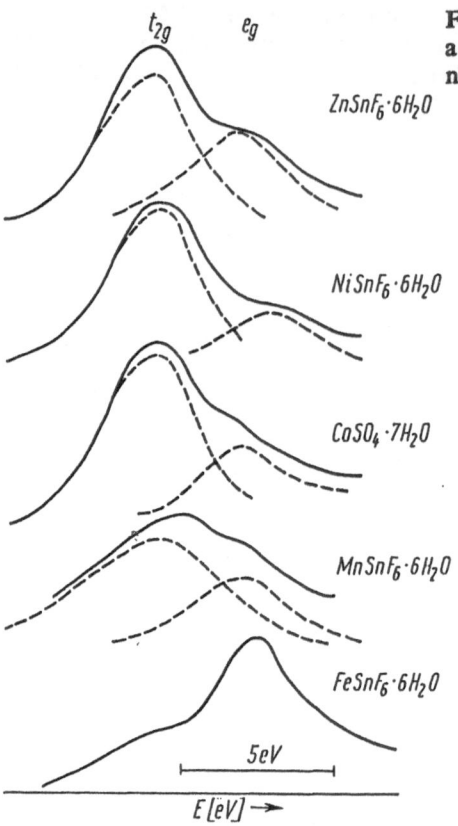

Fig.5.41. L emission bands of the cations in a series of complexes having six-fold coordinations [5.232]

single electron level, consequently cannot always be applied to paramagnetic compounds. Due to the smaller lifetime width of the $L_{II,III}$ levels compared to the width of the K level, multiplet effects should be seen more clearly in the L emission bands. In an effort to study these effects, the $L\ell,\eta$ lines (3s → 2p) of a series of hexaquo ions of 3d metals were recorded which, however, showed no change in the line shapes (disregarding a small broadening of about 0.5eV) upon transitions from diamagnetic to paramagnetic compounds. From these unexpected results – contrary to both theoretical predictions and the strong multiplet splitting of the 3s levels found experimentally [5.241] – it was concluded that the structure of the Lα emission band (3d → $2p_{3/2}$) is not produced by multiplet effects, i.e., by the splitting of terms of a configuration, but rather by MO and ligand effects [5.232]. The similar shapes of the Lα emission bands of the diamagnetic (M=Zn) and paramagnetic (M=Ni, Co, Mn) compounds $MSnF_6 \cdot H_2O$ (Fig.5.41) confirm these conclusions as far as conclusions are possible within the framework of the limited resolution achieved. The relative intensity of the t_{2g} and e_g components agrees with the occupancy (Mn: $t_{2g}^3 e_g^2$; Co: $t_{2g}^5 e_g^2$; Ni: $t_{2g}^6 e_g^2$; Zn: $t_{2g}^6 e_g^4$) caused by the weak crystal field in the hexaquo ions. The reversal of the inten-

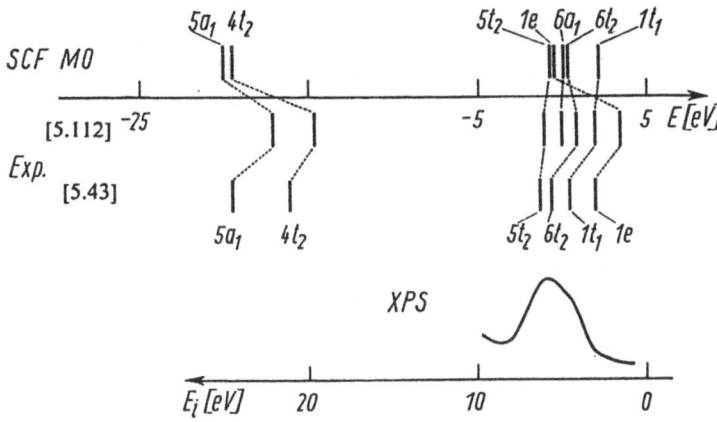

Fig.5.42. Theoretical (SCF MO) [5.43] and experimental energy levels together with the XPS spectrum in the valence electron region [5.43,112] for the anion CrO_4^{2-} and for chromates, respectively

sity ratio in the spectra of iron compounds can be explained by a change of the order of the t_{2g} and e_g levels.

The limited applicability of Koopmans' theorem for determination of ionization potentials for transition metal compounds using SCF MO calculations of the ground state is evident by a comparison of the orbital energies determined by X-ray and photoelectron spectroscopies for a series of compounds. Figure 5.42 shows a comparison of the ionization energies obtained from the components of the O Kα, Co Kβ_5 and Cr Lα emission bands [5.15,112,229,242-244] using the binding energies $E_{1s}(O)$ = 528.8 eV and $E_{2p3/2}(Cr)$ = 578.4 eV for K_2CrO_4 [5.112] and $E_{1s}(O)$ = 529.6 eV and $E_{2p3/2}(Cr)$ = 578.7 eV for Na_2CrO_4 [5.43] with the orbital energies obtained from theoretical calculations. While different calculations of the ground state suggest that the highest occupied MO of oxygen can be identified as the $1t_1$ orbital [5.43,201, 245,246], it follows from the relative position of the O Kα and Lα emission bands that the $1e$ orbital must have the lowest binding energy. Similar results have been obtained for the isoelectronic anions VO_4^{3-} and MnO_4^- [5.247]. Although relative intensities of individual components of the Kβ_5 and Lα emission bands of the metals have been satisfactorily reproduced by various calculations, the component with the shortest wavelength is seen in the Lα bands of all components. Thus, contrary to the theoretical results quoted above, a level with mostly 3d metal character, i.e., an e state, is associated with the first ionization potential. This identification depends on the assumption that disturbances from short-wavelength satellites in the Lα spectrum can be excluded, an assumption which must still be tested.

Analysis of the Kβ_5 emission band in a series of complex compounds with ligands CO, CN$^-$, NO$_2^-$, SCN$^-$, H$_2$O, and C$_6$H$_6$ (Fig.5.43 and below) confirms that the positions of the components in the emission band are de-

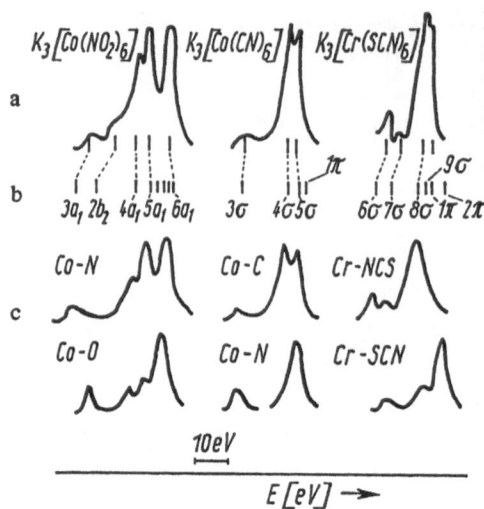

Fig.5.43a-c. Interpretation of the experimental Kβ emission bands (a) of Co and Cr in complex compounds using the energy level diagram of the free ligands (b); theoretical bands for various coordinations (c) [5.45]

termined mainly by the energy levels of the free ligands as predicted by the model developed first for Co and Cr complexes [5.45]. The assumption of only weak participation of (unoccupied) 4p orbitals of the metal in binding is convincingly supported by comparison of XPS curves and energy levels of the free ligands with the XPS and UPS curves together with the X-ray emission spectra from the complexes. The relative intensities of K and L emission bands of the metal atoms in the complexes follow from the number of 4d and 3d AO in the MO of the complexes (5.5); in contrast the XPS and UPS curves arise from a more complicated combination of the MO structure and cross-sections. Thus information obtained from X-ray spectra pertains to the local character of the structure, and thus nicely complements results from UPS and XPS curves where higher resolutions are available. The information obtainable by means of X-ray spectroscopy is rather clearly illustrated by the following example.

Figure 5.43 compares the experimental Kβ bands of cobalt and chromium complexes with theoretical curves in which the relative position of the calculated components are disposed according to the energy levels of the free ligands and the relative intensities set by the square of the amplitude of each MO function of the ligand in the region of the 1s function of the metal atom in the complex. This region consists predominantly of the extended tail of the wave function localized mainly on the ligands, to which only the AO's of the nearest neighbors contribute appreciably. Consequently, the shape of the calculated $K\beta_5$ band depends heavily on the assumed geometrical structure and hence the comparison in Fig.5.43 of the theoretical curves obtained for different coordinations with experimental results makes it possible to distinguish between different possible structural geometries. The coordination influences not only the shape of the $K\beta_5$ band of the central atom in the complex but also the shape of the band from the ligand atom (Fig.5.4) [5.22, 248, 249].

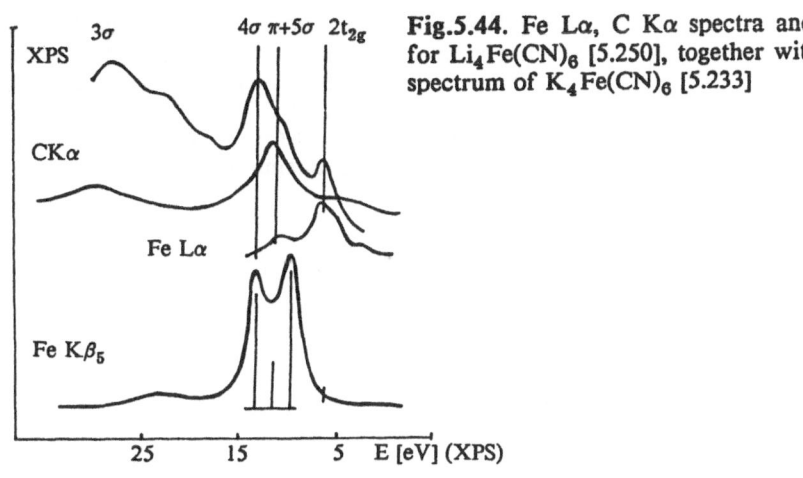

Fig.5.44. Fe Lα, C Kα spectra and XPS curves for $Li_4Fe(CN)_6$ [5.250], together with the Fe $K\beta_5$ spectrum of $K_4Fe(CN)_6$ [5.233]

A comprehensive picture of the electronic structure of the $Fe(CN)_6^{n-}$ (n=3,4) anions can be obtained from investigating the Fe $K\beta_5$ and Fe Lα emission bands, the C K emission band, the XPS curves of the valence electron region and from theoretical calculations (Fig.5.44) [5.137,233, 250]. The molecular orbitals of the octahedral anion are formed by interaction of the σ (a_{1g}, e_g, t_{1u}) and π (t_{1u}, t_{2g}, t_{1g}, t_{2u}) states of the ligand CN^- with 3d, 4s, and 4p orbitals of the central atom. The highest occupied $2t_{2g}$ level is dominated by Fe 3d orbitals. The relative order of the orbitals 3σ, 4σ, 1π, and 5σ of the CN^- ligand which is isoelectronic to N_2 and CO (Sect..3.1) is hardly influenced by formation of the complex and is evident in the fine structure of the Fe $K\beta_5$ emission band. The individual components determine the relative positions of the corresponding t_{1u} levels and give ionization potentials of 13.5 eV (4σ), 11.8 eV (1π), and 10.0 eV (5σ) based on the C 1s binding energy of 290 eV [5.233]. Analysis of the components in the Fe Lα spectrum then gives the value of the $2t_{2g}$ level to 6.6 eV [5.250]. Several long-wavelength structures in the Lα spectrum give evidence of the $2e_g$, $3e_g$ and $1t_{2g}$ states. Whilst the relative positions of the t_{1u} levels agree well with results from Xα SW calculations, a good result for the Lα emission band is obtained only for the relative intensities [5.137].

An analogous picture is obtained for the electronic structure of $Cr(CO)_6$. X-ray and photoelectron spectra of $Cr(CO)_6$ are juxtaposed in Fig.5.45. If one compares the XPS curves of CO with the XPS and UPS curves of chromium hexacarbonyl, it is clear that the levels between 13 and 16 eV are formed by combining 1π and 5σ MO of the free ligand and that the highest level $2t_{2g}$ at 8.4 eV is formed by Cr 3d electrons (3d part $\simeq$ 75% [5.178]). The levels $4e_g$, $6t_{1u}$ and $7a_{1g}$ correspond to 4σ of CO and the deeper lying levels are formed by 3σ MO of the ligand dominated by O 2s AO. The shift of the photoelectron spectra of the complexes compared to CO can be explained by a charge transfer towards the ligand

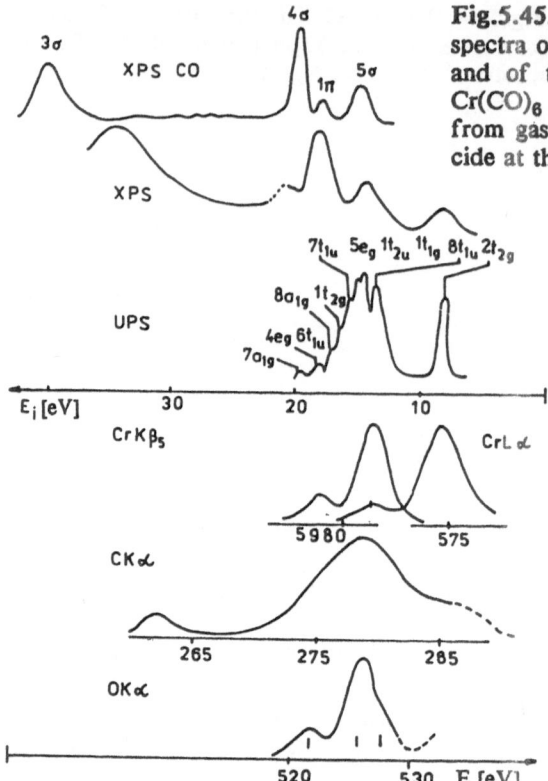

Fig.5.45. Interpretation of the photoelectron spectra of CO [5.7] and solid $Cr(CO)_6$ [5.226] and of the X-ray emission bands of solid $Cr(CO)_6$ [5.136, 233]. XPS and UPS curves from gaseous $Cr(CO)_6$ were drawn to coincide at the $2t_{2g}$ maximum

because of the dominating π acceptor character of CO, which is evident also from similar shifts of C 1s (2.7 eV) and O 1s (3.0 eV) [5.219]. If the $K\alpha_1$ energy $h\nu = 5414.7$ eV [5.183] (a small additional long-wavelength shift of the $K\alpha_1$ wavelength in the complex should not exceed 0.2 eV compared to the given value for the element) and $E_{2p3/2} = 581.8$ eV [5.219] are used to locate the Cr $K\beta_5$ emission band on the binding energy scale, then three components can be identified from the spectral analysis, i.e. $8t_{1u}$, $7t_{1u}$ and $6t_{1u}$ at the levels 4σ, 1π and 5σ of the ligands. The separation of the components $6t_{1u}$ and $8t_{1u}$, as well as $7t_{1u}$ and $8t_{1u}$ in $K\beta_5$ (4.4 and 1.7 eV, respectively) agrees well with the values obtained from the UPS (4.4 and 1.7 eV) and XPS curves (3.9 eV) and also with results from SCF Xα SW calculations (3.8 and 1.4 eV, respectively) [5.137].

The maxima of the Lα emission band give the positions of the $5e_g$ and $2t_{2g}$ levels, allowing a 3d electron population to be seen from the σ states by the electron transfer from ligand to metal following a σ donor mechanism. The components of the O Kα and the C Kα emission bands characterize the valence states formed by 4σ, 1π, and 5σ orbitals of the free CO group. Due to the limited resolution in this wavelength region it is not possible to separate individual levels [5.136]. Although satellites can occur in the short-wavelength region of the C $K\alpha$ band, structures in this region of the $2t_{2g}$ level can also arise from a metal-ligand π acceptor interaction which is related to a charge transfer of Cr 3d electrons into free π orbitals of the ligand. The 3d and 4p charges of the Cr sphere determined from SCF $X\alpha$ SW calculations satisfactorily account for relative intensities of the components in the Cr $K\beta_5$ and $L\alpha$ emission bands [5.137].

If the Cr $K\beta_5$ emission band of $Cr(CO)_6$ is compared with the Mn $K\beta_5$ band of the compound $Mn(CO)_5Cl$, a pronounced maximum is noted on the short-wavelength side of the 5σ maximum in the region of the ionization potential of the 3d electron of the metal, which indicates π and σ interaction of the 3p electron of chlorine with 4p and 3d electrons of manganese. In binuclear complexes, such as $Mn_2(CO)_{10}$, a similar shoulder gives evidence of the interaction between Mn 4p and 3d electrons of the two neighboring Mn atoms [5.251].

Figure 5.46 shows the photoelectron spectra of benzene and dibenzene chromium, chosen as an example of metallo-organic compounds with aromatic π ligands, the C K emission band of ligand and complex as well as the K and L emission bands of the metal. As for the carbonyl compounds already discussed, the main structures in the spectra from the free ligand in the complex remain largely the same except for a broadening and decrease in intensity for $Cr(C_6H_6)_2$ due to splitting of the levels attendant on altered symmetry and aggregation state, since the system of the MO levels in benzene remains largely unchanged in the coordination compound.

The main components of the $K\beta_5$ emission band (referred to the binding energy zero point via the binding energy $E_{1s}(Cr) = 5994.0$ eV [5.220]) involve transitions from MO's which represent the interaction between 4p AO of the metal and the π system of the ligands. The long-wavelength components, on the other hand, agree with the position of the corresponding component of the free ligand and arise from the σ levels of benzene. The shortest-wavelength component can be explained as quadrupole transition (3d $\rightarrow$ 1s). The level scheme obtained from the relative position of the $K\beta_5$ and $L\alpha$ emission bands (calculated using $h\nu_{K\alpha_1} = 5414.7$eV) agrees well with theoretical predictions [5.225]. A decrease in intensity of the C $K\alpha$ emission band of the complex in comparison with free ligands (compare dashed and solid Curve 3 in Fig.5.46) in the region of the C $2p\pi$ levels and the occurence of short-wavelength components, which is evident from a tail on the band in the region of Cr 3d (a_{1g}, e_{2g}) levels, confirm experimentally the combined donor-acceptor character of the π ligand. Through interaction of the $e_{1g}(\pi)$ orbital of benzene with 3d AO of the metal a charge transfer occurs from the ligand to the metal (evident from the decrease of intensity of the C K band in the region of the e_{1g} level and from a shoulder in the L emission band of the metal). However, in the energy region of a_{1g} and e_{2g} levels a charge transfer occurs from the metal to the ligand caused by mixing the 3d AO of the

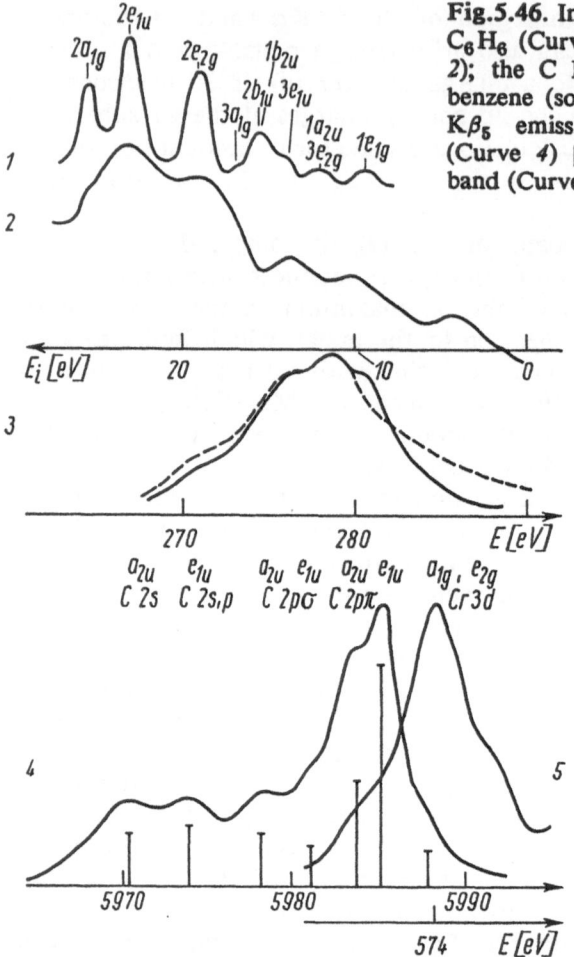

Fig.5.46. Interpretation of the XPS curves of C_6H_6 (Curve *1*) [5.1] and $Cr(C_6H_6)_2$ (Curve *2*); the C K emission bands (Curves *3*) of benzene (solid) and $Cr(C_6H_6)_2$ (dashed); Cr $K\beta_5$ emission band from $Cr(C_6H_6)_2BF_4$ (Curve *4*) [5.233]; and of the Lα emission band (Curve *5*) from $Cr(C_6H_6)_2$ [5.191]

metal with the π MO of the ligand, so that emission of the short-wavelength component of the C K band in the complex is possible [5.191].

From the similar positions and relative intensities of the components of the $K\beta$ emission bands of Cr and Fe in dibenzene chromium and ferrocene, respectively (compare Figs.5.46 and 40), it follows that the amounts of metal 4p AO in σ and π MO are almost equal in five- and six-membered rings.

Figure 5.47 shows the Cr $K\beta_5$ emission band of the compound $Cr(C_6H_6)(CO)_3$ compared with the Cr $K\beta_5$ band of dibenzene chromium and chromium hexacarbonyl. The correlation between the components of the $K\beta_5$ band, the UPS curve and the ionization potential (Table 5.8) confirms the congruence between the levels of the free ligand and those of the complex molecule. Disregarding small changes discussed below, contributions from the individual ligand levels add linearly. It thus follows

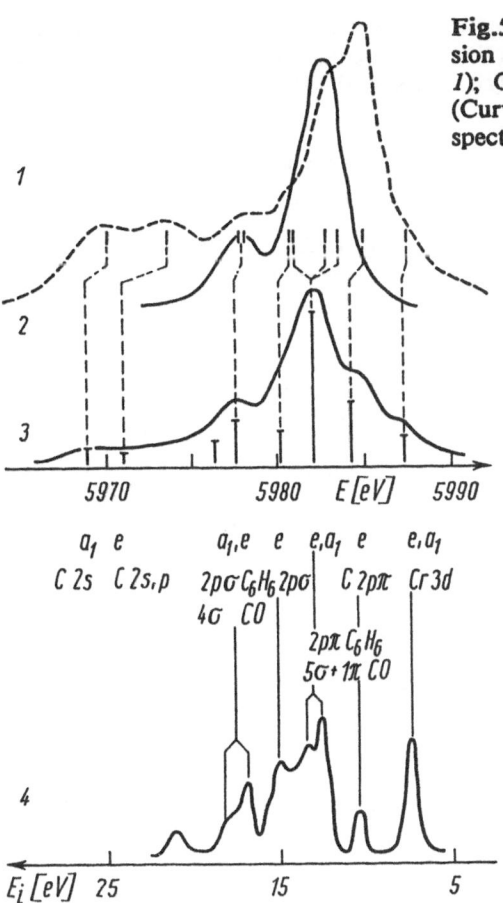

Fig.5.47. Interpretation of the Cr $K\beta_5$ emission bands [5.233] of $Cr(C_6H_6)_2BF_4$ (Curve *1*); $Cr(CO)_6$ (Curve *2*); and $Cr(C_6H_6)(CO)_3$ (Curve *3*); and the UV-excited photoelectron spectrum (Curve *4*) [5.225] of $Cr(C_6H_6)(CO)_3$

that bonding in coordination compounds of the type discussed with uniform ligands is generally similar to that in compounds with different ligands [5.45]. As in the cases discussed earlier, the shortest-wavelength component is related to the d levels of the metal. Due to the absence of a center of symmetry in the molecule, the corresponding e level obtains a relatively high admixture of 4p states of the metal, so that the relative intensity of this component is enhanced by 5-10 times compared to the neighboring C $2p\pi$ component.

Different coordination in the complex compounds $Mn(H_2O)_4Cl_2$ and $[Cr(H_2O)_4Cl_2]Cl\cdot 2H_2O$ is evident from the different shape of the $K\beta_5$ band of each central atom (Fig.5.48). The components of the Mn $K\beta_5$ band agree very well in position with the energy levels of the H_2O ligand if the main maximum is associated with transitions from the bonding a_1 MO of the water molecule (Sect.3.2). The origin of the second component (2) remains unclear (the satellite of the a_1 XPS maximum occurring at this position cannot be present in the X-ray spectrum) as does that of the sixth component. The presence of chlorine in the coordination sphere causes, as

Table 5.8. Energies of the components of the Cr $K\beta_5$ band and ionization potentials calculated using $E_{1s}(Cr) = 5995.1$ eV (XES), compared to UPS and XPS data for $Cr(C_6H_6)(CO)_3$ and C_6H_6, respectively (all values in eV) [5.233]

Photon energy	Level	Level energy		
		XES	UPS [5.225]	XPS [5.1]
5987.8	$17e+17a_1(Cr\ 3d)$	7.3	7.4	
5984.4	$16e(C\ 2p\pi)$	10.7	10.7	$9.2(1e_g)$
5982.0	$14e+16a_1+13e$	13.1	$12.7(5\sigma)$	
	$(C\ 2p\pi+CO\ 5\sigma+1\pi)$		$13.3(2p\pi+1\pi)$	$12.2(1a_{2u})$
5980.0	$11e(C\ 2p\sigma)$	15.1	15.05	$13.8(3e_{1u})$
5977.9	$10e+11a_1$	17.2	$16.6(4\sigma)$	
	$(C\ 2p\sigma+CO\ 4\sigma)$		$17.6(2p\sigma)$	$16.5(3a_{1g})$
5971.2	$8e(C\ 2s,p)$	23.9		$22.9(2e_{1u})$
5968.5	$10a_1(C\ 2s)$	26.6		$25.9(2a_{1g})$

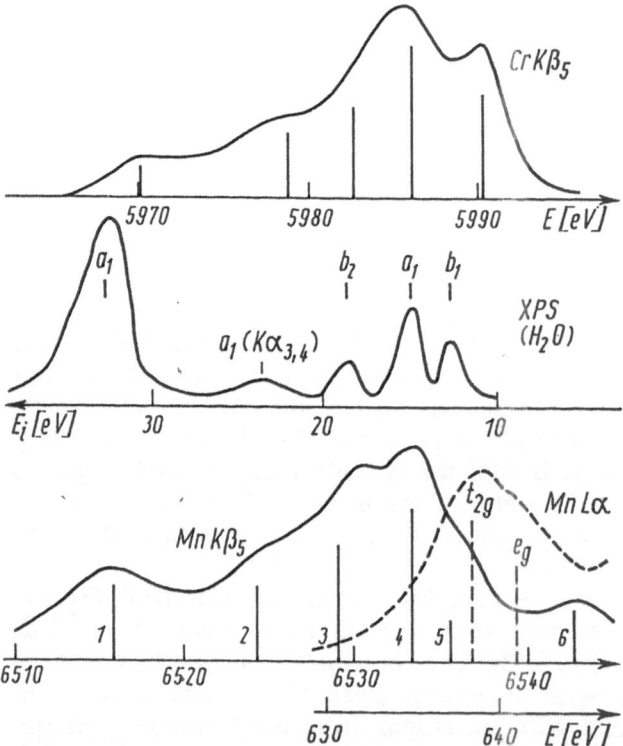

Fig.5.48. Interpretation of the $K\beta_5$ and $L\alpha$ emission bands from the cation in the compounds $Mn(H_2O)_4Cl_2$ and $[Cr(H_2O)_4Cl_2]\ Cl\cdot2H_2O$ [5.232, 233] and the XPS curve of H_2O [5.7]

shown in Fig.5.48, an additional component on the short-wavelength side of the maximum of the Cr $K\beta_5$ band which arises from transitions from MO with mostly Cl 3p character.

Electronic structure information on complex systems can also be obtained from X-ray absorption spectra. The pre-edge $1s \rightarrow nd$ peak of the K absorption spectra of transition metals with incompletely filled 3d shells and other features in the near-edge region are found to be intimately related to the local structure and the geometry for a great number of complexes, too [5.252-262].

5.5.3 Complex Compounds of the 4d and 5d Elements

X-ray emission bands and absorption spectra of elements with atomic numbers Z > 40 in complex compounds have little been studied up to now, since it is difficult to draw conclusions from the spectra, due to the increasing widths of the initial K and L levels. On the other hand, recordings of M spectra of 4d elements and N spectra of 5d elements in the ultrasoft X-ray region using fluorescence excitation and the corresponding absorption spectra are difficult, due to intensity and preparation conditions as found in other cases as well. Therefore, the results obtained give only qualitative information on the position of the MO levels. Some statements can be made regarding chemical bond on the basis of emission bands from ligand atoms such as phosphorus, sulfur, and chlorine which exhibit clearly resolved structures due to the smaller width of the core level participating in the X-ray transition. The type of results obtained will be discussed with a few selected examples.

X-ray emission and absorption spectra of diamagnetic 4d complexes with the isoelectronic anions RhX_6^{2-} and PdX_6^{2-} (X = F, Cl) and the configuration $4d^6$ of the central atom are shown together with the anticipated electronic structure in Fig.5.49 [5.263]. The positions of the MO levels were determined from a decomposition of the emission band into individual components. The relative intensities of these components are shown in the spectra as vertical lines. The splitting of the bonding and antibonding t_{2g} levels is observable through splitting of the $L\beta_2$ emission band ($4d \rightarrow 2p_{3/2}$) where an increasing population of the bonding t_{2g} states in Pd complexes is also clear. The reason for this is the higher binding energy of the 4d states of the metal. A decrease of the $t_{2g} \rightarrow t_{2g}^{*}$ splitting upon exchange of the F to Cl ligands is caused by strong interaction between the Me 4d and np_π orbital of the ligand combined with the different ionization energies of the np electrons of the ligand. The energy difference between short-wavelength emission maximum and absorption maximum ($2p_{3/2} \rightarrow e_g^{*}$) should correlate, subject to the limitations given in Sect.2.6.2, with the Δ parameter given from optical spectra (Δ is sometimes referred to as the ligand-field strength parameter).

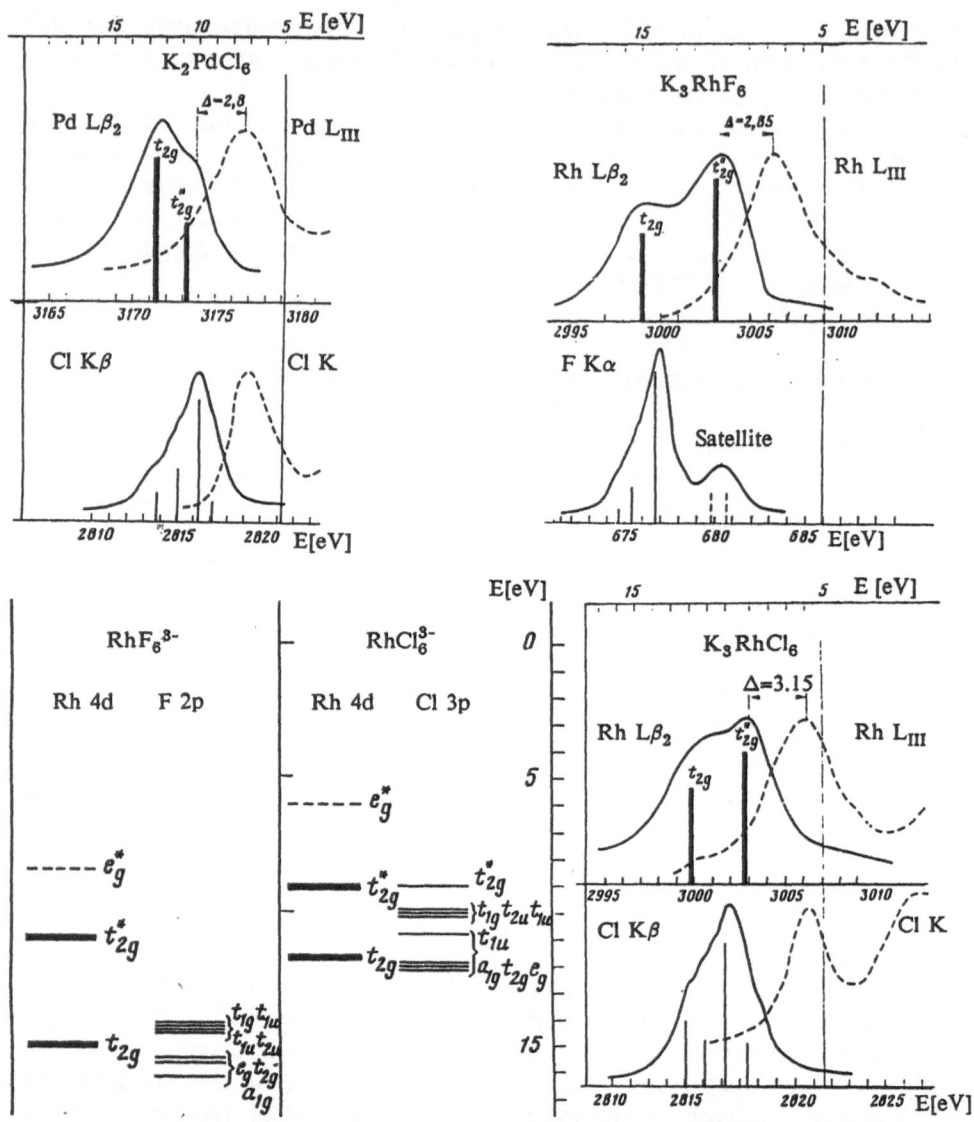

Fig.5.49. X-ray emission and absorption spectra of the compounds K_3RhF_6, K_3RhCl_6 and K_2PdCl_6 with energy level diagrams for the anions [5.263]

Analogous results have been obtained for the electronic structure of the complex anion RuX_6^{2-} with $4d^4$ configuration of the central atom [5.263]. In this case the L_{III} absorption curve shows two maxima (Fig.5.50) where the long-wavelength $(2p_{3/2} \rightarrow t_{2g}^*)$ component coincides with the short-wavelength maximum of the $L\beta_2$ emission band since the t_{2g}^* levels are only partly occupied in the ground state.

From a comparison of the Rh L and Cl K X-ray emission and absorption spectra for the anions $[Rh_2ac_4Cl_2]^{2-}$ and $RhCl_6^{2-}$, it follows that

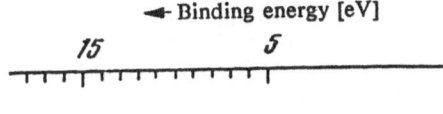

Fig.5.50. X-ray emission and absorption spectra of Cs_2RuCl_6 with XPS curve in the valence electron region (dotted) [5.263]

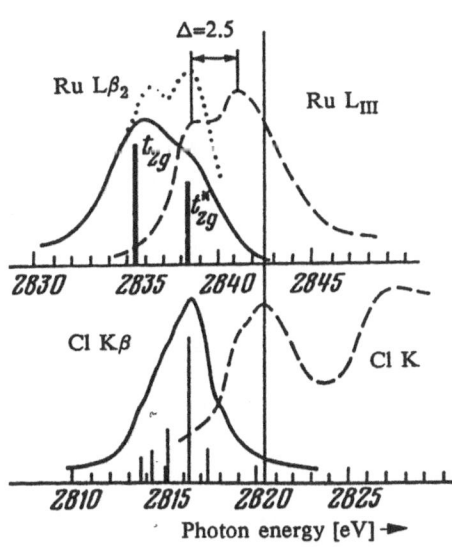

for the binuclear complex, the structure is Cl→Rh-Rh←Cl with a bond order larger than unity. Since the position of the e_g levels would change from the occupied to the unoccupied region of the orbitals if a structure such as Cl···Rh≡Rh···Cl is assumed, the intensity of Rh $L\beta_2$ must decrease in the short-wavelength region simultaneously with an increase of the intensity of the first Rh L_{III} absorption maximum. This behavior has, however, not been seen in experimental spectra [5.264].

Multiple metal-metal bonding is also seen for the anion $[Mo_2Cl_8]^{4-}$ from a superposition and component analysis of the Mo $L\beta_2$ and Cl $K\beta_{1,3}$ emission bands [5.265].

Figure 5.51 shows various experimental spectra and theoretical results on the MO structure of some $MeCl_4^{2-}$ complexes. Except for the MO formed mainly by Cl 3s orbitals, all the valence MO's (composed from Cl $3p\sigma$, $3p\pi$, and Me nd AO) occur in a relatively narrow energy region of about 5 eV so that unique decomposition of the emission band into individual components, particularly for the practically structureless Pd $L\beta_2$ emission band, is not possible. The relative position of the Cl $K\beta$ ($3p\sigma, \pi \to 1s$) and Pd $L\beta_2$ bands ($4d \to 2p$), as shown in the figure based on the binding energy scale of the photoelectron spectrum of K_2PtCl_4 according to the assumed level arrangement (see below) can be determined using a series of photon and binding energies [Pd $L\beta_2$, Pd $L\alpha_{1,2}$, Cl $K\beta_1$, Cl $K\alpha_1$, $E_{3d5/2}$(Pd), $E_{2p3/2}$(Cl)] [compare Fig. 2.3 and (5.3)]. Thus, as is shown in Fig.5.51, the maximum of the Pd $L\beta_2$ band lies in the region of smaller binding energy [5.21]. Consequently, the Pd 4d elec-

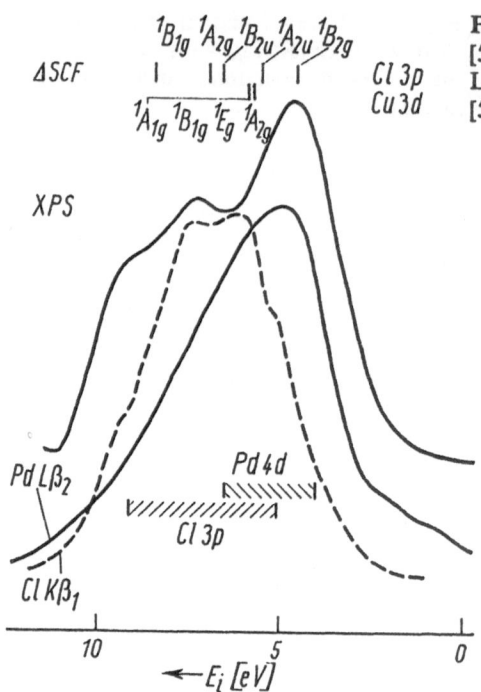

Fig.5.51 ΔSCF energy levels of $CuCl_4^{2-}$ [5.222], XPS curve of K_2PtCl_4 [5.266], Pd $L\beta_2$ [5.267] and Cl $K\beta_1$ emission bands [5.16] of K_2PdCl_4

trons in the complex should correspond to the lower ionization potential. Similar results were obtained from an analysis of the corresponding spectra of the compound K_3RhCl_6. From a decomposition of the emission band into its components and from theoretical results, the following qualitative picture emerges of the MO structure of the complex anions $MeCl_4^{n-}$ and $MeCl_6^{n-}$. In addition to the σ bonding MO states with about 75% Cl 3p character, a group of MO occurs toward lower binding energies which are dominated by the 3p AO of the ligand partly involving interaction with the nd AO of the metal. In the upper region there are only the d levels of the metal (d part exceeding 75%). The considerable Cl 3pπ and Me nd interaction in the halogen complexes results in a broadening of the Rh $L\beta_2$ band in K_3RhCl_6 compared to that from $Rh(NH_3)_6Cl_3$ [5.268].

The relative positions of the d levels of the metal and the 3pπ levels of the ligand are particularly interesting since different semiempirical [5.269-271] and ab-inito calculations [5.222] give contrary predictions. In contrast with the experimental situation shown in Fig.5.51, the ab-initio orbital energies of the 3d electrons are smaller (more negative) in the complex $CuCl_4^{2-}$ than those of the 3p electrons of the ligands so that as a result of Koopmans' theorem, the d electrons should have higher binding energies. As shown already in the previous section, omission of the relaxation energy when using Koopmans' theorem can lead to incorrect results. From ΔSCF calculations for the $CuCl_4^{2-}$ ion a comparatively high relax-

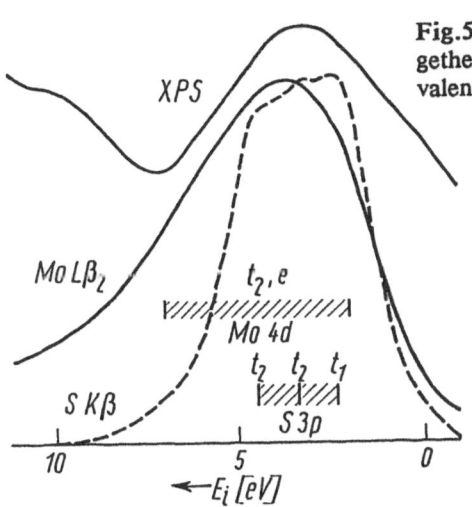

Fig.5.52. S Kβ and Mo Lβ_2 emission bands together with the photoelectron spectrum in the valence electron region of $(NH_4)_2MoS_4$ [5.272]

ation energy is obtained for the 3d orbitals while for the ligand orbitals Koopmans' theorem is applicable. In this way the 3d ionization potentials move in the upper region of the valence levels and become comparable with the ionization potentials for the 3p electrons of the ligands. Thus a picture of the MO levels emerges similar to that given by application of Koopmans' theorem to certain semiempirical calculations [5.269,270]. This agreement must, however, be accidental and the determination of the ionization potential of complex compounds using Koopmans' theorem is questionable.

As in the case just discussed, one also finds in the emission band of the ligand (S Kβ) in the tetrahedral complex $(NH_4)_2MoS_4$ that individual components are much more prominent than in the Lβ_2 band from the metal (Fig.5.52) [5.272]. If these emission bands are put on a consistent energy scale using the binding energies of $E_{2p}(S)$ and $E_{3d}(Mo)$ together with the energies of the L$\alpha_{1,2}(Mo)$ and K$\alpha_{1,2}(S)$ doublet in $(NH_4)_2MoS_4$, it can be seen from the relative position of the centers of gravity of the emission bands that the ionization energies are lower for the electrons localized mainly on the ligand than for the 4d electrons of the metal. From analysis of the intensity distribution, the shortest-wavelength component of the S Kβ spectrum is ascribed to the t_1 MO formed entirely by S 3p electrons (as is seen more clearly in WS_4^{2-} than in MoS_4^{2-}). If one assumes that no weaker components are hidden under the short-wavelength part of the spectrum, this maximum can be associated with the first ionization potential. Poor correspondence is found regarding the relative position of the MO levels among transition from tetrahedral MoS_4^{2-} ions in crystalline $(NH_4)_2MoS_4$ to the molecular gases RuO_4 and OsO_4. In the first case, accepting the limitations discussed above, the ionization potentials are obtained in the order $E_{t_1}(S\ 3p) < E_{t_2}(S\ 3p) < E_{t_2}, e(Me\ 4d, S\ 3p)$, while from the relative intensities and vibrational structures of He I

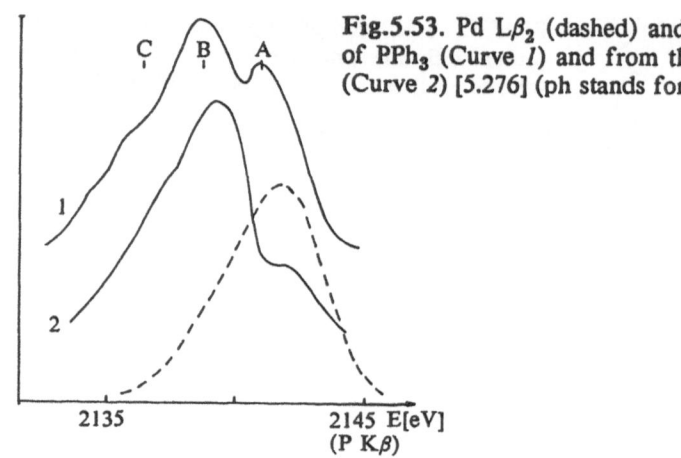

Fig.5.53. Pd $L\beta_2$ (dashed) and P $K\beta_{1,3}$ emission bands of PPh$_3$ (Curve 1) and from the complex $(PPh_3)_2PdCl_2$ (Curve 2) [5.276] (ph stands for phosphine)

photoelectron spectra and from the UV absorption spectra, the ordering $E_{t_2}(O\ 2p) < E_{t_1}(O\ 2p) < E_{t_{2,e}}(Me\ nd, O\ 2p)$ is obtained for the gases RuO$_4$ and OsO$_4$ [5.273]. Influence of the crystal field on the compound $(NH_4)_2MoS_4$ should be limited to a largely uniform stabilizing effect on all valence states in analogy with other cases [5.222], and no transposition in the ordering of the ionization potentials should be expected between isolated ions and molecules. If the results obtained are compared with studies of MeO_4^{n-} ions of the 3d metals (Sect.5.5.2) and theoretical estimates [5.273, 274], it is clear that a unified picture cannot as yet be obtained of the electronic structure of the tetrahedral oxy- and thio-anions of the transition metals.

Information regarding charge transfer in complex formation can be extracted from changes in shape of the emission bands of atoms in the ligand group in a series of compounds with Rh, Pd [5.16, 249, 275], Pt, and Au [5.250]. From the decrease in intensity of the short-wavelength maximum in the S $K\beta$ emission band of dimethyl sulfide and dimethyl sulfoxide when the molecule is coordinated to rhodium and palladium, delocalization of the outermost occupied orbitals, and charge transfer from ligand to metal can be inferred [5.275], in agreement with the results obtained for similar complex compounds of 3d metals. Analysis of the P $K\beta_{1,3}$ and Pd $L\beta_2$ X-ray emission bands indicates σ donor properties of the ligand in Pd complexes with triphenyl phosphine ligands (Fig.5.53). The short-wavelength maximum A of the P $K\beta_{1,3}$ emission band seems to arise from electron transitions from the lone-pair electrons on the P atom. If the P $K\beta$ and Pd $L\beta$ spectra from the complexes are compared on a common energy scale (Fig.5.53) using ESCA binding energies of the core levels, the maximum A and that of Pd $L\beta_2$ occur in the same energy region. The donor character of the ligand is obvious [5.276] from the decreasing relative intensity of the maximum A in complexes compared to free ligands, caused by transfer of electron density of lone-pair electrons from the P atom to the Pd ion.

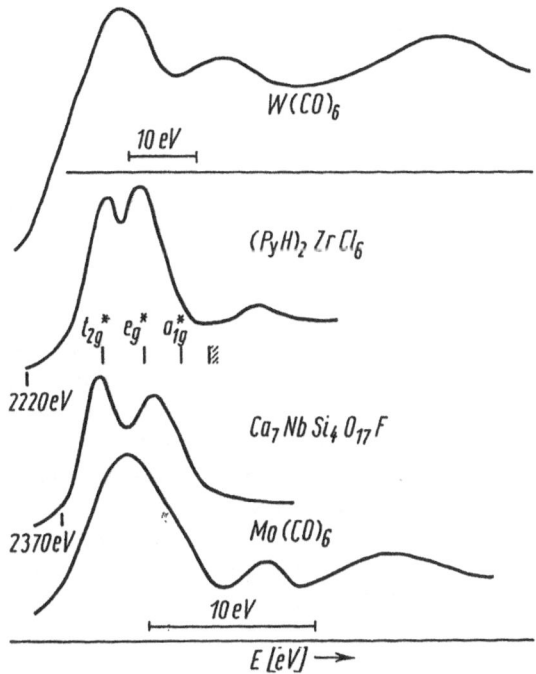

Fig.5.54. W, Zr, Nb, and Mo L_{III} absorption spectra from octahedral complexes [5.277,278]

The participation of the 3d AO of the ligand in the bond due to $d\pi$-$d\pi$ interaction, which would imply an electron transfer from the metal to the ligand, can be neglected, as shown by an analysis of the S L emission spectrum (no $3d \rightarrow 2p$ components are observed) [5.249]. Similar results are obtained from a study of Cl 3d influence on the bonding in $PtCl_4^{2-}$, $PtCl_6^{2-}$, and $AuCl_4^{2-}$. No structures can be identified as $3d \rightarrow 2p$ components in the L emission band and thus there is no population of Cl 3d [5.250].

The increasing widths of the core levels when going from atoms of the 3d series to those of the 4d and 5d series complicate the analysis of the experimental results also for the case of the absorption spectra. From Fig.5.54 it is obvious that the L_{III} absorption curves of octahedrally coordinated 4d and 5d metal atoms are dominated by an intense white line which is split in many compounds of Zr and Nb [5.277]. An unambiguous decomposition of the spectra of Mo and W carbonyls into the components $2p_{3/2} \rightarrow t_{2g}, e_g, a_{1g}$ is not possible. However, for other cases the energies of these excited states can be determined [5.277]. Information on the crystal field 10 Dq parameter and the d population at the metal site upon compound formation as a result of charge transfer has also been extracted [5.279].

6. X-Ray Spectroscopic Studies of the Band Structure of Solids

The theoretical basis for the determination of the density of states in solids and some experimental results are discussed. In Sect.6.1 different methods for the investigation of the band structure are compared. The correlation between the theoretical density of states and the experimental X-ray spectra is treated in Sect.6.2. Experimental results are described for most of the elements (Sect.6.3: main-group metals, d and f metals, alloys, nonmetallic elements) and for several compounds (Sect.6.4). It is tried to get a general classification for the electronic structure of compounds on the background of the total number of valence electrons (for example, $A^N B^{8-N}$, $A^N B_n^{16-N}$, and $A_m B_n$ compounds, transition metal compounds).

6.1 Experimental Methods for Studying the Band Structure

The electronic structure of a material is characterized by both the spatial distribution of the electrons (Sect.4.1) and by their energy distribution which for solids amounts to the band structure (Sect.2.3). Experimentally, it can be studied either by a determination of the energy surfaces E(k) of the valence and conduction bands and the Fermi surface through cyclotron resonance emission or the De Haas-Van Alphen effect [6.1], or by investigating the density of states N(E). Whilst measurements of the Knight shift, the specific heat and the magnetic susceptibility give information on the density of states in the vicinity of the Fermi edge only, studies of N(E) over a larger energy region are possible using photon spectroscopy in the visible, UV and X-ray regions, by studying electrons emitted by UV and X rays [6.2,3], and furthermore by using Auger [6.4] and ion neutralization spectroscopy [6.5]. The processes involved in these different kinds of electron spectroscopies are schematically presented in Fig.6.1. In this chapter we shall consider only a one-electron description and neglect additional effects from many-body interactions, excited states and experimental difficulties.

The optical properties of semiconductors and insulators in the UV-visible region are determined by electronic transitions between occupied and unoccupied bands, E_v and E_c (interband transitions). In metals, additional transitions are imaginable between occupied and unoccupied states of

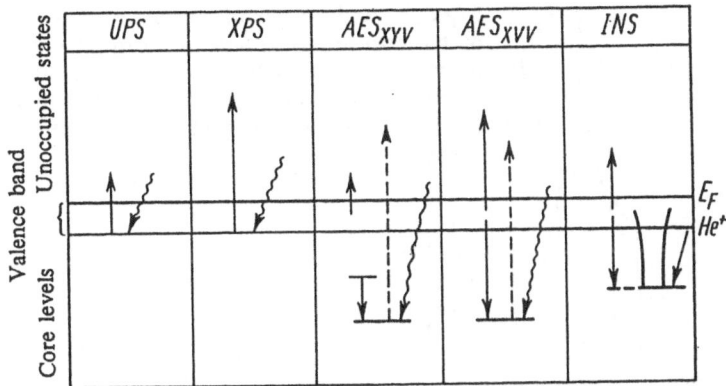

Fig.6.1. Scheme of the physical processes involved in several types of electron spectroscopy

a band (intraband transitions). The spectra are measured either in reflection $R = I_R/I_0$ (I_R is the intensity of the reflected light, I_0 the intensity of the incoming light) or in transmission $T = I_T/I_0 = (1-R)e^{-\mu x}$ (I_T is the intensity which is transmitted through a layer of thickness x); the latter method is applied for photon energies above 30 eV as here R < 2%. The reflectivity R_0 (R at normal incidence) and the absorption coefficient μ are related to the optical constants, the complex index of refraction $N = n+ik$ and the complex dielectric constant $\epsilon = \epsilon_1 + i\epsilon_2$, through

$$R_0 = \left| \frac{n-ik-1}{n-ik+1} \right|^2 \quad \text{and} \quad \mu = \frac{\omega \epsilon_2}{cn} \tag{6.1}$$

(ω is the angular frequency of the incoming light, c is the speed of light).

There is a direct relation between the imaginary part of the dielectric constant ϵ_2 and the electronic transitions. Neglecting intraband transitions and electron-photon interactions, in optically excited dipole transitions the k vector remains unchanged (direct transitions) and for ϵ_2 the following relation holds

$$\epsilon_2(\omega) = \frac{\hbar^2}{m^2 \omega^2 \pi} \int_S \frac{P(E,k)dS}{|\nabla_k(E_c - E_v)|} . \tag{6.2}$$

Here $P(E,k)$ is the transition probability and

$$N(E)_{vc} = \frac{V}{8\pi^3} \int_S \frac{dS}{|\nabla_k(E_c - E_v)|} \tag{6.3}$$

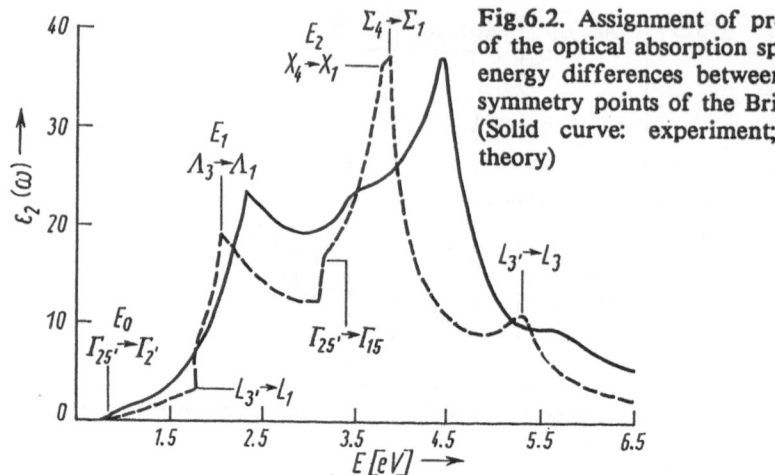

Fig.6.2. Assignment of prominent features of the optical absorption spectrum of Ge to energy differences between states at high-symmetry points of the Brillouin zone [6.6]. (Solid curve: experiment; dashed curve: theory)

is the joint density of states of the valence and conduction bands, as noted by E_v and E_c. The integration is carried out over the isoenergetic surface $S(E)$. For the spectra in the UV and visible regions, the individual densities of states of the valence or conduction bands are not the deciding factors, rather it is the density of states for constant energy difference between the bands. From (6.3) it is obvious that maxima occur in this joint density of states when the gradient of the energy difference goes to zero. This often occurs in the region of high-symmetry points of the Brillouin zone. It is thus possible to assign the structures in the $\epsilon_2(\omega)$ and also those in the experimental spectra to the energy differences of such points (Fig.6.2).

In contrast to spectroscopy in the UV-visible region, in X-ray spectroscopy inner electrons are included. These show very small k dependencies, i.e., the bands are flat and it is justified to assume sharp inner levels with electrons of a certain angular momentum. Selection rules for transitions from or in these flat bands correspond to those for free atoms. An important difference between UV-visible spectroscopy on the one hand, and X-ray spectroscopy on the other is that for the latter it is possible to determine the structure of both occupied and unoccupied bands from transitions between these bands and an inner level; i.e., $N(E)$ of both the valence and the conduction band can be individually studied as there is always only one final (emission spectrum) and initial (absorption spectrum) state, respectively, with known energy (Sect.6.2). The structures in the X-ray spectra can, to a first approximation, also be identified with the energy positions of bands in the high-symmetry points of the zone [6.7-9]. Subsequently, there should be a correspondence between structures obtained from the combination of X-ray emission and absorption spectra and the structures of the spectra in the UV and visible regions. This has been verified for $A^{III}B^{V}$ compounds [6.8] as well as for SiO_2 and MgO [6.7] (Fig.6.3).

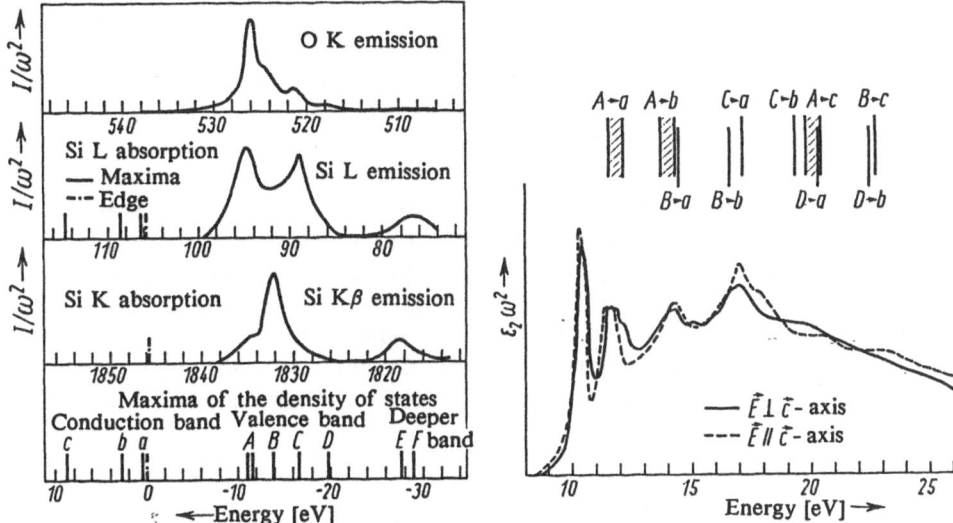

Fig.6.3. Comparison of X-ray spectra and optical absorption in SiO$_2$ [6.7]

Photoelectron spectroscopy with excitation by UV or X rays (UPS: UV photoelectron spectroscopy; XPS: X-ray photoelectron spectroscopy; ESCA: electron spectroscopy for chemical analysis) can also be used to get information about the band structure and density of states [6.2,3]. However, there are differences in the spectra obtained by UPS or XPS just as there are differences in the UV-visible and X-ray spectra. While UV photoelectron spectra are determined by both the occupied and unoccupied bands (i.e., through interband and intraband transitions), the X-ray photoelectron spectra are determined primarily only by the structure of the valence band.

Assuming only direct transitions to be allowed, the intensity in UPS is given by

$$I(E_{kin})_{UPS} \simeq \int_S \frac{P(E,k)' dS}{|\nabla_k (E_c - E_v)|} , \qquad (6.4)$$

where $P(E,k)'$ is the transition probability of the electrons between E_v and E_c for excitation with UV radiation. The measured kinetic energy of the electrons is given by

$$E_{kin} = h\nu - E_i - \Phi_{sp} \qquad (6.5)$$

($h\nu$: energy of the incoming light, E_i: binding energy of the electron, Φ_{sp}: electron work function of the spectrometer).

In contrast to photon spectroscopy, for $I(E_{kin})$ the escape probability and the inelastic scattering of the electrons play an important role as well

231

[6.10, 11]. Additionally, indirect electron transitions can contribute to the spectra [6.2]. A big advantage of UPS is that it is possible by varying the energy of the exciting monochromatic radiation to extract further information on the influence of the transition probability and the role of the initial and final states, thereby also providing some insight into the individual density of states of the valence and conduction bands.

When the energy of the exciting radiation is greater than about 20 eV, photoelectron spectra show only minor changes with excitation energy and resemble the X-ray photoelectron spectra, apart from the loss of resolution due to the spectral width of the frequently used Al or Mg Kα radiation. This is illustrated by Fig.6.4, where spectra of Au are presented. Obviously, the influence of the final state is rather small when the electrons are excited into states with sufficiently high energies. For XPS we have

$$I(E_{kin})_{XPS} \simeq \int_S \frac{P(E,k)''}{|\nabla_k E_v|} \, dS \; , \tag{6.6}$$

where $P(E,k)''$ is the transition probability of the electrons to be excited by X rays from the valence band into high-energy states.

The influence of the transition probability, which is obvious from the differences in the spectra of diamond, silicon, and germanium (Fig.6.42 below) [6.12] (these elements all have similar density-of-states structures but different widths of the valence bands), can be understood by considering the variation of the excitation probability for electrons with different symmetries. The excitation probabilities for the above examples of s and p valence electrons depend on the atomic number.

Consequently, the differences between X-ray photoelectron and X-ray emission spectroscopy can be explained using an LCAO model including different excitation and transition probabilities for electrons of different angular momentum. Thus, information on the "pure" symmetry components (for the elements from XES) and on the excitation probability from XPS

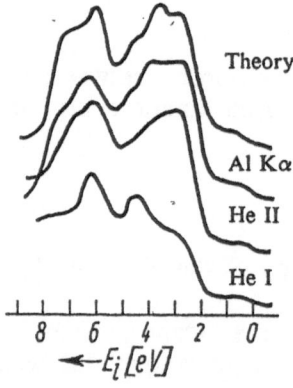

Fig.6.4. Theoretical spectrum of gold together with photoelectron spectra obtained using different excitation energies

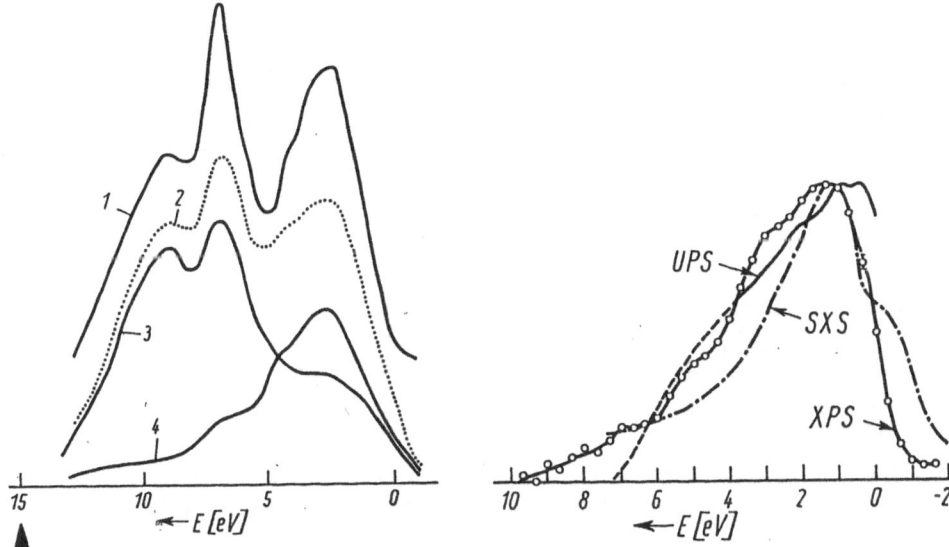

Fig.6.5. Comparison of the X-ray photoelectron spectrum of Si (Curve *1*) with the sum (Curve *2*) of L (Curve *3*) and K X-ray emission bands (Curve *4*) [6.13]

Fig.6.6. Comparison of the X-ray and UV excited photoelectron spectra with the Lα soft X-ray emission band (SXS) of Fe [6.14]

can be combined to estimate the degree of hybridization. It has been found, for example, that for Si the best combination of the s (L spectrum) and p parts (K spectrum) to restore the X-ray photoelectron spectrum is given by an area ratio of 2:1 (Fig.6.5) [6.13]. That means that with a theoretical ratio of the excitation probabilities $\sigma_s/\sigma_p = 6:1$ for XPS there is an s/p hybridization of 1:3, which agrees with the general description of the bonding in Si. Furthermore, (6.6) suggests that the X-ray photoelectron spectrum should be similar to the X-ray emission spectrum (2.30) when the difference of the influence of the transition probabilities is not too large. This has been verified for numerous cases (e.g., Fig.6.6).

Besides the role of the final states in XPS and UPS, differences can also arise from the transition probabilities [6.15] and from different escape depths of the electrons (or photons, if comparing with photon spectra), see Fig.6.7. The question then arises as to whether the photoelectron spectra are at all representative of the bulk solid. The answer can be found through a detailed comparison of the photoelectron spectra of electrons with kinetic energies around the minimum of the escape depth (50-100 eV). The influence of the escape depth must also be considered if results from the methods described above are compared with those from ion neutralization spectroscopy (Fig.6.8) which involves only the outermost surface layers.

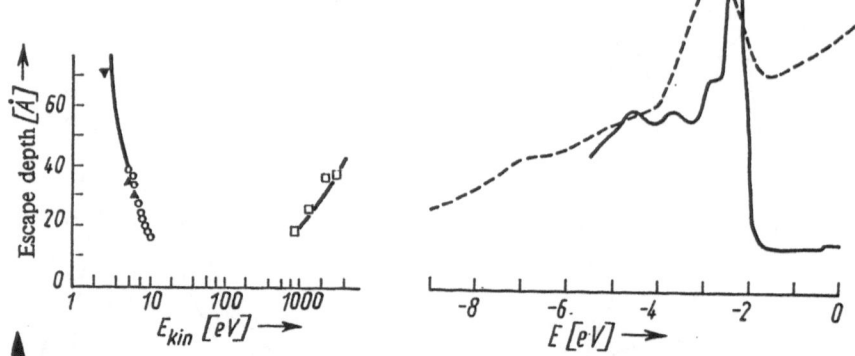

Fig.6.7. Dependence of the escape depth of photoelectrons on their kinetic energy E_{kin} for Au [6.16]

Fig.6.8. Comparison of the photoelectron spectrum (solid line) with the density of states obtained from ion-neutralization spectroscopy (dashed) for Cu [6.5]

6.2 Information Content of X-Ray Spectra

From the analysis in Sects.2.3,6, it would appear that X-ray emission bands and absorption spectra of solids arise from transitions from the occupied energy bands to inner vacancies and from inner energy levels to empty energy bands, respectively. Therefore it should be possible in principle to obtain information about the energy positions of the occupied and unoccupied states and their densities of states from the X-ray spectra. The basic advantages of X-ray spectroscopy (no important limitations as to elements or substances; small influence of impurities, surface states, and lattice defects on the spectra; coverage of a wide energy range; and ability to probe symmetries of the electronic states in the bands) must be opposed firstly to limitations caused by experimental distortions and secondly to difficulties introduced by different physical effects which restrict the ideal possibilities.

The specific distortions which are determined by instrumental effects are (Sect.3.6): finite spectral resolution of the instrument; the background produced by scattering of characteristic X-rays and bremsstrahlung; changes of the sample caused by the excitation procedure, the nature and the geometry of the sample, i.e., chemical and physical processes in the sample itself (decomposition, surface reactions, contamination processes) [6.17,18]; self-absorption in the sample which can result in a strong distortion of the spectrum [6.19] but can be mostly suppressed under favorable conditions [6.20]; the wavelength dependence of the reflectivity of the monochromator [6.21,22], the energy dependence of the quantum yield of the detector, and its nonlinearity with intensity [6.23].

234

Careful correction for all of these experimental factors requires great effort (Sect.3.6) and has been carried out more or less completely in only a few cases [6.20, 22, 23]. Even when careful corrections are applied, a comparison of the corrected spectrum with the results of theoretical calculations of the density of electronic states of solids is not directly possible. The reason for this lies within the one-electron model in that the distortion of the spectrum due to the width of the inner level is important and also that influences of transition probabilities have to be considered as in other spectroscopic methods (Sect.6.1). Besides, there are additional physical processes involved in the production of the excited state and the emission of radiation which go beyond the one-electron picture. (Whether a one-electron picture is applicable at all without including quasi-particle states has been discussed extensively in [6.24]).

The distortion of the spectrum caused by the inner level is readily understood since the width of each spectral line is the sum of the widths of the initial and final states, which are given again by the lifetimes of the states involved. To keep this influence small, transitions from and into narrow inner levels are needed such as those in the soft or ultrasoft X-ray region. When satisfactory values for the widths of these levels are available [6.25, 26], the effect can be corrected for through a deconvolution of the spectrum as was described for dealing with the window function of the spectrometer (Sect.3.6).

6.2.1 Interpretation of Emission Spectra in a Single-Particle Model

Complete calculation of a spectrum (2.30) including the transition probabilities has been done only for a few cases [6.27-37]. Following *Nikiforov* [6.30], the transition probability can be expressed as a product

$$P(E, k) \propto P_1(E) \cdot P_2(E, k) , \qquad (6.7)$$

where $P_1(E) = |\int_0^\infty R_{n\ell}(r) R_k(r, E) r \, dr|^2$ represents the radial factor ($R_{n\ell}$ and R_k is the radial part of the wave functions of the inner and valence electrons, respectively) and $P_2(E, k)$ the angular part. Presuming that the dipole selection rules are valid, the $P_2(E, k)$ factors are reduced for a particular atom in the solid to a calculation of the partial density of states of the valence electrons of corresponding angular momenta for the considered transition. If the energy dependence of the angular part is neglected, there remains a product of the partial density of states obtained by a series expansion of the wave functions for the valence electrons at a given mesh of k points and integration over the complete Brillouin zone and the energy-dependent radial factor. The influence of the radial factor is hardly noticeable for K spectra of heavier elements (Z>10) due to the small variation of $P_1(E)$ containing the sharply localized 1s function. Therefore it is possible to compare directly the partial densities of states

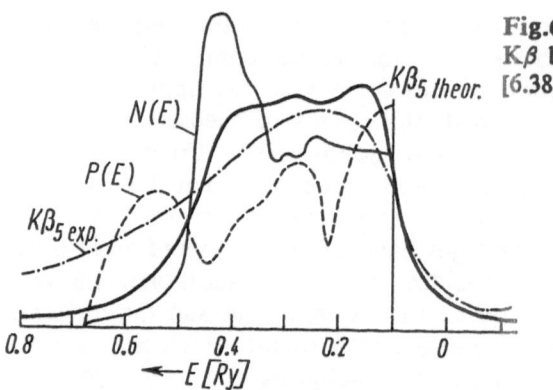

Fig.6.9. Comparison of the experimental Kβ band of iron with theoretical curves [6.38]

with the experimental spectra (Fig.6.9). Evidence of larger influence of this factor in transitions involving inner wave functions has been found in some cases [6.29, 33, 39, 40].

For alloys and compounds the question of the influence of the angular part of the wave function of the valence electrons on the calculation of $P_2(E, k)$ is especially important. As a result of the localized character of the inner function, this integral is different from zero only in a limited region. Thus, X-ray spectra are scanning only the part of the valence electron distribution in this region and therefore a local (partial) density of states near a certain component of the compound. This nearly corresponds to an LCAO expansion of the valence electron wave functions. In this way it is possible to compare locally and symmetry restricted partial

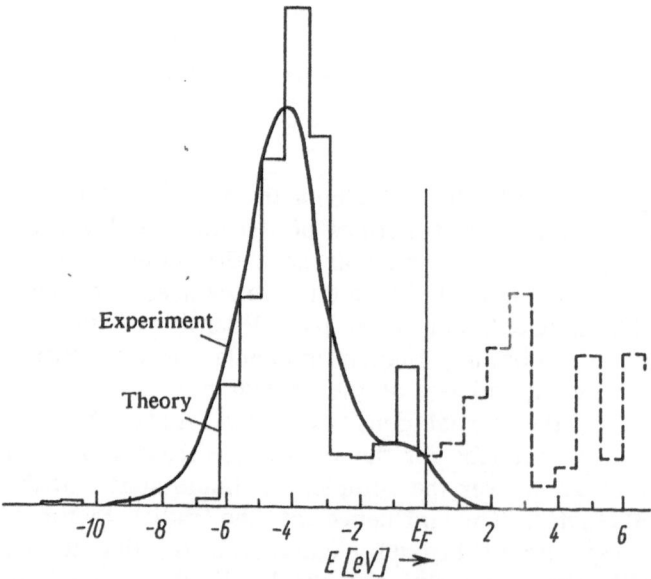

Fig.6.10. Comparison of the experimental C Kα band of NbC with the calculated partial density of states [6.41]

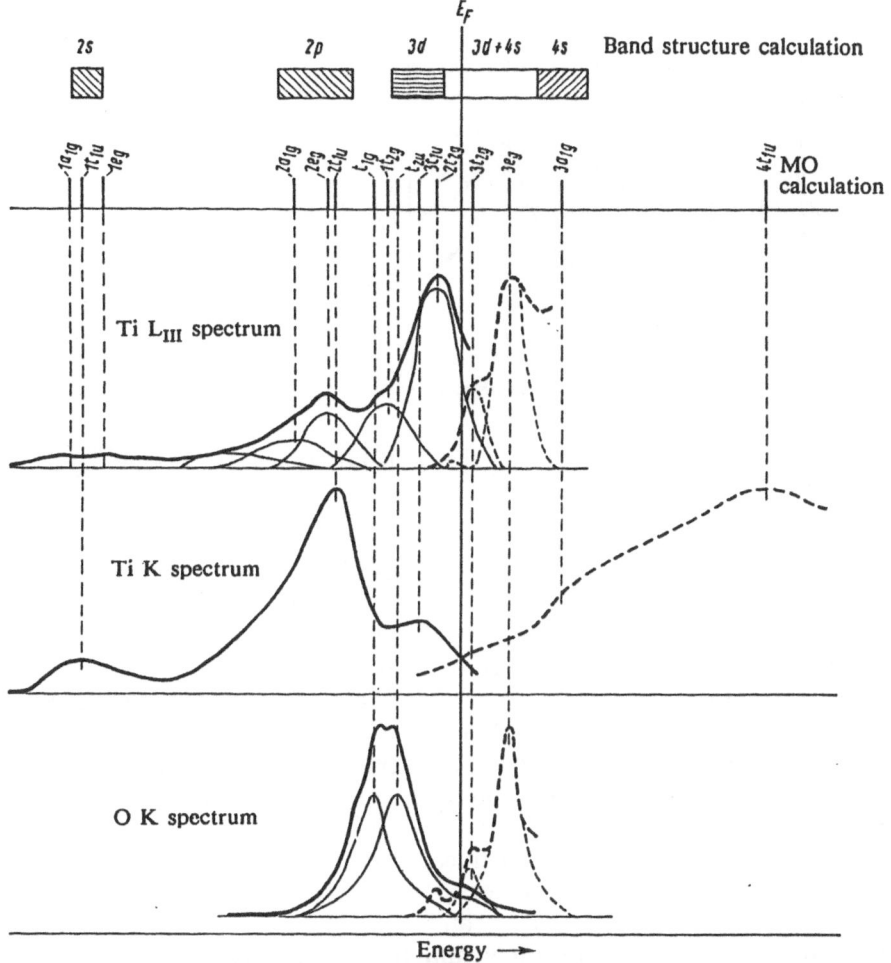

Fig.6.11. Comparison of the X-ray spectra of TiO (solid curve: emission; dashed curve: absorption) with theoretical results [6.42]

densities of states with X-ray spectra from compounds. Satisfactory agreement was thus found between theory and experiment for NbC and TiO (Figs.6.10, 11).

An exact inclusion of the transition probability requires an integration of $P(E,k)$ over the whole Brillouin zone, as calculating the density of states, i.e., a calculation of the matrix element $\langle \psi_{n\ell} | \nabla_k | \psi_k \rangle$ at a sufficiently large number of points in k space. Due to the influence of the transition probability $\dot{P}(k)$, there used to be a clear difference between the spectra of both components in a binary compound, as shown in Fig.6.12 for selected directions of k space for the Ga and As K emission bands in GaAs. Furthermore, the large differences for, e.g., X_3 and L_1 (the en- ergies in the $E(k)$ description differ only by 0.1 eV from each other) show that the approach of averaging the transition probability over k and therefore cal-

237

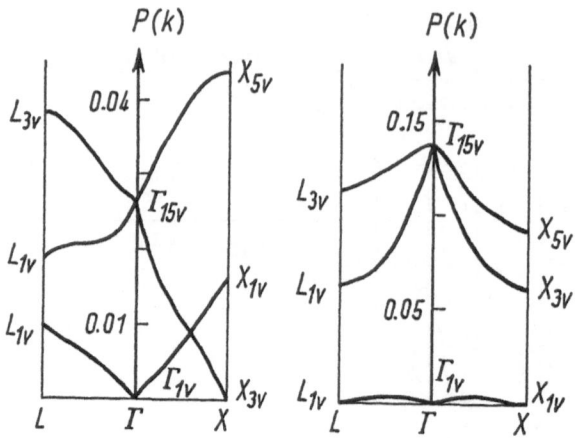

Fig.6.12. Calculated P(k) dependence for the Ga Kβ (left) and As Kβ emission bands (right) in GaAs [6.32]

culating the X-ray spectra from I(E) $\propto$ P(E)·N(E) is not always justified. Similar results have been obtained for the spectra of Al and P in AlSb and GaP [6.33]. A simple linear energy dependence of P(E), nearly corresponding to a partition of the total density of states into s and p parts, does not lead to satisfactory agreement with the experiment for compounds (Figs.6.13, 14). In particular, the difference between the components in binary A^{III}B^V compounds (difference between the A and B spectra) is not correctly described. Only a more complete calculation of P(E,k) produces agreement between the calculated and experimental spectra. The difference between the components, beyond a linear P(E) dependence, can be explained by the e^{ikr} factor in the valence electron wave function [6.33]. Thus, the different P(E) values (e.g., at X_1 and X_3) for the A and B components can then be ascribed to different symmetries of the valence electrons, a conclusion drawn already by *Skinner* [6.43] long ago.

Moreover, it follows from the calculations on binary compounds [6.33] that only wave functions orthogonalized to all inner states (thus not ones which merely correctly describe the valence band wave function) yield agreement between theory and experiment. This corresponds to the above-described conclusion that it is possible to obtain information on the localization of valence electron states from X-ray spectra. On the other hand, a small change of the inner wave function, e.g., application of the wave function for Al$^+$ instead of that for Al, has no major influence on the shape of the spectrum.

According to these statements, no information on the total density of states is to be obtained from X-ray spectra. On the other hand, the differences between X-ray spectra and densities of states due to the effects of transition probability can consist in both changes in the intensity contribution and simply suppression of structure. In most bands, even for

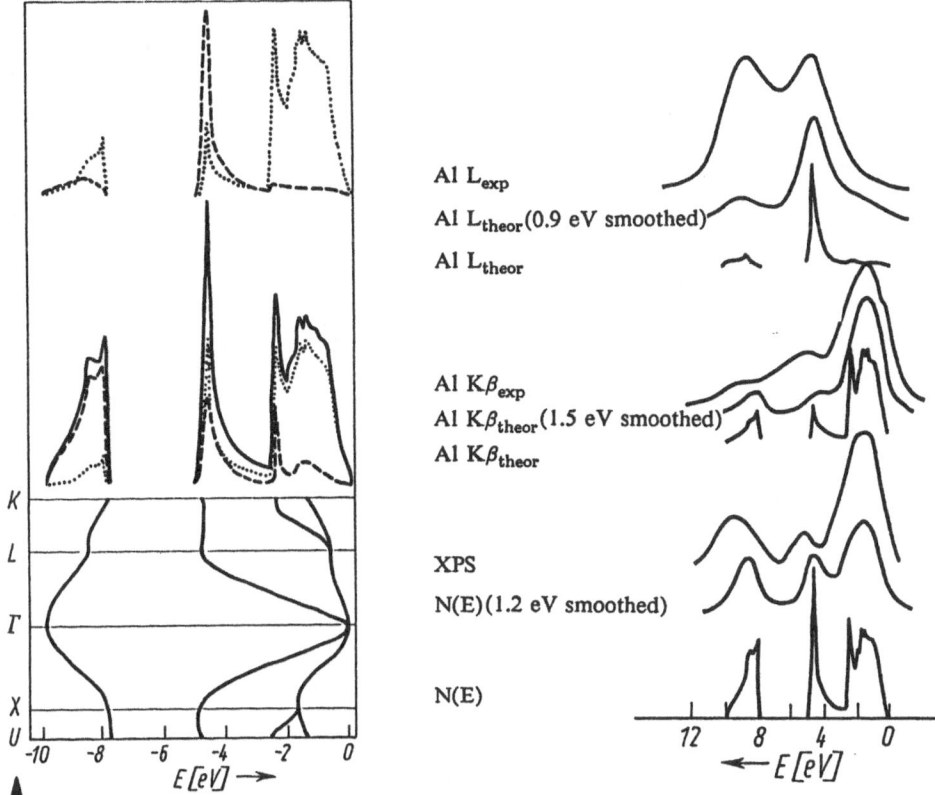

Al L$_{exp}$

Al L$_{theor}$(0.9 eV smoothed)

Al L$_{theor}$

Al Kβ_{exp}

Al Kβ_{theor}(1.5 eV smoothed)

Al Kβ_{theor}

XPS

N(E)(1.2 eV smoothed)

N(E)

Fig.6.13. Comparison of band structure (below), density of states (middle, solid curve) and calculated Al K (dotted) and L emission bands (dashed) in AlSb. Two calculations are shown: with exact transition probabilities included (top), and with a linear energy dependence of the transition probability assumed (middle) [6.33]

Fig.6.14. Comparison of the experimental XPS and XES curves with the calculated density of states and Al XES for AlSb [6.33]

crystals with high symmetry, there is a strong mixing of electronic states of different angular momenta, and most structures of the density of states can be identified in the spectra associated with the various inner levels, of course with very differing intensities (Fig.6.15). Structures in the density of states arise not only from individual bands in the E(k) description but also from the states at certain highly symmetric points in the zone (Van Hove singularities) allowing some information to be obtained about the energies of these states (accuracy is about ±0.5 eV, see Fig.6.16 and Table 6.1).

From the combination of emission bands with different initial states in a substance, the width of the valence band can be estimated, and a comparison of emission and absorption bands can give some information on the position of the Fermi level or the width of the energy gap. However, to achieve these results the relative energy positions of the inner

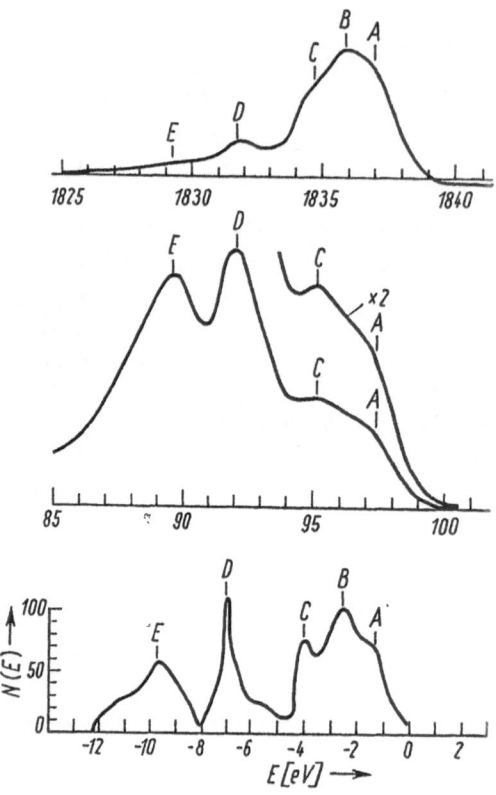

Fig.6.15. Comparison of K (top) and L (middle) emission bands of silicon with the density of states (bottom) [6.44]

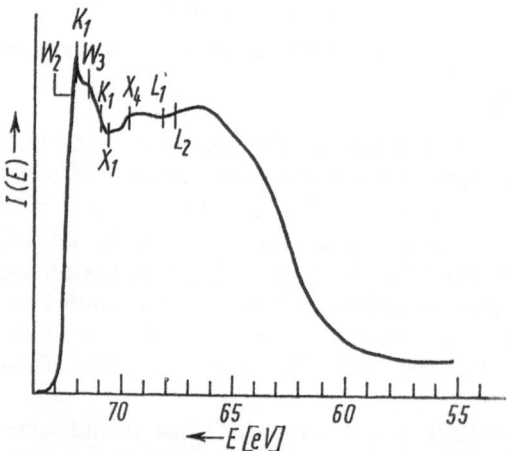

Fig.6.16. Kβ band of Al with energies corresponding to high-symmetry points in the zone indicated [6.9]

Table 6.1. Comparison of theoretical and experimental band structure data. For the experimental values (X-ray spectra) the components from whose spectra the data were drawn are given in parenthesis; since states $L_1(1)$ and X_1 are not resolved in the spectra the corresponding experimental values are placed between the last two columns [6.45] (values in eV). (DT: dielectric theory; XRS: X-ray spectra; OS: optical spectra)

Substance	Method	$\Gamma_{15}-X_5$	$X_5-L_1(2)$	$X_5-L_1(1)$	X_5-X_1
AlP	EPM-kp	2.04	3.48	9:19	8.64
	DT	2.6	-	-	-
	SCOPW	2.08	3.37	7.72	7.07
	XRS	2.3(Al)	3.3(Al)	7.4(Al)	
		3.3(P)	3.5(P)	7.27(P)	
GaP	EPM-kp	2.09	3.66	9.53	8.99
	DT	2.27	-	-	-
	OS	2.8	-	-	-
	XRS	2.85(P)	4.2(P)	7.88(P)	
InP	EPM-kp	1.69	3.49	8.32	7.80
	DT	1.85	-	-	-
	OS	2.5	-	-	-
	XRS	2.55(P)	3.42(P)	7.3(P)	
AlSb	EPM-kp	1.65	3.19	7.23	6.68
	DT	2.21	-	-	-
	OS	2.3	-	-	-
	XRS	2.2(Al)	3.5(Al)	7.23(Al)	

levels must be known. These data can be obtained either from the energy positions of X-ray emission lines (for the combination of the spectra from one component only) or from ESCA measurements (for the combination of spectra from different components of a compound). The determination of the valence band edge on the long-wavelength side is difficult due to Auger broadening (Sect.2.1) (correction by linear extrapolation from half-height or by assuming an $E^{1/3}$ or $E^{1/2}$ dependence, as appropriate [6.43]). By an accurate measurement of the intensity of such spectra it is possible to determine the number of electrons in certain states. However, there are difficulties caused by the necessary instrumental corrections (Sect.3.6) if spectra from different wavelength regions are to be compared. Nevertheless such studies have permitted estimates concerning changes of the number of electrons of a certain symmetry in alloys (Sect.6.3.3, Fig.6.39) [6.46] and on the charge distribution in $A^{III}B^V$ compounds (Sect.6.4.1) [6.47].

All these data contain direct information on the electronic structure of substances (local valence electron density, density-of-states structure, band separations, energies of states at high-symmetry points, valence-band widths, charge distributions). Looked at in another way, X-ray spectra are good experimental tests for band-structure calculations both with regard to the choice of the method and, especially, the choice of the potential. Thus, for example, the comparison of the calculated density of states of AlSb with the experimental spectra (Fig.6.14) shows that the pseudopotential used in the calculation and fitted to the optical spectra does not give entirely correct results; the distance between the energeti-cally deeper bands is larger in the experiment than obtained from theory. In this case X-ray emission and ESCA agree well with each other con-cerning the positions of the maxima. In such examples X-ray spectra can assist in the choice of pseudopotential form factors, which sometimes are obtained rather uncertainly from fits to other experiments.

Since band-structure calculations, especially for complicated struc-tures, are very cumbersome, attempts have been made to use MO calcula-tions including the nearest neighbors to interpret experimental spectra [6.42, 48-52]. As demonstrated in Fig.6.11 it is thus possible to explain the main features of certain experimental spectra. The scheme depicted in Fig.6.17 illustrates rather clearly the individual steps toward interpreting a spectrum. The intensity of an emission band depends directly on the an-gular-momentum parts of the LCAO expansion selected by the dipole selection rules for the corresponding X-ray transition, from this a simple explanation is also provided for the long-wavelength sidebands in the spectra of several compounds (Sects.2.4 and 6.4.2,3). It was shown through such a calculation for SiO_2 [6.53] that accurate predictions were

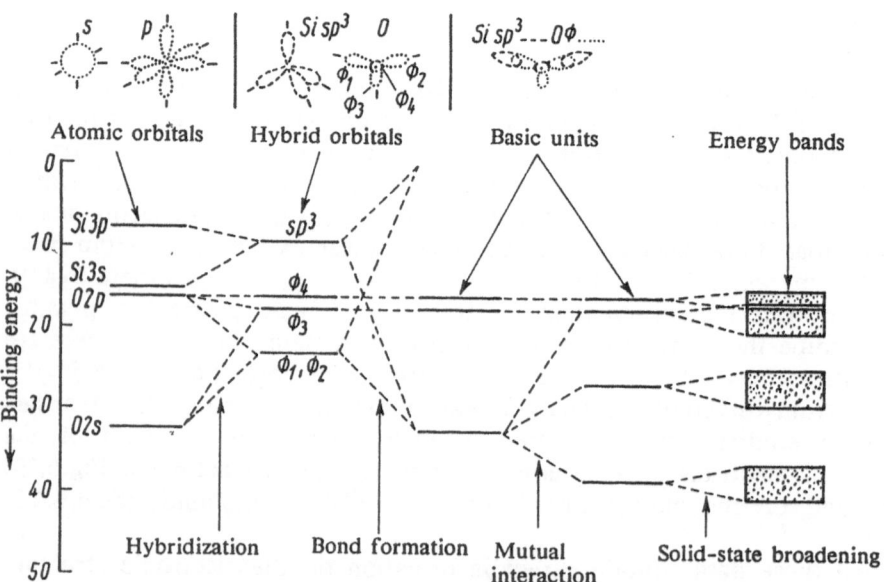

Fig.6.17. Scheme of the separate steps of an MO calculation for SiO_2 [6.50]

242

possible only by using an SCF formalism since the shape of the wave function and thereby the radial part of the transition probability can change appreciably. It was found that the ordering of the bands as calculated by MO methods and the resulting distribution of the valence electrons according to their angular momenta are reasonable. However, no estimates can be made from such simple MO calculations on, for example, bandwidth, band structure and density of states [6.50]. Such calculations can, however, serve very well as a first approximation to the electronic structure of a complicated compound. On the other hand, satisfactory agreement with experiment was found using calculations involving an LCAO expansion of the Bloch functions according to methods which are usually applied to calculations of molecules, e.g., HMO (Fig.6.41 below) [6.49].

Several researchers [6.18, 54] have attempted to correlate X-ray spectroscopic results with bonding parameters such as electronegativity and atomic charge. It is undoubtedly true that these parameters are closely related to the band structure [6.6, 33], but such relations, particularly with regard to the restricted quantum-mechanical definition of these parameters, could be of a qualitative nature only. Nevertheless it is not appropriate to confront the bonding theory with the band model (as was done in [6.18]) since these represent simply different stages of approximation. From this point of view the relation can be understood between the electronegativity and the shift of emission bands in oxides with similar crystal structures [6.54]. However, upon change of the crystal structure and, consequently, the band structure, such a correlation seems rather vague. Furthermore, the change in spectral profiles makes very uncertain the shift data. It is clear from Fig.6.18 that there is a dependence of the X-ray band spectra on the crystal structure which is related to the ionicity of the chemical bonding [6.6]. However, a change in the crystal structure is not necessarily connected with a variation of the spectrum since determining factors are the modification of the band structure and the density of states. Structural transitions do not necessarily show up as changes in X-ray spectra, in particular there is often only a small difference between the electronic structure for similar crystal structures and furthermore X-ray spectroscopy has rather limited resolving power. The role of the crystal structure becomes obvious if spectra are compared from crystalline and amorphous states (Fig.6.19) or from solid and liquid aggregation states (Fig.6.20).

By varying physical parameters which influence the X-ray spectra it is possible to obtain further information on certain details of the electronic structure. Thus a shift of the band and additional fine structure were observed for the L_{III} emission band of Cu by mechanical pressure variations on a polycrystalline copper sample (Fig.6.21). This experiment gave results concerning the deformation potential and the position of Van Hove singularities (since symmetry lowering can lift the degeneracy of otherwise highly-degenerate wave functions).

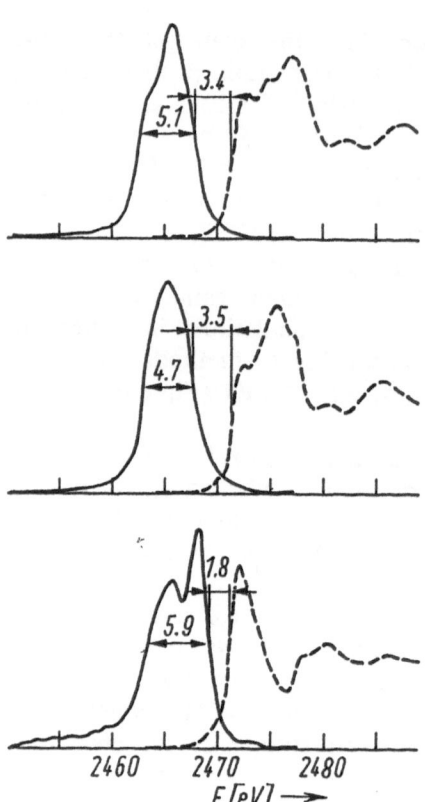

Fig.6.18. S K emission band (solid) and K absorption spectrum (dashed) of MnS in different crystal forms. Top: NaCl type; middle: ZnS type; bottom: NiAs type [6.55]

In strongly anisotropic crystals, especially layer crystals such as hexagonal Be, graphite and BN, additional band structure information can be obtained using polarized radiation [6.60]. By changing the angle of the outgoing radiation one obtains, for example, for a graphite crystal a clear change in the shape of the K emission band [6.61-64] which is caused by the σ and π bonding components (for a more detailed discussion, see Sect.5.2.2).

6.2.2 Interpretation of Absorption Spectra in a Single-Particle Model

All of these developments which relate the electronic structure of the occupied levels with the X-ray emission bands should be transferable to relations between the unoccupied electronic states and the X-ray absorption spectra. However, experience has shown that excitation states additionally can play an important role in the absorption spectra.

Good agreement has been obtained, e.g., for Ni metal between a careful study of the K absorption spectrum and the spectrum calculated including only transitions into p states of the conduction band [6.65,66].

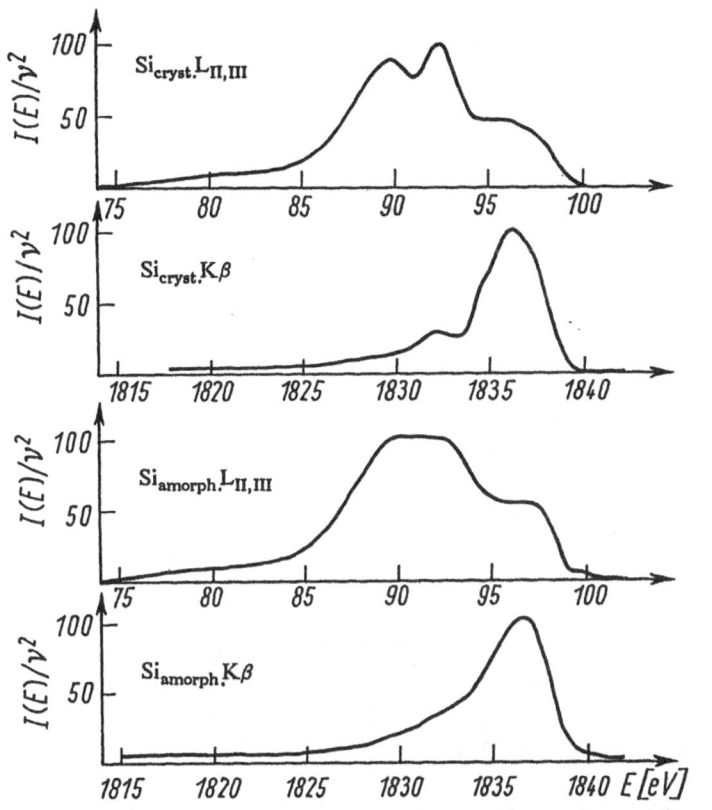

Fig.6.19. K and L emission bands of crystalline and amorphous Si [6.56]

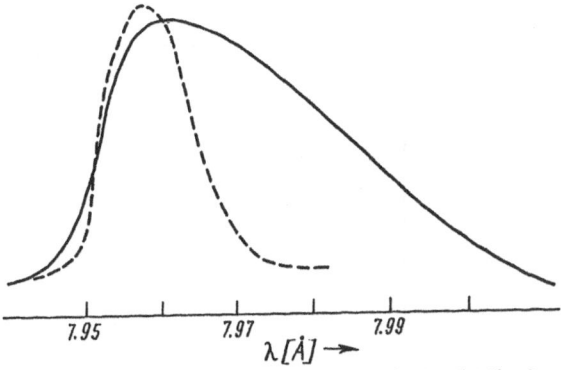

Fig.6.20. K emission band from solid (solid line) and liquid Al (dashed line) [6.57]

On the other hand, several investigations show that additional structures clearly exist for insulators. Lately, this has been shown also in experiments using synchrotron radiation for exciting inner levels [6.67]. The fine structure near the absorption edge arises not necessarily from transitions

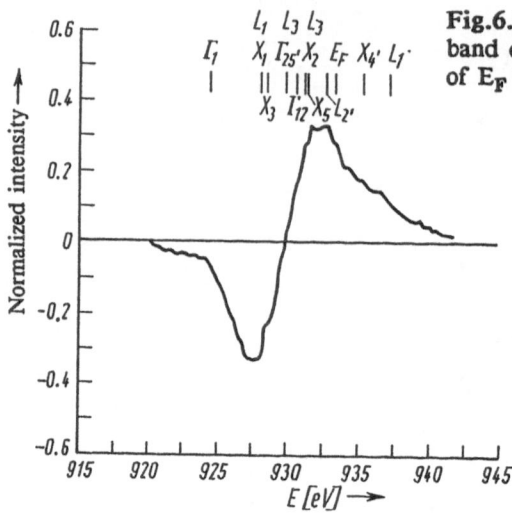

Fig.6.21. Modulation of the Cu L_{III} emission band due to alternating strain [6.58]; location of E_F and Van Hove singularities from [6.59]

to the conduction band, but may be due to the localization of the photo-electron on the corresponding hole-state (excitons, Sects. 2.7 and 6.2.3). This has caused rather different interpretations of the absorption spectra. On the one hand, the structure in the Li K absorption spectrum of LiF (Fig.6.22) was interpreted in terms of transitions to the conduction band alone [6.68,69]. According to this version the electron–hole interaction produces a shift of the spectrum as a whole which would be sufficient to explain the discrepancy between theory and experiment. On the other hand, the fine structure at the absorption threshold was interpreted as arising from exciton states [6.71,72] as has been borne out by later studies [6.70,73]. Through a combination of photoelectron and absorption spectra using $(E_{1s}-E_v)+E_g = E_c$ (E_{1s}: energy of the Li 1s level; E_v: energy of the valence-band edge; E_c: energy of the conduction band edge, Fig.

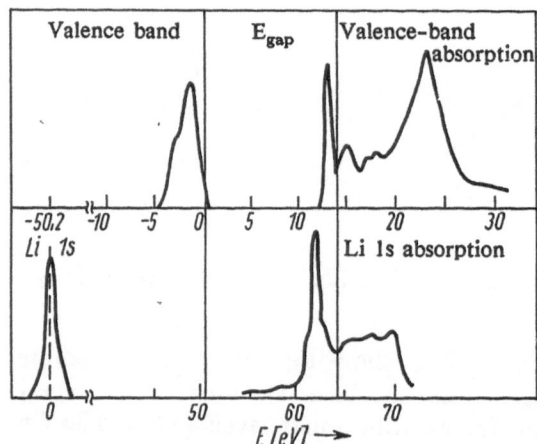

Fig.6.22. Photoelectron and absorption spectra of LiF [6.70]

6.22), with the width of the optical gap E_g = 13.6 eV [6.74] one obtains E_c = 63.8 eV (with structure in the range 54-60eV assigned to the forbidden s-s excitons, and at 61.9eV assigned to allowed s-p excitons). Furthermore, inclusion of X-ray emission data using $E_{1s}-E_v = E_b$ (E_b: short-wavelength limit of the X-ray emission band) supports such an interpretation [6.75].

Nefedov [6.76] and later *Åberg* and *Dehmer* [6.77] attributed the existence of such states to the formation of potential barriers for the photoelectron caused by the anions surrounding the positive ion - a situation analogous to the cases of SF_6 and BF_3 (Sect.2.7). The energies of these states nearly agree with those of the free ion since its potential nearly corresponds to the one of the ion in the lattice with the inner vacancy taken into account and the reference level changed. This conception is confirmed by differences found between the F and Li K absorption spectra: In the F K absorption spectrum there are no obvious structures before the beginning of the conduction band which can be associated with excitonic states [6.72]. The interpretation of the absorption spectra on the one hand by an atomic model and on the other hand by band-structure states can be understood in terms of inner- and outer-well states.

The approach to consider the surrounding ions in an ionic solid as a potential barrier was developed further by *Pavlychev* and *Kondrateva* [6.78] who use orbitals localized at the absorbing ion and calculate the changes in the absorption cross-section caused by the potential of the neighboring ions to interpret the K absorption spectra of NaF and NaCl.

Furthermore, delayed transitions, as in atoms and molecules (Sect. 2.7), are observed for solids too to continuum states with higher quantum numbers ℓ. Such transitions occur in particular in the spectra of the rare earths [6.79,80]. Scattering of the photoelectron, autoionization processes, multiplet splitting, configuration interaction, and multiple excitation can also influence the absorption spectra.

With proper allowance for these many-particle effects information can be obtained from the X-ray absorption spectra of solids analogously to that described for X-ray emission bands. If the same limitations are considered (transition probability, symmetry selection, local character), statements can be made about the electronic structure of the conduction band (density of states, see, e.g., Figs.2.45,46, energies of high-symmetry states in the conduction bands, see Fig.6.49). Additional results can be obtained from studies of the angular dependence of the absorption spectra of crystals of noncubic symmetry using polarized radiation [6.81,82].

Among the other X-ray spectroscopic methods for studying the conduction band (isochromat spectroscopy, appearance potential spectroscopy, radiative Auger effect, Raman scattering, Sects.2.4,5) especially isochromat and appearance potential spectroscopies are frequently used. Isochromat spectroscopy was developed mainly by *Ulmer* and co-workers [6.83] and applied to metals and alloys (Fig.6.31 and Sect.6.3.2 below) and later on also to compounds [6.84]. Isochromat spectra from 3d metal borides

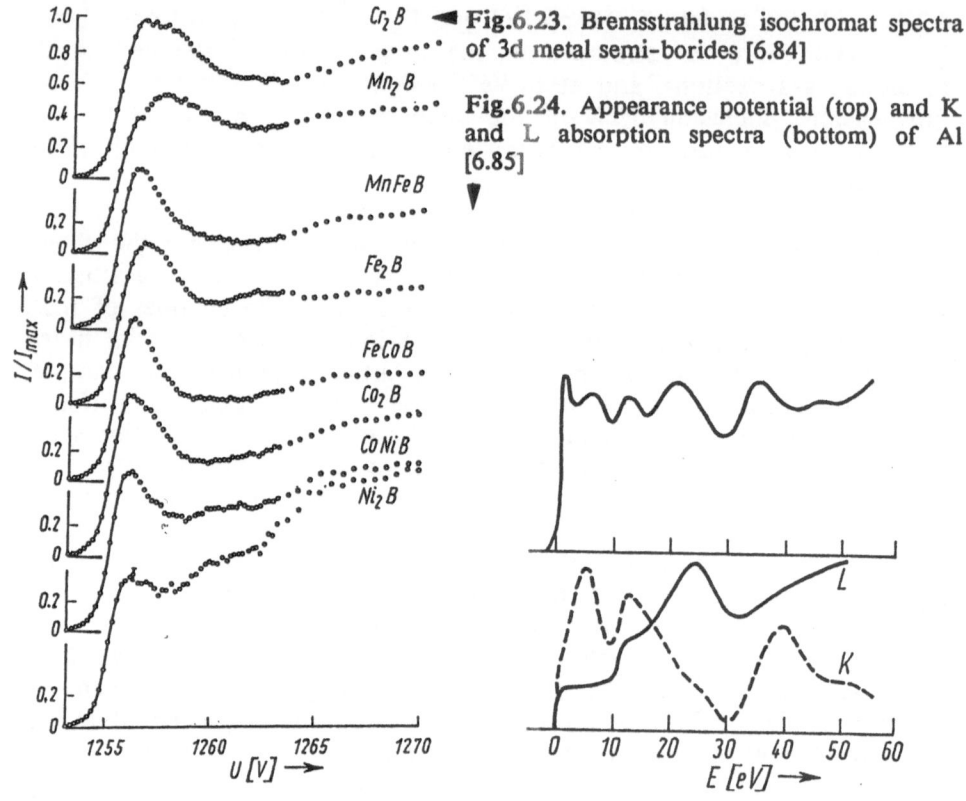

◄ **Fig.6.23. Bremsstrahlung isochromat spectra of 3d metal semi-borides [6.84]**

Fig.6.24. Appearance potential (top) and K and L absorption spectra (bottom) of Al [6.85]
▼

(Fig.6.23) give information on the width of the unoccupied d band and the number of electrons transferred from the metal to the boron atom.

The relations between X-ray absorption spectra, appearance potential spectra and the radiative Auger effect are shown in Figs.6.24,25 (see also Fig.2.30).

6.2.3 Many-Body Interactions

In the following section only the most important processes are briefly discussed qualitatively, since a quantitative description is not possible without major theoretical background [6.24]. Apart from the many-body interactions (e.g., electron correlation) which were neglected in the discussion of densities of states and X-ray spectra in Sect.6.2.1, there are to be considered further such effects predominantly in the X-ray production process.

Considering the time relations among the ionization, orbital relaxation, and deexcitation processes [6.50] (Fig.2.2), one states that in some cases the relaxation times are longer for the outer electrons than the ionization times (especially for electrons with high kinetic energies) and also than the lifetimes of some excited states (especially for heavier ele-

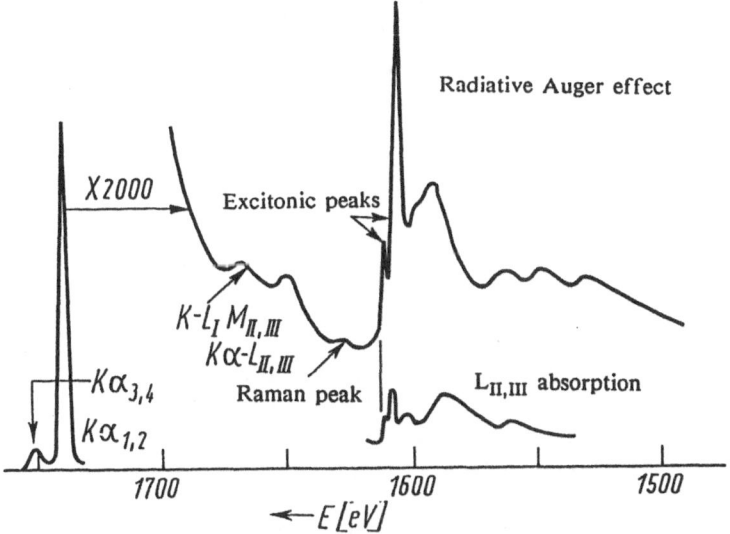

Fig. 6.25. Comparison of the Si $L_{II,III}$ absorption with the radiative Auger effect in the K shell emission for SiO_2 [6.86, 87]

ments). Therefore, excited states with incomplete relaxation of the valence electrons can be important and can differ for different spectra of an element [6.88].

Numerous investigations have dealt with the influence of the interaction between an X-ray produced inner vacancy and the remaining electrons in a solid [6.24, 89–99]. In metals the inner hole is screened by a localized electron cloud of conduction electrons, which causes a change of the density of states especially in the region of the Fermi level compared to the non-excited state. The shape of the emission band and the absorption edge in light metals was thus explained based on calculations by *Mahan* [6.96], and *Nozieres* and *Dominicis* [6.98] (MND theory, see, e.g., [6.100]). More accurate studies [6.101] showed, however, that such many-body effects cannot fully account for the shape of these spectra. Careful analysis of the emission and absorption spectra of Li, Na, Al, and Mg showed [6.90] that for certain cases the influence of these many-electron effects could be clearly proven. However, in each special situation it must be considered whether a one-electron picture suffices for the interpretation of the spectrum and where many-electron effects play an important role (Sect. 6.3.1).

For semiconductors and insulators the interaction of the inner hole with the excited electron leads to the existence of excitons. In a spectroscopic sense this means discrete transitions into energy states, which mostly lie below the conduction band. One distinguishes between strongly bound (Frenkel exciton) and weakly bound excitons (Wannier exciton: electron and hole are not in the same lattice cell). These excitonic states are especially important in X-ray absorption spectra of substances with a

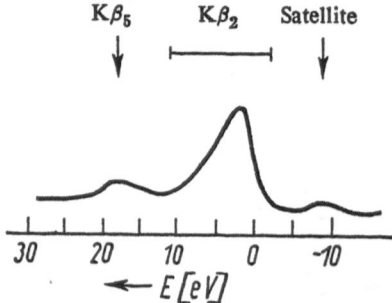

Fig.6.26. X-ray emission band for Ga [6.102]. The zero of the energy scale corresponds to the valence band edge [6.103]

large energy-gap width (Fig.6.22). This is one of the reasons why it is complicated to get precise conduction-band information on these substances from the X-ray absorption spectra (see also Sect.6.4).

Besides, through different processes (as, e.g., shake-off, Sect.2.4) additional vacancies can be created which (generally) lead to short-wavelength satellites of the X-ray emission bands. Such multiple-ionization satellites influence the spectra considerably (Fig.6.26) and often lead to erroneous interpretations. These effects can also be important in X-ray absorption spectra [6.104].

A possibility to get X-ray emission bands free from multiple-ionization satellites is the fluorescence excitation by monochromatized radiation of threshold energy. In the past, this was not applicable because of too low intensities. The situation has changed now with the use of electron storage rings and, especially, wigglers and undulators as sources of a high-intensity X-ray continuum [6.105]. Recently, the pure N K and Ti L_{III} emission spectra of TiN have been isolated for the first time, and interesting insight was got into the influence of excited states on the spectra [6.106].

A further distortion of the emission spectrum arises on the long-wavelength side caused by both Auger effect and plasmon excitation. The

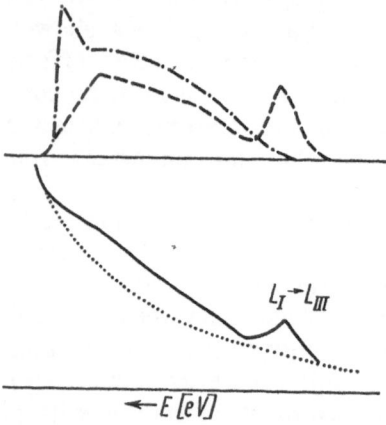

Fig.6.27. Satellite in the low-energy region of the Al K emission band associated with plasmon excitation. Solid line: experimental spectrum; dotted: extrapolation of the Al K band; dashed: difference between experimental and extrapolated spectra; dashed-dotted: shape of the Al K band for comparison with the shape of the plasmon satellite [6.31]

reason for the former is a decrease of the lifetime of the final state of the X-ray emission through Auger electron emission at the bottom of the valence band [6.107, 108]. Through excitation of collective plasmon oscillations of the electron gas additional long-wavelength tails arise, which can also be observed experimentally (Fig.6.27). Calculations show an influence of the inner hole for these effects too [6.24, 109]. A uniform theory to explain all observations from the Fermi edge to the long-wavelength structures is not available [6.89].

6.3 Investigation of Elements

6.3.1 Metals from the Main Groups of the Periodic Table

Metals from the main groups of the periodic table, especially those with lower Z, have been extensively studied both theoretically and experimentally. Experiments using ultrahigh-vacuum conditions and calculations including many-body effects have given new insights. Measurements published by different researchers usually agree; sometimes, in addition to small quantitative differences there are also qualitative deviations in the measured band shapes (e.g., Al $K\beta$ [6.110, 111]). Furthermore, it should be emphasized that the spectra have been presented very differently (intensity I in arbitrary units, $I(E)/\nu^3$ [6.112], $I(E)/\nu^2$ [6.23]).

From the measured spectra (Fig.6.28) and their theoretical interpretation the following statements can be derived.

1) Apart from the long-wavelength tails of the emisson bands (Sect.6.2.3 and below), there is in general good qualitative agreement of the width and shape of the emission bands with N(E) curves calculated using various methods (an exception is Li) [6.9, 114, 119] (Fig.6.28 and Table 6.2). The agreement between the experimental bandwidth and the bandwidth calculated using the free-electron model, $E_0-E_F = 50.05/r_s^2$ (with r_s values taken from [6.9]), is rather good [6.120]. In many cases structures in the emission spectra can be assigned to Van Hove singularities [6.9, 31, 114] (Fig.6.15). Such structures are not expected to arise for the alkali metals since, for these elements, the Fermi surface does not touch the boundaries of the Brillouin zone [6.9]. However, for example, structures in the Mg L bands found by *Watson* et al. [6.121] and attributed to Van Hove singularities were only partly confirmed by *Neddermeyer* [6.23] in an experiment with higher resolution. An investigation of the spectrum of liquid Al showed a very strong narrowing of the $K\beta$ band and a vanishing of its asymmetry (Fig.6.20).

2) The plasmon satellite in the Al L emission spectrum was found experimentally by *Rooke* [6.9]. A satellite of this type has also been identified in the emission bands of Be, Na, and Mg (Table 6.2). Furthermore, in the K and L bands of Be, Mg, and Al short-wavelength satellites due to

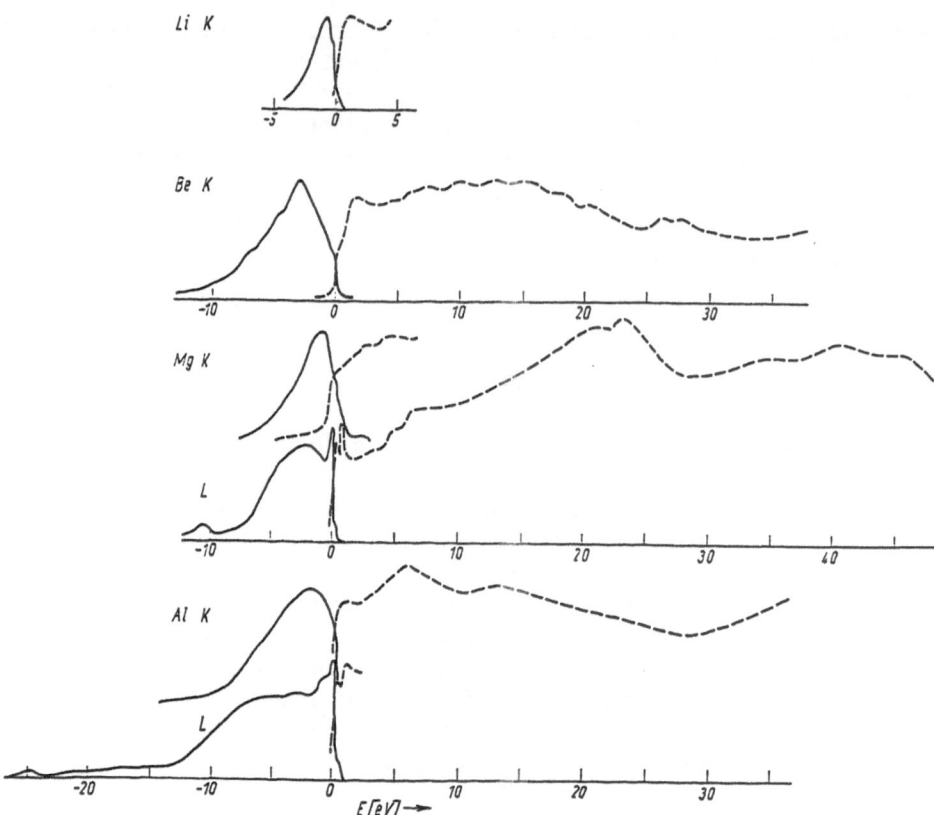

Fig.6.28. X-ray emission bands (solid) and absorption edges (dashed) of Li [6.100, 113], Be [6.114, 115], Mg [6.100, 116, 117], and Al [6.100, 110, 116, 118] in a uniform energy scale

multiple ionization have been found [6.122] whose intensity and shape are dependent on experimental conditions due to self-absorption effects [6.23]. The position of the L_I level can be determined in good agreement with [6.123] from the energy position of the line emission superposed with the long-wavelength tail of the L bands of Al and Mg, which according to [6. 124] arises from $L_{II,III} \rightarrow L_I$ transitions.

3) The K emission band of Li is particularly interesting. Assuming the dipole selection rule and the free-electron model, the K band should have a sharp cut-off on the short-wavelength side. Calculations of the density of states predicted similar behavior near the Fermi level [6.125]. However, the experimental result departs appreciably from these predictions. In addition, the K band appears to be independent of the crystal structure for lithium metal [6.113]. According to *Goodings* [6.126] and *Allotey* [6. 127] these discrepancies are to be attributed to effects caused by the inner vacancy. On the contrary, *Stott* and *March* [6.99] explain the "anomaly" by the influence of the transition probability. The particular shape of the Li K band compared to the emission bands of other alkali metals probably

Table 6.2. Comparison of plasmon energies and valence band widths determined from X-ray spectra and those determined by other methods (in eV) [6.31]. (EELS: Electron energy loss spectra; OS: Optical spectra; XRS: X-ray spectra; FEM: Free-electron model; BSC: Band structure calculations; PP: Pseudopotential)

		Li	Be	Na	Mg	Al	K
Plasmon energy	Theory	7.979	18.104	5.90	10.884	15.772	4.4
	EELS	7.12	18.7	5.85	10.6	15.3	4.05
	OS	8	18.4	5.49	10.34	14.9	4.25
	XRS		18.2±0.5	5.5±0.2	10.6±0.5	14.9±0.1	
Band width	FEM	4.691	13.987	3.139	7.097	11.638	2.14
	BSC	3.42	11.38	3.31	–	11.28	2.20
	PP	3.95	10.94	3.14	7.03	11.19	2.16
	XRS	3.0±0.1	13.8±1.0	2.6±0.3	6.84±0.05	11.3±0.5	1.62±0.04

results from the absence of p electrons in Li. Investigations of the temperature dependence of emission band edges showed a clear dependence of the shape of the K band of the alkali metals on the hole state-phonon interaction [6.128].

4) The particular shape of the L absorption edges of Na, Mg, and Al up to a distance of about 0.5 eV above threshold can be explained largely through the hole-conduction electron interaction using the MND theory (Sect.6.2.3) [6.100]. Corresponding structures are not present in the spectrum from Na vapor [6.129]. Besides this effect, influences of the density of states (as seen particularly clearly in the Al and Mg L emission bands), the spin-orbit interaction (especially for the Na L absorption) and the hole-phonon interaction (particularly for the Li absorption) as well as the ubiquitous smearing of the spectrum due to the finite lifetime of the inner hole are important for the interpretation of these spectra. The influence of the individual factors is rather different in each case [6.90].

5) The low-energy region of the K absorption spectra can be partially explained in terms of the density of states of the conduction band (Be [6.130], Mg [6.131]). Peaks within a distance of 20 eV from the absorption edge were explained by *Sagawa* and coworkers [6.115] as 2p → nd transitions.

6) The L absorption spectra of Na, Mg, and Al are very similar. The general shape of these spectra can be obtained from atomic calculations [6.132]. For Al a satisfactory explanation of this spectrum resulted from the assumption of plasmarons [6.24].

6.3.2 Transition Metals

X-ray spectra from 3d transition metals have been extensively studied. Due to improvements in experimental techniques, it has been possible to measure more accurately both the L and M spectra including, in particular, corrections for self-absorption and satellite structures. Typical spectra have been collected in Fig.6.29. Comparing these spectra one must take into account that intensity changes in the individual spectra due to experimental conditions can be rather significant and differ from case to case and, also, that different correction procedures have been utilized. As a consequence of the dipole selection rules the K, L, and M spectra of each element differ strongly from each other. General correlations or trends among the spectra of different elements are not evident, especially when comparing spectra from metals at the beginning and at the end of the 3d series.

The K emission bands in 3d elements were interpreted in older papers as due to quadrupole transitions 3d → 1s [6.146]. However, a first argument to the contrary is that there are strong shape differences between the K, $L_{II,III}$ (3d4s → 2p), and $M_{II,III}$ (3d4s → 3p) bands. Secondly, the widths of the bands differ not only by their absolute values but also the trend of the bandwidths within the 3d series is different. The K band widths increase from Sc to Zn whilst the L and M bandwidths go through a maximum somewhere in the middle of the 3d series [6.135]. For the intensities of the K and L bands the opposite is true: The $K\beta_5$ intensity is largest in the middle of the 3d series whilst the L intensity increases steadily in going from Sc to Zn [6.135]. Even if the L and M spectra in [6.135] are partly based on older work, the conclusions do not change when more recent investigations as in Fig.6.29 are considered. Thirdly, calculations [6.147, 148] predict that the quadrupole transitions are about two orders of magnitude lower in intensity than experimental intensity ratios appear to suggest. Furthermore, more recent measurements give both smaller widths compared with earlier and lower intensity of the K emission band in the vicinity of the Fermi edge where predominantly d states are located [6.149]. For these reasons, it must be assumed that the K bands of the 3d metals arise from transitions of p electrons (see below); however, for heavier d elements, due to the larger statistical weight of the electrons with d symmetry, a contribution from quadrupole transitions (for example, the high-energy structure of the $K\beta_{2,5}$ bands of Co and Ni) cannot be excluded [6.135]. Whilst the $K\beta_{2,5}$ bands of Ti to Fe are very similar, the bands of Co and the following elements differ rather strongly from each other. The different appearance of $K\beta_{2,5}$ bands recorded under different experimental conditions, e.g., the bands from Ni and Cr [6.150] (in particular in the higher-energy region), can largely be explained as produced by strong influences of self-absorption and multiple-ionization satellites.

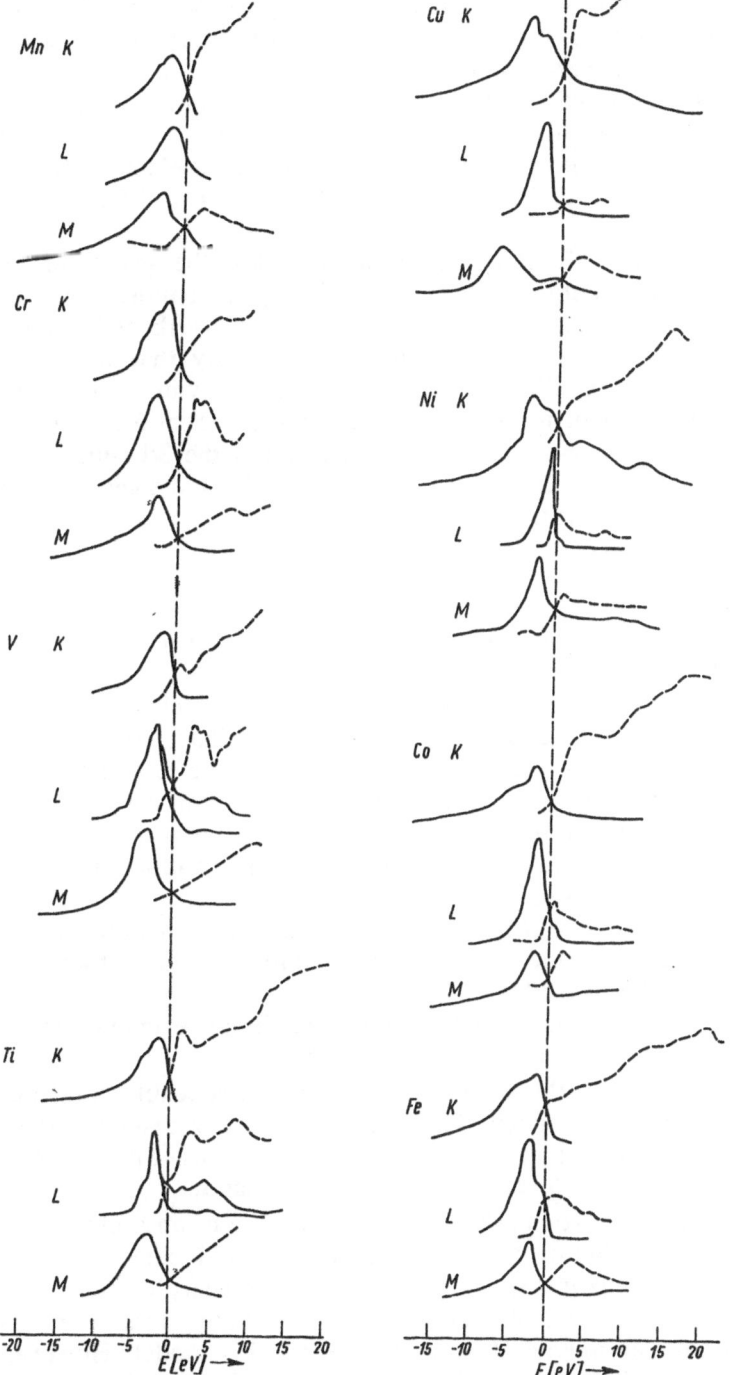

Fig.6.29. X-ray emission bands (solid) and absorption edges (dashed) of 3d metals; for Ti and V, two L bands are shown with different amounts of self-absorption [6.133-145]

The L emission bands were measured carefully [6.138-140, 144] with particular attention paid to the effects of self-absorption and satellite emission. Even narrower bands than suggested by $K\beta_{2,5}$ were got. The structures of the L bands of Ti, V, and Cr are very similar to those of the corresponding $K\beta_{2,5}$ bands [6.151]. The L bands of the elements Co, Ni, Cu, and Zn show, besides the main peak caused by d electron transitions, a tail on the high-energy side apparently due to transitions of the 4s electrons.

Considering only the effect of the selection rules, the shapes of the $M_{II,III}$ bands should be identical with those of the $L_{II,III}$ bands. However, it is observed that the M bands usually have a larger width and often a structure differing from the L bands. This shows clearly that the influence of the transition probability can exceed that of the symmetry restriction. It can be understood in terms of the radial factors (Sect. 6.2.1). Using atomic wave functions for the 2p and 3p electrons and the 3d function of a band-structure calculation, *Cuthill* and coworkers [6.39] showed that for Ni the radial parts of the transition probability

$$\left| \int R_{np}(r) r R_{3d}(r) dr \right|^2$$

for the upper and lower edges of the valence band differ by 70% for $L_{II,III}$, whereas the difference is only 10% for $M_{II,III}$.

The fine structure near the K absorption edge of the elements at the beginning and the end of the 3d series is rather similar. On the other hand, it is remarkable that differences appear between the L and M edge regions. The white lines in the $L_{II,III}$ absorption spectra are generally attributed to transitions to the unoccupied d states and the widths of these lines are of the same order as those expected for the conduction bands. The shape of the M absorption spectrum is explained within the framework developed by *Fano* and *Cooper* [6.152] as due to interference between the intense background and the superimposed line absorption (interchannel interaction) [6.145].

The following conclusions can be drawn from the experimental spectra about the electronic structure of the 3d metals:

1) In the valence bands of the 3d metals there is a noticeable contribution of electrons with p symmetry. Their electron density extends over a wider energy range than that of the d-symmetry electrons. The electrons with s symmetry form a broad, not sharply localized band.

2) Uniform variations, particularly of the K emission and absorption spectra, were taken as evidence for the validity of a rigid-band model by *Finkelshtein* and *Nemnonov* [6.136] (Fig. 6.30). The similarity of the K, L, and M emission bands of Ti, V, and Cr supports these ideas. Furthermore, isochromat measurements (Fig. 6.31) confirm this model. However, one has to have in mind that the identical lattice structure of these metals is taken as a presupposition. On the other hand, significant variations are observed in going from the 3d to the 4d and 5d metals [6.153, 154].

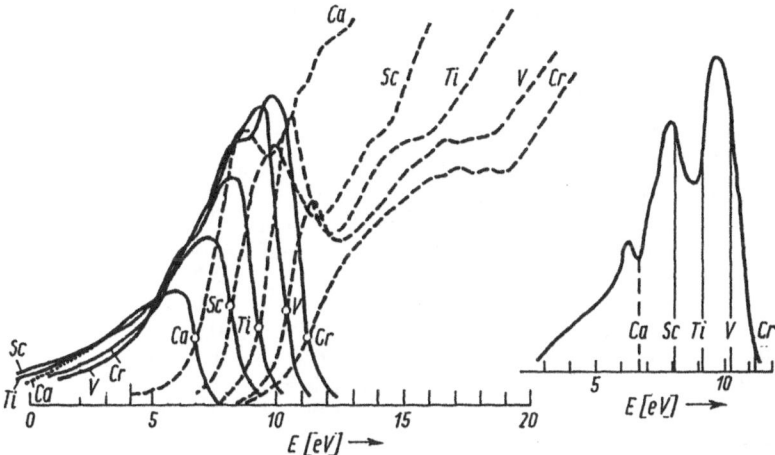

Fig.6.30. K spectra of 3d metals (emission bands - solid lines; absorption edges - dashed lines); as common origin the long-wavelength rise of the emission bands was chosen. On the right are shown the partial densities of states for these metals obtained from the emission spectra assuming the rigid-band model [6.136]

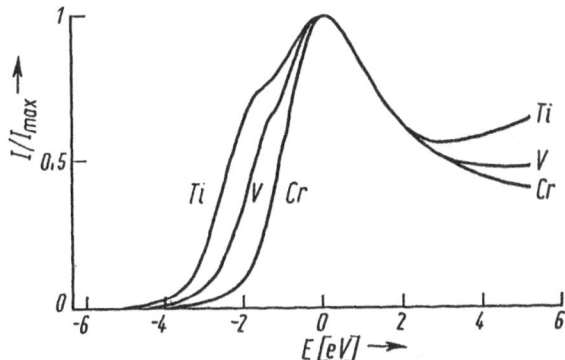

Fig.6.31. Isochromat spectra of 3d metals [6.153] aligned so that their maxima coincide

3) From the normalized intensities of the K and L emission bands the electronic configurations in the metal crystals were determined by *Dekhtyar* and *Nemoshkalenko* [6.135] (Table 6.3). In this approach it was assumed that the transition probabilities do not change for electrons of all angular momenta within the 3d series, which seems somewhat dubious considering the arguments given above. Taking those into account, *Blokhin* et al. [6.155] obtained differing results. In Fig.6.32 the data are compared.

4) There is a semi-quantitative agreement between the spectra and the calculated densities of states. The bandwidths [6.138, 139, 149, 158] and also the shape of the spectra [6.39, 144], particularly if later data are included [6.140] (Figs.6.33, 34), are related to the densities of states in a generally

Table 6.3. Valence electron configuration in metal crystals determined from X-ray spectroscopic data [6.135]

	$d_{3/2}$	$d_{5/2}$	d	sp	s	p
Ti	-	-	1.0	-	1.0	2.0
V	-	-	1.7	-	0.7	2.6
Cr	-	-	2.4	-	0.5	3.1
Fe	-	-	6.0	-	0.5	1.5
Co	-	-	7.5	-	0.5	1.0
Ni	-	-	8.6	-	0.5	0.9
Zr	0.5	0.5	1.0	3.0	1.6...2.0	1.0...1.4
.Nb	1.0	1.0	2.1	2.9	0.6...1.4	1.6...2.3
Mo	1.5	1.2	2.7	3.3	0.0...1.0	2.3...3.3
Ru	1.4	2.6	4.0	4.0	-	-
Rh	2.8	3.6	6.4	2.6	-	-
Pd	3.1	3.9	7.0	3.0	-	-
Ag	4.9	5.9	10.8	1.0	-	-
In	4.0	5.8	9.8	3.0	0.8	2.3
Sn	4.0	6.4	10.4	4.0	1.4	2.7
Sb	3.3	6.2	9.5	5.0	2.2	2.8

obvious way. Especially those calculations of the spectra including exact transition probabilities as presented by *Nikiforov* et al. [6.38, 161] agree well with experiment. Therefore, it is possible to prove the interpretation of the spectra (e.g., $K\beta_5$) and to use spectra as testing ground for band-structure calculations [6.162]. *Blokhin* et al. [6.163] compared the cellular, APW, and Green functions method.

It is possible only in a still more limited way to draw conclusions on the electronic structure of the 4d and 5d metals from their spectra.

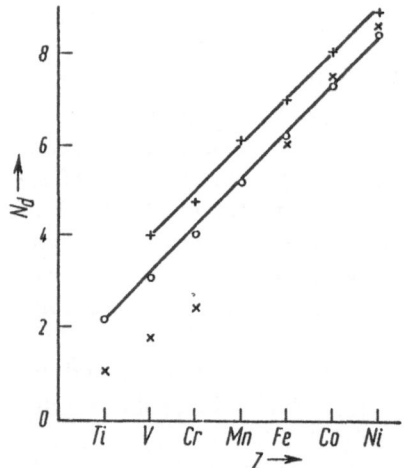

Fig.6.32. Number of 3d electrons N_d in the metals determined from X-ray spectra. + from [6.155], o from [6.156], × from [6.157]

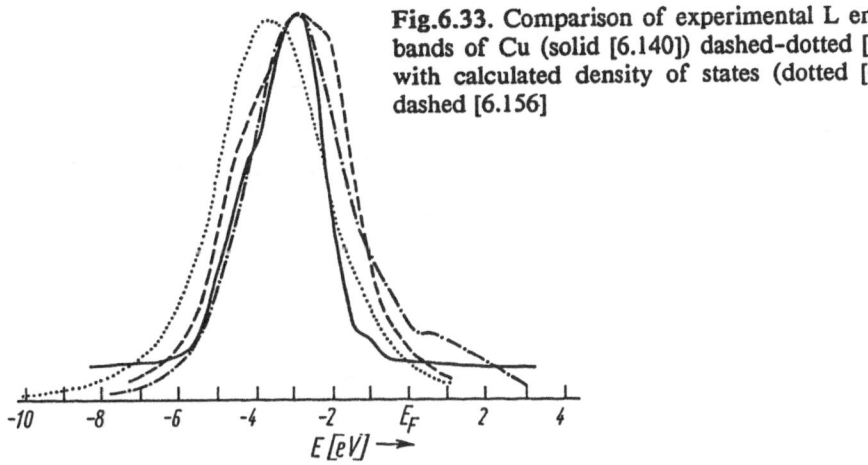

Fig.6.33. Comparison of experimental L emission bands of Cu (solid [6.140]) dashed-dotted [6.158]; with calculated density of states (dotted [6.159]) dashed [6.156]

Linewidths and shapes of the emission bands $L\beta_2$ $(4d_{5/2} \rightarrow 2p_{3/2})$ and $L\gamma_1$ $(4d_{3/2} \rightarrow 2p_{1/2})$ differ considerably in some cases [6.135, 164] (Fig. 6.35). Furthermore, the $L\beta_2$ band is superposed with the $L\beta_{15}$ $(4d_{3/2} \rightarrow 2p_{3/2})$ and $L\beta_7$ bands $(5s \rightarrow 2p_{3/2})$; however, the contribution from $L\beta_{15}$ is less than 10% and from $L\beta_7$ about 7%. The $L\gamma_1$ line is superimposed by the $L\gamma_8$ line $(5s \rightarrow 2p_{1/2})$ with an intensity contribution of only 3%.

The results from [6.165] appear quite contrary to those in [6.135]. After corrections, the L emission bands of Zr, Nb, and Mo were modeled as lines with two components which were explained as caused by localized and collective electrons in the metal lattice [6.165]. The L emission bands of Zr, Nb, Mo, Rh, and Pd were compared with band-structure data by *Shveitser* et al. [6.164]. The existence of M_V bands of Zr, Nb, and Mo (Fig.6.36) points towards the presence of p-symmetry electrons in the valence band [6.135, 166, 167]. The similarity of these bands to the $K\beta_5$ bands of the corresponding 3d metals supports this assumption. Following *Nemoshkalenko*'s analysis of the 3d metals, he and *Dekhtyar* [6.135] also deter-

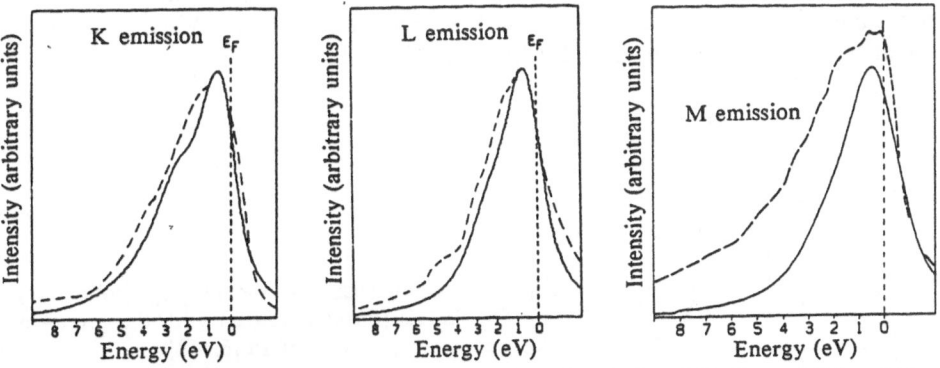

Fig.6.34. Comparison of experimental (dashed) and calculated K, L, and M emission bands (solid) for V [6.160]

259

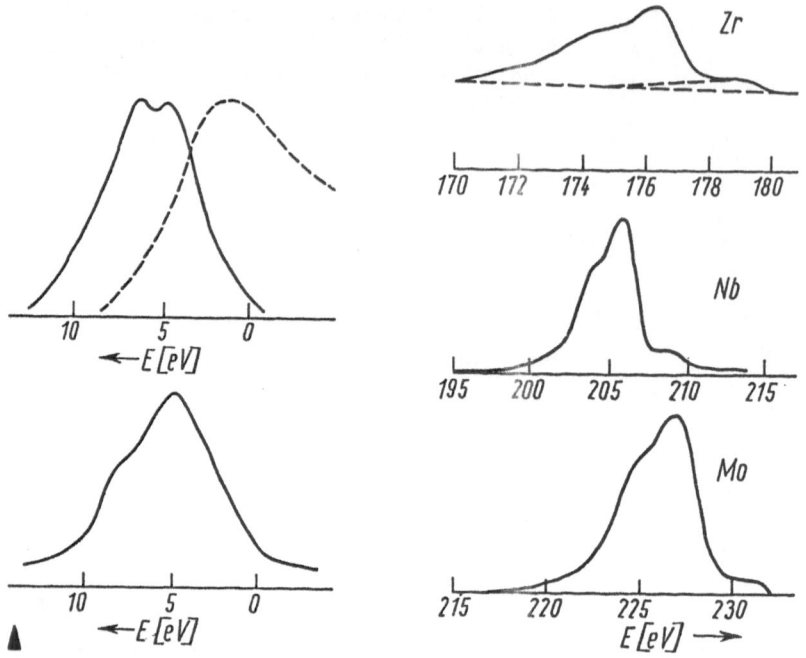

Fig.6.35. $L\beta_2$ (top) and $L\gamma_1$ X-ray emission bands (bottom) and L_{III} absorption edge (dashed) of Ru [6.135, 164]

Fig.6.36. M_V emission bands of 4d metals [6.135]

mined the electron configuration in the lattice of these metals using relative intensities of the bands (Table 6.3).

In some investigations, the $N_{II,III}$, $N_{VI,VII}$, and $O_{II,III}$ spectra were associated solely with dipole-allowed X-ray transitions [6.168, 169]. The agreement with ESCA results (Fig.6.37) [6.170] is not satisfactory. However, dividing the calculated density of states into partial densities of states and considering the experimental distortions and the broadening by the inner level and the Auger effect, it was possible to obtain rather good agreement between theoretical and experimental L_{III}, M_V, and $N_{II,III}$ spectra of Nb, Mo, Rh, and Pd [6.171] (Fig.6.38).

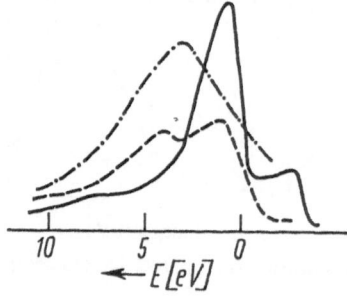

Fig.6.37. Comparison of the $N_{VI,VII}$ (solid curve) and O_{III} bands (dashed-dotted) with the photoelectron spectrum (dashed) of Pt [6.168]

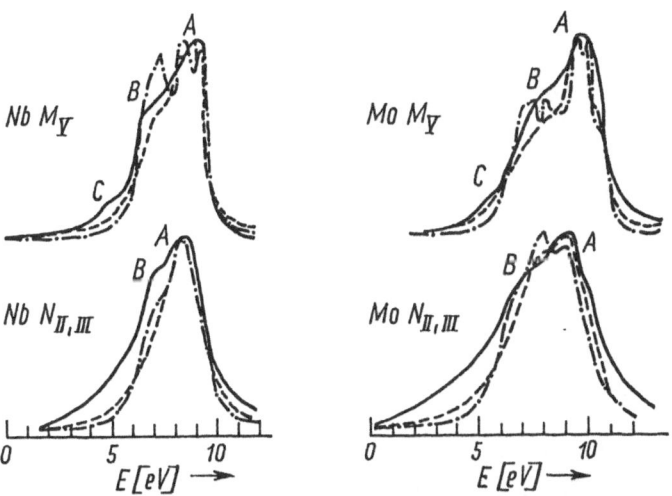

Fig.6.38. Experimental (solid curve) and theoretical X-ray emission bands (dashed-dotted without, dashed with Auger broadening) for Nb and Mo [6.171]

6.3.3 Alloys

X-ray emission spectroscopy has been applied in investigations of the electronic structure of alloys up to recent time [6.34, 172-176]. Here we will discuss briefly only some major results from the studies of alloys. An extensive discussion of alloys would go beyond the intention of this book; for reviews, see [6.135, 177-180].

Numerous investigations of the X-ray emission and absorption spectra and the isochromat spectra were carried out for alloys composed of transition metals and for alloys of transition metals with Al and Mg. Such measurements have some importance for the development of an electronic theory for alloys and also in explaining relations between the electronic structure and practically important properties of alloys.

Emission bands have been studied most frequently. The results show that the bands change their shapes considerably upon alloying. These changes can be explained by variations of the band structure (for example, caused by new lattice structures) and also from shifts in the electronic density, i.e., an electron transfer between the metals in the alloy. As a result there is a gain in intensity on the short-wavelength side of the band of the metal A when it forms an alloy with a metal B with higher valency [6.181]. The opposite is true for the metal A with higher valency.

The positions and widths of the bands in alloys are similar to those of the pure metals [6.181]. It was shown, however, by some investigations [6.116, 180] that there are experimental difficulties in determining bandwidths and the position of the low-energy limit of the band, so that to draw conclusions from different bandwidths of the components of an alloy requires very careful measurements. No conclusions can be extracted from

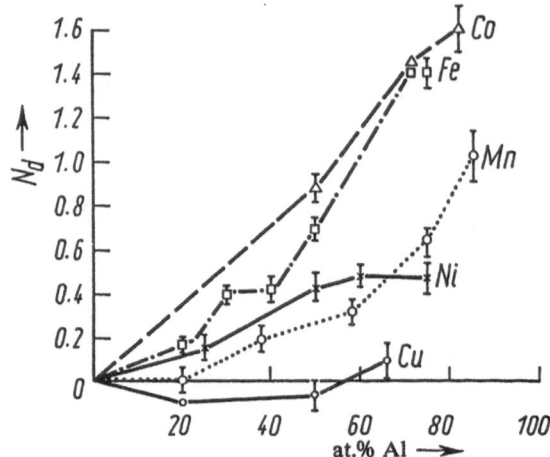

Fig.6.39. Number N_d of 3d electrons in Al alloys obtained from intensity measurements of emission bands [6.46]

shifts in the emission bands due to the lack of shift data for the inner levels.

Investigations of X-ray absorption spectra particularly by *Azaroff* and co-workers [6.65, 182] have led to certain conclusions about the validity of different models for the band structure of alloys. The following conclusions regarding the electronic structure of alloys can be drawn from the X-ray spectra:

1) The rigid-band model is applicable only for alloys of neighboring elements in the periodic table, but even this rule has exceptions [6.183-185]. For alloys of metals whose atomic numbers differ widely, good predictions can sometimes be made with models based on the coherent potential approximation [6.116, 180, 186]. In contrast to results from photoelectron spectra [6.186], the X-ray spectra have given no experimental proof for the virtual-bond-state model [6.180]. However, in some alloys the hybridization of electrons leads to a localization of the electronic states of a certain symmetry (see below).

2) For alloys with transition metals only, no essential changes were found in the spectra compared to those from the pure metals.

3) From investigations of alloy systems composed of Al and transition metals there are two important results. Measurements of the intensities of the bands of Al and the 3d metals have shown changes of the number of 3p and 3d electrons, respectively (Fig.6.39). A low-energy peak was observed in the L spectra of Al in alloys with noble metals and d metals, which agrees with theoretical predictions of a hybridization of the d states with the s states of Al and the formation of a semilocal state with s symmetry around the Al atom [6.180, 185]. In the K spectra of some alloys (for example, Al-Pd and V-3d metal [6.141]), peaks associated with such a localization of certain electronic states arise too.

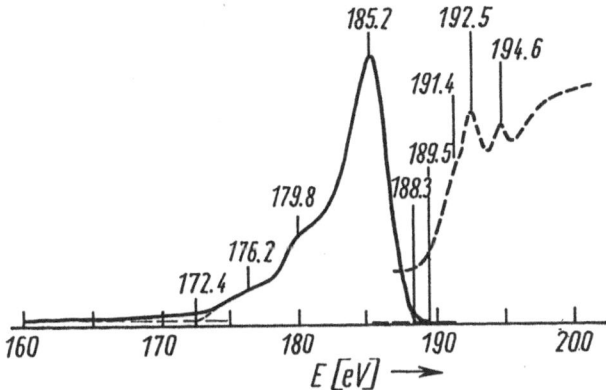

Fig.6.40. Emission (solid) and absorption spectra (dashed) of B [6.187]

6.3.4 Nonmetals

The X-ray spectra of the elements B, C, Si, Ge, P, and S are discussed in this section.

Figure 6.40 presents the K emission and absorption spectra of boron by *Fomichev* [6.187]. The shape of the K band agrees with the measurements by *Holliday* [6.188]; *Aita* et al. [6.114]; and *Hoffmann* et al. [6.189]. Full-width values (at the base) from 14.1 eV [6.190] to 15.9 eV [6.187] have been reported; at half-height, the corresponding values were 4.1 eV [6.187] to 5 eV [6.191]. Differences in energy positions (main peak located between 66.92 Å [6.187] and 66.56 Å [6.188]) are probably caused by the calibration wavelengths used and not by differences in the crystal structure [6.189]. The energy gap between the valence and conduction bands obtained from the spectra is 1.2±0.2 eV, in reasonable agreement with the results of electrical measurements [6.187].

For boron there is no detailed comparison with theoretical predictions because of the lack of suitable band-structure calculations. However, for C, Si, and Ge there are numerous calculations available which allow an exact interpretation of the X-ray spectra (Fig.6.41). The spectra of graphite and diamond differ considerably due to the striking differences in the structure of these carbon modifications. On the other hand, the spectra published by *Holliday* [6.18] are different from those by other researchers [6.44, 192–195]. The electronic structure of graphite can be understood within an LCAO model [6.196]. The σ bonds formed by overlapping sp² orbitals within the graphite layers form three subbands and the π bonds directed perpendicularly to the layers a further band. As a result of the different angular dependences of the σ → 1s and π → 1s transitions, the widths of the σ and π bands can be determined from angular resolved measurements with polarized X rays (Fig.6.41). Widths of 16.9 and 9.0 eV have thus been obtained; the total width of the valence band is 21.7 eV [6.63]. This appears roughly consistent with the results of MO calculations

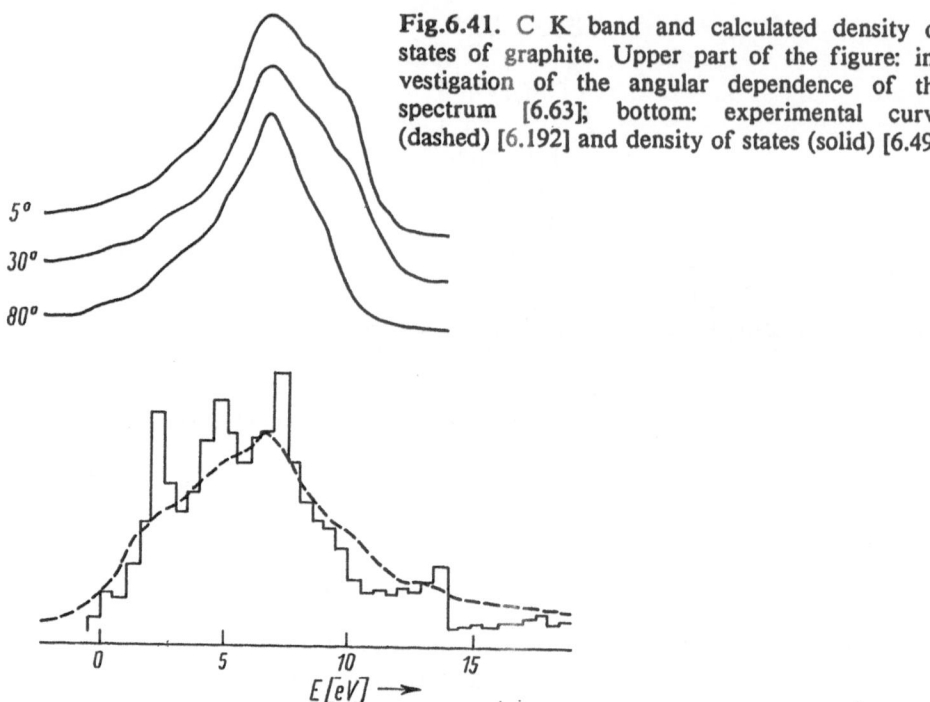

Fig.6.41. C K band and calculated density of states of graphite. Upper part of the figure: investigation of the angular dependence of the spectrum [6.63]; bottom: experimental curve (dashed) [6.192] and density of states (solid) [6.49]

by *Nefedov* et al. [6.197] (20 and 10 eV, respectively) but disproves the interpretation given by *Sagawa* [6.192] who claims clearly separated σ and π bands.

Satellites were found above and below the band at separations of 39.2 and 29.8 eV, respectively [6.198]. The high-energy satellite appears to arise from multiple-ionization states, while the low-energy satellite is thought to involve plasmon excitation, which earlier was assumed to exist only in metals. *Blokhin* et al. [6.155] also reported on the observation of a long-wavelength satellite (31 eV) and a short-wavelength satellite at a separation of 58 eV from the maximum. These values agree with calculations based on a many-body theory [6.199]. The K emission band of a-morphous carbon is similar to that of graphite; however, structural details are very faint [6.200].

With regard to crystal and electronic structure, diamond resembles silicon and germanium. This is also evident from both X-ray spectra and photoelectron spectra (Fig.6.42). The K band of diamond - the results of different measurements are very similar [6.18, 196] - shows clear similarities to those of Si and Ge. The somewhat less pronounced features of these bands as compared to the C K band in diamond follow from the increased smearing due to the larger K level widths in Si and Ge. Comparisons with the densities of states for all of those elements show considerable similarities, except for their different bandwidths. The observed features in the K bands have correspondences in the density of states

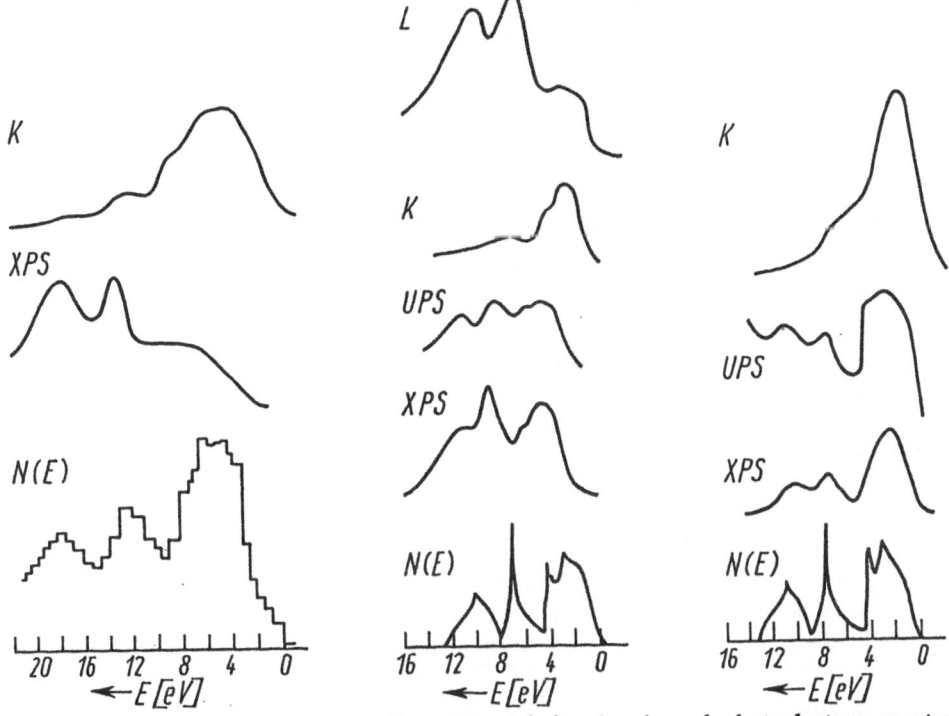

Fig.6.42. Comparison of X-ray K and L emission bands and photoelectron spectra (XPS, UPS) with the density of states N(E) of diamond (left), silicon (middle), and germanium (right) [6.45]

which can be related to the four valence bands, V_1 to V_4. The main maximum of the K band can be associated with V_3+V_4, the long-wavelength shoulder being explained as the density-of-states maximum related to V_3. The structures in the density of states caused by V_1 and V_2 are suppressed in the K bands due to the influence of the transition probability (Sect.6.2) - the number of p states decreases strongly from V_4 to V_1, and V_1 has almost pure s symmetry. Therefore, the shape of the Si L spectrum is determined mostly by V_1 and V_2. *Klima* [6.29] has reported results of a calculation of the spectra using more accurate transition probabilities which quantitatively confirms these qualitative interpretations; the agreement between theory and experiment is rather good. Together with the photoelectron spectra [6.12] shown in Fig.6.42 the relations between the different methods and the density of states are obvious (Sect.6.1).

Measurements of the Si K and L bands (Fig.6.19) and the L absorption spectrum of amorphous silicon [6.56] show appreciable smearing of the fine structure in comparison with results from crystalline material even for the energetically deeper lying valence bands; this has been confirmed by ESCA measurements [6.201]. Furthermore, small shifts of the spectra are reported. Measurements of the Si L absorption spectra with higher resolution using synchrotron radiation [6.202] have shown that the fine structure of the spectra from crystalline silicon can be associated with the density of states of the conduction band and that this structure disappears in amorphous silicon. Furthermore, there is a shift of the edge of 0.5 eV towards lower energies.

Recent investigations of the $L_{II,III}$ emission [6.203] and absorption spectra [6.204] of Si have shown that it is possible to derive even surface properties like filled surface/interface states and the influence of surface reconstruction on the partial density of unoccupied states from the X-ray spectra.

A comparison of the Ge $L_{II,III}$ and $M_{II,III}$ spectra calculated by *Klima* [6.29] with experiment is not meaningful as yet. These spectra differ, of course, because of the influence of the radial part of the transition probability (compare Sect.6.2), but the experimental results are not sufficient for comparison [6.44,205,206] due to the weak intensity and the appreciable spin-orbit splitting of the Ge $M_{II,III}$ levels (about 3.5eV) [6.44,206]. The Ge $M_{IV,V}$ band is similar to the Ge K band [6.206].

The X-ray spectra of phosphorus are shown in Fig.6.43. If the long-wavelength maximum in the L band is attributed to a satellite transition [6.207], then one can interpret the spectra on the same base as the ESCA observations on As, Sb, and Bi [6.201]. Thus, the double structure in the L main band should be due to transitions to valence states with mainly s symmetry and the short-wavelength structure in the K and L emission bands due to transitions from the three upper valence bands which have mainly p symmetry. Contrary to the group IV elements, in phosphorus there are available two extra s electrons which are inserted between the s and p bands as an additional band. In the $A^{III}B^V$ compounds there should not be two s bands due to the sp^3 hybridization.

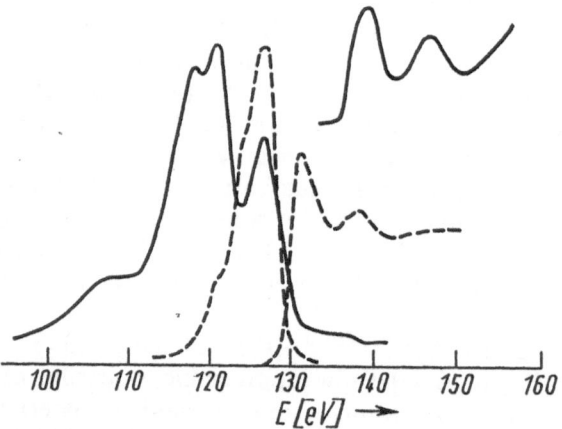

Fig.6.43. X-ray emission (left) and absorption spectra (right) of red phosphorus. (K spectra: dashed; L spectra: solid)

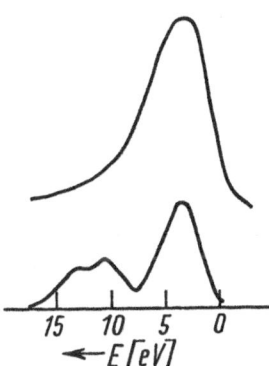

Fig.6.44. K emission band (top) and X-ray photoelectron spectrum (below) of As [6.102, 201]

Indeed, the P L bands in these compounds show only one prominent L maximum (Sect.6.4.1). The spectra of red and black phosphorus differ especially in the upper part of the valence band [6.208] where the influence of the lattice symmetry is particularly important (Sect.6.4.2).

The As $K\beta$ spectrum from the pure element can be explained similarly (Fig.6.44): The three upper valence bands have mainly p symmetry and form the main band. The upper of the two bands with predominantly s symmetry results only in a weak shoulder on the main band due to the weak hybridization with the p states. The lower s band has so little p symmetry that it cannot be clearly identified in the $K\beta$ emission band.

Experimental studies of the spectra of pure sulfur have been reported by numerous authors (the results of the different measurements agree well with each other) but there have been only few theoretical interpretations of these spectra. The agreement of the positions of the MO levels of the S_8 molecule with the S K and L emission bands and the K absorption spectrum [6.209] is not satisfactory. A comparison of the spectra with band-structure calculations by MO LCAO approaches is shown in Fig. 6.45. The experimental bandwidth (15eV) agrees better with the results of the EHT (16.5eV) than those of the CNDO calculation (21.5eV) but the shape of the $K\beta$ band is better described by the CNDO approximation. The upper of the six valence bands have mainly p symmetry and the lower mainly s symmetry, nevertheless hybridization occurs over all six bands.

6.4 Compounds

X-ray spectra are very well suited to give a survey of the electronic structure of compounds (Sect.6.2) because of their insensitivity to impurities and defects. This is especially important for compounds where (apart from a few simple cases), due to their complicated crystal and electronic struc-

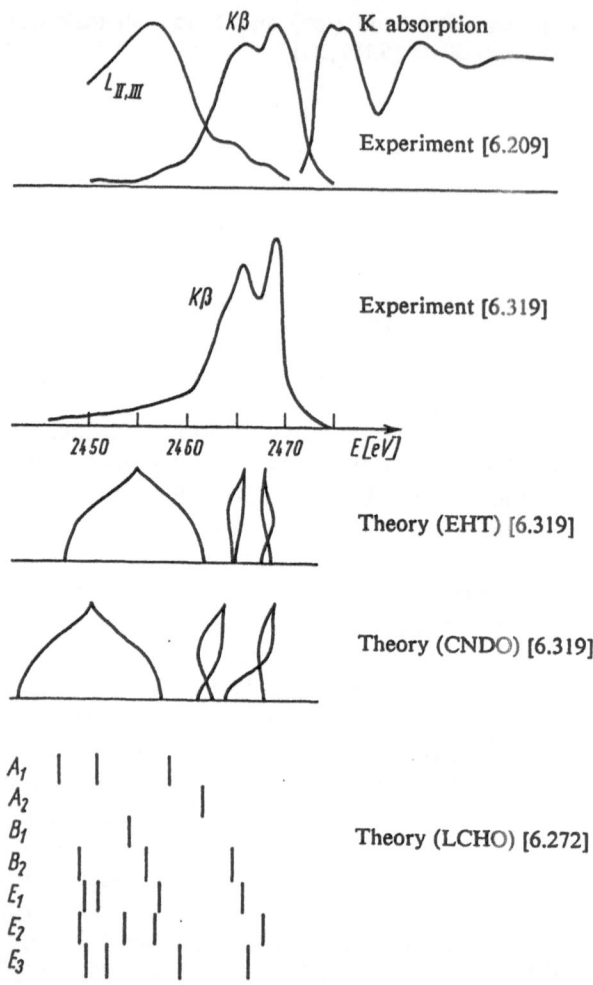

Fig.6.45. Comparison of the X-ray spectra of sulfur with the results of MO calculations

ture, calculations are possible only with radically simplified approximations. One should attempt to come to general conclusions on the basis of homologous series combined with detailed interpretations of spectra of simple compounds, where good calculations are possible. So one can get some conclusions regarding the electronic structure of complicated compounds. Thus X-ray spectroscopy can be of significant help in the development of general theories of the electronic structure of inorganic compounds. It should agree with current models of the chemical bonding in crystals [6.6, 210] or lead to a refinement of these models.

The scheme according to which the topic will be treated, follows from the above consideration. Starting with the electronic structure of the elements in the IVth group of the periodic table, binary compounds with eight valence electrons, the $A^N B^{8-N}$ compounds, can be discussed. In this way, an attempt is made to draw general conclusions on the electronic structure of inorganic crystals and to use these results as described in Sect.6.4.2 for other compounds of the elements of the main groups. As will be seen from the analysis given in Sect.6.4.1, the number of valence electrons, the kind of the bonding partner and the crystal structure of the compound each are essential for the spectrum. Since the d electrons are particularly important due to their strong localization, compounds with d electrons are discussed separately in Sect.6.4.3, where the great amount of experimental data on binary compounds of d metals warrants special attention.

In this connection, the difficulty to represent spectra on a uniform energy scale must be pointed out. While different spectra of one component of a compound can be brought into correspondence using measured values for X-ray spectra (e.g., K and L emission bands or absorption spectra using the $K\alpha$ lines), difficulties arise when attempting to compare spectra from different components. Then one has to resort to more or less well-founded assumptions [6.211]. A unique assignment is possible, however, using ESCA data for the energies of the initial states.

6.4.1 $A^N B^{8-N}$ Compounds

The $A^N B^{8-N}$ compounds can be systematized according to their crystal structures starting with zinc blende (most of the $A^{III} B^V$ and $A^{II} B^{VI}$ compounds have this structure) which is the binary analogue to the diamond structure of the elements of the IVth group. The essential qualitative difference in the band structure and density of states of such a binary substance compared to the group IV elements is the existence of an energy gap between the two deepest-lying valence bands V_1 and V_2, the "antisymmetric" energy gap (Fig.6.13). This implies that the X-ray emission bands of the substances with zinc-blende structure should be qualitatively similar to the spectra from the group IV elements; this agreement has been confirmed by experimental results for both K and L spectra (compare Figs.6.13, 15, 42, and 46).

The interpretation of these spectra is identical to that given in Sect.6.3.4 for the elements of the IVth main group, i.e., the individual peaks of the emission bands are identified with the valence bands V_1 to V_4. The structure connected with V_1 is separated from those dependent on V_2 to V_4 with increasing ionicity. This is because the antisymmetric energy gap between the valence bands – the denotation comes from the separation of the crystal potential into a symmetric and an antisymmetric part [6.6, 212] – also increases (Fig.6.61 below). For compounds with very high ionicity a separate long-wavelength side band originates, which has often been denoted a satellite and has been interpreted very differently (see the discussion of MgO). For BN [6.211] the so-called long-wave-

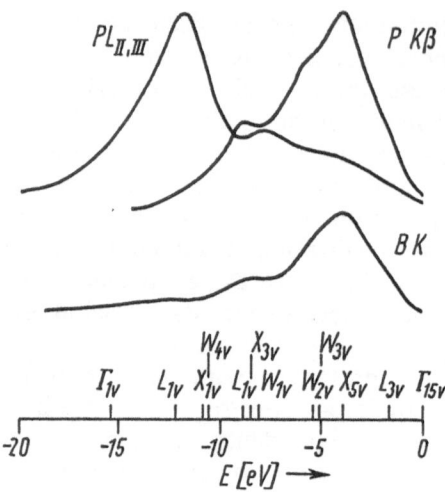

Fig.6.46. Comparison of experimentally obtained K and L emission bands of BP with energies of high-symmetry states in the band structure [6.8]

length satellite of the B K band corresponds to the valence band V_1. From the positions of structures in the spectra, band separations can be obtained for different $A^{III}B^V$ compounds which satisfactorily agree with data obtained from other experimental and theoretical methods (Table 6.1) [6.213].

A careful study of the spectra shows that there are characteristic differences between the spectra from the A and B components. In the K spectrum of the A components, e.g., of B in BP (Fig.6.46) [6.211] and of Al in AlP, AlSb and AlN [6.110], there is a clear long-wavelength side maximum caused by V_1. Furthermore, in the L spectrum of the A component the maximum caused by V_2 is much more intense than in the L spectrum of the B component, for example, for Al L in AlSb (Fig.6.14) [6.214] compared to P L in BP, GaP and InP (Fig.6.47) [6.208], see also below for SiC. Calculations of the spectra using an empirical pseudopotential method including the correct transition probabilities [6.33] gave particularly good agreement with experiment for the K spectra (Fig.6.47). The calculated intensity of the long-wavelength side band in the Ga K spectrum of GaAs agrees well with experimental results [6.32]. A discrepancy between theory and experiment for the L spectra (Fig.6.47) can be attributed to their larger sensitivity to transition probability effects against changes in the valence electron distribution, which is a consequence of the different spatial extension of the K and L shell wave functions. The same was proved by calculations of the spectra using nonorthogonalized valence electron wave functions (Sect.6.2.1) and also by the differences in the Sb Lγ spectra (5p → 2s transition) for AlSb, GaSb, and InSb [6.40]. On the other hand, the K spectra of the same element in different $A^{III}B^V$ compounds are very similar. Summarizing, it can be stated that the differences in the spectra of the A and B components can be explained by considering not only a P(E) dependence but the full P(E,k) one of the transition probability, as discussed in Sect.6.2.1.

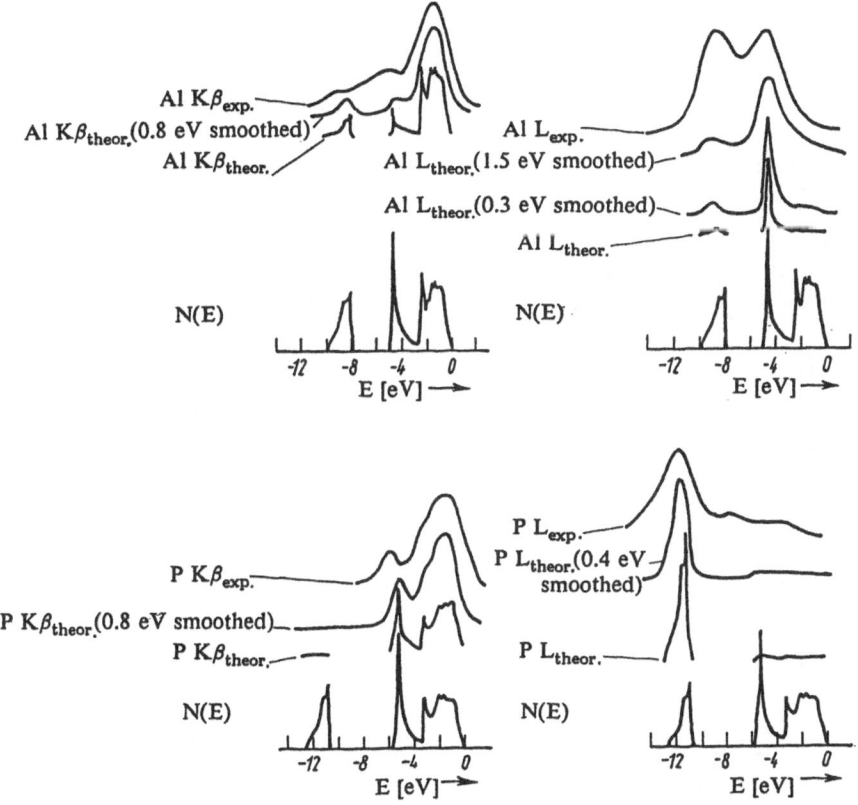

Fig.6.47. Experimental K and L X-ray emission bands with calculated density of states N(E) and calculated X-ray spectra for AlSb (top) and GaP (bottom). The theoretical spectra were folded with a Lorentzian curve [6.33]

The experimental bandwidths can be explained using Penn's model [6.8]. Thus, the change in bandwidth arises mainly from a change in the lattice dimensions. With fixed lattice parameters the bandwidth decreases with increasing ionicity.

Among the $A^{II}B^{VI}$ and $A^{I}B^{VII}$ compounds with zinc-blende structure ZnS, ZnSe, CdS, CuCl, and CuBr have been particularly well studied [6.28, 209, 214-216]. A special feature of these compounds compared to the $A^{III}B^{V}$ compounds is the existence of d electrons between the V_1 to V_4 valence bands. This is shown clearly in Fig.6.48 by the separation of the structure named $K\beta_5$ from $K\beta_2$. The former is due either to quadrupole transitions $(3d' \rightarrow 1s)$ [6.28] or to the 4p part of a band build up from the 3d atomic states, while the latter is simply a dipole-allowed transition $(4p \rightarrow 1s)$. Whereas the 3d level is separated by about 25 eV from the valence band in Ge, the 3d states lie within the valence band for ZnSe and CuBr. This is also confirmed by ESCA measurements (Fig.6.48) [6.217, 218]. The existence of a spectral structure in the region of the d states in the spectra

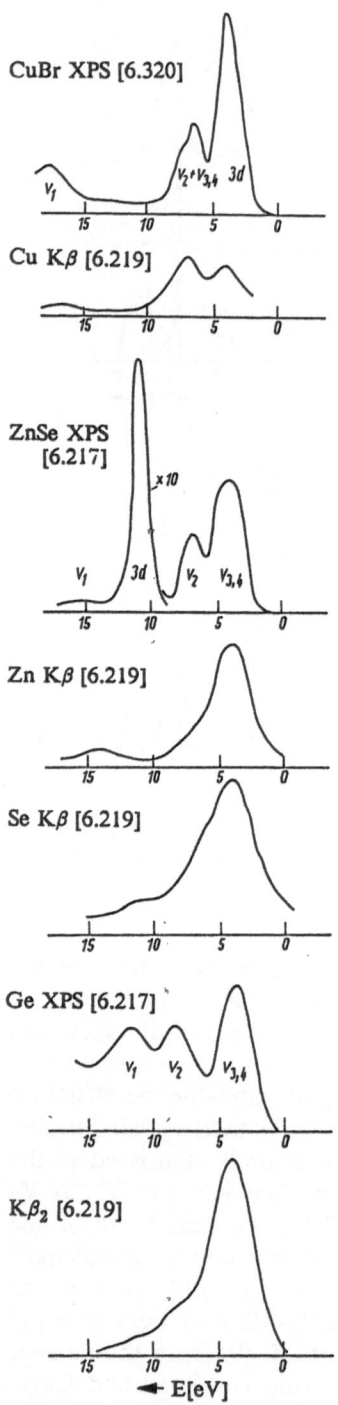

Fig.6.48. X-ray and photoelectron spectra of Ge (bottom), ZnSe (middle), and CuBr (top)

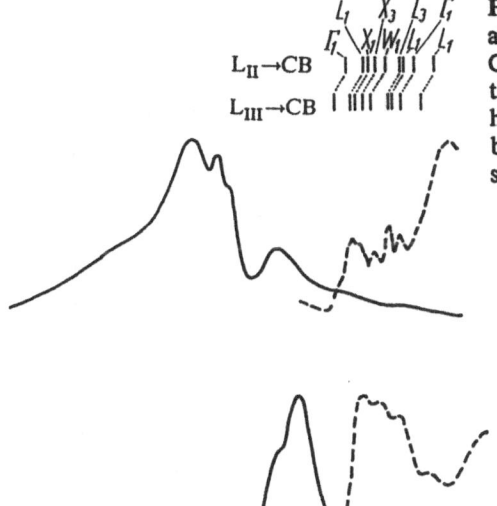

Fig.6.49. X-ray emission bands (solid) and absorption edges (dashed) of sulfur in CdS (bottom: K spectrum; top: L spectrum) and assignment of the structures to high-symmetry states of the conduction band, taking account also of the L_{II}-L_{III} spin-orbit splitting [6.220]

of the A and B components of ZnSe and CuBr indicates a hybridization of these states [6.219]. If the d bands are considered to be "inserted" between the other bands, then the spectra of ZnSe and CuBr can be explained similarly to those of $A^{III}B^{V}$ compounds in terms of the V_1 to V_4 bands. In this way, for example, the long-wavelength structure of the S L band in ZnS and CdS can also be identified by the d states (Fig.6.49) [6.215, 216]. Additional structures appear due to the spin-orbit splitting of the S $L_{II,III}$ levels. The energy positions of the 3d states determined by the X-ray spectra agree well with those determined by ESCA measurements and show that the calculations using an SC-OPW method [6.221] give too large values. Calculations using a relativistic SCF-OPW method [6.222] agree better with experiment for the positions of the 3d levels.

The structures in the vicinity of the S K and L absorption edges in CdS were explained by the energy values of high-symmetry states in the band structure (Fig.6.49) [6.220]. Consequently, the 4s states of sulfur are dominant in the lower conduction band. The strong increase of the absorption coefficient at about 170 eV can be explained as a delayed contribution of 2p → nd transitions [6.223].

Studies of substances with wurtzite structures have been carried out for SiC [6.224, 225], BN [6.211], AlN [6.214, 226], BeO [6.206], and ZnO [6. 227] (Sect.6.4.3). From a comparison of the Si K and L bands in SiC with those of pure silicon one clearly perceives the structures in the valence bands assigned to the V_1 to V_4 bands but the distance between V_1 and V_2 is appreciably increased in SiC. Furthermore, in the K spectrum of SiC

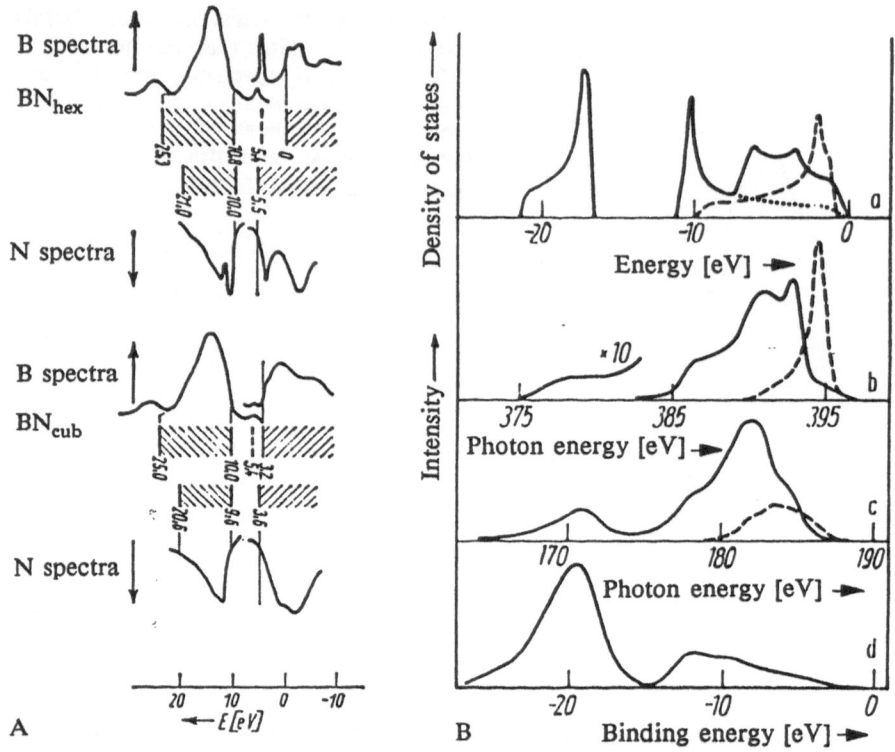

Fig.6.50. Spectra of BN: A) Comparison of the spectra of cubic and hexagonal BN [6.228] (left: K emission band; right: K absorption edge; hatched: experimentally determined bandwidths). B) σ (solid curve) and π bands (dashed) of hexagonal BN [6.64]. a) Theory, b) N K band, c) B K band, d) photoelectron spectrum

there is an intense side band, connected with V_1, and in the L spectrum of SiC the intensity ratio V_2/V_1 is considerably larger, too. Both these results lead to the conclusion that there is an electron transfer from the Si to the C atom in SiC.

For BN there is a clear difference between the wurtzite and zinc-blende structures especially in the N K band and in the absorption spectra (Fig.6.50). The spectra were aligned to a uniform energy scale using ESCA values; in this way, good agreement of all structures in the valence band was obtained. The spectra of hexagonal BN show a more pronounced structure in the upper valence bands V_3 and V_4 compared to the cubic structure. Similar features are found for both crystal structures of SiC [6.110], in qualitative agreement with band-structure calculations for the wurtzite structure [6.229], which show that the difference from the zinc-blende structure is mainly caused by the large splitting of the upper valence bands due to the decrease in symmetry. The short-wavelength maximum observed in the B K band of BN appears to arise from a re-emission process analogous to that found in the spectra of other boron

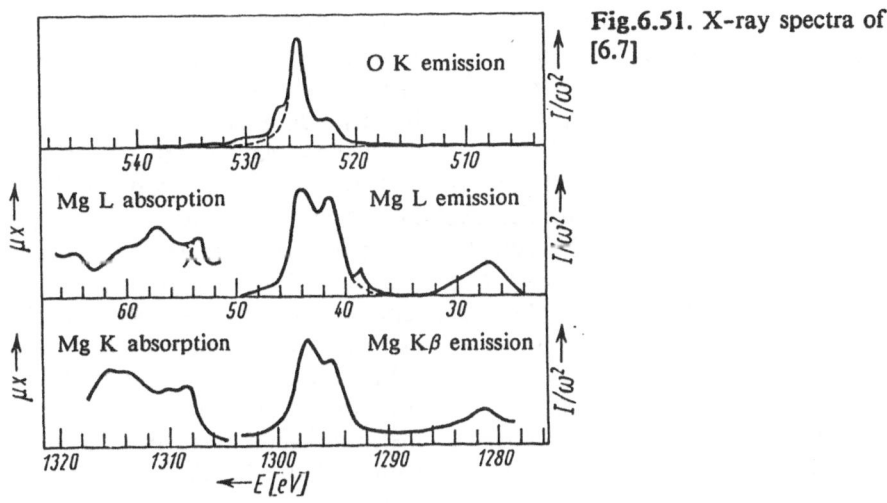

Fig.6.51. X-ray spectra of MgO [6.7]

compounds (Sect.6.4.2) [6.207]. The investigation of the anisotropy of the K spectra has permitted to determine the σ and π bands of hexagonal BN (Fig.6.50). Available band-structure calculations do not agree satisfactorily with experiment [6.64].

The Al spectra of AlN are similar to those of Mg in MgO and also show a structured region in the upper valence bands and a deeper lying band clearly separated from this. However, for both the Al K and L bands a larger intensity occurs in the short-wavelength region than in the spectra discussed earlier (for discussion of MgO, see below). Furthermore, structures are observed which obviously are due to Al_2O_3 [6.226].

The O K band of the BeO spectra is particularly interesting [6.230]. A comparison with the Be K band shows that the long-wavelength structures of the O K emission belong to the upper valence bands and that the short-wavelength structure is a satellite emission.

The energy band structure of substances with NaCl structure is composed of three upper and very flat bands formed mainly by states of p symmetry and one flat band separated from these which shows mainly s symmetry. Extensive X-ray spectroscopic studies exist for MgO and the alkali halides. The emission bands of MgO are shown in Fig.6.51. The O K band can be interpreted in a similar way to that of BeO, i.e., the splitting of the main band should be due to the structure of the upper valence bands of MgO. *Chun* and *Hendel* [6.230] did not find good correspondence of the O K bands among oxides with the same crystal structure (NaCl structure: MgO, BaO, NiO) but did find it when comparisons were made according to electronegativity. Three reasons could be suggested: Firstly, the band structures of substances with zinc-blende, wurtzite and NaCl structures are basically not different considering the moderate resolving power of X-ray spectroscopy. Thus, electronegativity influences can be more evident, since, for example, with increasing ionicity the bands get flatter. Secondly, the influence of the d electrons is important for NiO (Sect.6.4.3), and possibly for BaO, too. Thirdly, changes of the satellites caused by different chemical bonding can probably not be neglected.

As observed for the B K band of BN, the Be K band of BeO, and the Al K and L bands in AlN, well-resolved long-wavelength side bands of lower intensity are found also in the Mg K and L emission bands of MgO. *O'Bryan* and *Skinner* [6.122] explained these bands as cross-transitions, e.g., O 2s → Mg 1s. Different symmetry behavior of the O 2s wave function as seen from the O and Mg cores was invoked to suggest the possibility of such transitions. This concept was later developed further [6.231-233] to the point that, in one case, even the main K band was explained as a cross-transition (O 2p → Mg 1s) assuming a completely ionic bonding. However, calculations of the transition probabilities [6.52,53] have shown that the matrix elements of such transitions are too small to be able to account for the intensities of the bands in the X-ray spectra. According to later qualitative and quantitative studies [6.48,51-53,131], these long-wavelength bands are explained, as for molecules (Sect.2.4), in terms of dipole transitions of electrons of the more electropositive components from the valence band, which in turn is formed mainly by the valence electrons of the more electronegative components, i.e., as Mg 3p → 1s and Mg 3s → 2p. Thus, these side bands correspond to the valence band V_1 whose energy position depends on the kind of electronegative bonding partner. The position of V_1 is, to a first approximation, given by the energy of the corresponding atomic level (i.e., by the energy of the 2s level of N, O, F in all nitrides, oxides, fluorides). Due to the large binding energies of these levels and the large electronegativities of these elements the overlap with the orbitals of the electropositive partner will be less, i.e., the intensity of V_1 in the spectrum of the A component is distinctly smaller in AlN, MgO, and BeO compared to GaP and AlSb.

The structure of the upper valence bands, which occurs in the K and L bands of Mg and in the K band of O in MgO, was satisfactorily identified with the energy separation between the σ and π orbitals obtained from MO considerations [6.48]. Calculations for an AO_6 cluster [6.234] gave a value of about 3 eV which agrees rather well with experimental results. Band-structure calculations in [6.234] starting from the same potential gave a bandwidth from V_2 to V_4 of about 3 eV and a separation of the two maxima in the density of states in this region of about 1.6 eV.

The X-ray spectra of the alkali halides can be generally understood in terms of the band structure, where transitions from the valence band are expected to be seen only in the X-ray spectra of the anions. As an example, Fig.6.52 shows the Cl K spectrum of NaCl. As shown by different researchers [6.235-238], the Cl K emission band consists of a main peak associated with the valence band and short-wavelength satellite features that arise from multiple-ionization processes. The width of the upper, p-like main part of the valence band derived from the Cl K emission band

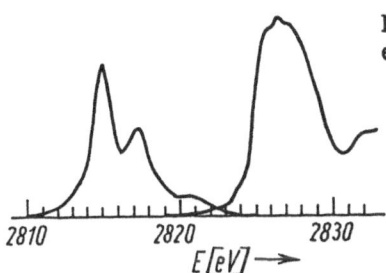

Fig.6.52. Cl K emission band (left) and absorption edge (right) of NaCl [6.235]

2810   2820   2830

$E [eV] \longrightarrow$

(about 3eV) is in agreement with photoelectron spectra [6.239] but is smaller than the calculated value (4.38eV) [6.213]. At the same time, in agreement with density-of-states calculations, the photoelectron spectrum shows structure in the valence band (also, for example, for LiF [6.70]), which cannot be discerned in the X-ray spectrum due to insufficient resolution.

Structure analogous to that found in NaCl is also present in the Cl bands of the heavier alkali halides [6.235,237] as well as in the F and Cl K bands of LiF [6.72] and LiCl [6.209]. However, newer investigations have shown that the short-wavelength structure reported earlier for the Li K band can be explained as caused by decomposition of the sample [6.75,240].

Although no transitions from the valence band to inner states of the cation are expected in the completely ionic picture, such transitions have been observed. A theoretical interpretation of the K and Cl K bands of KCl [6.241] suggests that the hole in the valence shell is located on a single atom during the lifetimes of the K and Cl 1s vacancies. This is said to lead to a real cross-transition Cl 3p $\rightarrow$ K 1s which is, in fact, seen; well resolved from the K $K\beta_{1,3}$ emission a short-wavelength feature (historically called $K\beta_5$) is found whose energy position corresponds exactly to that of the valence band. On the same basis, it is obvious that the width of the $K\beta_{1,3}$ line, which is a core transition 3p $\rightarrow$ 1s, has nothing to do with the valence band and the width of the $K\beta_5$ side band. The energy position of the Na $K\beta_1$ line in alkali halides suggested that this line is also due to a cross-transition [6.238,242]; a calculation of the transition probability [6.242] has shown, however, that the line's intensity is determined by the portion of Na 3p states in the valence band. Therefore, the intensity of this band is a measure of the degree of covalency of these compounds. Possibly, the same explanation holds for the Li K band in the halides. *Fomichev* [6.240], and *Elango* and *Maisie* [6.75] (using primary excita- tion),· besides bands from decomposition products, have found a Li K band, which, on one hand, does not agree in position with the F K band but, on the other hand, gives a satisfactory explanation of the absorption spectra. However, *Kosuch* et al. [6.243] could not find such a band with fluorescence excitation.

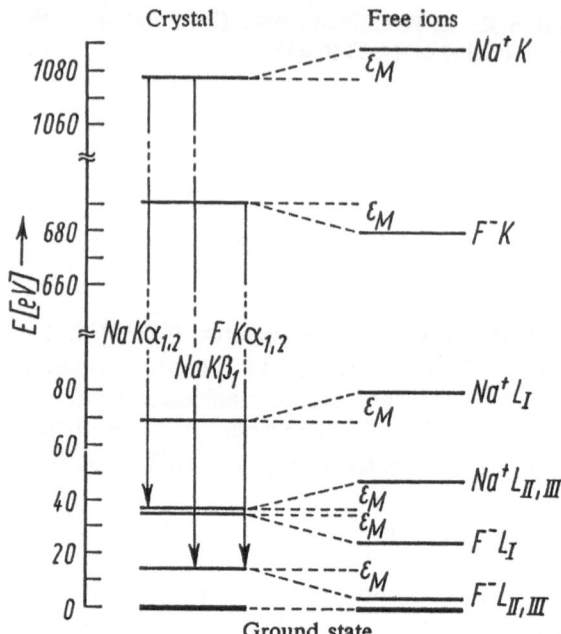

Fig.6.53. Energy level diagram of NaF determined by X-ray spectroscopy and theoretical calculations [6.242]

The energy intervals ascertained from X-ray spectra for the upper occupied anion and cation levels for ionic crystals can be explained using a point-charge model [6.242, 244]. By this model the energies E_v of the valence electrons of an ionic crystal can be predicted, neglecting the band broadening in the solid and the polarization following the interaction between the inner hole and the surrounding ions - for KCl, e.g., both these contributions are equal in magnitude and cancel each other [6.245]

$$E_v = E_i \pm \epsilon_M$$

(E_i ionization energy of the free ion, ϵ_M the Madelung energy). For NaF, for example, the calculated energy positions of the Na and F $K\alpha_{1,2}$ and Na $K\beta_1$ lines satisfactorily agree with experimental results (Fig.6.53). Moreover, results from photoelectron spectroscopy also agree with such a model [6.246-248].

Extensive investigations have also been carried out of the absorption spectra of alkali halides [6.235-238, 249], especially the utilization of synchrotron radiation has given numerous results in the ultrasoft X-ray region. In this way it is possible to link the X-ray region immediately with investigations of the optical properties of solids in the visible and UV regions. An important series of questions in such studies considers whether the absorption spectra merely reflect the density of states of the

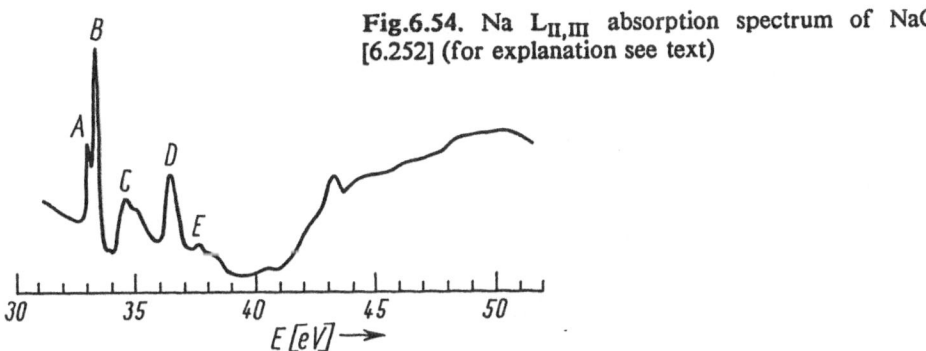

Fig.6.54. Na $L_{II,III}$ absorption spectrum of NaCl [6.252] (for explanation see text)

conduction band or what the influences of transitions into discrete excitation levels are.

As for LiF, fine structures of the Na $L_{II,III}$, K $M_{II,III}$, Rb $N_{II,III}$, and Cs $O_{II,III}$ absorption spectra are understandable in terms of the influence of potential barriers due to the anions surrounding the cation in an atomic or molecular picture [6.76, 77, 250, 251]. The energy position of the absorption edge E_c can be accurately estimated from electrostatic considerations as was done for E_v, viz.,

$$E_c = E_i \pm \epsilon_M - \epsilon_a$$

(ϵ_a is the electron affinity) [6.77]. The values obtained from this formula show, as is also evident from the experimental determination using $E_c = E_v + E_g$, that E_c lies considerably higher than the onset of the absorption spectra. The existence of excitons is thereby required; their characterization as localized states (Frenkel exciton) at the cation can be attributed to the presence of a potential barrier. Thus, for example, the structure in the Na $L_{II,III}$ absorption spectrum (Fig.6.54) can be associated with $2p^6 \rightarrow 2p^5 3\ell$ transitions ($\ell' = s, p, d$). The energy positions of the discrete levels in the free ion and in the crystal are shown in Fig.6.55. As a result of the localization of the 2p, 3s, and 3p states their energy separations are maintained and they are all shifted by the same amount ϵ_M. The Madelung energy for the delocalized states of the conduction band is equal to zero. The localization and thereby the shift of the d states falls inbetween these two values. Thus, it follows that the $2p^5 3p$ and $2p^5 3d$ levels represent autoionization states.

Quantitative estimates of energy values as required for the explanation of the fine structure must include, in addition to the electron-hole interaction, also the spin-orbit and crystal-field splittings. The differing interaction with the inner vacancy can also explain differences of the Na spectrum from the K, Rb, and Cs spectra (transitions to d states with different and equal principal quantum number as the vacancy in the latter, therefore multiplet structure [6.251]). By describing the crystal field using the ligand-field theory, one can also interpret the chemical shifts of the exci-

279

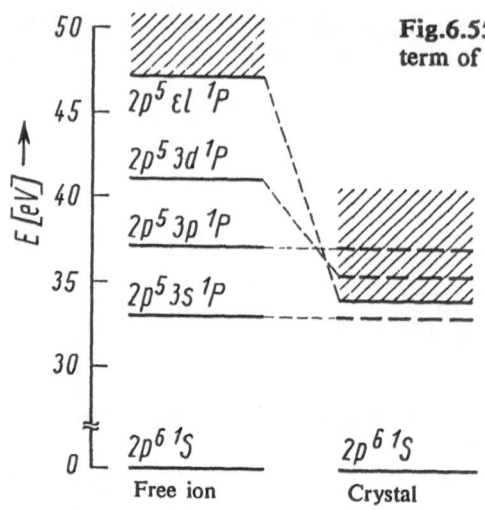

Fig.6.55. Influence of the crystal lattice on the ^1P term of Na$^+$ [6.77]

tonic levels taking refuge with the spectrochemical series [6.251]. Further-more, the line intensities are also influenced by electron-phonon and con-figuration interactions [6.77, 251]. Thus, the structures A, B, C, D, and E in the Na absorption spectrum can be explained as 2p → 3s (splitting into A and B due to spin-orbit interaction of the Na L$_{II,III}$ levels), 2p → 3d (splitting into C and D by the crystal field) and 2p→3p transitions (break-ing of the atomic selection rules through electron-phonon interaction). A further comparison between theory and experiment is shown in Fig.6.56.

This interpretation of absorption spectra with the help of Frenkel excitons is, according to *Satoko* and *Sugano* [6.251], equivalent to a de-

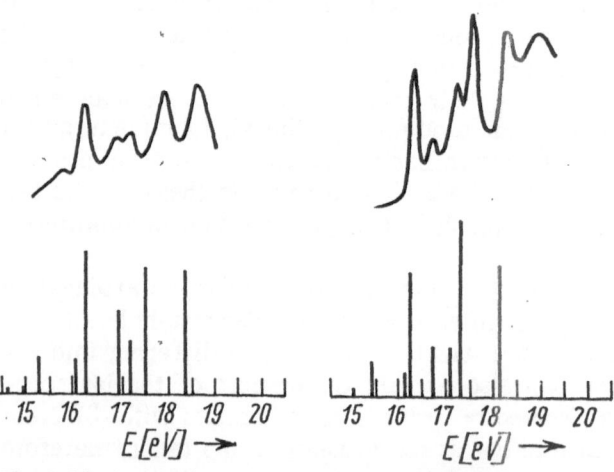

Fig.6.56. Comparison of the N$_{II,III}$ absorption spectra of Rb (top) in RbF (left) and RbCl (right) with calculations (bottom) [6.251]

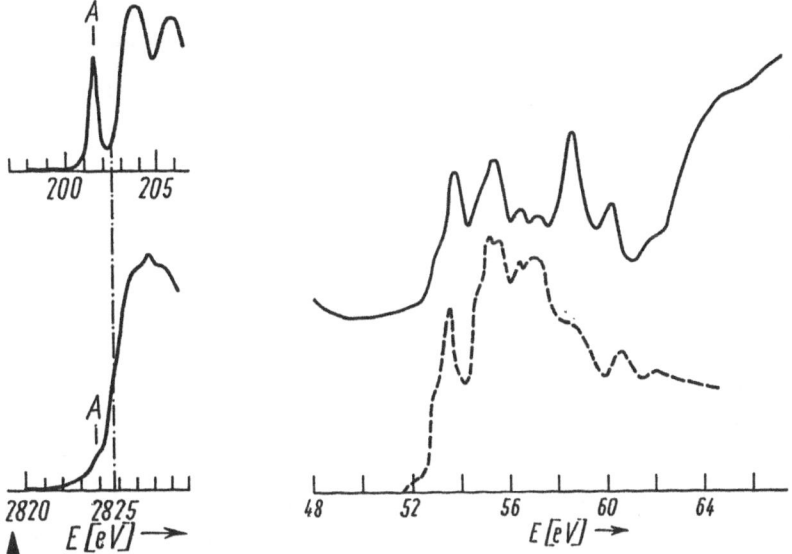

Fig.6.57. Comparison of Cl K (bottom) and $L_{II,III}$ absorption spectra (top) of NaCl [6.235]

Fig.6.58. Comparison of the I $N_{IV,V}$ absorption spectrum (solid) of KI with the calculated joint density of states (dashed) [6.253]

scription of $np^6 \rightarrow np^5(n+1)s$ transitions as delocalized Wannier excitons. This could possibly explain the existence of exciton-like structures in the spectra of the anions. An example of this case is shown in Fig.6.57 in which the structure A is referred to as a Γ exciton [6.235].

As was shown above for LiF, some interpretations of absorption spectra rely exclusively on the one-electron model and the conduction band structure. A particularly good agreement between the density of states of the conduction band and the absorption spectrum was found, for example, for the $N_{IV,V}$ spectrum in KI (Fig.6.58) [6.253,254]. The Cl $L_{II,III}$ spectrum in NaCl and the absorption spectra of LiCl could also be explained in this way [6.213,235] assuming additional structures caused by plasmons [6.249,255].

The shape of absorption spectra with different symmetries of the initial state was studied for the I and Cs L_I and $L_{II,III}$ absorption spectra [6.254,256]. The L_I spectrum is clearly different from the L_{II} and L_{III} spectra, the latter being very similar. The L_I and $L_{II,III}$ spectra were interpreted using band-structure data [6.257] as transitions of the 2s and 2p electrons, respectively, to maxima in the density of states of the conduction band with corresponding symmetries (Fig.6.59). The I L_I spectrum shows some similarity to the $N_{IV,V}$ spectrum (both are transitions into p states) if the $N_{IV,V}$ spin-orbit splitting is taken into account. Also the Cs N_{III} absorption spectrum is similar to the Cs L_{III} spectrum. The clear differences particularly in the $L_{II,III}$ spectra of both components of CsI can

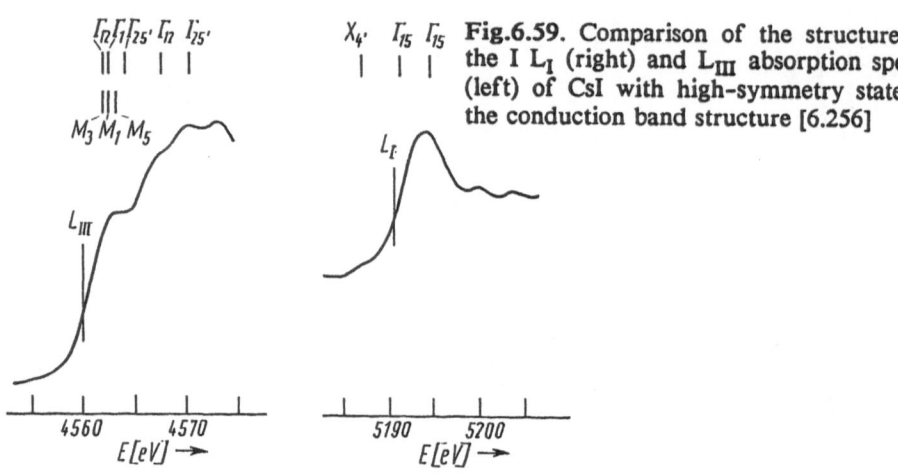

Fig.6.59. Comparison of the structures of the I L_I (right) and L_{III} absorption spectra (left) of CsI with high-symmetry states in the conduction band structure [6.256]

be attributed to different charge distributions of the states at the Γ, X, and M points in the Brillouin zone. Influences of the lattice type are evident from the similarity of the Cs L_I and K K spectra of CsF and KF and the differences between the same spectra of CsCl, CsBr, and CsI from that of KCl, KBr, and KI (CsF, KF, KCl, KBr, KI all have the NaCl structure while CsCl, CsBr, and CsI have the CsCl structure).

From the interpretation of the X-ray spectra of $A^N B^{8-N}$ compounds the following conclusions on their electronic structure can be drawn.

1) The valence band structure can be satisfactorily understood in terms of the ground state of the system of electrons. Excited states and many-electron effects do not have appreciable influence. The whole valence band of $A^N B^{8-N}$ compounds consists of four valence bands whose energy positions are determined by the kind of bonding partners and the interactions between them. Thereby, hybridization of the electronic states takes place in all four valence bands, even for the highly ionic compounds.

2) The energy separation of the bands for ionic crystals is governed by electrostatic relations, while for more covalent compounds the energetic order of the bands can be obtained to a first approximation from atomic data. In particular, the lowest valence band V_1 is mainly formed from s states of the B component. The overlap with the states of the A component is dependent on the kind of the B component and the electronegativity difference, i.e., the bond ionicity (Fig.6.60).

3) The splitting of the valence band into a lower (V_1) and an upper part (V_2 to V_4) and the widths of these bands can be directly related to the antisymmetric part of the crystal potential and to the ionicity of substances using the dielectric theory (Fig.6.61). For the bond ionicity (4.9) can be used [6.33, 258]. By means of this relation results of band-structure calculations can be compared with experimental values of the widths of the valence bands determined by X-ray spectra (Table 6.1).

4) The influence of the crystal structure consists in a change of the shape

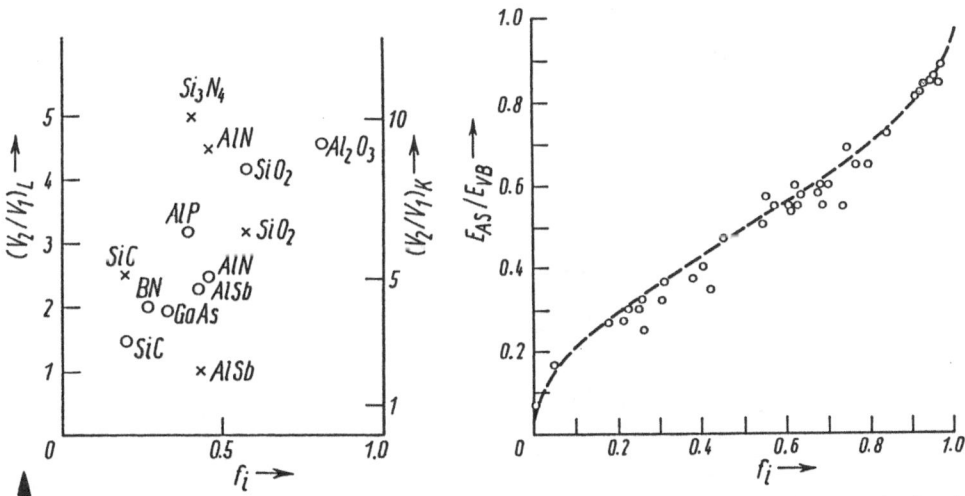

Fig.6.60. Variations of intensity ratios of the valence bands V_1 and V_2, obtained from the K (circles) and L (crosses) emission bands, as a function of ionicity f_i

Fig.6.61. Relation between the relative width of the antisymmetric energy gap E_{AS}/E_{VB} and the ionicity f_i of various compounds [6.258]

of the upper valence bands V_2 to V_4 additionally to the ionicity effects described in (2.) and (3.) and determining mainly the separation of V_1 and V_2. In the amorphous state there is an obvious loss of structural details in the spectra.

5) Thus, essential classification parameters for the electronic structure of the valence bands of inorganic solids are: the number of valence electrons (determines the number of bands), the kind of bonding partner (determines the relative energy position of the valence bands), and the lattice type of the substance (determines the splitting, in particular, of the upper valence bands).

6) If many-body effects (especially excitonic levels) are taken into account statements about the conduction-band structure are possible. The origin of excitons can be understood with the aid of the potential-barrier picture (the barrier arises from the negative ions surrounding the cation permitting states of the photoelectron to be considered as Frenkel excitons). Quantitative values are obtained if effects of the electron-hole interaction, spin-orbit coupling and crystal-field splitting are included.

6.4.2 Other Compounds of Main-Group Elements

In this section the validity of the conclusions obtained in Sect.6.4.1 for $A^N B^{8-N}$ compounds will be tested for other compounds of the main-group elements. This is possible only qualitatively due to the limited number of systematic studies of such compounds. Following the conclu-

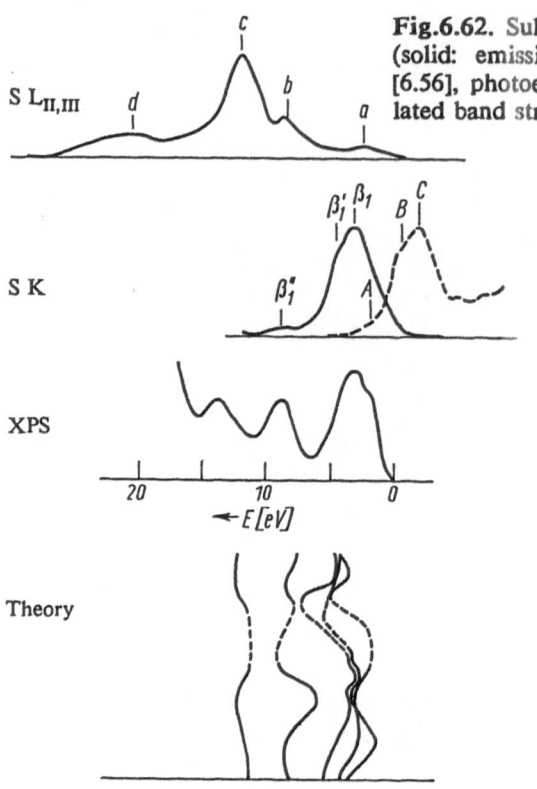

Fig.6.62. Sulfur $L_{II,III}$ emission band, K spectra (solid: emission band; dashed: absorption edge) [6.56], photoelectron spectrum [6.259], and calculated band structure [6.260] of PbS

sions in Sect.6.4.1 (Point 5) the compounds are systematized first according to the number of valence electrons in three groups: AB compounds with ten valence electrons, $A^N B_n^{16-N}$, and $A_m B_n$ compounds.

Of the binary compounds with ten valence electrons, X-ray spectroscopic studies have been presented for TlCl [6.198] and PbS [6.55]. The Cl $L_{II,III}$ emission spectrum in TlCl differs from those of the alkali halides by a small intensity in the region of the upper valence bands which points to s and d symmetry relative to the Cl position in this energy region. This should be expected as a result of the two additional electrons. The width of the band and its shape agree well with the fine structure of the theoretically predicted band.

The S spectra of PbS closely resemble those of the $A^{II} B^{VI}$ compounds (Fig.6.62). If the spectra are compared with band-structure calculations, however, the short-wavelength bands (a or β_1 and β_1') do not correspond to the notion of two mainly p-symmetry bands as for the $A^{III} B^V$ and $A^{II} B^{VI}$ compounds but rather suggest three such bands. The maxima β_1'' or b and c can be associated with s-symmetry bands which are dominantly formed by the Pb 6s and S 3s electrons. The maximum d can be correlated to transitions from states having mainly Pb 5d symmetry which overlap with S 3s states and then participate in the bonding. These interpretations and also the energy separations of the individual bands agree well with ESCA measurements (Fig.6.62). In the S K absorption spectrum, the shoulder A can be assigned to the onset of the conduction band, the structures B and C to states in the band structure with p symmetry. A generalization of these results suggests that the valence band structure of $A^{IV}B^{VI}$ compounds is characterized by two rather strongly bound s bands (mainly from atomic s states of the VI and IV components, respectively) and three p-symmetric bands (mainly from the p states of the VI component) at the upper valence band edge. Starting with the models of chemical bonding in $A^N B^{8-N}$ compounds it is possible to understand the electronic structure of $A^{IV}B^{VI}$ compounds as binary analogues of the elements of the Vth main group, whose band structures are rather similar, as shown by ESCA measurements [6.201]. The two additional electrons added to the existing octet are "inserted" as an additional s band between V_1 and V_2 to V_4.

X-ray spectroscopic studies of compounds of the type $A^N B_n^{16-N}$ have been reported for SiO_2, GeO_2, and MeF_2 (Me: Ca, Sr, Ba). Most of the results have been obtained for SiO_2. Figure 6.3 summarizes these spectra. The long-wavelength side bands in the Si spectra can be understood similarly to those for MgO (Sect.6.4.1). Certainly, for SiO_2 the cross-transition picture presuming fully ionic bonding [6.232] is less probable, a conclusion which is supported by calculations of the transition probabilities for SiO_2 [6.52,53]. Interpretations of the spectra based on an LCAO-MO model (including dipole-selection rules) have been reported by several researchers [6.48,52,53,261]. This model predicts that the long-wavelength side bands in the Si K and L spectra reflect the fractions of Si 3p and 3s wave functions in the lowest energy band. It also was found that both structures in the upper valence bands can be explained as due to the σ-π splitting of the corresponding orbitals. Since this splitting is dependent on the geometrical arrangement of the atoms, an influence of the lattice structure should be expected (compare also Point 4 in Sect.6.4.1). *Läuger* [6.110] found a clearly changed profile of the Si $K\beta$ band in stishovite (a high-pressure form of SiO_2 with 6-fold coordination) compared to the tetrahedrally coordinated compounds of Si. *Brytov* et al. [6.262] generalized these differences between the tetrahedral and octahedral coordination using an MO model to explain Si L and O K emission and absorption spectra and also the corresponding spectra for Al-O compounds.

Breeze and *Perkins* [6.263] calculated the density of states based on Bloch functions with the approximations of the HMO theory. Not to mention quantitative differences of the energy positions of individual bands, no calculations have yet given a complete even qualitative explanation of the spectra (e.g., in [6.263] there are no Si 3p states predicted in the lower valence band; in [6.261] the upper valence band is dominated by O 2p functions; cf. Fig.6.3). From the extensive SCF calculation of the O K band performed by *Gilbert* et al. [6.53], it follows that in contrast to results of other MO calculations [6.52, 261] the wave functions of oxygen are not strongly hybridized and d contributions are not important. Furthermore, these calculations show the clear influence of the change of the shape of the wave function on the intensity of the spectra (e.g., the long-wavelength Si bands), which in contrast to the usual approaches (see above) cannot be simply determined by the eigenvectors from the LCAO approach. The calculations also show an influence of the bending of the Si-O-Si bond on the σ-π splitting. Calculations of the multiplet structure of atomic oxygen also gave satisfactory agreement with experiment. *Ruffa* [6.264] explained the separation and intensity ratio of the σ and π parts in the O K band considering excited states (disruption of the Si-O bonding by transition of the bonding O σ electrons to the 1s level).

The Ge $M_{II,III}$ emission spectrum in GeO_2 [6.265] can be understood in a similar way as the SiO_2 spectrum. The spectrum is composed of a region containing the upper valence bands, the lower valence band, and $M_{IV,V} \rightarrow M_{II,III}$ transitions (which have been erroneously explained as transitions from MO levels by *Iguchi*). The upper valence bands show three maxima, which agree with results of an APW calculation [6.266], even though the experimentally determined full widths of these bands are larger than those expected from the calculations. The width of the forbidden gap between valence and conduction band nearly corresponds to the theoretical value.

Both the width of the energy gap between the upper and lower valence bands and the relative intensities in the spectrum of SiO_2 (compare Figs.6.60, 61) conform to the expectations. The smaller widths of the upper valence bands in MeF_2 compounds [6.267] (after subtracting the satellites at about 3-4 eV instead of 9.5 eV in SiO_2) also agree with these estimations. The increasing energy gap between the bands from SiO_2 to MeF_2 cannot, however, be directly inferred from the available X-ray spectra of alkaline earth fluorides, but photoelectron spectra [6.239.267] give indications thereof. These spectra can, at least, be qualitatively interpreted by means of a theoretical picture according to which the Me np levels lie in the energy gap [6.268].

The existence of a Ca L emission band in the region of the upper valence bands can be explained as arising from a certain covalency in the bonding of the alkaline earth fluoride, but it can, however, partially be attributed to decomposition of the sample as was the case for alkali halides (Sect.6.4.1).

The L absorption spectra of Mg halides [6.269] show, as did those of the alkali halides, exciton levels which are linked to the conduction band minimum at the Γ point, where the band is mainly formed by s states of Mg (corresponding to a transition $2p^6 \rightarrow 2p^5 3s$ in the free Mg^{2+} ion). This structure is followed by structures that originate from interband transitions to final states with s and d symmetries, and at about 20 eV above the $L_{II,III}$ absorption edges by delayed $2p \rightarrow d$ transitions, as has also been found for metallic Mg. At still higher energies transitions from

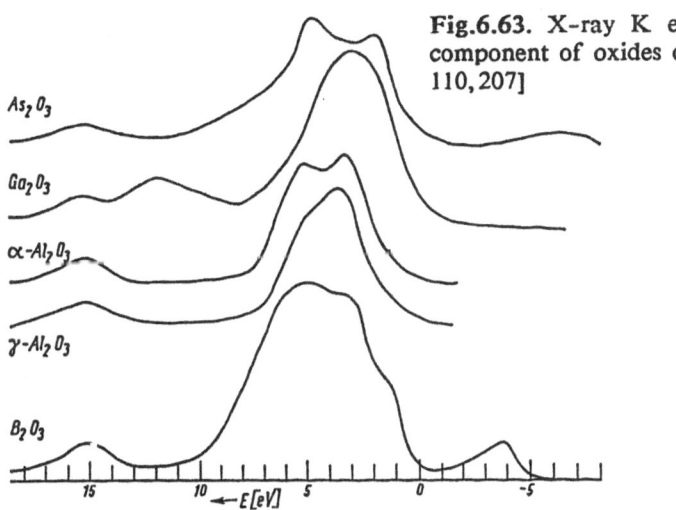

Fig.6.63. X-ray K emission band of the A component of oxides of the type A_2B_3 [6.102, 110, 207]

the 2s state of Mg to p states in the conduction band occur (L_I edge). In the Ca L absorption spectrum of CaF_2 [6.267] sharp maxima have been observed which can be assigned to transitions to empty d states.

Investigations of compounds of the type A_2B_3 have been carried out for B_2O_3, Al_2O_3, Ga_2O_3, As_2O_3, Sb_2O_3, As_2S_3, Sb_2S_3, and As_2Se_3. Parallel investigations of both X-ray and photoelectron spectra showed that a consistent interpretation was possible for all the samples [6.270, 271] (Fig.6.63). According to this, these compounds form three bands with binding energies E_B < 10eV arising mainly from non-bonding electron pairs of oxygen or sulfur (bands with the smallest E_B) and from p_σ bonds between the bonding partners. The structure of these upper valence bands should depend on the lattice type as stated in Sect.6.4.1. This is clearly demonstrated, e.g., from the differences of the Al $K\beta$ bands in α- and γ-Al_2O_3 [6.110] and has been explained using MO theories [6.262]. Furthermore, the particular spectrum of B_2O_3 supports these assumptions (Fig.6.63) [6.207].

The existence of long-wavelength side bands in the oxides and sulfides is analogous to MgO and SiO_2 and can be understood in terms of a band produced mainly by 2s and 3s states of oxygen or sulfur. In compounds of group Va elements an additional valence band should appear as compared to the analogous compounds of group IIIa elements. From the models described in Sect.6.4.1, and applied successfully to AB compounds with ten valence electrons, there should be one band formed mainly by ns valence electrons, which according to ESCA data [6.203] should be situated about 10-14 eV below the Fermi level and therefore in the energy gap between the upper and lower valence bands. Indeed, the As $K\beta$ bands of all As compounds have a shoulder in this region, which is not present in the corresponding compounds of group III elements (compare Fig.6.64). This band is clearly visible in the photoelectron spectra due to its high photoelectric cross-section.

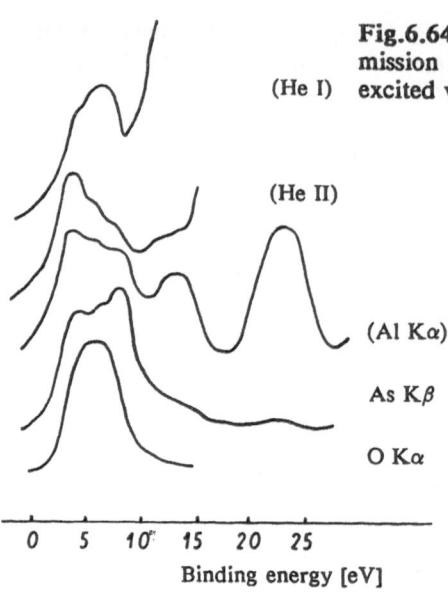

(He I)

(He II)

(Al Kα)

As Kβ

O Kα

Binding energy [eV]

The $M_{IV,V}$ spectra of both components of As_2Se_3 have been studied in emission and absorption. It is interesting to note that in the spectrum of As, there is a long-wavelength maximum which cannot be observed in the spectrum of Se, so agreeing with results from the $A^{III}B^{V}$ compounds. The splitting of the upper valence bands (ca. 2.5 eV) and the separation of the maxima of the upper and lower valence bands (4-5eV), determined from the spectra, differ appreciably from the predictions of an MO calculation [6.272] (ca. 0.5 and 9eV, respectively).

All compounds described so far are characterized by only one type of bonding between the atoms. The observed relations between electronic structure and chemical bonding cannot automatically be transferred to compounds with mixed bonding. Such a special case is given, for example, for most of the borides, silicides, and germanides, which have been frequently studied [6.21,273-276]. A reevaluation of predictions from the dielectric theory regarding additivity of the bonding parameters [6.210] could elucidate further the relations between band structure and chemical bonding, considering those experimental results.

6.4.3 Transition-Metal Compounds

Compounds of transition metals have special magnetic and electric properties due to the presence of d electrons. Accordingly, studies of the electronic structure of these substances have been of great interest. Following the order of treatment in Sect.6.4.1, the binary compounds will be discussed first.

Compounds of d metals from the IIIrd, IVth, and Vth groups with O, C, and N have been studied frequently with X-ray spectroscopy. The emission spectra of different substances from these groups are rather similar: The K spectrum of the nonmetal represents a partially structured band, the K and L main bands of the metals are more or less splitted with long-wavelength side bands (e.g., TiO, Fig.6.11). This splitting and also the separation of the side band from the main band decrease clearly in the order MeO, MeN, MeC.

Different models have been proposed to explain the existence of the long-wavelength side band. Already *Vainshtein* et al. [6.277] interpreted this $K\beta''$ band in the oxide as due to a cross-transition O 2s → Me 1s as was suggested above for the $K\beta'$ satellite in the $K\beta$ bands of compounds of main-group elements (compare Sects.2.4 and 6.4.1). Here, for the transition elements analogously to the main-group elements, the energy position of $K\beta''$ depends basically on the nonmetallic component. *Fischer's* interpretation [6.42] based on MO theory corresponds to the models used for the main-group elements (see MgO and SiO_2, and Fig.6.65), according to which the emission bands of transition-metal compounds are caused solely by dipole transitions and the intensity of the bands reflects the number of appropriate electronic states in the corresponding parts of the valence band. Therefore, the long-wavelength side bands correspond to the deepest-lying valence bands in agreement with conclusions from Sect.6.4.1. The energy position of this band is to a first approximation determined by the atomic s states of the nonmetal.

An extensive discussion giving all details for the spectra of TiO, TiN, and TiC based on band theory is possible from the results of the APW calculations by *Ern* and *Switendick* [6.279], and has been pursued in particular by *Nemnonov* and co-workers [6.151,278,280]. Lately, mixed crystals of these compounds and defect structures have been investigated both theoretically and experimentally [6.281-283]. From these investigations it follows that in these substances there are three main valence regions: a low-energy band mainly formed by the nonmetal 2s states, a band formed by the 2p states of the non-metal and 3d4p states of the metal, and finally a band mainly formed by 3d states (Fig.6.11). The two latter bands can be considered to arise from metal-nonmetal and metal-metal electron interaction processes. An analysis of the electron distribution based on dipole selection rules gives rather good agreement between X-ray spectra and the partial density of states. The intensity of the long-wavelength side bands of the K and L bands of the metals can be referred to as 4p and 3d4s states (as also MO calculations predict) and their energy separations from the main band are basically determined by the 2p-2s separation of the nonmetal. The splitting of the main band can be estimated from the distance of the 2p(3d4sp) and 3d(4s) bands. Thus, the intensities of the subbands in the K and L spectra reflect the decreasing number of 4p states and the related increases in the number of 3d states with increasing energy; their relative intensities reflect the proportions of the correspond-

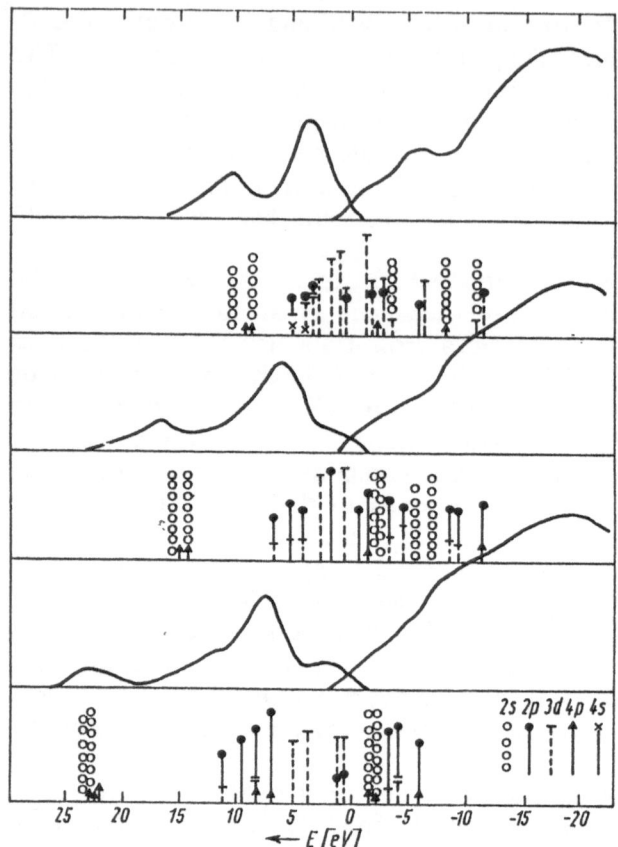

Fig.6.65. Comparison of the Ti K emission band (left) and absorption edge (right) of TiO (bottom), TiN (middle) and TiC (top) with theoretical results [6.278]

ing electronic states. The ratio of the two subbands of the Ti K band in TiO, TiN, and TiC corresponds to the statistical weight of the Ti 4p states [6.280].

Calculations of the band structure and the partial densities of states for the carbides of Ti, V, Zr, Hf, Nb, and Ta of different researchers [6.41, 284-286] support the above discussion on the correlation between X-ray spectra and partial densities of states (Figs.6.10,65). For heavier elements, however, relativistic calculations are needed [6.287]. The calculations by *Ern* and *Switendick* [6.279] suggest that at an occupation of the valence band by eight electrons the band formed by metal-nonmetal interaction should be completely filled. At still higher electron concentration (TiN, TiO), occupation of the next band starts. Accordingly, only for TiN and TiO a clear splitting of the Ti main emission bands can be observed and, in particular, an increase in intensity of the short-wavelength component of the Ti L band occurs. From investigations of the X-ray spectra of corresponding compounds of the group IIIb to Vb metals it follows that a

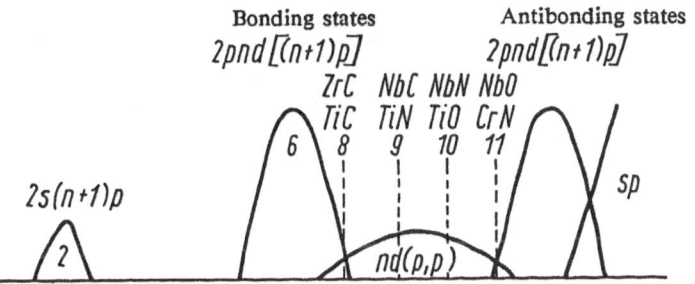

Fig.6.66. Schematic band-structure model for compounds with NaCl structure with numbers of valence electrons [6.151]

similar picture holds for all compounds with NaCl structure [6.280]. Occupation of the d band starts at an electron concentration of eight electrons (ScN, YN, TiC, ZrC, HfC) and at an occupation number of eleven electrons per unit cell (CrN, WN, VO, NbO) the d band is filled (Fig.6.66). Thus, the increase of the intensity of the short-wavelength component of the main band in the K and L spectra in the series MeC < MeN < MeO and similarly in the series TiN < VN < CrN (Fig.6.67) and TiN ≃ ZrN < NbN, etc., can be understood. Such a formation of additional d bands at a number of electrons in excess of the octet corresponds to the features observed for the $A^N B^{8-N}$ compounds. In this way some of the fine-structure components in the near-edge absorption structure can also be understood as transitions into unoccupied states of the d band and the following antibonding 4sp band. The separation of these bands is clearly seen, for example, in the K absorption spectra [6.151, 278].

Brytov and *Kurmaev* [6.227], in particular, discussed the binary oxides of all 3d elements (Fig.6.68). It is especially interesting to consider whether the differences of the electrical properties of the oxides are reflected in the spectra; for example, TiO and VO are electrical conductors in contrast to all other oxides [6.288]. According to assumptions from different researchers [6.289], such metal-insulator transitions can arise from changes in the overlap of the 3d wave functions and, therefore, by a formation of delocalized bands from local levels at a critical value of the lattice constant. Indeed, the shapes of particularly the Ti and V $L_{II,III}$ bands of TiO and VO are clearly different from the analogous bands of the other 3d elements in their oxides, but a unique statement about the local character of the 3d

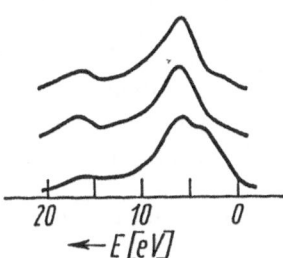

Fig.6.67. Kβ bands of Ti (top), V (middle), and Cr (bottom) in the series TiN, VN, CrN with increasing valence-electron concentration [6.151]

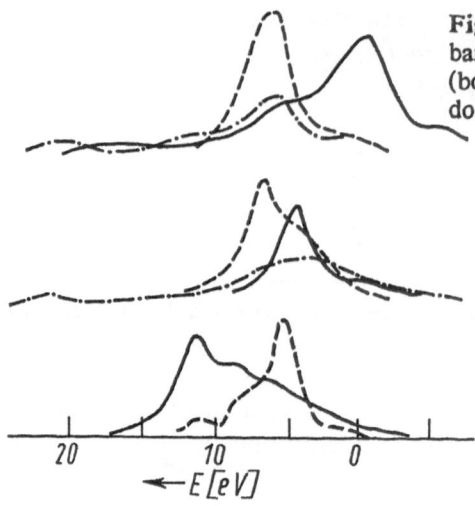

Fig.6.68. Comparison of the X-ray emission bands of TiO (top), CuO (middle), and ZnO (bottom). Solid: Ti, Cu, Zn $L\alpha$; dashed-dotted: Ti, Cu $K\beta_5$; dashed: O $K\alpha$ [6.227]

bands is not possible. The L emission and absorption spectra overlap in all oxides, i.e., there is no energy gap between the occupied and unoccupied d states. The energy position of this overlap is at the Fermi level, and with increasing atomic number of the metal the d states become localized at the metal and do not contribute to the conductivity [6.262].

From the above discussion it can be concluded merely that in TiO and VO in addition to the metal-nonmetal interactions there are also strong metal-metal interactions. With oxides of heavier 3d metals, judged by their electrical properties, probably the first kind of interaction is dominant. The agreement between the energetic positions of the bands determined experimentally and the theoretical values is not very good. On the other hand, a calculation of the crystal field splitting of the 3d states (TiO: 4.8eV; VO: 4.6eV; FeO: 0.65eV; NiO: 0.72 eV) agrees well with experiment [6.290] and also agrees with the conclusions from Sect.6.4.1.

Figure 6.68 shows that from TiO to CuO the separation between the 2p and 3d bands decreases. For ZnO the 3d band is strongly localized and situated well below the 2p band. Thus, there is a natural connection to $A^N B^{8-N}$ compounds (see Sect.6.4.1 and below). This also follows from a comparison of the band structures of these compounds (e.g., TiO [6.279]). The O K band of ZnO resembles the K band of the B component of $A^N B^{8-N}$ compounds. The S $K\beta$ spectra of NiS and CuS [6.291,292] can be understood from a comparison with ZnS by the shift of the 3d band toward the upper valence band edge in the series ZnS-CuS-NiS. In CuS and NiS the d states of the metal are above the 3p states of sulfur in contrast to ZnS. Furthermore, the influence of the lattice symmetry on the splitting of the upper valence bands is very clear, for example, in the spectra of MnS, compare Fig.6.18 [6.293].

Recently, the discovery of the high-temperature superconducting ceramics has strongly attached the interest of the whole scientific community [6.294]. These compounds are derived from La_2CuO_4 with a perovskite-like lattice. After starting with $La_{2-x}Ba_xCuO_{4-y}$ [6.294], the compounds $La_{2-x}Sr_xCuO_{4-y}$, and especially $YBa_2Cu_3O_{7-y}$ [6.295] with a much higher transition temperature T_c are the main field of investigation now.

X-ray emission [6.296-301] and absorption spectroscopy [6.296,300, 302-310] combined with photoelectron spectroscopy [6.307,311-313], bremsstrahlung isochromat spectroscopy (now called widely "inverse photoemission" if excitations in the vacuum UV are considered) [6.312] and theoretical investigations [6.36,301,314,315] have yielded much information on important questions to be solved - the electronic band structure and the Cu valency in these compounds. A review on this was given by *Wendin* [6.316].

The electronic structure of these compounds is closely related to that of CuO, since the main contributions to the valence bands come from the O 2p and Cu 3d states forming the Cu-O bonds. From O $K\alpha$ and Cu $L\alpha$ emission bands, the position below the Fermi energy of the maxima of p and d partial densities of states was determined at $\simeq 4eV$ and $\simeq 2.5eV$, respectively [6.296,298]. This is contradictory to photoemission results ($\simeq 3eV$ and $\simeq 4.5eV$, respectively [6.316]). The reasons for these discrepancies seem to be strong correlation effects in the Cu d bands and charge-transfer screening of Cu 3d holes by O 2p states - both effects going beyond the one-electron picture used in this monograph.

From the X-ray absorption spectra, the Cu valency was derived. It was found that Cu^{3+}, if at all, is present much less than follows from the stoichiometry of the compounds. This is explained again by charge-transfer screening at the Cu sites (formally, $Cu^{3+}-O^{2-}$ pairs change over into $Cu^{2+}-O^-$ pairs).

In conclusion, the following statements can be made concerning the spectra of binary transition-metal compounds and their interpretation:
1) From the discussion on the $A^N B^{8-N}$ compounds it was concluded that the electronic structure of compounds with d electrons could be understood provided that the additional electrons in excess of the electron octet per unit cell must have additional bands. These bands lie either between the upper and lower valence bands according to their atomic energies or even above the upper valence band (ZnSe $\rightarrow$ CuBr). It is possible to arrange these compounds according to their valency in $A^{III}B^V$, $A^{II}B^{VI}$, or A^IB^{VII} compounds. Thus, the electronic structure of a transition-metal compound is given by an addition of d electron states to the valence band of the corresponding compound of main-group elements (see ZnS, ZnSe, CdS, CuCl, CuBr in Sect.6.4.1; binary compounds with C, N, O see above). For more complicated compounds, the experimental material is too scarce to be able to make detailed statements. At an electron occupancy of $n \le 8$ (for example, TiC)) the same partition holds of the valence electrons into an energetically deeper lying valence band V_1 and the higher valence

Table 6.4. Comparison of ionicities, f_i, determined from X-ray spectra, band-structure calculations and the dielectric theory (E_{AS} is the antisymmetric valence band gap, E_{VB} is the full valence band width) [6.318]

	X-ray spectra			Band-structure calculations			Dielectric theory
	E_{AS}	E_{VB}	f_i	E_{AS}	E_{VB}	f_i	f_i
CaO	–	–	–	15.2	16.3	0.99	0.92
TiO	11	25	0.41	12.5	26.6	0.45	–
VO	11.5	24	0.47	12.6	26.0	0.48	–
MnO	–	–	–	13.4	23.4	0.61	0.89
FeO	–	–	–	13.1	22.9	0.61	0.87
CoO	13	18.5	0.80	12.9	22.2	0.63	0.86
NiO	14	20	0.79	12.8	21.7	0.64	0.84
CuO	13.5	19	0.81	–	–	–	–
ZnO	3.5	6	0.63	–	–	–	0.65
NbN	–	–	–	8.4	18.7	0.42	–
Ti_2O_3	13	23	0.60	–	–	–	–
V_2O_3	13	22	0.64	–	–	–	–
Cr_2O_3	13.5	21	0.72	–	–	–	0.78
Mn_2O_3	12	19	0.70	–	–	–	–
Fe_2O_3	13.5	22	0.67	–	–	–	0.68

bands V_2 to V_4. Concerning V_1 the same assertion holds, as shown in Sect. 6.4.1.

2) The electronic structure of binary compounds of the group IIIb to Vb elements with C, N, and O of NaCl structure - disregarding certain changes in the energy positions - can be explained using a rigid-band model (Fig. 6.66).

3) The electronic states of the individual bands can be understood as more or less strongly hybridized from atomic wave functions of the bonding partners.

4) In agreement with Point 1 one gets a consistent relationship between the separation of the bands and the ionicity f_i also for compounds of the 3d elements. Table 6.4 compares f_i values obtained from the spectra and from theoretical band-structure calculations [6.317] using (4.9) with the corresponding values from the dielectric theory [6.210]. It becomes clear that (4.9) is applicable also to transition-metal compounds. Furthermore, the substances studied can be divided into the following three groups [6.318]:

a) $f_i = 0.8 - 0.9$: CaO, MnO, FeO, CoO, NiO, CuO
b) $f_i = 0.6 - 0.7$: Ti_2O_3, V_2O_3, Cr_2O_3, Mn_2O_3, Fe_2O_3, ZnO
c) $f_i = 0.4 - 0.5$: TiO, VO, NbN.

This classification closely corresponds to electrical properties: The compounds of the first group are all insulators, those of the last group are all metals. The substances from the second group should be semiconducting compounds as seems, in fact, to be the case.

References

Chapter 1

1.1 M. Siegbahn: *Spektroskopie der Röntgenstrahlen* (Julius Springer, Berlin 1931)

1.2 A.E. Lindh: In *Handbuch der Experimentalphysik*, ed. by W. Wien, F. Harms, Vol.24 II (Akademische Verlagsgesellschaft, Leipzig 1930) p.281–323

1.3 A.H. Compton, S.K. Allison: *X-Rays in Theory and Experiment* (Van Nostrand, London 1935)

1.4 Y. Cauchois: *Les Spectres de Rayons X et la Structure Electronique de la Matiere* (Gauthier-Villards, Paris 1948)
Y. Cauchois: *Atomes, Spectres, Matiere* (Albin Michel, Paris 1952)

1.5 M.A. Blokhin: *Physik der Röntgenstrahlen* (Verlag Technik, Berlin 1957)

1.6 M.A. Blokhin: *Methods of X-Ray Spectroscopic Research* (Pergamon, Oxford 1965)

1.7 S. Flügge (ed.): *Handbuch der Physik*, Vol.30 (Springer, Berlin, Göttingen 1957)

1.8 E.F. Kaelble (ed.): *Handbook of X-Rays* (McGraw-Hill, New York 1965)

1.9 H. Niehrs: Ergeb. Exakt. Naturwiss. **23**, 359 (1950)

1.10 L.V. Azaroff: *X-Ray Spectrometry* (McGraw-Hill, New York 1974)

1.11 B.K. Agarwal: *X-Ray Spectroscopy*, 2nd. ed., Springer Ser. Opt. Sci., Vol.15 (Springer, Berlin, Heidelberg 1989)

1.12 C. Bonnelle, C. Mande (eds.): *Advances in X-Ray Spectroscopy* (Pergamon, Oxford 1982)

1.13 R.L. Barinskii, V.I. Nefedov: *Röntgenspektroskopische Bestimmung der Atomladungen in Molekülen* (Akademische Verlagsgesellschaft, Leipzig 1969)

1.14 V.I. Nefedov: *Valence Electron Levels in Chemical Compounds* (VINITI, Moscow 1975) (in Russian)

1.15 T.M. Zimkina, V.A. Fomichev: *Ultra Soft X-Ray Spectroscopy* (University of Leningrad 1971) (in Russian)

1.16 V.V. Nemoshkalenko: *X-Ray Emission Spectra of Metals and Alloys* (Naukova Dumka, Kiev 1972) (in Russian)

1.17 V.V. Nemoshkalenko, V.G. Aleshin: *Theory of X-Ray Emission Spectroscopy* (Naukova Dumka, Kiev 1974) (in Russian)

1.18 L.N. Mazalov, V.D. Yumatov, V.V. Murakhtanov, F.Kh. Gelmukhanov, G.N. Dolenko, E.S. Gluskin, A.V. Kondratenko: *X-Ray Spectra of Molecules* (Nauka, Novosibirsk 1977) (in Russian)
L.N. Mazalov: *X-Ray Spectroscopy and X-Ray Electron Spectroscopy of Molecules, Theoretical Principles* (Nauka, Novosibirsk 1979) (in Russian)
L.N. Mazalov: *X-Ray Spectra and Chemical Bond* (Nauka, Novosibirsk 1982) (in Russian)
A.V. Kondratenko, L.N. Mazalov, I.A. Topol: *Highly Excited States of Molecules* (Nauka, Novosibirsk 1982) (in Russian)
B.Yu. Khelmer, A.T. Shuvaev, O.E. Shelepin, L.N. Mazalov: *X-Ray and Pho-*

toelectron Spectra of Compounds Containing Oxygen (University of Rostov (Don) 1983) (in Russian)

1.19 R. Karazija: *The Theory of X-Ray and Electronic Spectra of Free Atoms* (Mokslas, Vilnius 1987) (in Russian)

1.20 B.K. Teo, D.C. Joy: *EXAFS Spectroscopy, Techniques and Applications* (Plenum, New York 1981)
B.K. Teo: *EXAFS, Basic Principles and Data Analysis* (Springer, Berlin, Heidelberg 1986)

1.21 A. Faessler: Röntgenspektrum und Bindungszustand, in *Landolt-Börnstein*, Vol.I/4, 6th ed. (Springer, Berlin, Göttingen 1955) p.769

1.22 A.J. McAlister, R.C. Dobbyn, J.R. Cuthill, M.L. Williams: Soft X-Ray Emission Spectra of Metallic Solids; Critical Review of Selected Systems and Annotated Spectral Index. NBS (US) Spec. Publ. 369 (Washington, DC 1974)

1.23 H. Yakowitz, J.R. Cuthill: Annotated Bibliography on Soft X-Ray Spectroscopy. NBS (US) Monogr.52 (Washington, DC 1962)

1.24 A. Meisel: Phys. Stat. Sol. 10, 365 (1965)

1.25 A.P. Hitchcock: Bibliography of atomic and molecular inner-shell excitation studies. J. Electron Spectrosc. 25, 245 (1982)

1.26 J.A. Bearden: X-Ray Wavelengths (US Atomic Energy Commission, Oak Ridge, TN 1964)
J.A. Bearden, A.F. Burr: Atomic Energy Levels (US Atomic Energy Commission, Oak Ridge, TN 1965)

1.27 M.A. Blokhin, I.G. Shveitser: *X-Ray Spectroscopic Tables* (Nauka, Moscow 1982) (in Russian)

1.28 A. Faessler: Angew. Chem. 84, 51 (1972)

1.29 L.G. Parratt: Rev. Mod. Phys. 31, 616 (1959)

1.30 D.J. Fabian: Crit. Rev. Solid State Sci. 2, 255 (1971)

1.31 D.J. Fabian, L.M. Watson, C.A.W. Marshall: Rep. Prog. Phys. 34, 601 (1972)

1.32 Y. Gohshi: Bunko Kenkyu 18, 235 (1969)

1.33 R.L. Barinskii: Zh. Strukt. Khim. 8, 897 (1967)

1.34 Y. Cauchois, C. Bonnelle: Colloq. Int. CNRS 191, 125 (1969)

1.35 U.C. Srivastava, H.L. Nigam: Coord. Chem. Rev. 9, 275 (1973)

1.36 V.I. Nefedov: Zh. Strukt. Khim. 13, 352 (1972)

1.37 W.H.E. Schwarz: Angew. Chem. 86, 505 (1974)

1.38 E.W. White: In *Microprobe Analysis*, ed. by C.A. Andersen (Wiley, New York 1973) p.349

1.39 M. Grasserbauer: Anal. Chem. 273, 401 (1975)
M. Grasserbauer: Mikrochim. Acta 1, 145, 563, 597 (1975)
S. Kosina, M. Grasserbauer, H. Drack: Talanta 26, 765 (1979)

1.40 V.I. Nefedov: Mod. Phys. Chem. 1, 1 (1976)

1.41 O.I. Sumbaev: Mod. Phys. Chem. 1, 31 (1976)

1.42 D.S. Urch: In *Electron Spectroscopy, Theory, Techniques and Applications*, Vol.3, ed. by G.R. Brundle, A.D. Baker (Academic, London 1979)

1.43 J. Nordgren, H. Agren: Interpretation of Ultra-Soft X-Ray Emission Spectra. (UUIP 1091) (Uppsala University, Institute of Physics 1983)

1.44 J.P. Pendry: Com. Solid State Phys. 10, 219 (1983)

1.45 P. Rabe, R. Haensel: *Festkörperprobleme* 20, 43 (Vieweg, Braunschweig 1980)

1.46 P.A. Lee, P.H. Citrin, P. Eisenberger, B.M. Kincaid: Rev. Mod. Phys. 53, 769 (1981)

1.47 C.D. Garner, S.S. Hasnain: EXAFS for Inorganic Systems. DL/SCI/R 17 (Daresbury 1981)

1.48 T.M. Hayes, J.B. Boyce: *Solid State Physics. Advances in Research and Applications* **37**, 173-351 (Academic, New York 1982)

1.49 C. Kunz (ed.): *Synchrotron Radiation, Techniques and Applications*, Topics Curr. Phys., Vol. 10 (Springer, Berlin, Heidelberg 1979)

1.50 H. Winick, S. Doniach (eds.): *Synchrotron Radiation Research* (Plenum, New York 1980)

1.51 E.E. Koch (ed.): *Handbook on Synchrotron Radiation*, Vols.1a,b (North-Holland, Amsterdam 1983)

1.52 S. McGlynn, G. Findley, R. Huebner (eds.): *Photophysics and Photochemistry in the Vacuum Ultraviolet* (Reidel, Dordrecht 1983)

1.53 R. Prins, D. Koningsberger (eds.): *X-Ray Absorption. Principles, Applications, Techniques of EXAFS, SEXAFS, and XANES* (Wiley, New York 1985)

1.54 Y. Cauchois, C. Bonnelle: In *Atomic Inner-Shell Processes*, ed. by B. Crasemann (Academic, New York 1975) Vol.2, p.83

1.55 Y. Cauchois: In *Soft X-Ray Band Spectra and the Electronic Structure of Metals and Materials*, ed. by D.J. Fabian (Academic, London 1968) p.71

1.56 Bull. Am. Phys. Soc. **10**, 4, 1218-1225 (1965)

1.57 A. Meisel (ed.): *Röntgenspektren und chemische Bindung* (Karl-Marx-Universität, Leipzig 1966)

1.58 V.V. Nemoshkalenko (ed.): *X-Ray Spectra and Electronic Structure of Matter*, Vols. 1 and 2 (Institute of Metal Physics, Kiev 1969)

1.59 J. Physique **32**, C4, 1-344 (1971)

1.60 A. Faessler, G. Wiech (eds.): *X-Ray Spectra and Electronic Structure of Matter*, Vols. 1 and 2 (Universität München 1973)

1.61 Phys. Fenn. **9**, 3-437 (1974)

1.62 R.D. Deslattes (ed.): Int'l Conf. Physics of X-Ray Spectra (NBS, Gaithersburg, MD 1976) (extended abstracts)

1.63 Jpn. J. Appl. Phys. **17**, S2, 1-516 (1978)

1.64 D.J. Fabian, H. Kleinpoppen, L.M. Watson (eds.): *Inner-Shell and X-Ray Physics* (Academic, New York 1981)

1.65 B. Crasemann (ed.): *X-Ray and Atomic Inner-Shell Physics* (American Institute of Physics, New York 1982)

1.66 A. Meisel, J. Finster (eds.): *X-Ray and Inner-Shell Processes in Atoms, Molecules and Solids* (Karl-Marx-Universität, Leipzig 1984)

1.67 J. Physique **48**, C9, 1-1243 (1987) (2 vols.)

1.68 Izv. Akad. Nauk SSSR, Ser. Fiz. **28**, 758-938 (1964)

1.69 Izv. Akad. Nauk SSSR, Ser. Fiz. **31**, 874-1034 (1967)

1.70 Izv. Akad. Nauk SSSR, Ser. Fiz. **36**, 226-450 (1972)

1.71 Izv. Akad. Nauk SSSR, Ser. Fiz. **38**, 426-662 (1974)

1.72 Izv. Akad. Nauk SSSR, Ser. Fiz. **40**, 226-444 (1976)

1.73 Izv. Akad. Nauk SSSR, Ser. Fiz. **46**, 720-828 (1982)

1.74 Izv. Akad. Nauk SSSR, Ser. Fiz. **49**, 1458-1566 (1985)

1.75 Izv. Sib. Otd. AN SSSR, Ser. Khim. Nauk 1975, No.4, 1-154 (1975)

1.76 Adv. X-Ray Anal. **9**, 1-544 (1966)

1.77 Adv. X-Ray Anal. **13**, 1-681 (1970)

1.78 D.J. Fabian (ed.): *Soft X-Ray Band Spectra and the Electronic Structure of Metals and Materials* (Academic, London 1968)

1.79 D.J. Fabian, L.M. Watson (eds.): *Band Structure Spectroscopy of Metals and Alloys* (Academic, London 1973)

1.80 *Structure of Non-Crystalline Materials*, Conf. Proc. Daresbury 1982 (Taylor and Francis, London 1983)

1.81 J. Auleytner, W. Zahorowski (eds.): Proc. Conf. Phys., Vol.5 (Ossolineum, Wroclav 1983)

1.82 A. Bianconi, L. Incoccia, S. Stipcich (eds.): *EXAFS and Near-Edge Structure*, Springer Ser. Chem. Phys., Vol.27 (Springer, Berlin, Heidelberg 1983)

1.83 K.O. Hodgson, B. Hedman, J.E. Penner-Hahn (eds.): *EXAFS and Near Edge Structure III*, Springer Proc. Phys., Vol.2 (Springer, Berlin, Heidelberg 1984)

1.84 J. Physique 47, C8, 3-1209 (1986) (2 Vols.)

1.85 Y. Nakai (ed.): Proc. 3rd Int'l Conf. Vacuum Ultraviolet Radiation Physics (Physical Society of Japan, Tokyo 1971)

1.86 E.E. Koch, R. Haensel, C. Kunz (eds.): *Vacuum Ultraviolet Radiation Physics* (Vieweg, Braunschweig 1974)

1.87 M.C. Castex, M. Pouey, N. Pouey (eds.): 5th Int'l Conf. Vacuum Ultraviolet Radiation Physics (Meudon 1977) (extended abstracts)

1.88 Proc. 7th Int'l Conf. Vacuum Ultraviolet Radiation Physics. Ann. Isr. Phys. Soc. 6 (1984)

1.89 Phys. Scripta T17, 7-239 (1987)

1.90 N.N. Sirota (ed.): *Chemical Bond in Semiconductors and Solids* (Consultants Bureau, New York 1967)

1.91 N.N. Sirota (ed.): *Chemical Bond in Semiconductors and Thermodynamics* (Consultants Bureau, New York 1968)

1.92 N.N. Sirota (ed.): *Chemical Bond in Solids* (Consultants Bureau, New York 1972)

1.93 N.N. Sirota (ed.): *Chemical Bond in Semiconductors* (Nauka i Tekhnika, Minsk 1972)

1.94 N.N. Sirota (ed.): *Chemical Bond in Semiconductors and Semimetal Crystals* (Nauka i Tekhnika, Minsk 1973)

1.95 I.Ya. Dekhtyar (ed.): *Electronic Structure of Transition Metals and Their Alloys* (Institute of Metal Physics, Kiev 1968)

1.96 I.Ya. Dekhtyar, V.V. Nemoshkalenko (eds.): *Electronic Structure of Transition Metals and Their Alloys* (Institute of Physics, Kiev 1970)

1.97 I.Ya. Dekhtyar, V.V. Nemoshkalenko: *Electronic Structure and Electronic Properties of Transition Metals and Their Alloys* (Naukova Dumka, Kiev 1971)

1.98 I.Ya. Dekhtyar (ed.): *Electronic Structure of Transition Metals, Their Alloys and Compounds* (Naukova Dumka, Kiev 1974)

1.99 V.V. Nemoshkalenko (ed.): *Some Questions of the Electronic Structure of the Transition Elements* (Voic, Kiev 1968) (in Russian)

1.100 V.V. Nemoshkalenko (ed.): *Electron Structure of Transition Metals, Their Alloys and Intermetallic Compounds* (Naukova Dumka, Kiev 1979)

1.101 L.H. Bennett (ed.): Electronic Density of States (NBS US) Spec. Publ. No.323 (Washington, DC 1971)

1.102 R.W. Fink, S.T. Manson, J.M. Palms, P.V. Rao (eds.): Proc. Int'l Conf. Inner-Shell Ionization Phenomena Future Appl. (US Atomic Energy Commission, Oak Ridge, TN 1973)

1.103 Proc. Int'l Conf. X-Ray and VUV Synchrotron Radiation Instrumentation. Nucl. Instrum. Methods Phys. Res. 208, 1-866 (1983)

Chapter 2

2.1   D.J. Nagel, D.A. Papaconstantopoulos, J.W. Caffrey, J.W. Criss: In *X-Ray Spectra and Electronic Structure of Matter*, Vol.2, ed. by A. Faessler, G. Wiech (Universität München 1973) p.51

2.2   D.J. Nagel: Adv. X-Ray Anal. 13, 183 (1970)
2.3   L.G. Parratt: Rev. Mod. Phys. 31, 616 (1959)
2.4   H.W. Schnopper, L.G. Parratt: In *Röntgenspektren und chemische Bindung*, ed. by A. Meisel (Karl-Marx-Universität, Leipzig 1966) p.314
2.5   M.A. Blokhin: *Physik der Röntgenstrahlen* (Verlag Technik, Berlin 1957)
2.6   A. Faessler, M. Goehring: Naturwiss. 31, 567 (1943)
2.7   M. Fichter: In *Röntgenspektren und chemische Bindung*, ed. by A. Meisel (Karl-Marx-Universität, Leipzig 1966) p.112; Spectrochim. Acta B 30, 417 (1975)
2.8   H. Krämer: Ann. Physik 4, 263 (1959)
2.9   K. Läuger: Über den Einfluß der Bindungsart und der Kristallstruktur auf das K-Röntgenemissionsspektrum von Aluminium und Silizium. Dissertation (Universität München 1968)
2.10 E.W. White: Nature 200, 649 (1963); Mat. Res. Bull. 2, 395 (81967)
2.11 G. Leonhardt: Zur theoretischen Interpretation röntgenspektroskopischer Untersuchungen an Verbindungen der 3d-Übergangselemente. Dissertation (Karl-Marx-Universität, Leipzig 1969)
2.12 M. Malkowska: Exp. Techn. Physik 6, 130 (1958)
2.13 A. Meisel: In *Röntgenspektren und chemische Bindung*, ed. by A. Meisel (Karl-Marx-Universität, Leipzig 1966) p.212
2.14 A.T. Shuvaev, G.M. Kulyabin: Izv. Akad. Nauk SSSR, Ser. Fiz. 27, 322 (1963)
2.15 J. Finster, G. Leonhardt, A. Meisel: J. Physique 32, C4-218 (1971)
2.16 J. Finster: Untersuchung des Mo $K\beta_{1,3}$-Dubletts im Metall und in einigen Verbindungen mit einer neuen, automatischen Zählrohrregistriereinrichtung. Dissertation (Karl-Marx-Universität, Leipzig 1967)
2.17 A. Meisel, R. Szargan: Z. Phys. Chemie 236, 113 (1967)
2.18 T. Mukoyama, H. Kaji, K. Yoshihara: Phys. Lett. A 11, 44 (1986)
2.19 I.M. Band, A.P. Kovtun, M.A. Listengarten, M.B. Trzhaskovskaya: J. Electron Spectrosc. 3, 59 (1985)
2.20 G. Brunner, M. Nagel, E. Hartmann, E. Arndt: J. Phys. B 1, 4517 (1982)
2.21 J.L. Campbell, A. Perujo, W.J. Teesdale, B.M. Millman: Phys. Rev. A3, 2410 (1986)
2.22 K.E. Collins, C.H. Collins, C. Heitz: Radiochim. Acta 28, 7 (1981)
2.23 L. Obert, J.A. Bearden: Phys. Rev. 54, 1000 (1938)
2.24 T.M. Snyder: Phys. Rev. 59, 168 (1941)
2.25 E.E. Vainshtein: C.R. Acad. Sci. URSS 40, 102 (1943)
2.26 R.L. Barinskii, V.I. Nefedov: *Röntgenspektroskopische Bestimmung der Atomladungen in Molekülen* (Akademische Verlagsgesellschaft, Leipzig 1969)
2.27 B. Ekstig, E. Källne, E. Noreland, R. Manne: Phys. Scr. 2, 38 (1970)
2.28 B. Ekstig, E. Källne, E. Noreland, R. Manne: J. Physique 32, C4-214 (1971)
2.29 Z. Horak: Czech. J. Phys. 10, 405 (1960)
2.30 L.K. Izraileva: Izv. Akad. Nauk SSSR, Ser. Fiz. 25, 954 (1961); Dokl. Akad. Nauk SSSR 168, 777 (1966)
2.31 V.I. Nefedov: Zh. Strukt. Khim. 7, 549,672,719 (1966)
2.32 K. Tsutsumi: J. Phys. Soc. Jpn. 14, 1696 (1959); also in *Röntgenspektren und chemische Bindung*, ed. by A. Meisel (Karl-Marx-Universität, Leipzig 1966) p.336
2.33 K. Tsutsumi, H. Nakamori: J. Phys. Soc. Jpn. 25, 1418 (1968); and in *X-Ray Spectra amd Electronic Structure of Matter*, Vol. 1, ed. by A. Faessler, G. Wiech (Universität München 1973) p.100
2.34 C.S. Fadley, D.A. Shirley: Phys. Rev. Lett. 23, 1397 (1969)
2.35 D.R. Hartree: *The Calculation of Atomic Structures* (Wiley, New York 1957)

2.36 C.C.J. Roothaan: Rev. Mod. Phys. **23**, 69 (1951); ibid. **32**, 179 (1960)
C.C.J. Roothaan, P.S. Bagus: *Methods in Computational Physics*, Vol.2, ed. by B. Alder, S. Fernbach, M. Rotenberg (Academic, New York 1963) p.47

2.37 O. Goscinski, B.T. Pickup. G. Purris: Chem. Phys. Lett. **22**, 167 (1973)
O. Goscinski, G. Howat, T. Åberg: J. Phys. B **8**, 11 (1975)

2.38 J.C. Slater, J.B. Mann, T.M. Wilson, J.H. Wood: Phys. Rev. **184**, 672 (1969)
J.C. Slater: In *Advances in Quantum Chemistry*, Vol.6, ed. by P.O. Löwdin (Academic, New York 1972)

2.39 P.S. Bagus, H.F. Schaefer: J. Chem. Phys. **55**, 1474 (1971); ibid. **56**, 224 (1972)

2.40 P.S. Bagus, M. Krauss, R.E. LaVilla: Chem. Phys. Lett. **23**, 13 (1973)

2.41 P.S. Bagus: in *X-Ray Spectra and Electronic Structure of Matter*, Vol.1, ed. by A. Faessler, G. Wiech (Universität München 1973) p.256

2.42 T.L. Gilbert, W.J. Stevens, H. Schrenk, M. Yoshimine, P.S. Bagus: Phys. Rev. B **8**, 5977 (1973)

2.43 H. Schrenk: Röntgenzustände symmetrischer Moleküle. SCF-Rechnungen an einigen Schwefel- und Chlorverbindungen. Dissertation (Universität München 1969)

2.44 M.E. Schwartz: Chem. Phys. Lett. **5**, 50 (1970)

2.45 E.U. Condon, C. Shortley: *The Theory of Atomic Spectra* (Cambridge Univ. Press, Cambridge 1950)

2.46 J.C. Slater: *Quantum Theory of Atomic Structure* (McGraw-Hill, New York 1960)

2.47 P.S. Bagus: Phys. Rev. A **139**, 619 (1965)

2.48 F.A. Gianturco, C.A. Coulson: Mol. Phys. **14**, 223 (1968)

2.49 I. Lindgren: In *Röntgenspektren und chemische Bindung*, ed. by A. Meisel (Karl-Marx-Universität, Leipzig 1966) p.182

2.50 A. Sureau, G. Berthier: J. Physique **24**, 672 (1963)

2.51 I.M. Band, M.B. Trzhaskovskaya: *Eigenvalues, Electronic Charge Densities in the Vicinity of the Nuclei and Expectation Values in Self-Consistent Potentials for Free Atoms and Ions*, $37 \leq Z \leq 64$ and $2 \leq Z \leq 94$ (Institute of Nuclear Physics, Academy of Sciences of USSR, Leningrad 1971 and 1974)

2.52 E. Clementi: *Tables of Atomic Functions*, Suppl. to IBM J. Res. Develop. **9** (1965)

2.53 A.T. Shuvaev: Izv. Akad. Nauk SSSR, Ser. Fiz. **28**, 758,934 (1964)

2.54 R.E. Watson: *Iron Series Hartree-Fock Calculations*. Techn. Rep. Nr. 12 (MIT, Cambridge, MA 1959)

2.55 G. Leonhardt, A. Meisel: J. Chem. Phys. **52**, 6189 (1970)

2.56 S.M. Karalnik: Izv. Akad. Nauk SSSR, Ser. Fiz. **20**, 815 (1956)

2.57 O.I. Sumbaev, E.V. Petrovich, Yu.P. Smirnov, A.I. Egorov, V.S. Zykov, A.I. Grushko: Zh. Eksp. Teor. Fiz. **53**, 1545 (1967)

2.58 P.S. Bagus, A.J. Freeman, F. Sasaki: Phys. Rev. Lett. **30**, 850 (1973)

2.59 R.L. Barinskii, I.M. Kulikova: Zh. Strukt. Khim. **14**, 372 (1973); Izv. Akad. Nauk SSSR, Ser. Fiz. **38**, 444 (1974)

2.60 M. Siegbahn: Ergebn. Exakt. Naturwiss. **16**, 104 (1937)

2.61 H.W.B. Skinner: Phil. Trans. Roy. Soc. A **239**, 95 (1940)

2.62 J. Farineau: Ann. Physique **10**, 20 (1938)

2.63 A. Neckel, P. Rastl, P. Weinberger, R. Mechtler: Theor. Chim. Acta **24**, 170 (1972)
A. Neckel, P. Rastl, K. Schwarz, R. Eibler-Mechtler: Z. Naturforsch. **29a**, 107 (1974)
A. Neckel, K. Schwarz, R. Eibler, P. Rastl, P. Weinberger: Mikrochim. Acta Suppl. **6**, 257 (1975)

K. Schwarz, A. Neckel: Ber. Bunsenges. Phys. Chemie 79, 1071 (1975)

P. Weinberger: Theor. Chim. Acta 41, 169 (1976)

K. Schwarz: J. Phys. C 10, 195 (1977)

P. Weinberger, F. Rosicky: Theor. Chim. Acta 48, 349 (1978)

2.64 J.C. Slater: *Quantum Theory of Molecules and Solids* (McGraw-Hill, New York 1965) Vol.2

2.65 J.M. Ziman: *Principles of the Theory of Solids* (Cambridge Univ. Press, Cambridge 1965)

2.66 U. von Darth, G. Grossmann: Phys. Scr. 21, 580 (1980)

2.67 P. Nozieres, C.T. Dominicis: Phys. Rev. 178, 1097 (1969)

2.68 D.A. Papaconstantopoulos, J.W. McCaffrey, D.J. Nagel: J. Phys. F3, 126 (1973)

2.69 L.O. Werme, B. Grennberg, J. Nordgren, C. Nordling, K. Siegbahn: J. Electron Spectrosc. 2, 435 (1973)

L.O. Werme, J. Nordgren, H. Ågren, C. Nordling, K. Siegbahn: Z. Physik A 272, 131 (1975)

H. Ågren, J. Müller: Phys. Scr. 20, 627 (1979)

H. Ågren, J. Müller: J. Chem. Phys. 72, 4078 (1980)

2.70 D.T. Clark, J. Müller: Theor. Chim. Acta 41, 193 (1976)

H. Ågren, J. Müller, J. Nordgren: J. Chem. Phys. 72, 4078 (1980)

2.71 F.K. Gelmukhanov, L.N. Mazalov, A.V. Kondratenko: Chem. Phys. Lett. 46, 133 (1977)

A.V. Kondratenko, L.N. Mazalov, F.K. Gelmukhanov, V.I. Avdeev, E.A. Saprykina: Zh. Strukt. Khim. 18, 622 (1977)

2.72 J. Nordgren, H. Ågren, L. Selander, C. Nordling, K. Siegbahn: Phys. Scr. 16, 280 (1977)

H. Ågren, J. Nordgren, L. Selander, C. Nordling, K. Siegbahn: Phys. Scr. 18, 499 (1978); Chem. Phys. 37, 161 (1979)

L. Petterson, N. Wassdahl, M. Bäckström, J.E. Rubensson, J. Nordgren: J. Phys. B 18, L125 (1985)

2.73 H. Adachi, K. Taniguchi: Technol. Rep. Osaka Univ. 30, 365 (1980); J. Phys. Soc. Jpn. 49, 1944 (1980)

2.74 B.L. Henke, K. Taniguchi: J. Appl. Phys. 47, 1027 (1976)

K. Taniguchi, B.L. Henke: J. Chem. Phys. 64, 3021 (1976)

2.75 K.H. Johnson, F.C. Smith: Chem. Phys. Lett. 7, 541 (1970); ibid. 10, 219 (1971)

K.H. Johnson: In *Advances in Quantum Chemistry*, ed. by P.-O. Löwdin (Academic New York 1973) Vol.7, p.143

2.76 V.A. Gubanov, J. Weber, J.W.D. Connolly: Chem. Phys. 11, 319 (1975)

V.A. Gubanov: J. Electron Spectrosc. 9, 85 (1976)

2.77 R. Szargan, E. Hartmann, H. Sommer, A. Meisel: In *Inner-Shell and X-Ray Physics*, ed. by D. Fabian, H. Kleinpoppen, L.M. Watson (Academic, New York 1981) p.717

2.78 J.A. Tossell: Geochim. Cosmochim. Acta 37, 583 (1973); J. Phys. Chem. Solids 34, 307 (1973)

D.J. Vaughan, J.A. Tossell: Am. Mineral. 58, 765 (1973)

J.A. Tossell: J. Phys. Chem. Solids 36, 1273 (1975); ibid. 37, 1043 (1976); Chem. Phys. 15, 303 (1976)

J.A. Tossell: Am. Mineral. 62, 136 (1977)

2.79 J.A. Tossell: Inorg. Chem. 19, 3328 (1980)

2.80 P. Weinberger: Ber. Bunsenges. Phys. Chemie 81, 459 (1977)

2.81 V.A. Gubanov, N.I. Lazukova, E.Z. Kurmaev: Izv. Sib. Otd. Akad. Nauk SSSR, Ser. Khim. Nauk (1975) Nr.9, H.4, 18

V.A. Gubanov, E.Z. Kurmaev: Zh. Strukt. Khim. **16**, 731 (1975)
V.A. Gubanov, B.G. Kasimov, E.Z. Kurmaev: J. Phys. Chem. Solids **36**, 861 (1975)
V.A. Gubanov, E.Z. Kurmaev, G.P. Shveikin: J. Phys. Chem. Solids **38**, 201 (1977)

2.82 D. Coster, R. de L. Kronig: Physica 2, 13 (1935)
2.83 M.J. Druyvesteyn: Z. Physik **43**, 707 (1928)
2.84 G. Wentzel: Ann. Physik **65**, 437 (1921)
2.85 F.K. Richtmyer: Phil. Mag. **6**, 645 (1928)
2.86 A.I. Kostarev: Zh. Eksp. Teor. Fiz. 11, 60 (1941); ibid. 22, 628 (1952)
2.87 J. Valasek: Phys. Rev. **52**, 250 (1937); ibid. **53**, 274 (1938)
2.88 T. Åberg: in *X-Ray Spectra and Electronic Structure of Matter*, ed. by A. Faessler and G. Wiech (Universität München 1973) Vol.1, p.1
In: *Atomic Inner-Shell Processes*, Vol.1, ed. by B. Crasemann (Academic, New York 1975) p.353
2.89 A.R. Knudson, D.J. Nagel, P.G. Burkhalter, R.L. Dunning: Phys. Rev. Lett. **26**, 1149 (1971)
2.90 J.R. McDonald, M.D. Brown: Phys. Rev. Lett. **29**, 4 (1972)
2.91 P. Richard, I.L. Morgan, T. Furuta, D. Burch: Phys. Rev. Lett. **23**, 1009 (1969)
2.92 F.W. Saris, W.F. van der Weg, H. Tawara, R. Laubert: Phys. Rev. Lett. **28**, 717 (1972)
2.93 D. Coster, M.J. Druyvesteyn: Z. Physik 40, 765 (1926)
2.94 D. Coster: Phil. Mag. **44**, 546 (1922)
2.95 P.U. Sakellaridis: J. Physique **16**, 422 (1955); C.R. Acad. Sci. 247, 876 (1958)
2.96 S.I. Salem, C.W. Schultz, B.A. Rabbani, R.T. Tsutsui: Phys. Rev. Lett. **27**, 447 (1971)
2.97 J.L. Dehmer, A.F. Starace, U. Fano, J. Sugar, J.W. Cooper: Phys. Rev. Lett. **26**, 1521 (1971)
2.98 C.S. Fadley, D.A. Shirley: Phys. Rev. A2, 1109 (1970); and in *Electronic Density of States*, ed. by L.H. Bennett, NBS Spec. Publ. 323 (Washington 1971) p.163
2.99 J.P. Coad: Phys. Lett. A 37, 437 (1971)
2.100 H.W. Schnopper: Phys. Rev. **154**, 118 (1967)
2.101 M. Linkoaho, T. Åberg, G. Graeffe, J. Utriainen: Z. Naturforsch. 24a, 775 (1969)
2.102 H. Mendel: Proc. Kon. Ned. Akad. Wetensch. B 70, 276 (1967)
2.103 D. Urch: J. Phys. C 3, 1275 (1970)
2.104 M.O. Krause, M.L. Vestal, V.H. Johnston, T.A. Carlson: Phys. Rev. A 133, 385 (1964)
T.A. Carlson, M.O. Krause: Phys. Rev. A 137 (1965); ibid. A 140, 1057 (1965)
2.105 R.D. Richtmyer: Phys. Rev. **49**, 1 (1926)
2.106 L.G. Parratt: Phys. Rev. **50**, 598 (1936); ibid. **54**, 99 (1938); ibid. **56**, 285 (1939)
2.107 F.K. Richtmyer, E.G. Ramberg: Phys. Rev. **51**, 925 (1937)
2.108 Y. Cauchois: J. Physique 5, 1 (1944)
2.109 L. Pincherle: Phys. Rev. **61**, 225 (1942)
2.110 N. Wassdahl, J.-E. Rubensson, G. Bray, N. Martensson, J. Nordgren, R. Nyholm, S. Cramm, K.-L. Tsang, T.A. Callcott, D.L. Ederer, C.W. Clark: UUIP-1177 (1987)

N. Wassdahl, J.-E. Rubensson, G. Bray, N. Martensson, J. Nordgren, R. Nyholm, S. Cramm: UUIP-1178 (1987)

2.111 T.A. Carlson, C.W. Nestor, T.C. Tucker, F.B. Malik: Phys. Rev. 169, 27 (1968)

2.112 R.D. Deslattes: Phys. Rev. A 133, 390, 395, 399 (1964)

2.113 T. Åberg: Phys. Rev. 156, 35 (1967)

2.114 V.P. Sachenko, V.F. Demekhin: Zh. Eksp. Teor. Fiz. 49, 765 (1965)

2.115 M.O. Krause: J. Physique 32, C4-67 (1971)

2.116 J.-E. Rubensson, L. Pettersson, N. Wassdahl, M. Bäckström, J. Nordgren, O.M. Kralheim, R. Manne: J. Chem. Phys. 82, 4486 (1985)

R. Brammer, N. Wassdahl, J.-E Rubensson, J. Nordgren: Phys. Scr. 36, 262 (1987)

2.117 M.O. Krause, F. Wuilleumier, C.W. Nestor: Phys. Rev. A6, 871 (1972); F. Wuilleumier, M.O. Krause: In *X-Ray Spectra and Electronic Structure of Matter*, ed. by A. Faessler, G. Wiech (Universität München 1973) Vol.1, p.397

2.118 H. Hartmann, D. Hendel: Theoret. Chim. Acta 15, 303 (1969)

2.119 F.J. McGuire: Scandia Res. Rep. 70, 721 (1970); Phys. Rev. A 5, 2313 (1972)

2.120 M.A. Blokhin, A.I. Platkov: Izv. Akad. Nauk SSSR, Ser. Fiz. 36, 325 (1972)

2.121 T. Åberg: Phys. Lett. A 26, 515 (1968); Ann. Acad. Sci. Fenn. A 6, 308 (1969)

2.122 T. Åberg: In *Proc. Int'l Conf. Inner-Shell Ionization Phenomena Future Appl.*, ed. by R.W. Fink (US Atomic Energy Commission, Oak Ridge 1973) p.1509

2.123 W.L. Baun, D.W. Fischer: Nature 204, 642 (1964); Spectrochim. Acta 21, 1471 (1965)

2.124 W.L. Baun, D.W. Fischer: Adv. X-Ray Anal. 8, 371 (1965)

D.W. Fischer, W.L. Baun: J. Appl. Phys. 36, 534 (1965); ibid. 38, 2404 (1967)

2.125 H.-U. Chun: in *X-Ray Spectra and Electronic Structure of Matter*, ed. by A. Faessler, G. Wiech (Universität München 1973) Vol.1, p.426; Phys. Fenn. 9, S1-144 (1974)

2.126 V.F. Demekhin, V.P. Sachenko: Izv. Akad. Nauk SSSR, Ser. Fiz. 31, 900, 907 (1967); Bull. Acad. Sci. USSR 31, 921 (1967)

2.127 F.A. Gianturco: J. Phys. B 1, 614 (1968)

2.128 L. Papula, W. Strehl, H.-U. Chun: Theoret. Chim. Acta 22, 149 (1971)

2.129 K. Läuger: J. Phys. Chem. Solids 32, 609 (1971); ibid. 33, 1343 (1972)

2.130 R.L. Watson, A. Langenberg, F.E. Jenson, R.M. Hedges: Jpn. J. Appl. Phys. 17, S2-93 (1978)

2.131 R. Szargan, A. Meisel. E. Hartmann, G. Brunner: Jpn. J. Appl. Phys. 17, S2-174 (1978)

E. Hartmann, R. Szargan: Chem. Phys. Lett. 68, 175 (1979)

2.132 R.L. Watson, T.Chiao, F.E. Jenson: Phys. Rev. Lett. 35, 254 (1975)

R.L. Watson, A.K. Leeper, B.I. Sonobe, T. Chiao, F.E. Jenson: Phys. Rev. A 15, 914 (1977)

J.A. Demarest, R.L. Watson: Phys. Rev. A 17, 1302 (1978)

2.133 J. Kawai, C. Satoko, Y. Gohshi: J. Phys. C 20, 69 (1987)

2.134 M. Uda, A. Koyama, K. Maeda, Y. Sasa: In *X-Ray and Inner-Shell Processes in Atoms, Molecules and Solids X84*, ed. by A. Meisel, J. Finster (Karl-Marx-Universität, Leipzig 1984) p.307

2.135 J. Valjakka, J. Utriainen, T. Åberg, J. Tulkki: Phys. Rev. B 32, 6892 (1985)

2.136 L.N. Mazalov, B.A. Treiger: Zh. Strukt. Khim. 24, 128 (1983)

2.137 T. Maekawa, T. Yokokawa: Spectrochim. Acta B 37, 713 (1982)

2.138 T. Åberg, J. Utriainen: Phys. Rev. Lett. 22, 1346 (1969)

2.139 T. Åberg: Phys. Rev. A 4, 1735 (1971)

2.140 R.L. Martin, S.P. Kowalczyk, D.A. Shirley: J. Chem. Phys. 68, 3829 (1978)

2.141 S. Asada, C. Satoko, S. Sugano: J. Phys. Soc. Jpn. **38**, 855 (1975)
2.142 R.E. LaVilla: Phys. Rev. A **4**, 476 (1971)
2.143 L.O. Werme, B. Grennberg, J. Nordgren, C. Nordling, K. Siegbahn: Nature **242**, 453 (1973)
2.144 M.A. Blokhin, V.P. Sachenko, I.G. Shveitser: In *Proc. Int'l Conf. Inner-Shell Ionization Phenomena Future Appl.*, ed. by R.W. Fink (US Atomic Energy Commission, Oak Ridge, TN 1973) p.1696
2.145 M.V. Linkoaho: In *Chemical Bond in Semiconductors and Semimetals*, ed. by N.N. Sirota (Nauka i Tekhnika, Minsk 1972) p.146
2.146 J. Utriainen: In *X-Ray Spectra and Electronic Structure of Matter*, ed. by A. Faessler, G. Wiech (Universität München 1973) Vol.1, p.382
2.147 M. Sawada, K. Taniguchi, H. Nakamura: In *X-Ray Spectra and Electronic Structure of Matter*, ed. by V.V. Nemoshkalenko (Institute of Metal Physics, Kiev 1969) Vol.2, p.122; J. Phys. Soc. Jpn. **33**, 1496 (1972)
2.148 M. Linkoaho, J. Utriainen: Phys. Fenn. **8**, 67 (1973)
2.149 J.W. Cooper, R.E. LaVilla: Phys. Rev. Lett. **25**, 1745 (1970)
2.150 L.O. Werme, B. Grennberg, J. Nordgren, C. Nordling, K. Siegbahn: Phys. Lett. A **41**, 113 (1972)
      L.O. Werme, J. Nordgren, C. Nordling, K. Siegbahn: C.R. Acad. Sci. B **279**, 119 (1974)
2.151 C. Nordling: Jpn. J. Appl. Phys. 17, S 2-7 (1978)
      J. Nordgren, H. Ågren, L. Pettersson, L. Selander, S. Griep, C. Nordling, K. Siegbahn: Phys. Scr. **20**, 623 (1979)
2.152 R.C. Ehlert, R.A. Mattson: J. Chem. Phys. **48**, 5471 (1968)
2.153 T. Åberg, J.J. Utriainen: J. Physique 32, C4-295 (1971)
2.154 G. Böhm, K. Ulmer: Z. Angew. Physik **29**, 287 (1970); J. Physique **32**, C4-241 (1971)
2.155 D.J. Nagel: In *Band Structure Spectroscopy of Metals and Alloys*, ed. by D.J. Fabian, L.M. Watson (Academic, London 1973) p.457
2.156 R.I. Liefeld: In *Soft X-Ray Band Spectra and the Electronic Structure of Metals and Materials*, ed. by D.J. Fabian (Academic, London 1968) p.133
2.157 R.L. Park, J.E. Houston, D.G. Schreiner: Rev. Sci. Instr. **41**, 1810 (1970)
2.158 J.E. Houston, R.L. Park: In *Electron Spectroscopy*, ed. by D.A. Shirley (North Holland, Amsterdam 1972) p.895
2.159 J. Kanski, P.O. Nilsson: Phys. Fenn. **9**, S1-68 (1974)
2.160 J.E. Houston, R.L. Park, G.E. Laramore: Phys. Rev. Lett. **30**, 846 (1973)
2.161 J.E. Holliday: In *Soft X-Ray Band Spectra and the Electronic Structure of Metals and Materials*. ed. by D.J. Fabian (Academic, London 1968) p.101
2.162 A.P. Lukirskii, I.A. Brytov: Fiz. Tverd. Tela **6**, 43 (1964)
2.163 A.P. Lukirskii, T.M. Zimkina: Izv. Akad. Nauk SSSR, Ser. Fiz. **28**, 765 (1964)
      A.S. Vinogradov, T.M. Zimkina, J.F. Maltsev: Fiz. Tverd. Tela **11**, 3354 (1969)
2.164 A.P. Lukirskii, E.P. Savinov, I.A. Brytov, Yu.F. Shepelev: Izv. Akad. Nauk SSSR, Ser. Fiz. **28**, 866 (1964)
      E.P. Savinov, A.P. Lukirskii: Opt. Spektrosk. **23**, 303 (1967)
2.165 A.P. Lukirskii, T.M. Zimkina, O.A. Ershov, E.P. Savinov: Fiz. Tverd. Tela **8**, 1787 (1966)
2.166 T.M. Zimkina, V.A. Fomichev: *Ultra Soft X-Ray Spectroscopy* (University of Leningrad 1971) (in Russian)
2.167 G.A. Sawatzky, J.W. Allen: Phys. Rev. Lett. **53**, 2339 (1984)
2.168 W. Speier, J.C. Fuggle, R. Zeller, B. Ackermann, K. Szot, F.H. Hillebrecht, M. Campagna: Phys. Rev. B **30**, 6921 (1984)

2.169 V. Dose, H.J. Gossmann, D. Straub: Phys. Rev. Lett. **47**, 608 (1981)

2.170 J.K. Lang, Y. Baer, P.A. Cox: J. Phys. F **11**, 121 (1981)

2.171 F.U. Hillebrecht, J.C. Fuggle, G.A. Sawatzky, R. Zeller: Phys. Rev. Lett. **51**, 1187 (1983)

2.172 F.U. Hillebrecht, J.C. Fuggle, G.A. Sawatzky, M. Campagna, O. Gunnarsson, K. Schönhammer: Phys. Rev. B **30**, 1777 (1984)

2.173 P.O. Nilsson, A. Kovacs: Phys. Scr. T **4**, 61 (1983)

2.174 N.V. Smith: Phys. Scr. T **17**, 214 (1987)

2.175 H. Ågren, N. Mårtensson, R. Manne: Chem. Phys. **99**, 357 (1985)

2.176 E.A. Stern: J. Physique **47**, C8-3 (1986)

2.177 W. Speier, T.M. Hayes, J.W. Allen, J.B. Boyce, J.C. Fuggle, M. Campagna: Phys. Rev. Lett. **55**, 1693 (1985)

2.178 H.A. Bethe, E.E. Salpeter: *Quantum Mechanics of One and Two Electron Atoms* (Springer, Berlin, Göttingen 1957)

2.179 W. Heitler: *The Quantum Theory of Radiation* (Clarendon, Oxford 1954)

2.180 A. Sommerfeld: *Atombau und Spektrallinien* (Vieweg, Braunschweig 1960)

2.181 R. Haensel, G. Keitel, C. Kunz, P. Schreiber: Phys. Rev. Lett. **25**, 208, 1281 (1970)
R. Haensel, G. Keitel, E.E. Koch, N. Kosuch, M. Skibowski: Phys. Rev. Lett. **25**, 1281 (1970)
R. Haensel, G. Keitel, N. Kosuch, U. Nielson, P. Schreiber: J. Physique **32**, C4-236 (1971)
R. Haensel, N. Kosuch, U. Nielsen, B. Sonntag, U. Rössler: Phys. Rev. B **7**, 1577 (1973)

2.182 K. Codling: J. Electron Spectrosc. **17**, 279 (1979)

2.183 H. Petersen, K. Radler, B. Sonntag, R. Haensel: J. Phys. B **8**, 31 (1975)
H. Petersen: Phys. Stat. Sol. B **72**, 591 (1972)
K. Radler, B. Sonntag: Chem. Phys. Lett. **39**, 371 (1976)
K. Radler, B. Sonntag, H.W. Wolff: In *Int'l Conf. on the Physics of X-Ray Spectra, Extended Abstracts*, ed. by R.D. Deslattes (NBS, Gaithersburg 1976) p.54

2.184 R.D. Deslattes: Phys. Rev. Lett. **20**, 483 (1968)

2.185 R. Haensel, G. Keitel, P. Schreiber, C. Kunz: Phys. Rev. **188**, 1375 (1969)

2.186 U. Fano, J.W. Cooper: Rev. Mod. Phys. **40**, 441 (1968); ibid. **41**, 724 (1969)

2.187 D.C. Griffin, K.L. Andrew, R.D. Cowan: Phys. Rev. **177**, 62 (1969)

2.188 J.L. Dehmer: Phys. Fenn. **9**, S1-60 (1974)

2.189 P. Rabe: DESY F 41-74/2 (1974)

2.190 J.W. Cooper: Phys. Rev. **165**, 681 (1962)

2.191 D. Ottewell, E.A. Stewardson, J.E. Wilson: J. Phys. B **6**, 2184 (1973)

2.192 R. Haensel, P. Rabe, B. Sonntag: Solid State Commun. **8**, 1845 (1970)

2.193 H.W. Wolff, R. Bruhn, K. Radler, B. Sonntag: Phys. Lett. A **59**, 67 (1976)

2.194 B.J. Thole, G. van der Laan, J.C. Fuggle, G.A. Sawatzky, R.C. Karnatak, J.-M. Esteva: Phys. Rev. B **32**, 5107 (1985)

2.195 R. Bruhn, B. Sonntag, H.W. Wolff: J. Phys. B **12**, 203 (1979)

2.196 M. Ya. Amusia, N.A. Cherepkov, L.V. Chernysheva: Zh. Eksp. Teor. Fiz. **60**, 160 (1971)

2.197 M. Ya. Amusia: In *Vacuum Ultraviolet Radiation Physics*, ed. by E.E. Koch, R. Haensel, C. Kunz (Vieweg, Braunschweig 1974) p.205

2.198 C. Bonnelle, R.C. Karnatak, J. Sugar: Phys. Rev. A **9**, 1920 (1974)

2.199 D.J. Kennedy, S.T. Manson: Phys. Rev. A **5**, 227 (1972)

2.200 A.F. Starace: J. Phys. B **7**, 14 (1974)

2.201 J. Sugar: Phys. Rev. B **5**, 1785 (1972)

2.202 G. Wendin: J. Phys. B **6**, 42 (1973)

2.203 P.J. Durham, J.B. Pendry, C.H. Hodges: Sol. State Commun. **38**, 159 (1981)

2.204 J.L. Dehmer: Phys. Rev. Lett. **35**, 213 (1975)

2.205 F.W. Kutzler, C.R. Natoli, D.K. Misemer, S. Doniach, K.O. Hodgson: J. Chem. Phys. **73**, 3274 (1980)

2.206 A.P. Hitchcock: J. Electron Spectrosc. **25**, 245 (1982)

2.207 J. Physique **47**, C8, 3–1209 (1986) (2 vols.)

2.208 J.C. Fuggle: Phys. Scr. T **17**, 64 (1987)

2.209 H. Friedrich B. Sonntag, P. Rabe, W. Butcher, W.H.E. Schwarz: Chem. Phys. Lett. **64**, 360 (1979)

H. Friedrich, B. Pittel, P. Rabe, W.H.E. Schwarz: J. Phys. B **13**, 25 (1980)

2.210 C.H. Shaer: Phys. Rev. **57**, 877 (1940)

2.211 J. Backovsky: Czech. J. Phys. **4**, 118 (1954)

2.212 D. Blechschmidt, R. Haensel, E.E. Koch, U. Nielsen, T. Sagawa: Chem. Phys. Lett. **14**, 33 (1972)

2.213 J.L. Dehmer: J. Chem. Phys. **56**, 4496 (1972)

2.214 D. Dill, J.L. Dehmer: J. Chem. Phys. **61**, 692 (1974)

J.L. Dehmer, D. Dill: Phys. Rev. Lett. **35**, 213 (1975); J. Chem. Phys. **65**, 5327 (1976).

2.215 A.S. Vinogradov, T.M. Zimkina, V.N. Akimov, B. Schlarbaum: Izv. Akad. Nauk SSSR, Ser. Fiz. **38**, 508 (1974)

A.S. Vinogradov, V.N. Akimov, T.M. Zimkina, E.B. Dobryakova: Izv. Sib. Otd. Akad. Nauk SSSR, Ser. Khim. Nauk 1975, Nr.9, H.4, 88

2.216 G.R. Wight, C.E. Brion, M.J. van der Wiel: J. Electron Spectrosc. **1**, 457 (1973)

G.R. Wright, C.E. Brion: **3**, 191 (1974); ibid. **4**, 25 (1974)

M.J. van der Wiel, Th.M. El-Sherbini, C.E. Brion: Chem. Phys. Lett. **7**, 161 (1970)

2.217 R.L. Barinskii: In *X-Ray Spectra and Electronic Structure of Matter*, ed. by V.V. Nemoshkalenko (Institute of Metal Physics, Kiev 1969) Vol.2, p.222

2.218 V.A. Fomichev, R.L. Barinskii: Zh. Strukt. Khim. **11**, 875 (1970)

2.219 V.I. Nevedov: Zh. Strukt. Khim. **11**, 292, 299 (1970)

2.220 W.H.E. Schwarz: Angew. Chemie **86**, 505 (1974)

2.221 E.E. Vainshtein, R.L. Barinskii, K.I. Narbutt: Dokl. Akad. Nauk SSSR **78**, 39 (1951); Zh. Eksp. Teor. Fiz. **23**, 593 (1952)

2.222 M.J. Hanus, E. Gilberg: In *X-Ray Spectra and Electronic Structure of Matter*, ed. by A. Faessler, G. Wiech (Universität München 1973) Vol.1, p.338; J. Phys. B **9**, 137 (1976)

2.223 W. Butscher, R.J. Buenker, S.D. Peyerimhoff: Chem. Phys. Lett. **52**, 449 (1977)

R.J. Buenker, S.D. Peyerimhoff, W. Butscher: Mol. Phys. **35**, 771 (1978)

2.224 H.-U. Chun, H. Gebelein: Z. Naturforsch. 22a, 1813 (1967)

H.-U. Chun: Phys. Lett. A 30, 445 (1969)

2.225 G. van der Laan, J. Zaanen, G.A. Sawatzky, R. Karnatak, J.-M. Esteva: Sol. State Commun. **56**, 673 (1985); Phys. Rev. B **33**, 4253 (1986)

2.226 A. Fujimori, F. Minami, S. Sugano: Phys. Rev. B **30**, 957 (1984)

2.227 I. Davoli, A. Marcelli, A. Bianconi, M. Tomellini, M. Fafoni: Phys. Rev. B **33**, 2979 (1986)

2.228 W.H.E. Schwarz: In *Int'l Conf. on the Physics of X-Ray Spectra, Extended Abstracts*, ed. by R.D. Deslattes (NBS, Gaithersburg 1976) p.49

2.229 W.H.E. Schwarz: Chem. Phys. **9**, 157 (1975); ibid. **11**, 217 (1975)

2.230 V.I. Baranovskii, T.M. Zimkina, V.A. Fomichev: Teor. Eksp. Khim. 3, 354 (1967)

2.231 R.A. Mattson, R.C. Ehlert: J. Chem. Phys. 48, 5465, 5471 (1968)

2.232 L.N. Mazalov, A.P. Sadovskii, V.M. Bertenev, V.V. Murakhtanov, E.A. Galtsova, L.I. Chernyavski: Teor. Eksp. Khim. 7, 46 (1971)

2.233 A.P. Sadovskii, L.N. Mazalov, V.M. Bertenev: Teor. Eksp. Khim. 6, 502 (1970)

2.234 L.N. Mazalov, A.P. Sadovskii, P.I. Vadash, F.Kh. Gelmukhanov: Zh. Strukt. Khim. 14, 262 (1973)
L.N. Mazalov, V.V. Murakhtanov, T.I. Guzhavina, A.P. Sadovskii: Zh. Strukt. Khim. 16, 262, 267 (1975)

2.235 M.L. Sink, G.E. Jura: Chem. Phys. Lett. 20, 474 (1973)

2.236 L.N. Mazalov, V.D. Yumatov, V.V. Murakhtanov, F.Kh. Gelmukhanov, G.N. Dolenko, E.S. Gluskin, A.V. Kondratenko: *X-Ray Spectra of Molecules* (Nauka, Novosibirsk 1977) (in Russian)
L.N. Mazalov: *X-Ray Spectroscopy and X-Ray Electron Spectroscopy of Molecules, Theoretical Principles* (Nauka, Novosibirsk 1979) (in Russian)

2.237 A.A. Pavlychev, A.S. Vinogradov, D.E. Onopko, S.A. Titov: Fiz. Tverd. Tela 20, 3674 (1978)
A.A. Pavlychev, A.S. Vinogradov, T.M. Zimkina, D.E. Onopko, R. Szargan: Opt. Spektrosk. 48, 192 (1980)

2.238 V.P. Sachenko, E.V. Polozhentsev, A.P. Kovtun, Yu.F. Migal, V.V. Kolesnikov, R.V. Vedrinskii: Phys. Fenn. 9, S 1-129 (1974)
V.P. Sachenko, E.V. Polozhentsev, A.P. Kovtun, Yu.F. Migal, R.V. Vedrinski, V.V. Kolesnikov: Phys. Lett. A 48, 169 (1974)

2.239 R.V. Vedrinskii, A.P. Kovtun, V.V. Kolesnikov, Yu.F. Migal, E.V. Polozhentsev, V.P. Sachenko: Izv. Akad. Nauk SSSR, Ser. Fiz. 38, 434 (1974)
R.V. Vedrinskii, V.L. Kraizman: Zh. Eksp. Teor. Fiz. 74, 1215 (1978)

2.240 M. Nakamura: Phys. Rev. 178, 80 (1969)
Y. Moriota, M. Nakamura, E. Ishiguro, M. Sasanuma: J. Chem. Phys. 61, 1426 (1974)

2.241 W.H.E. Schwarz, T.C. Chang: Chem. Phys. Lett. 49, 207 (1977); J. Phys. B 11, 591 (1978)

2.242 R. Haensel, B. Sonntag: In *Computational Solid State Physics*, ed. by F. Herman, N.W. Dalton, T.R. Koehler (Plenum, New York 1972) p.43

2.243 P. Rabe, B. Sonntag, T. Sagawa, R. Haensel: Phys. Stat. Sol. B 50, 559 (1972)

2.244 R. Haensel, C. Kunz, B. Sonntag: Phys. Rev. Lett. 20, 262 (1968)
R. Haensel: DESY F 41-70/1 (1970)

2.245 R.S. Knox: *The Theory of Excitons* (Academic, New York 1969)

2.246 K. Codling, R.P. Madden: Phys. Rev. Lett. 12, 106 (1964); Appl. Optics 4, 1431 (1965)

2.247 J. Barth, F. Gerken, C. Kunz: Phys. Rev. B 28, 3608 (1983)

2.248 R.D. Leapman, L.A. Grunes: Phys. Rev. Lett. 45, 397 (1980)
L.A. Grunes: Phys. Rev. B 27, 2111 (1983)

2.249 W.G. Waddington, P. Rez, I.P. Grant, C.J. Humphreys: Phys. Rev. B 34, 1467 (1986)

2.250 H. Nakamura, K. Ichikawa, Y. Watanabe, K.J. Tsutsumi: J. Phys. Soc. Jpn. 52, 4014 (1983)

2.251 J. Fink, Th. Müller-Heinzeling, B. Scheerer, W. Speier, F.U. Hillebrecht, J.C. Fuggle, J. Zaanen, G.A. Sawatzky: Phys. Rev. B 32, 4899 (1985)

2.252 J. Zaanen, G.A. Sawatzky, J. Fink, W. Speier, J.C. Fuggle: Phys. Rev. B 32, 4905 (1985)

2.253 T. Fransen, P.J. Gellings, J.C. Fuggle, G. van der Laan, J.-M. Esteva, R.C. Karnatak: Appl. Surf. Sci. 20, 257 (1985)

2.254 G. Kaindl, G. Kalkowski, W.D. Drewer, B. Perscheid, F. Holtzberg: J. Appl. Phys. 55, 910 (1984)

2.255 J.C. Fuggle, F.U. Hillebrecht, J.-M. Esteva, R.C. Karnatak, O. Gunnarsson, K. Schönhammer: Phys. Rev. B 27, 4637 (1983)

2.256 G. van der Laan, B.T. Thole, G.A. Sawatzky, J.C. Fuggle, R.C. Karnatak, J.-M. Esteva, B. Lengeler: J. Phys. C 19, 817 (1986)

2.257 G. van der Laan, J. Zaanen, G.A. Sawatzky, R.C. Karnatak: Phys. Rev. B 33, 4253 (1986)

2.258 N. Kosugi, T. Yokoyama, K. Asakura, H. Kuroda: Chem. Phys. 91, 249 (1984)

2.259 R.A. Bair, W.A. Goddard: Phys. Rev. B 22, 2767 (1980)

2.260 G. van der Laan, B.T. Thole, G.A. Sawatzky, J.B. Goedkoop, J.C. Fuggle, J.-M. Esteva, R.C. Karnatak, J.P. Remeika, H.A. Dabkowska: Phys. Rev. B 34, 6529 (1986)

2.261 J. Stöhr: Z. Physik B 61, 453 (1985)

2.262 B.M. Kincaid, P. Eisenberger: Phys. Rev. Lett. 34, 1361 (1975)

2.263 D.E. Sayers, E.A. Stern, F.W. Lytle: Pys. Rev. Lett. 27, 1204 (1971)
     E.A. Stern, D.E. Sayers: Phys. Rev. Lett. 30, 174 (1973); Phys. Rev. B 10, 3027 (1974)
     F.W. Lytle, D.E. Sayers, E.A. Stern: Phys. Rev. B 11, 4825 (1975)
     E.A. Stern, D.E. Sayers, F.W. Lytle: Phys. Rev. B 11, 4836 (1975)
     D.E. Sayers, E.A. Stern, F.W. Lytle: Phys. Rev. Lett. 35, 584 (1975)

2.264 C.A. Ashley, S. Doniach: Phys. Rev. B 11, 1279 (1975)

2.265 P.A. Lee, J.B. Pendry: Phys. Rev. B 11, 2795 (1975)
     B.-K. Teo, P.A. Lee: J. Am. Chem. Soc. 101, 2815 (1979)

2.266 A. Bianconi, L. Incoccia, S. Stipcich (eds.): EXAFS and Near Edge Structure, Springer Ser. Chem. Phys., Vol.27 (Springer, Berlin, Heidelberg 1983)

2.267 H. Winick, S. Doniach: Synchrotron Radiation Research (Plenum, New York 1980)

2.268 P. Rabe, R. Haensel: Festkörperprobleme 20, 43 (Vieweg, Braunschweig 1980)

2.269 C.D. Garner, S.S. Hasnain: EXAFS for Inorganic Systems, Daresbury publication DL/SCI/R17 (1980)

2.270 J. Stöhr: In Chemistry and Physics of Solid Surfaces V, ed. by R. Vanselow, R. Howe, Springer Ser. Chem. Phys., Vol.35 (Springer, Berlin, Heidelberg 1984) p.231

2.271 R. Prins, D. Koningsberger (eds.): X-Ray Absorption: Principles, Applications, Techniques of EXAFS, SEXAFS, and XANES (Wiley, New York 1985)

2.272 B.K. Teo, C. Joy: EXAFS Spectroscopy, Techniques and Applications (Plenum, New York 1981)
     B.K. Teo: EXAFS, Basic Principles and Data Analysis (Springer, Berlin, Heidelberg 1986)

2.273 J.D. Hanawalt: Z. Physik 70, 293 (1931)

2.274 R. de L. Kronig: Z. Physik 70, 317 (1931); ibid. 75, 191 (1932)

2.275 T. Hayasi: Sci. Rep. Tohoku Univ. 33, 123 (1949); ibid. 44, 87 (1960)

2.276 F.W. Lytle: Developm. Appl. Spectrosc. 12, 275 (1963); and in Physics of Non-Crystalline Solids, ed. by J.A. Prins (North Holland, Amsterdam 1965) p.12

2.277 H. Petersen: Z. Physik 80, 258 (1933); ibid. 98, 569 (1936)
     D.R. Hartree, R. de L. Kronig, H. Petersen: Physica 1, 895 (1934)

2.278 A.I. Kozlenkov: Izv. Akad. Nauk SSSR, Ser. Fiz. **25**, 957 (1961); ibid. **27**, 364 (1963)

2.279 V.V. Shmidt: Izv. Akad. Nauk SSSR, Ser. Fiz. **25**, 977 (1961); ibid. **27**, 384 (1963)

2.280 T. Shiraiwa, T. Ishimura, M. Sawada: J. Phys. Soc. Jpn. **12**, 788 (1957); ibid. **13**, 847 (1958); ibid. **15**, 240 (1960)

2.281 H. Keilacker: Der Einfluß der Teilchengröße auf die kantenferne Feinstruktur von Röntgenspektren. Dissertation (Karl-Marx-Universität, Leipzig 1973)

2.282 A.I. Kostarev, W.M. Weber: Phys. Rev. B **3**, 4124 (1971)

2.283 M. Sawada: Rep. Sci. Works Fac. Sci. Osaka Univ. **7**, 1 (1959)

2.284 R.A. Van Nordstrand: In *Non-Crystalline Solids*, ed. by V.D. Frechette (Wiley, New York 1960) p.168

2.285 O. Brümmer, G. Dräger: Phys. Stat. Sol. **27**, 513 (1968)

2.286 W.F. Nelson, J. Siegel, R.W. Wagner: Phys. Rev. **127**, 2025 (1962)

Chapter 3

3.1 G. Brogren: Ark. Fysik **6**, 321 (1953); ibid. **8**, 391 (1954)

3.2 V.A. Trapeznikov, V.A. Trofimova: Prib. Tekh. Eksp. **6**, 191 (1967)

3.3 A.N. Nigam: Indian J. Theor. Phys. **2**, 175 (1955)
   D.C. Purkayastha: Indian J. Phys. **39**, 250 (1956)

3.4 T. Hayasi, T. Nishimura, M. Suzuki: Sci. Rep. Tohoku Univ. **41**, 183 (1958)

3.5 L.S. Birks, R.E. Seebold, A.P. Batt, J.S. Grosso: J. Appl. Phys. **35**, 2578 (1964)
   A.A. Sterk: Adv. X-Ray Anal. **8**, 189 (1965); ibid. **9**, 410 (1966)

3.6 F.W. Saris, W.F. van der Weg, I. Kistenmacher: Phys. Lett. A **26**, 592 (1968)

3.7 A.G. Stadnikov, A.P. Nikolskii: Dokl. Akad. Nauk SSSR **191**, 315 (1970)

3.8 C.R. Worthington, S.G. Tomlin: Proc. Phys. Soc. **69**, 401 (1956)

3.9 H. Bethe: Ann. Physik **5**, 325 (1930)

3.10 V.E. Cosslett, R.N. Thomas: Brit. J. Appl. Phys. **15**, 1283 (1964)

3.11 L. Hoffmann, G. Wiech, E. Zöpf: Z. Physik **229**, 131 (1969)

3.12 C. Feldman: Phys. Rev. **117**, 455 (1960)

3.13 D.W. Fischer, W.L. Baun: J. Appl. Phys. **39**, 4757 (1968)
   D.W. Fischer: J. Appl. Phys. **41**, 3561 (1970)

3.14 I.B. Borovskii, V.I. Rydnik: Izv. Akad. Nauk SSSR, Ser. Fiz. **31**, 1009 (1967)

3.15 A. Vignez, G. Dez: J. Phys. D **1**, 1309 (1968)

3.16 M. Green, V.E. Cosslett: J. Phys. D **1**, 425 (1968)

3.17 L.S. Birks, R.E. Seebold, J.S. Grosso: J. Appl. Phys. **36**, 699 (1965)

3.18 J.E. Holliday: Rev. Sci. Instr. **31**, 891 (1960); J. Appl. Phys. **33**, 3259 (1962)

3.19 R.D. Deslattes, B.G. Simson: Rev. Sci. Instr. **37**, 753 (1966)

3.20 E. Gilberg: Untersuchung der Bindungsabhängigkeit von Röntgenemissionsspektren an freien Molekülen mit Fluoreszenzanregung. Dissertation (Universität München 1969)

3.21 R.A. Mattson, R.C. Ehlert: Adv. X-Ray Anal. **9**, 471 (1966)

3.22 M.A. Blokhin: *Methods of X-Ray Spectroscopic Research* (Pergamon, Oxford 1965)

3.23 W. Schaafs: In *Handbuch der Physik*, ed. by S. Flügge, Vol.30 (Springer, Berlin, Göttingen 1957) p.1

3.24 A. Faessler: In *Soft X-Ray Band Spectra and the Electronic Structure of Metals and Materials*, ed. by D.J. Fabian (Academic, London 1968) p.93

3.25 G. Wiech: Z. Physik **193**, 490 (1966)

3.26 K. Tögel: Siemens Z. **8**, 597 (1962)

3.27 J.C. Slater: *Quantum Theory of Atomic Structure* (McGraw-Hill, New York 1960)

3.28 N. Spielberg: Philips Res. Rep. 14, 215 (1959); Adv. X-Ray Anal. 10, 534 (1967)

3.29 R.O. Müller: *Spektrochemische Analysen mit Röntgenfluoreszenz* (Oldenbourg, München 1967)

3.30 E. Alexander, A. Faessler: Z. Physik 68, 260 (1931)

3.31 D. Heintz: Über den Einfluß der chemischen Bindung auf das Kβ-Emissionsspektrum von Cobalt. Dissertation (Universität München 1969); Z. Angew. Physik 27, 98 (1969)

3.32 M.I. Korsunkii, Ya.E. Genkin, M.M. Omarov: Zavodskaya Lab. 32, 381 (1966); in *Electronic Structure of Transition Metals and Their Alloys*, ed. by I.Ya. Dekhtyar (Institute of Metal Physics, Kiev 1968) p.61

3.33 K. Läuger: Über den Einfluß der Bindungsart und der Kristallstruktur auf das K-Röntgenemissionsspektrum von Aluminium und Silizium. Dissertation (Universität München 1968)

3.34 B.L. Henke: Adv. X-Ray Anal. 6, 361 (1963); ibid. 8, 269 (1964)

3.35 C. Kunz (ed.): *Synchrotron Radiation, Techniques and Applications*, Top. Curr. Phys., Vol.10 (Springer, Berlin, Heidelberg 1979)

3.36 E.-E. Koch (ed.): *Handbook on Synchrotron Radiation*, Vol.1 (North-Holland, Amsterdam 1983)

3.37 R. Haensel, C. Kunz, T. Sasaki, B. Sonntag: Appl. Opt. 7, 301 (1968)

3.38 U. Bonse: NATO Adv. Study Inst. Ser. B 63, 298 (1980)

3.39 N. Schwentner, U. Hahn, D. Einfeld, G. Mühlhaupt: Nucl. Instrum. Meth. 167, 499 (1979)

3.40 H. Winick: Nucl. Instr. Meth. A 261, 9 (1987); see also Synchrotron Radiation News 2, No.2, 25 (1989)

3.41 N. Kosuch, J. Müller, G. Wiech, A. Faessler: Phys. Fenn. 9, S1–189 (1974)
N. Kosuch, E. Tegeler, G. Wiech, A. Faessler: J. Electron Spectrosc. 13, 263 (1978)

3.42 O. Aita, K. Tsutsumi, K. Ichikawa, M. Kamada, M. Okusawa, H. Nakamura, T. Watanabe: Phys. Rev. B 23, 5676 (1981)

3.43 A. Bianconi: In *EXAFS and Near Edge Structure III*, ed. by K.O. Hodgson, B. Hedman, J.E. Penner-Hahn, Springer Proc. Phys., Vol.2 (Springer, Berlin, Heidelberg 1984) p.167

3.44 E.A. Stern, S.M. Heald: In [Ref.3.36, p.955]

3.45 W. Greulich, E. Dreisigacker: Phys. Bl. 43, 80 (1987)

3.46 E.-E. Koch: Phys. Bl. 40, 324 (1984)

3.47 A. Gaupp, E.-E. Koch, R. Maier, W. Peatman, A.M. Bradshaw: BESSY II - Eine optimierte Undulator/Wiggler-Speicherring-Lichtquelle für den VUV- und XUV-Spektralbereich, BESSY (1986)

3.48 D.W. Fischer, W.L. Baun: J. Appl. Phys. 38, 4830 (1967)

3.49 M.A. Blokhin: *Physik der Röntgenstrahlen* (Verlag Technik, Berlin 1957)

3.50 A.P. Lukirskii: Opt. Spektrosk. 19, 800 (1965)

3.51 H.W.B. Skinner, J.E. Johnston: Proc. Roy. Soc. A 161, 420 (1937); Proc. Cambridge Phil. Soc. 34, 109 (1938)

3.52 T.M. Zimkina, V.A. Fomichev: *Ultra Soft X-Ray Spectroscopy* (University of Leningrad 1971) (in Russian)

3.53 R.I. Liefeld: In *Soft X-Ray Band Spectra and the Electronic Structure of Metals and Materials*, ed. by D.J. Fabian (Academic, London 1968) p.133

3.54 E.E. Koch, R. Haensel, C. Kunz (eds.): *Vacuum Ultraviolet Radiation Physics* (Vieweg, Braunschweig 1974)

3.55 V.A. Fomichev: Fiz. Tverd. Tela 9, 3034, 3167 (1967)

3.56 A.P. Lukirskii, I.A. Brytov: Fiz. Tverd. Tela 6, 43 (1964)

3.57 A.P. Lukirskii, T.M. Zimkina: Izv. Akad. Nauk SSSR, Ser. Fiz. 28, 765 (1964)
      A.S. Vinogradov, T.M. Zimkina, J.F. Maltsev: Fiz. Tverd. Tela 11, 3354 (1969)

3.58 S. Kiyono, Y. Hayasi, T. Muranaka: In *Int'l Conf. on the Physics of X-Ray Spectra*, ed. by R.D. Deslattes (NBS, Gaithersburg, MD 1976) Extended Abstracts, p.241
      S. Kiyono, T. Muranaka, K. Aota: Tech. Rep. Tohoku Univ. 43, 423 (1978)
      S. Kiyono, T. Muranaka, T. Watanabe: Jpn. J. Appl. Phys. 18, 1865 (1979)

3.59 A.K. Baird: Adv. X-Ray Anal. 13, 26 (1969)

3.60 F.D. Davidson, R.W.G. Wyckhoff: Adv. X-Ray Anal. 9, 344 (1966)

3.61 R.C. Ehlert: Adv. X-Ray Anal. 8, 325 (1964)

3.62 D.W. Fischer, W.L. Baun: Adv. X-Ray Anal. 7, 489 (1963); Spectrochim. Acta 21, 443 (1965)

3.63 B.L. Henke: Adv. X-Ray Anal. 9, 430 (1966)

3.64 M. Singh, S. Singh: Appl. Opt. 19, 3313 (1980)

3.65 J.B. Nicholson: Adv. X-Ray Anal. 7, 497 (1963)

3.66 D.W. Fischer: J. Appl. Phys. 36, 2048 (1965); Adv. X-Ray Anal. 13, 159 (1970)

3.67 A. Fingerland, L. Cervinka: Acta Crystallogr. 17, 508 (1964)

3.68 A.H. Compton, S.K. Allison: *X-Rays in Theory and Experiment* (Van Nostrand, London 1935)

3.69 G. Brogren: Ark. Fysik 22, 535 (1962); ibid. 23, 81,219 (1963)

3.70 D. Chopra: Rev. Sci. Instr. 41, 1004 (1970)

3.71 J. Backovsky: Izv. Fiz. Inst. Bulg. Akad. Nauk 10, 5 (1962)

3.72 A. Meisel, To ba Trong: Z. Kristallogr. 112, 148 (1965)

3.73 L.S. Birks, R.T. Seal: J. Appl. Phys. 28, 541 (1957)
      J. Vierling, J.V. Gittrich, L.S. Birks: Appl. Spectrosc. 23, 342 (1968)

3.74 L.G. Parratt: Rev. Sci. Instr. 5, 395 (1934)

3.75 S.S. Lenin, I.V. Serikov, A.I. Spichkin, A.N. Meshevich: App. Met. Rentgen. Anal. 10, 102 (1972)

3.76 R. Bubakova, J. Drahokoupil, A. Fingerland: Czech. J. Phys. B 17, 657 (1967)

3.77 K. Feser, A. Faessler: Z. Physik 209, 1 (1968)

3.78 G. Brogren: Ark. Fysik 3, 507,515 (1951)

3.79 A.B. Gilvarg: Dokl. Akad. Nauk SSSR 72, 489 (1950)

3.80 E. Gilberg: Rev. Sci. Instr. 42, 1189 (1971)

3.81 G. Leonhardt, A. Kopczynski, H. Ehrhardt, A. Meisel: Exp. Techn. Physik 20, 221 (1972)

3.82 J. Eggs, K. Ulmer: In *Röntgenspektren und chemische Bindung*, ed. by A. Meisel (Karl-Marx-Universität, Leipzig 1966) p.96

3.83 M. Siegbahn: *Spektroskopie der Röntgenstrahlen* (Julius Springer, Berlin 1931)

3.84 G. Spraque, D. Tomboulian, D. Bedo: J. Opt. Soc. Am. 45. 756 (1955)

3.85 A.P. Lukirskii, E.P. Savinov, O.A. Ershov, I.I. Zhukova, V.A. Fomichev: Opt. Spektrosk. 19, 425 (1965)

3.86 R.L. Johnson: Nucl. Instr. Meth. A 246, 303 (1986)

3.87 R.L. Johnson: In [Ref.3.36, p.173]

3.88 J.E. Holliday: Adv. X-Ray Anal. 9, 38, 365 (1966)

3.89 H. Neddermeyer: Beiträge zur Spektroskopie der ultraweichen Röntgenstrahlung. Dissertation (Universität München 1969)

3.90 J.E. Mack, J.R. Stehn, B. Edlen: J. Opt. Soc. Am. 22, 245 (1932)

3.91 G. Wiech, E. Zöpf: In *Electronic Density of States*. NBS Spec. Publ. 323, ed. by L.H. Bennett (Washington 1971) p.335

3.92   H.I. Israel, D.W. Lier, E. Storm: Nucl. Instr. Meth. **91**, 141 (1971)
3.93   S.A. Baldin, L.M. Joannesyants: Prib. Tekhn. Eksp. (1972) p.7
3.94   U. Fano: Phys. Rev. **72**, 26 (1947)
3.95   L. Kuhn: Glas Instr. Techn. **12**, 91 (1968)
3.96   A.E. Sandström: In *Handbuch der Physik*, ed. by S. Flügge, Vol.30 (Springer, Berlin, Göttingen 1957) p.78
3.97   A.P. Lukirskii, V.A. Fomichev, I.A. Brytov: Opt. Spektrosk. **20**, 366,368 (1966)
3.98   A.J. Caruso, H.H. Kim: Rev. Sci. Instr. **39**, 1059 (1968)
3.99   A.P. Lukirskii, I.A. Brytov: Prib. Tekhn. Eksp. (1966) p.66
3.100 R.D. Deslattes, B.G. Simson, R.E. LaVilla: Rev. Sci. Instr. **37**, 596 (1966)
3.101 B.L. Henke, R.E. Lent: Adv. X-Ray Anal. **12**, 480 (1969); ibid. **13**, 1 (1970)
3.102 J.B. Birks: *The Theory and Practice of Scintillation Counting* (Pergamon, London 1964)
3.103 A.P. Lukirskii, E.P. Savinov, I.A. Brytov, Yu.F. Shepelev: Izv. Akad. Nauk SSSR, Ser. Fiz. **28**, 866 (1964)
       E.P. Savinov, A.P. Lukirskii: Opt. Spektrosk. **23**, 303 (1967)
3.104 W.L. Baun: J. Appl. Phys. **40**, 4210 (1969); Adv. X-Ray Anal. **13**, 49 (1970)
3.105 C. Nordling: Jpn. J. Appl. Phys. **17**, S 2-7 (1978)
       J. Nordgren, H. Ågren, L. Pettersson, L. Selander, S. Griep, C. Nordling, K. Siegbahn: Phys. Scr. **20**, 623 (1979)
3.106 A. Boksenberg: Anal. Chem. **41**, 87 (1969)
3.107 F.J. Zutavern, S.E. Schnatterly, E. Källne, C.P. Franck, T. Aton: Nucl. Instr. Meth. **172**, 351 (1980)
3.108 E. Suoninen, M. Pessa: Phys. Scr. **7**, 89 (1973)
3.109 M.A. Blokhin, E.Ya. Ovcharenko, P.I. Myakov, V.A. Sotnikov, Yu.M. Mamonov, G.L. Belkina: Zavodskaya Lab. **31**, 423 (1965)
3.110 O.I. Sumbaev, A.F. Mezentsev: Zh. Eksp. Teor. Fiz. **48**, 445 (1965); ibid. **49**, 459 (1965)
3.111 A. Meisel, J. Finster, A. Kopczynski: Exp. Techn. Physik **16**, 396 (1968)
3.112 R. Plesch: Archiv Techn. Mess. **412**, R49 (1970)
3.113 N. Spielberg, W. Parrish, K. Lowitsch: Spectrochim. Acta **15**, 564 (1959)
3.114 M. Romand, R. Bador, M. Charbonnier, F. Gaillard: X-Ray Spectrometry **16**, 7 (1987)
3.115 H.H. Johann: Z. Physik **69**, 185 (1931)
3.116 R.C. Ehlert, R.A. Mattson: Adv. X-Ray Anal. **9**, 456 (1966)
3.117 Y. Cauchois: J. Physique Radium **3**, 320 (1932); ibid. **4**, 61 (1933)
3.118 R.W. Jewell, W. John, R. Massey, B.G. Saunders: Nucl. Instr. Meth. **62**, 68 (1968)
3.119 E. Källne, J. Källne: Phys. Scr. T **17**, 152 (1987)
3.120 T. Johansson: Z. Physik **82**, 507 (1933)
3.121 J. Cermak: J. Phys. E **3**, 615 (1970)
3.122 J. Drahokoupil: Czech. J. Phys. **12**, 752 (1962)
3.123 W. Ehrenberg, H. Mark: Z. Physik **42**, 807 (1927)
3.124 M.A. Blokhin, I.Ya. Nikiforov: App. Met. Rentgen. Anal. **10**, 89 (1972)
3.125 J. Drahokoupil, A. Fingerland: Czech. J. Phys. B **18**, 1034 (1968)
3.126 L.V. Azaroff: Adv. X-Ray Anal. **9**, 251 (1965)
3.127 H.W. Schnopper: J. Appl. Phys. **36**, 1423 (1965)
3.128 R.D. Deslattes: Rev. Sci. Instr. **38**, 815 (1967)
3.129 I.A. Brytov, M.S. Goldenberg, L.E. Mstibovskaya, N.N. Danilova: App. Met. Rentgen. Anal. **10**, 95 (1972)
3.130 A.S. Bhalla, E.W. White: Adv. X-Ray Anal. **13**, 272 (1970)

3.131 J.A. Golovchenko, R.A. Levesque, P.L. Cowan: Rev. Sci. Instr. **52**, 509 (1981)
3.132 W. Braun, H. Petersen, J. Feldhaus, A.M. Bradshaw, E. Dietz, J. Haase, I.T. McGovern, A. Puschmann, A. Reimer, H.H. Rotermund, R. Unwin: Proc. SPIE **447**, 117 (1984)
3.133 J. Barth, F. Gerken, C. Kunz, J. Schmidt-May: Nucl. Instr. Meth. A **208**, 307 (1983)
3.134 T. Magnusson: Nova Acta Reg. Soc. Sci. Upsal. IV **11**, 3 (1938)
3.135 P. Fischer: J. Opt. Soc. Am. **44**, 665 (1954)
3.136 A.P. Lukirskii, I.A. Brytov, N.I. Komyak: App. Met. Rentgen. Anal. **2**, 4 (1967)
3.137 G. Wiech: In *Röntgenspektren und chemische Bindung*, ed. by A. Meisel (Karl-Marx-Universität, Leipzig 1966) p.343
3.138 A.P. Lukirskii: Izv. Akad. Nauk SSSR, Ser. Fiz. **25**, 910,913 (1961)
     A.P. Lukirskii, V.A. Fomichev, A.V. Rudnev: App. Met. Rentgen. Anal. **6**, 89 (1970)
3.139 G. Andermann, L. Bergknut, M. Karras, G. Griesehaber: Spectrosc. Lett. **11**, 571 (1978)
     G. Andermann, L. Bergknut, M. Karras, G. Griesehaber, J. Smith: Rev. Sci. Instr. **51**, 814 (1980)
3.140 M. Sawada, K. Tsutsumi: J. Appl. Phys. **40**, 1950 (1969)
3.141 P. Jaegle: C.R. Acad. Sci. **250**, 3620 (1960)
3.142 L.O. Werme, B. Grennberg, J. Nordgren, C. Nordling, K. Siegbahn: Phys. Lett. A **41**, 113 (1972)
     L.O. Werme, J. Nordgren, C. Nordling, K. Siegbahn: C.R. Acad. Sci. B **279**, 119 (1974)
3.143 E. Gilberg, M.J. Hanus, B. Föltz: Jpn. J. Appl. Phys. **17**, S2-101 (1978)
3.144 E. Dietz, W. Braun, A.M. Bradshaw, R.L. Johnson: Nucl. Instr. Meth. A **239**, 359 (1985)
3.145 J.A. Bearden: *X-Ray Wavelengths* (US Atomic Energy Commission, Oak Ridge, TN 1964)
     J.A. Bearden, A.F. Burr: *Atomic Energy Levels* (US Atomic Energy Commission, Oak Ridge, TN 1965)
3.146 J. Backovsky: Czech. J. Phys. B **15**, 783 (1965)
3.147 J.S. Thomsen, F.Y. Yap: J. Res. NBS A **72**, 187 (1968)
3.148 G.C. Nelson, W. John, B.G. Saunders: Phys. Rev. **187**, 1 (1969)
3.149 K.D. Sevier: *Low Energy Electron Spectrometry* (Wiley Interscience, New York 1972)
3.150 I.Ya. Nikiforov: Izv. Akad. Nauk SSSR, Ser. Fiz. **21**, 1362 (1957)
3.151 J.O. Porteus: J. Appl. Phys. **33**, 700 (1962); ibid. **39**, 163 (1968)

Chapter 4

4.1 A. Faessler: Z. Physik **72**, 734 (1931); Proc. 10th Coll. Spectrosc. Internationale (Spartan Books, Washington, DC 1963) p.307
4.2 O. Stelling: Z. Physik **50**, 506 (1928)
4.3 N.G. Johnson: Nature **138**, 1056 (1936)
4.4 V. Kunzl: Coll. Czechoslov. Chem. Commun. **4**, 213 (1932)
4.5 A. Meisel, E. Döring: Z. Phys. Chemie **220**, 397 (1962)
4.6 A. Meisel: Izv. Akad. Nauk SSSR, Ser. Fiz. **28**, 811 (1964)
4.7 A. Meisel, G. Leonhardt: Z. Anorg. Allg. Chemie **339**, 1 (1965)
4.8 A. Meisel, M. Köstler, A. Merkel: J. Prakt. Chemie **34**, 112 (1966)

4.9   O.I. Sumbaev, Yu.P. Smirnov, E.V. Petrovich, V.S. Sykov, A.I. Egorov, A.I.
      Grushko: Zh. Eksp. Teor. Fiz. **56**, 536 (1969)
      O.I. Sumbaev: Zh. Eksp. Teor. Fiz. **57**, 1716 (1969); Phys. Lett. **30 A**, 129
      (1969)
      A.I. Grushko, T.B. Mezentseva, N.M. Miftakhov, Yu.P. Smirnov, A.E. Sovest-
      nov, O.I. Sumbaev, V.A. Shaburov: Zh. Eksp. Teor. Fiz. **74**, 501 (1978)
4.10 A.K. Dey, B.K. Agarwal: Lett. Nuovo Cim. **1**, 803 (1971)
4.11 K. Böke: Die K-Kantenfeinstruktur von weiteren Komplexen der Übergangs-
      elemente Chrom bis Zink. Dissertation (Technische Universität München 1956)
4.12 I.A. Ovsyannikova, S.S. Batsanov, L.I. Nasonova, L.R. Batsanova, E.A. Nekra-
      sova: Izv. Akad. Nauk SSSR, Ser. Fiz. **31**, 922 (1967)
4.13 N. Okamoto, M. Kajikawa, K. Hasegawa: J. Chem. Soc. Jpn., Pure Chem.
      Sect. **87**, 363 (1966); ibid. **88**, 165 (1967)
4.14 E.P. Domashevskaya, Ya.A. Ugai: In [Ref.1.57, p.70]
4.15 D.W. Fischer: J. Chem. Phys. **42**, 3814 (1965)
4.16 D.W. Fischer, W.L. Baun: Analyt. Chem. **37**, 902 (1965)
4.17 L.L. Makarov, Yu.P. Kostikov, G.P. Kostikova: Teor. Eksp. Khim. **8**, 403
      (1972)
4.18 D.W. Fischer, W.L. Baun: J. Chem. Phys. **43**, 2075 (1965); Phys. Rev. **145**, 555
      (1966)
4.19 J. Drahokoupil: J. Phys. C **5**, 2259 (1972)
4.20 A. Meisel, H. Keilacker: Z. Phys. Chemie **247**, 321 (1971)
4.21 D. Heintz: Z. Angew. Physik **27**, 98 (1969); Über den Einfluß der chemischen
      Bindung auf das $K\beta$-Emissionsspektrum von Kobalt. Dissertation (Universität
      München 1969)
4.22 R.A. Van Nordstrand: In [Ref.1.57, p.255]
4.23 U.C. Srivastava, H.L. Nigam: Coord. Chem. Rev. **9**, 275 (1973)
4.24 M. Malkovska: Exp. Techn. Physik **6**, 130 (1958); Intensive X-ray lines of the
      K series of the transition elements of the iron group (in Czech) (CSAV,
      Prague 1963)
4.25 H.-U. Chun, D. Hendel: Z. Naturforsch. **22a**, 1401 (1967)
4.26 V.F. Demekhin, V.P. Sachenko: Izv. Akad. Nauk SSSR, Ser. Fiz. **31**, 900, 907
      (1967); Bull. Acad. Sci. USSR **31**, 921 (1967)
4.27 M.A. Blokhin: Zh. Teor. Eksp. Fiz. **9**, 1515 (1939)
4.28 G. Böhm, A. Faessler, G. Rittmayer: Z. Naturforsch. **9b**, 509 (1954)
4.29 E.E. Vainshtein, R.M. Ovrutskaya, B.I. Kotlyar, V.R. Linde: Fiz. Tverd. Tela
      5, 2955 (1963); ibid. **7**, 2120 (1965)
4.30 P.P. Kirichok: Izv. Vyssh. Uchebn. Zaved., Fiz. **11**, 67 (1965)
4.31 E.E. Vainshtein, S.M. Blokhin, Yu.B. Paderno: Fiz. Tverd. Tela **6**, 2909 (1964)
      E.E. Vainshtein, S.M. Blokhin, M.N. Bril, I.B. Stary, Yu.B. Paderno: Zh.
      Neorg. Khim. **10**, 121 (1965)
4.32 A. Meisel, D. Köstler: Z. Phys. Chemie **231**, 99 (1966)
4.33 A.N. Gusatinskii, S.A. Nemnonov: Izv. Akad. Nauk SSSR, Neorg. Mater. **1**,
      838 (1965); in [Ref.1.57, p.124]
      A.N. Gusatinskii, S.A. Ishchenko: Izv. Akad. Nauk SSSR, Ser. Fiz. **31**, 1002
      (1967)
4.34 H. Krämer: Ann. Physik **4**, 263 (1959); Die K-Emissions- und Absorptions-
      spektren von Silizium. Dissertation (Universität München 1960)
4.35 E.E. Vainshtein, M.N. Bril, I.B. Stary, M.E. Kost: Fiz. Met. Metalloved. **21**,
      138 (1966)
4.36 E.E. Vainshtein, S.M. Blokhin, V.M. Bertenev: Izv. Sib. Otd. Akad. Nauk
      SSSR, Ser. Khim. Nauk **1966**, 59

4.37 E.E. Vainshtein, M.N. Bril, I.B. Stary, E.I. Gladyshevskii, P.I. Kripyakevich: Izv. Akad. Nauk SSSR, Neorg. Mater. 3, 644, 1685 (1967)

4.38 M.A. Blokhin, A.T. Shuvaev, V.V. Gorski: Izv. Akad. Nauk SSSR, Ser. Fiz. 28, 801 (1964)

4.39 A. Faessler: Proc. Int'l Conf. Semiconductor Physics, Prague 1960, p.914

4.40 Ya.A. Ugai, E.P. Domashevskaya, T.A. Marshakova: Zh. Strukt. Khim. 4, 250 (1963); in [Ref.1.92, p.347]

4.41 M. Fichter: Über die Bindungsabhängigkeit des K-Röntgenemissionsspektrums von Phosphor. Dissertation (Universität München 1966); in [Ref.1.57, p.112]; Spectrochim. Acta B 30, 417 (1975)

4.42 S. Aoyama, K. Kimura, Y. Nishina: Z. Physik 44, 810 (1927)

4.43 E.W. White, R. Roy: Solid State Commun. 2, 151 (1964)
      E.W. White, G.V. Gibbs: Am. Mineral. 52, 985 (1967)

4.44 E.W. White, G.V. Gibbs: Am. Mineral. 54, 931 (1969)

4.45 F. Freund, M. Hamich: Z. Anorg. Allg. Chemie 385, 209 (1971); Fortschr. Mineral. 48, 243 (1971)

4.46 I.A. Ovsyannikova: Izv. Sib. Otd. Akad. Nauk SSSR, Ser. Khim. Nauk 1966, 151

4.47 D.E. Day: Nature 200, 649 (1963)

4.48 A.T. Shuvaev, M.A. Blokhin, E.A. Israilevich: Izv. Akad. Nauk SSSR, Ser. Fiz. 31, 919 (1967)

4.49 T. Iseki, H. Tagai: J. Am. Ceram. Soc. 53, 582 (1970)

4.50 Y. Suwa, S. Naka, T. Noda: Kogyo Kagaku Zasshi 74, 845 (1971)

4.51 K. Läuger: J. Phys. Chem. Solids 32, 609 (1971); ibid. 33, 1343 (1972)

4.52 A. Wolberg, J.F. Roth: J. Catalysis 15, 250 (1969)

4.53 R.O. Keeling: J. Chem. Phys. 31, 279 (1959); Develop. Appl. Spectrosc. 2, 263 (1963)

4.54 F.W. Lytle: Acta Crystallogr. 22, 321 (1967)

4.55 F.A. Cotton, C.J. Ballhausen: J. Chem. Phys. 25, 617 (1956)
      F.A. Cotton, H.P. Hanson: J. Chem. Phys. 25, 619 (1956); ibid. 26, 1758 (1957); ibid. 28, 83 (1958)

4.56 G.R. Mitchell, W.W. Beeman: J. Chem. Phys. 20, 1298 (1952)
      G.R. Mitchell: J. Chem. Phys. 37, 216 (1962)

4.57 G. Rittmayer: Valenzelektronenübergänge im K-Röntgenspektrum von Chrom. Dissertation (Universität München 1960)

4.58 R.M. Agarwal, A.N. Nigam: Proc. Indian Acad. Sci. 63, 200 (1966)

4.59 B.D. Padalia, C.S. Gupta, A. Paigankar, B.C. Halder: Curr. Sci. 38, 490 (1969)

4.60 B.D. Padalia, V. Krishnan: Indian J. Pure Appl. Phys. 9, 813 (1971)

4.61 F.W. Lytle: Appl. Phys. Lett. 24, 45 (1974)

4.62 E.A. Stern, D.E. Sayers: Phys. Rev. Lett. 30, 174 (1973); Phys. Rev. B 10, 3027 (1974)
      F.W. Lytle, D.E. Sayers, E.A. Stern: Phys. Rev. B 11, 4825 (1975)
      E.A. Stern, D.E. Sayers, F.W. Lytle: Phys. Rev. B 11, 4836 (1975)
      D.E. Sayers, E.A. Stern, F.W. Lytle: Phys. Rev. Lett. 35, 584 (1975)

4.63 S.P. Cramer, K.O. Hodgson, E.O. Stiefel, W.E. Newton: J. Am. Chem. Soc. 100, 2748 (1978)
      S.P. Cramer, K.O. Hodgson, W.O. Gillum, L.E. Mortenson: J. Am. Chem. Soc. 100, 3398 (1978)
      S.P. Cramer, H.B. Gray, Z. Dori, A. Bino: J. Am. Chem. Soc. 101, 2770 (1979)
      S.P. Cramer, H.B. Gray, K.V. Rajagopalan: J. Am. Chem. Soc. 101, 2772 (1979)

S.P. Cramer, K.O. Hodgson: Progr. Inorg. Chem. 25, 1 (1979)
4.64 T.M. Hayes, P.N. Sen, S.H. Hunter: J. Phys. C 9, 4357 (1976)
T.M. Hayes: J. Non-Cryst. Solids 31, 57 (1978)
G. Lucovsky, T.M. Hayes: Topics Appl. Phys. 36, 275 (1979)
4.65 P. Rabe, R. Haensel: Festkörperprobleme 20, 43 (1980)
4.66 D.R. Sandström, F.W. Lytle: Ann. Rev. Phys. Chem. 30, 215 (1979)
4.67 R.G. Shulman, P. Eisenberger, B.-K. Teo, B.M. Kincaid, G.S. Brown: J. Mol. Biol. 124, 305 (1978)
R.G. Shulman, P. Eisenberger, B.M. Kincaid: Ann. Rev. Biophys. Bioeng. 7, 559 (1978)
R.G. Shulman: Trends Biochem. Sci. 3, N 282 (1980)
4.68 B.-K. Teo, R.G. Shulman, G.S. Brown, A.E. Meixner: J. Am. Chem. Soc. 101, 5624 (1979)
B.-K. Teo: Accounts Chem. Res. 13, 412 (1980)
B.-K. Teo, C. Joy: *EXAFS Spectroscopy, Techniques and Applications* (Plenum, New York 1981)
B.-K. Teo: *EXAFS, Basic Principles and Data Analysis* (Springer, Berlin, Heidelberg 1986)
A. Bianconi, L. Incoccia, S. Stipcich (eds.): *EXAFS and Near Edge Structure*, Springer Ser. Chem. Phys. Vol.27 (Springer, Berlin, Heidelberg 1983)
K.O. Hodgson, B. Hedman, J.E. Penner-Hahn (eds.): *EXAFS and Near Edge Structure III*, Springer Proc. Phys., Vol.2 (Springer, Berlin, Heidelberg 1984)
J. Stöhr: In *Chemistry and Physics of Solid Surfaces V*, ed. by R. Vanselov, R. Howe, Springer Ser. Chem. Phys., Vol.35 (Springer, Berlin, Heidelberg 1984) p.231
R. Prins, D. Koningsberger (eds.): *X-Ray Absorption: Principles, Applications, Techniques of EXAFS, SEXAFS, and XANES* (Wiley, New York 1985)
P. Lagarde, O. Raoux, J. Petiau (eds.): J. Physique 47, C 8 (1986)
D.A. Outka, J. Stöhr: In *Chemistry and Physics of Solid Surfaces VII*, ed. by R. Vanselow, R. Howe, Springer Ser. Surf. Sci., Vol.10 (Springer, Berlin, Heidelberg 1988) Chap.6
4.69 S.I. Chan, V.W. Hu, R.C. Gamble: J. Mol. Struct. 45, 239 (1978)
S.I. Chan, R.C. Gamble: Methods Enzym. 54, 323 (1980)
4.70 F.W. Lytle, G.H. Via, J.H. Sinfelt: Prepr. Div. Pet. Chem. Am. Chem. Soc. 21, 366 (1976); J. Chem. Phys. 67, 3831 (1977); in [1.50] p.401
4.71 J.B. Boyce, T.M. Hayes: In *Physics of Superionic Conductors*, ed.by M.B. Salamon, Topics Curr. Phys., Vol.15 (Springer, Berlin, Heidelberg 1979) p.5
J.B. Boyce, T.M. Hayes, C.J. Mikkelsen, Jr., W. Stutius: Solid State Commun. 33, 183 (1980)
4.72 P. Rabe, G. Tolkiehn, A. Werner: J. Phys. C 12, 899 (1979)
4.73 S.P. Cramer, T.K. Eccles, F. Kutzler, K.O. Hodgson, S. Doniach: J. Am. Chem. Soc. 98, 8059 (1976)
4.74 P. Rabe, G. Tolkiehn, A. Werner, R. Haensel: Z. Naturforsch. 34a, 1528 (1979)
4.75 G. Martens, P. Rabe, N. Schwentner, A. Werner: Phys. Rev. Lett. 39, 1411 (1977); Phys. Rev. B 17, 1481 (1978); J. Phys. C 11, 3125 (1978)
G. Martens, P. Rabe: Phys. Stat. Sol. (a) 57, K 31 (1980)
4.76 E.D. Crozier, F.W. Lytle, D.E. Sayers, E.A. Stern: Can. J. Chem. 55, 1968 (1977)
4.77 E.D. Crozier, A.J. Seary: Can. J. Phys. 58, 1388 (1980)
4.78 J.J. Jaklevic, J.A. Kirby, M.P. Klein, A.S. Robertson, G.S. Brown, P. Eisenberger: Solid State Commun. 23, 679 (1977)

4.79 P.H. Citrin, P. Eisenberger, R.C. Hewitt: Phys. Rev. Lett. **41**, 309 (1978); ibid. **45**, 1948 (1980)

4.80 P.R. Sarode, S. Ramasesha, W.H. Madhusudan, C.N.R. Rao: J. Phys. C **12**, 2439 (1979)

4.81 G. Martens, P. Rabe, G. Tolkiehn, A. Werner: Phys. Stat. Sol. A **55**, 105 (1979)
P. Rabe, G. Tolkiehn, A. Werner: J. Phys. C **12**, L545 (1979)

4.82 J. Stöhr, D. Denley, P. Perfetti: Phys. Rev. B **18**, 4132 (1978)
L.I. Johansson, J. Stöhr: Phys. Rev. Lett. **43**, 1882 (1979)
J. Stöhr, L.I. Johansson, S. Brennan, M. Hecht, J.N. Miller: Phys. Rev. B **22**, 4052 (1980)

4.83 J. Stöhr: Jpn. J. Appl. Phys. **17**, S 2-217 (1978); J. Vac. Sci. Technol. **16**, 37 (1979)
J. Stöhr, L. Johansson, I. Lindau, P. Pianetta: Phys. Rev. B **20**, 664 (1979)

4.84 C.A. Ashley, S. Doniach: Phys. Rev. B **11**, 1279 (1975)

4.85 P.A. Lee, G. Beni: Phys. Rev. B **15**, 2862 (1977)

4.86 P.H. Citrin, P. Eisenberger, B.M. Kincaid: Phys. Rev. Lett. **36**, 1346 (1976)

4.87 B.-K. Teo, P.A. Lee, A.L. Simons, P. Eisenberger, B.M. Kincaid: J. Am. Chem. Soc. **99**, 3854 (1977)
B.-K. Teo, K. Kijima, R. Bau: J. Am. Chem. Soc. **100**, 621 (1978)
B.-K. Teo, P. Eisenberger, J. Reed, J.K. Barton, S.J. Lippard: J. Am. Chem. Soc. **100**, 3225 (1978)

4.88 D.E. Sayers, E.A. Stern, F.W. Lytle: Phys. Rev. Lett. **27**, 1204 (1971)
E.A. Stern, D.E. Sayers: Phys. Rev. Lett. **30**, 174 (1973)

4.89 H. Keilacker, A. Meisel: Wiss. Z. Karl-Marx-Univ. Leipzig, Math.- Naturwiss. R. **22**, 585 (1973)

4.90 G.S. Brown, P. Eisenberger, P. Schmidt: Solid State Commun. **24**, 201 (1977)

4.91 S.M. Heald, E.A. Stern: Phys. Rev. B **16**, 5549 (1977)

4.92 P. Rabe, G. Tolkiehn, A. Werner: J. Phys. C **13**, 1857 (1980)

4.93 P.J. Mallozzi, R.E. Schwerzel, H.M. Epstein, B.E. Campbell: Science **206**, 353 (1979)

4.94 R.L. Barinskii, V.I. Nefedov: *Röntgenspektroskopische Bestimmung der Atom-ladungen in Molekülen* (Akademische Verlagsgesellschaft, Leipzig 1969)
V.I. Nefedov, V.G. Yarzhemsky, A.V. Chuvaev, E.M. Trishkina: J. Electron Spectrosc. **46**, 381 (1988)

4.95 A.E. Attard: Solid State Commun. **5**, 360 (1972)

4.96 K. Hübner, G. Leonhardt: Wiss. Z. Karl-Marx-Univ. Leipzig, Math.-Naturwiss. R. **20**, 21 (1971)

4.97 N.N. Sirota: In [Ref.1.90, p.12]

4.98 C.V. de Álvarez, M.L. Cohen: Phys. Rev. B **8**, 1603 (1973)

4.99 O. Madelung: *Solid State Theory* (Springer, Berlin, Heidelberg 1972)

4.100 J.P. Walter, M.L. Cohen: Phys. Rev. Lett. **26**, 17 (1971)

4.101 C.J. Ballhausen, H.B. Gray: *Molecular Orbital Theory* (Benjamin, New York 1965)

4.102 L.C. Cusachs, P. Politzer: Chem. Phys. Lett. **1**, 529 (1968)

4.103 R.S. Mulliken: J. Chem. Phys. **23**, 1833, 1841 (1955)

4.104 L. Pauling: *The Nature of the Chemical Bond* (University, New York 1960)

4.105 H. Preuss: Angew. Chemie **77**, 666 (1965)

4.106 Ya.K. Syrkin: Zh. Fiz. Khim. **37**, 1422 (1963)

4.107 B.F. Levine: J. Chem. Phys. **59**, 1463 (1973); Phys. Rev. B **7**, 2591 (1973)

4.108 J.C. Phillips: Rev. Mod. Phys. **42**, 317 (1970)
J.C. Phillips: *Bands and Bonds in Semiconductors* (Academic, New York 1973)

4.109 R.L. Mössbauer: Angew. Chemie **83**, 524 (1971)
U. Gonser (ed.): *Mössbauer Spectroscopy II*, Topics Current Phys., Vol.25 (Springer, Berlin, Heidelberg 1981)

4.110 A. Lösche, S. Grande: In [Ref.1.92, p.208]
C.P. Slichter: *Principles of Magnetic Resonance*, 4th ed., Springer Ser. Solid-State Sci., Vol.1 (Springer, Berlin, Heidelberg 1989)

4.111 K. Siegbahn, C. Nordling, A. Fahlman, R. Nordberg, K. Hamrin, J. Hedman, G. Johansson, T. Bergmark, S.-E. Karlsson, I. Lindgren, B. Lindberg: *ESCA (Atomic, Molecular and Solid State Structure Studied by Means of Electron Spectroscopy* (North Holland, Amsterdam 1967)

4.112 K. Siegbahn, C. Nordling, G. Johansson, J. Hedman, P.F. Heden, K. Hamrin, U. Gelius, T. Bergmark, L.O. Werme, R. Manne, Y. Baer: *ESCA Applied to Free Molecules* (North Holland, Amsterdam 1969)

4.113 G. Leonhardt, A. Meisel: J. Chem. Phys. **52**, 6189 (1970)

4.114 G. Leonhardt, J. Hedman, C. Nordling, A. Meisel: In [Ref.1.60, Vol.II, p.313]

4.115 L. Ley, R.A. Pollak, F.R. McFeely, S.P. Kowalczyk, D.A. Shirley: Phys. Rev. B **9**, 600 (1974)

4.116 N.J. Shevchik, J. Tejeda, M. Cardona: Phys. Rev. B **9**, 2627 (1974)

4.117 U. Gelius: Phys. Scr. **9**, 133 (1974)

4.118 M.E. Schwartz, J.D. Switalski, R.E. Stronski: In *Electron Spectroscopy* ed. by D.A. Shirley (North Holland, Amsterdam 1972) p.605

4.119 D.A. Shirley: Adv. Chem. Phys. **23**, 85 (1973)

4.120 H. Basch: Chem. Phys. Lett. **5**, 337 (1970)

4.121 T.K. Ha, L.C. Allen: Int'l J. Quant. Chem. S **1**, 199 (1967)

4.122 F.A. Ellison, L.L. Larcom: Chem. Phys. Lett. **10**, 580 (1971)

4.123 A.T. Shuvaev, V.V. Krivitskii, A.P. Semlyanov: Izv. Akad. Nauk SSSR, Ser. Fiz. **36**, 259 (1972)

4.124 V.I. Nefedov: Phys. Stat. Sol. **2**, 904 (1962)

4.125 E. Clementi: Tables of Atomic Functions, Suppl. to IBM J. Res. Develop. (1965) No.9

4.126 C.A. Coulson, C. Zauli: Mol. Phys. **6**, 525 (1963)

4.127 A.T. Shuvaev: Izv. Akad. Nauk SSSR, Ser. Fiz. **28**, 758, 934 (1964)

4.128 I.M. Band, M.B. Trzhaskovskaya: Eigenvalues, Electron Charge Densities in the Vicinity of the Nuclei and Expectation Values in Self-consistent Potentials for Free Atoms and Ions, $37 \leq Z \leq 64$, Institute of Nuclear Physics, Academy of Sciences of USSR, Leningrad (1971); ibid. $2 \leq Z \leq 94$, Leningrad (1974)

4.129 R.E. Watson: Iron Series Hartree-Fock Calculations, Techn. Rep. No. 12, MIT Cambridge, MA (1959)

4.130 G. Leonhardt: Informationsgehalt der Röntgen- und Photoelektronenspektren binärer Festkörper. Dissertation (Karl-Marx-Universität, Leipzig 1976)

4.131 W.L. Jolly, D.N. Hendrickson: J. Am. Chem. Soc. **92**, 1863 (1970)

4.132 D.A. Shirley: Chem. Phys. Lett. **16**, 220 (1972)

4.133 S.M. Karalnik: Izv. Akad. Nauk SSSR, Ser. Fiz. **20**, 815 (1956)

4.134 J. Blomqvist, B. Roos, M. Sundbom: Chem. Phys. Lett. **9**, 160 (1971)

4.135 E.V. Petrovich, Yu.P. Smirnov, V.S. Sykov, A.I. Grushko, O.I. Sumbaev, I.M. Band, M.B. Trzhaskovskaya: Zh. Eksp. Teor. Fiz. **61**, 1756 (1971)

4.136 V.M. Mikhailov, M.A. Khanonkind: Izv. Akad. Nauk SSSR, Ser. Fiz. **35**, 86 (1971)

4.137 R. Friedemann, W. Gründler: Z. Chemie **13**, 308 (1973)

4.138 G. Leonhardt, P. Pelowa, A. Meisel: Z. Anorg. Allg. Chemie **397**, 209 (1973)

4.139 R. Manne: J. Chem. Phys. **46**, 4645 (1967)
A. Stogard, R. Manne: Chem. Phys. **8**, 348 (1975)

4.140 L.E. Harris, E.A. Boudreaux: Chem. Phys. Lett. 23, 434 (1973)

4.141 J. Tilgner, I. Topol, G. Leonhardt, A. Meisel: J. Phys. Chem. Solids 36, 27 (1975)

4.142 J. Finster, P. Müller, N. Meusel, A. Meisel: In [Ref.1.60, Vol.II, p.467]

4.143 J. Finster, N. Meusel, A. Meisel: Phys. Fenn. 9, S 1-425 (1974)

4.144 P. Müller, J. Finster, A. Meisel: Izv. Akad. Nauk SSSR, Ser. Fiz. 40, 373 (1976)

4.145 V.M. Vdovenko, L.L. Makarov, I.G. Suglobova, V.A. Volkov, N.P. Chibisov: Dokl. Akad. Nauk SSSR 202, 868 (1972)
L.L. Makarov, I.G. Suglobova, R.I. Karaziya, Yu.M. Zaitsev, Yu.F. Batrakov, N.P. Chibisov: Vestn. Leningr. Univ. 1975, Nr.16, 87
L.L. Makarov, R.I. Karaziya, Yu.F. Batrakov, N.P. Chibisov, A.N. Mosevich, Yu.M. Zaitsev, A. Udris, L.V. Shishkunova: Radiokhimiya 20, 116 (1978)
L.L. Makarov, B.F. Myasoedov, Yu.P. Novikov, Yu.F. Batrakov, R.I. Karaziya, A.N. Mosevich, V.B. Gliva: Zh. Neorg. Khim. 24, 1014 (1979)

4.146 K. Läuger: Über den Einfluß der Bindungsart und der Kristallstruktur auf das K-Röntgenemissionsspektrum von Aluminium und Silizium. Dissertation (Universität München 1968)

4.147 G. Leonhardt, J. Tilgner, I. Topol, R. Szargan, A. Meisel: In [Ref.1.60, Vol.I, p.38]

4.148 G. Leonhardt, H. Sommer, A. Meisel: Phys. Fenn. 9, S 1-316 (1974)

4.149 R. Szargan, G. Leonhardt, F.-U. Flöther, A. Meisel: Spectrochim. Acta 28 B, 359 (1973)

4.150 P. Politzer, K.C. Daiker: Chem. Phys. Lett. 20, 309 (1973)

4.151 L. Papula, W. Strehl, H.-U. Chun: Theoret. Chim. Acta 22, 149 (1971)

4.152 F.A. Gianturco: J. Phys. B 1, 614 (1968)

4.153 A. Meisel, R. Szargan, G. Leonhardt, H.-J. Köhler: J. Physique 32, C4-301 (1971)

4.154 W.H.E. Schwarz: Angew. Chemie 86, 505 (1974)

Chapter 5

5.1 U. Gelius: J. Electron Spectrosc. 5, 985 (1974)

5.2 W.C. Price, A.W. Potts, D.G. Streets: In *Electron Spectroscopy*, ed. by D.A. Shirley (North-Holland, Amsterdam 1972) p.187

5.3 K. Siegbahn: In *Electron Spectroscopy*, ed. by D.A. Shirley (North-Holland, Amsterdam 1972) p.15; J. Electron Spectrosc. 5, 3 (1974)

5.4 D.W. Turner, C. Baker, A.D. Baker, C.R. Brundle: *Molecular Photoelectron Spectroscopy* (Wiley, London 1970)

5.5 H. Fellner-Feldegg: Wiss. Z. Karl-Marx–Univ. Leipzig, Math-Naturwiss. R. 25, 355 (1976)

5.6 G. Leonhardt: Wiss. Z. Karl-Marx–Univ. Leipzig, Math.-Naturwiss. R. 25, 439 (1976)

5.7 K. Siegbahn, C. Nordling, G. Johansson, J. Hedman, P.F. Heden, K. Hamrin, U. Gelius, T. Bergmark, L.O. Werme, R. Manne, Y. Baer: *ESCA Applied to Free Molecules* (North-Holland, Amsterdam 1969)

5.8 W. Gudat, C. Kunz: Phys. Rev. Lett. 29, 169 (1972)
W. Gudat: Photoelektrische Ausbeutespektroskopie und Spektroskopie der Photoelektronen bei Anregung im extremen Vakuum Ultraviolett. Dissertation (Universität Hamburg 1974)

5.9 A.P. Lukirskii, M.A. Rumsh, I.A. Karpovich: Opt. Spektrosk. 9, 653 (1960)

5.10 T.M. Zimkina, V.A. Fomichev: *Ultra Soft X-Ray Spectroscopy* (State University, Leningrad 1971) (in Russian)

5.11 A.P. Hitchcock, C.E. Brion: J. Electron Spectrosc. **14**, 417 (1978)

5.12 M. Tronc, G.C. King, F.H. Read: J. Phys. B **9**, L555 (1976); ibid. **12**, 137 (1979)

5.13 G.R. Wight, C.E. Brion, M.J. van der Wiel: J. Electron Spectrosc. **1**, 457 (1973); ibid. **3**, 191 (1974); ibid. **4**, 25 (1974); Chem. Phys. Lett. **7**, 161 (1970)

5.14 G. Andermann, R.A. Lynch, H.C. Whitehead, K. Myers: In [Ref.1.60, p.283]
R.A. Lynch, G. Andermann: Phys. Fenn **9**, S 1-138 (1974)
D.R. Phillips, G. Andermann: Phys. Fenn. **9**, S 1-141 (1974)

5.15 D.W. Fischer: In [Ref.1.79, p.669]

5.16 V.I. Nefedov: Zh. Strukt. Khim. **12**, 1019 (1971)
K.I. Narbutt, V.I. Nefedov, M.A. Porai-Koshits, A.P. Kochetkova: Zh. Strukt. Khim. **13**, 451 (1972)

5.17 L.O. Werme, G. Grennberg, J. Nordgren, C. Nordling, K. Siegbahn: J. Electron Spectrosc. **2**, 435 (1973); Z. Physik A **272**, 131 (1975)
H. Ågren, J. Müller: Phys. Scr. **20**, 627 (1979); J. Chem. Phys. **72**, 4078 (1980)

5.18 W.H.E. Schwarz: Angew. Chemie **86**, 505 (1974)

5.19 V.I. Nefedov: Zh. Strukt. Khim. **13**, 309, 352 (1972)

5.20 V.I. Nefedov: *Valence Electron Levels in Chemical Compounds* (VINITI, Moscow 1975) (in Russian)

5.21 V.I. Nefedov, V.I. Vovna: *Electronic Structure of Chemical Compounds* (Nauka, Moscow 1987) (in Russian)

5.22 M.A. Blokhin, A.T. Shuvaev, E.I. Fedorov, V.P. Kurbatov, O.A. Osipov, A.V. Kozinkin: Izv. Akad. Nauk SSSR, Ser. Fiz. **38**, 544 (1974)
A.T. Shuvaev, M.A. Blokhin, E.I. Fedorov, A.V. Kozinkin: Izv. Sib. Otd. Akad. Nauk SSSR, Ser. Khim. Nauk 1975, No.9, H.4, 86

5.23 C. Beyreuther, G. Wiech: In [Ref.1.86, p.517]
G. Wiech: In [Ref.1.62, p.195]

5.24 I.B. Borovskii, V.I. Matyskin: Dokl. Akad. Nauk SSSR **192**, 63 (1970); ibid. **195**, 1072 (1970)

5.25 O. Brümmer, G. Dräger, W. Starke: J. Physique **32**, C 4-169 (1971)

5.26 V.I. Nefedov, V.I. Matyskin, I.B. Borovskii: Zh. Strukt. Khim. **12**, 893 (1971)

5.27 E. Tegeler, N. Kosuch, G. Wiech, A. Faessler: Jpn. J. Appl. Phys. **17**, S2-97 (1978)
E. Tegeler, M. Iwan, E.E. Koch: J. Electron Spectrosc. **22**, 297 (1981)
E. Tegeler, G. Wiech, A. Faessler: J. Phys. B **14**, 1273 (1981)

5.28 E. Tegeler, N. Kosuch, G. Wiech, A. Faessler: J. Electron Spectrosc. **18**, 23 (1980)

5.29 G. Dräger, O. Brümmer: In [Ref.1.66, p.337]
G. Dräger, W. Czolbe, G. Schulz, A. Simunek, J. Drahokoupil, Yu.N. Kucherenko, V.V. Nemoshkalenko: Phys. Stat. Sol. B **131**, 183 (1985)

5.30 G. Dräger, O. Brümmer: Phys. Stat. Sol. B **124**, 11 (1984)

5.31 R. Eisberg, G. Wiech: J. Physique **48**, C9-1133 (1987)

5.32 R.A. Rosenberg, P.J. Love, V. Rehn: Phys. Rev. B **33**, 4034 (1986)

5.33 A. Simunek, G. Wiech: J. Physique **48**, C9-1129 (1987)

5.34 R. Eisberg, P. Josuks, G. Wiech, R. Schlögl: Solid State Commun. **60**, 827 (1986)

5.35 A. Mansour, S.E. Schnatterly, J.J. Ritsko: Phys. Rev. Lett. **58**, 614 (1987)

5.36 R. Manne: J. Chem. Phys. **52**, 5733 (1970)

5.37 V.I. Nefedov: Zh. Strukt. Khim. **8**, 686, 1037 (1967)

5.38 V.I. Nefedov: Zh. Strukt. Khim. **12**, 303, 521 (1971)

5.39 H. Adachi, K. Taniguchi: Technol. Rep. Osaka Univ. **30**, 365 (1980); J. Phys. Soc. Jpn. **49**, 1944 (1980)

5.40 P.E. Best: J. Chem. Phys. **49**, 2797 (1968)

5.41 A. Meisel, R. Szargan, G. Leonhardt, H.-J. Köhler: J. Physique **32**, C 4-301 (1971)

5.42 A.T. Shuvaev: Izv. Akad. Nauk SSSR, Ser. Fiz. **25**, 986, 992 (1961)

5.43 J.A. Connor, I.H. Hillier, V.R. Saunders, M. Barber: Mol. Phys. **23**, 81 (1972); ibid. **24**, 497 (1972)

5.44 J.A. Connor, I.H. Hillier, M.II. Wood, M. Barber: J. Chem. Soc., Faraday Trans. II **70**, 1040 (1974)

5.45 V.I. Nefedov, E.Z. Kurmaev, M.A. Porai-Koshits, S.A. Nemnonov, G.V. Tsintsadze: Zh. Strukt. Khim. **13**, 637 (1972)

5.46 Y. Takahashi: Bull. Chem. Soc. Jpn. **44**, 586 (1971); ibid. **46**, 2039 (1973)

5.47 A.P. Sadovskii, G.N. Dolenko, L.N. Mazalov, V.D. Yumatov, E.S. Gluskin, Yu.I. Nikonorov, E.A. Galtsova: Izv. Akad. Nauk SSSR, Ser. Fiz. **78**, 606 (1974)
    G.N. Dolenko, A.P. Sadovskii, L.N. Mazalov: Izv. Sib. Otd. Akad. Nauk SSSR, Ser. Khim. Nauk (1974) Nr.14, H.6, p.157

5.48 E. Gilberg: Untersuchung der Bindungsabhängigkeit von Röntgenemissionsspektren an freien Molekülen mit Fluoreszenzanregung. Dissertation (Universität München 1969)

5.49 E. Gilberg: In [Ref.1.58, Vol.1, p.277]; Z. Physik **236**, 21 (1970)

5.50 E. Gilberg, M.J. Hanus, B. Föltz: Jpn. J. Appl. Phys. **17**, S2-101 (1978)

5.51 E.S. Gluskin, A.P. Sadovskii, L.N. Mazalov: Zh. Strukt. Khim. **14**, 739 (1973)
    E.S. Gluskin, L.N. Mazalov, A.P. Sadovskii, D.A. Zhogolev: Zh. Strukt. Khim. **16**, 1061 (1975)

5.52 A.P. Lukirskii: Izv. Akad. Nauk SSSR, Ser. Fiz. **25**, 910, 913 (1961)
    A.P. Lukirskii, V.A. Fomichev, A.V. Rudnev: App. Met. Rentgen. Anal. **6**, 89 (1970)

5.53 A.P. Sadovskii, L.N. Mazalov, V.M. Bertenev: Teor. Eksp. Khim. **6**, 502 (1970)

5.54 J. Nordgren, R. Nyholm: Nucl. Instr. Methods A **246**, 242 (1986)

5.55 J. Nordgren: J. Physique **48**, C9-693 (1987)

5.56 E.S. Gluskin, L.N. Mazalov, A.A. Krasnoperova, V.A. Kochubei, S.I. Mishnev, A.N. Skrinskii, E.M. Trakhtenberg, G.M. Tumaikin: Izv. Akad. Nauk SSSR, Ser. Fiz. **40**, 226 (1976)

5.57 R. Haensel, C. Kunz: Z. Angew. Physik **23**, 276 (1967)

5.58 R. Haensel, B. Sonntag: In *Computational Solid State Physics* ed. by F. Herman, N.W. Dalton, T.R. Koehler (Plenum, New York 1972) p.43

5.59 W. Hayes: Contemp. Phys. **13**, 441 (1972); Phys. Rev. A **6**, 21 (1972)

5.60 M. Nakamura: Phys. Rev. **178**, 80 (1969)
    Y. Morita, M. Nakamura: J. Chem. Phys. **61**, 1426 (1974)

5.61 H.-U. Chun, H. Gebelein: Z. Naturforsch. **22a**, 1813 (1967)
    H.-U. Chun: Phys. Lett. A **30**, 445 (1969)

5.62 R.D. Deslattes, R.E. LaVilla: Appl. Optics **6**, 39 (1967)

5.63 R.A. Mattson, R.C. Ehlert: Adv. X-Ray Anal. **9**, 471 (1966)

5.64 R.A. Mattson, R.C. Ehlert: J. Chem. Phys. **48**, 5465, 5471 (1968)

5.65 R. Brammer: Spectroscopic and Computational Studies of Soft X-Ray Emission in Free Molecules. Dissertation (University of Uppsala 1986)

5.66 A. Bianconi, H. Petersen, F.C. Brown, R.Z. Bachrach: Phys. Rev. A**17**, 1907 (1978)

5.67 D. Dill, J.L. Dehmer: J. Chem. Phys. 61, 692 (1974)
J.L. Dehmer, D. Dill: Phys. Rev. Lett. 35, 213 (1975); J. Chem. Phys. 65, 5327 (1976)

5.68 B.L. Henke, R.C.C. Perera, E.M. Gullikson, M.L. Schattenburg: J. Appl. Phys. 49, 480 (1978)
B.L. Henke, R.C.C. Perera, D.S. Urch: J. Chem. Phys. 68, 3692 (1978)
B.L. Henke: Nucl. Instrum. Methods 177, 161 (1980)

5.69 S. Iwata: Nippon Kessho Dakhaishi 11, 102 (1969)
S. Iwata, N. Kosugi, O. Nomura: Jpn. J. Appl. Phys. 17, S2-109 (1978)

5.70 R.B. Kay, Ph.E. van der Leeuw, M.J. van der Wiel: J. Phys. B 10, 2513 (1977)

5.71 A.V. Kondratenko, L.N. Mazalov, F.Kh. Gelmukhanov, V.I. Avdeev, E.A. Saprykina: Zh. Strukt. Khim. 18, 546 (1977)
A.V. Kondratenko, L.N. Mazalov, K.M. Neiman: Zh. Strukt. Khim. 20, 203 (1979); Opt. Spektrosk. 48, 1072 (1980); ibid. 49, 488 (1980)

5.72 A.V. Kondratenko, L.N. Mazalov, F.Kh. Gelmukhanov, V.I. Avdeev, E.A. Saprykina: Zh. Strukt. Khim. 18, 622 (1977)
A.V. Kondratenko, L.M. Mazalov, B.A. Korney: Theor. Chim. Acta 52, 311 (1979)

5.73 R.E. LaVilla: J. Chem. Phys. 56, 2345 (1972)

5.74 J. Nordgren, H. Ågren, L. Selander, C. Nordling, K. Siegbahn: Phys. Scr. 16, 280 (1977)
H. Ågren, J. Nordgren, L. Selander, C. Nordling, K. Siegbahn: Phys. Scr. 18, 499 (1978); Chem. Phys. 37, 161 (1979)

5.75 A.S. Vinogradov, T.M. Zimkina, V.N. Akimov, B. Schlarbaum: Izv. Akad. Nauk SSSR, Ser. Fiz. 38, 508 (1974)
A.S. Vinogradov, V.N. Akimov, T.M. Zimkina, E.B. Dobryakova: Izv. Sib. Otd. Akad. Nauk SSSR, Ser. Khim. Nauk Nr.9, H.4, 88 (1975)

5.76 J. Nordgren, H. Ågren: Comm. Atomic Molecular Phys. 14, 203 (1984)

5.77 R.A. Rosenberg, P.J. Love, P.R. LaRoe, V. Rehn, C.C. Parks: Phys. Rev. B 31, 2634 (1985)

5.78 V.N. Akimov, A.S. Vinogradov, T.M. Zimkina: Opt. Spektrosk. 53, 109 (1982)

5.79 D.M. Barrus, R.L. Blake, A.J. Byrek, K.C. Chambers, A.L. Pregenter: Phys. Rev. A 20, 1045 (1979)

5.80 S. Bodeur, C. Senemaud, C. Bonnelle, J.P. Connerade: In [Ref.1.86, p.94]

5.81 R.E. LaVilla: J. Chem. Phys. 58, 3841 (1973); ibid. 63, 2733 (1975)

5.82 M. Nakamura, T. Hayashi, E. Ishiguro, M. Sasanuma: Proc. 3rd Int'l Conf. VUV Rad. Phys. (Tokyo 1971) 1pA1.

5.83 E.S. Gluskin, A.A. Krasnoperova, L.N. Mazalov: Zh. Strukt. Khim. 18, 185, 665 (1977)

5.84 L.N. Mazalov, A.V. Nikolaev, E.A. Galtsova, A.P. Sadovskii, V.G. Torgov, G.K. Parygina: Izv. Sib. Otd. Akad. Nauk SSSR, Ser. Khim. Nauk (1971) Nr.2, H.6, p.3
L.N. Mazalov, A.P. Sadovskii, E.A. Galtsova, V.V. Murakhtanov, V.G. Torgov, V.M. Bertenev, A.P. Zeif: Zh. Strukt. Khim. 14, 76 (1973)

5.85 F.J. Comes, U. Nielsen, W.H.E. Schwarz: J. Chem. Phys. 58, 2230 (1973)

5.86 F.Kh. Gelmukhanov, L.N. Mazalov, A.V. Kondratenko: Chem. Phys. Lett. 46, 133 (1977)
A.V. Kondratenko, L.N. Mazalov, F.Kh. Gelmukhanov, V.I. Avdeev, E.A. Saprykina: Zh. Strukt. Khim. 18, 622 (1977)

5.87 R. Manne: J. Chem. Phys. 52, 5733 (1970)

5.88 A. Flores-Riveros, N. Correia, H. Ågren, L. Pettersson, M. Bäckström, J. Nordgren: J. Chem. Phys. 83, 2053 (1985)

5.89 L.N. Mazalov, A.P. Sadovskii, E.S. Gluskin, G.N. Dolenko, A.A. Krasnoperova: Zh. Strukt. Khim. 15, 800, 805 (1974)
F.Kh. Gelmukhanov, L.N. Mazalov, A.V. Nikolaev, A.V. Kondratenko, V.G. Smirnyi, P.I. Vadash, A.P. Sadovskii: Dokl. Akad. Nauk SSSR 225, 597 (1975)
E.S. Gluskin, L.N. Mazalov, A.P. Sadovskii, D.A. Zhogolev: Zh. Strukt. Khim. 16, 1061 (1975)

5.90 C. Nordling: Jpn. J. Appl. Phys. 17, S 2-7 (1978)
J. Nordgren, H. Ågren, L. Pettersson, L. Selander, S. Griep, C. Nordling, K. Siegbahn: Phys. Scr. 20, 623 (1979)

5.91 R.C.C. Perera, B.L. Henke: Jpn. J. Appl. Phys. 17, S2-112 (1978); J. Chem. Phys. 70, 5398 (1979); X-Ray Spectrom. 9, 81 (1980)

5.92 J. Nordgren, L. Selander, L. Pettersson, C. Nordling, K. Siegbahn, H. Ågren: J. Chem. Phys. 76, 3928 (1982)

5.93 Y. Morioka, M. Nakamura, E. Ishiguro, M. Sasanuma: In [Ref.1.86, p.92]

5.94 W.H.E. Schwarz, T.C. Chang: Chem. Phys. Lett. 49, 207 (1977)
W.H.E. Schwarz, W. Butscher, D.L. Ederer, T.B. Lucatorto, B. Ziegenbein, W. Mehlhorn, H. Prömpeler: J. Phys. B 11, 591 (1978)

5.95 L. Pettersson, M. Bäckström, R. Brammer, N. Wassdahl, J.-E. Rubensson, J. Nordgren: J. Phys. B 17, L 279 (1984)

5.96 F.P. Larkins, R.A. Phillips: J. Physique 48, C9-729 (1987)

5.97 B.L. Henke, K. Taniguchi: J. Appl. Phys. 47, 1027 (1976)
K. Taniguchi, B.L. Henke: J. Chem. Phys. 64, 3021 (1976)

5.98 L.N. Mazalov, V.M. Bertenev, A.P. Sadovskii, T.I. Guzhavina: Zh. Strukt. Khim. 13, 855, (1972)
L.N. Mazalov, A.P. Sadovskii, V.M. Bertenev, K.E. Mironov, T.I. Guzhavina, L.I. Chernyarskii: Zh. Strukt. Khim. 13, 859 (1972)

5.99 L.N. Mazalov, A.P. Sadovskii, P.I. Vadash, F.Kh. Gelmukhanov: Zh. Strukt. Khim. 14, 262 (1973)
L.N. Mazalov, V.V. Murakhtanov, T.I. Guzhavina, A.P. Sadovskii: Zh. Strukt. Khim. 16, 262, 267 (1975)

5.100 A.V. Nikolaev, A.P. Sadovskii, L.N. Mazalov, G.N. Dolenko, E.S. Gluskin: Dokl. Akad. Nauk SSSR 212, 1149 (1973)

5.101 H. Schrenk: Röntgenzustände symmetrischer Moleküle, SCF-Rechnungen an einigen Schwefel- und Chlorverbindungen. Dissertation (Universität München 1969)

5.102 R.V. Vedrinskii, S.A. Prosandeev, A.N. Pavlov, A.P. Kovtun: Teor. Eksp. Khim. 16, 19 (1980)

5.103 T.M. Zimkina, A.S. Vinogradov: J. Physique 32, C4-3, 278 (1971)
A.S. Vinogradov, T.M. Zimkina: Zh. Strukt. Khim. 31, 685 (1971)

5.104 S. Bodeur, J.M. Esteva: Chem. Phys. 100, 415 (1985)

5.105 V.N. Akimov, A.S. Vinogradov, T.M. Zimkina: Opt. Spektrosk. 53, 918 (1982)

5.106 T.I. Guzhavina, L.N. Mazalov, V.V. Murakhtanov: Izv. Sib. Otd. Akad. Nauk SSSR, Ser. Khim. Nauk Nr.9, H.4, 14 (1975)

5.107 L.N. Mazalov, E.A. Kravtsova, S.V. Zemskov, Yu.I. Nikonorov: Zh. Strukt. Khim. 18, 565 (1977)

5.108 L.A. Demekhina, V.L. Sukhorukov, V.F. Demekhin, V.A. Yarna: Opt. Spektrosk. 49, 861 (1980)

5.109 H. Ågren, S. Svensson, U.I. Wahlgren: Chem. Phys. Lett. 35, 336 (1975)

5.110 R.E. LaVilla: J. Chem. Phys. 62, 2209 (1975)

5.111 J. Nordgren, L.O. Werme, H. Ågren, C. Nordling, K. Siegbahn: J. Phys. B8, L18 (1975)

5.112 A.P. Sadovskii, E.A. Kravtsova, L.N. Mazalov: Izv. Sib. Otd. Akad. Nauk SSSR, Ser, Khim. Nauk Nr.9, H.4, 62 (1975)

5.113 J.E. Rubensson, H. Ågren, R. Manne: J. Electron Spectrosc. **36**, 307 (1985)

5.114 J.-E. Rubensson, N. Wassdahl: J. Physique **48**, C9-793 (1987)

5.115 V.N. Akimov, A.S. Vinogradov, T.M. Zimkina: Opt. Spektrosk. **53**, 476 (1982)

5.116 E.S. Gluskin, L.N. Mazalov, A.P. Sadovskii, G.N. Dolenko, A.L. Shor: Izv. Sib. Otd. Akad. Nauk SSSR, Ser. Khim. Nauk (1974) Nr.7, H.3, p.3
E. Gluskin, A.P. Sadovskii, L.N. Mazalov, G.N. Dolenko: Zh. Strukt. Khim. **15**, 304 (1974)

5.117 R.E. LaVilla, R.D. Deslattes: J. Chem. Phys. **44**, 4399 (1966)

5.118 R.E. LaVilla: Phys. Fenn. **9**, S1-126 (1974)

5.119 L.N. Mazalov, A.P. Sadovskii, V.M. Bertenev, V.V. Murakhtanov, E.A. Galtsova, L.I. Chernyarskii: Teor. Eksp. Khim. **7**, 46 (1971)

5.120 W.H.E. Schwarz: Chem. Phys. **9**, 157 (1975); ibid. **11**, 217 (1975)

5.121 H. Ågren, J. Nordgren, L. Selander, C. Nordling, K. Siegbahn: J. Electron Spectrosc. **14**, 27 (1978)
H. Ågren, J. Muller: J. Chem. Phys. **72**, 4078 (1980)

5.122 G.N. Dolenko, L.I. Nasonova, L.N. Mazalov, G.G. Furin, G.G. Yakobson: Zh. Strukt. Khim. **17**, 435 (1976)
G.N. Dolenko, S.A. Krupoder, L.N. Mazalov: Zh. Strukt. Khim. **20**, 334 (1979)
G.N. Dolenko, S.A. Krupoder, L.N. Mazalov, L.I. Nasonova, G.G. Furin, G.G. Yakobson: Izv. Akad. Nauk SSR, Ser. Khim. **2**, 343 (1979)
G.N. Dolenko, A.I. Kholkin, A.S. Chernobrov, E.A. Galtsova, L.M. Gindin, L.M. Kuznetsova, N.K. Kalish: Izv. Sib. Otd. Akad. Nauk SSSR, Ser. Khim. Nauk (1979) Nr.2, H.1, p.32
M.G. Voronkov, G.N. Dolenko, L.N. Mazalov, M.S. Sorokin, N.V. Bausk: Dokl. Akad. Nauk SSSR **248**, 897 (1979)
A.V. Zibarev, G.N. Dolenko, S.A. Krupoder, L.N. Mazalov, A.I. Rezvukhin, G.G. Furin, G.G. Yakobson: Izv. Sib. Otd. Akad. Nauk SSSR, Ser. Khim. Nauk (1980) Nr.4, H.2, p.73

5.123 H. Friedrich, B. Sonntag, P. Rabe, W. Butscher, W.H.E. Schwarz: Chem. Phys. Lett. **64**, 360 (1979)
H. Friedrich, B. Pittel, P. Rabe, W.H.E. Schwarz, B. Sonntag: J. Phys. B13, 25 (1980)

5.124 F.C. Brown, R.Z. Bachrach, A. Bianconi: In [Ref.1.87, p.17]; Chem. Phys. Lett. **54**, 425 (1978)
R.Z. Bachrach, A. Bianconi, F.C. Brown: Nucl. Instrum. Meth. **152**, 53 (1978)
A. Bianconi, H. Petersen, F.C. Brown, R.Z. Bachrach: Phys. Rev. A **17**, 1907 (1978)

5.125 A.A. Pavlychev, A.S. Vinogradov, D.E. Onopko, S.A. Titov: Fiz. Tverd. Tela **20**, 3671 (1978)
A.A. Pavlychev, A.S. Vinogradov, T.M. Zimkina, D.E. Onopko, R. Szargan: Opt. Spektrosk. **48**, 192 (1980)

5.126 M.L. Sink, G.E. Jura: Chem. Phys. Lett. **20**, 474 (1973)

5.127 R. Szargan, A. Meisel, E. Hartmann, G. Brunner: Jpn. J. Appl. Phys. **17**, S2-174 (1978)
E. Hartmann, R. Szargan: Chem. Phys. Lett. **68**, 175 (1979)

5.128 R.L. Watson, T. Chiao, F.E. Jenson: Phys. Rev. Lett. **35**, 254 (1975)
R.L. Watson, A.K. Leeper, B.I. Sonobe, T. Chiao, F.E. Jenson: Phys. Rev. A **15**, 914 (1977)
J.A. Demarest, R.L. Watson: Phys. Rev. A17, 1302 (1978)

5.129 A.A. Krasnoperova, E.S. Gluskin, L.N. Mazalov, V.A. Kochubei: Zh. Strukt. Khim. **17**, 1113 (1976)
5.130 R.C.C. Perera, R.E. LaVilla: J. Chem. Phys. **81**, 3375 (1984)
5.131 F.J. Comes, R. Haensel, U. Nielsen, W.H.E. Schwarz: J. Chem. Phys. **58**, 516 (1973)
5.132 U. Nielsen, R. Haensel, W.H.E. Schwarz: J. Chem. Phys. **61**, 3581 (1974)
5.133 W.H.E. Schwarz: Ber. Bunsenges. Phys. Chemie **78**, 1206 (1974)
5.134 R.L. Barinskii, I.M. Kulikova: Zh. Strukt. Khim. **14**, 372 (1973); Izv. Akad. Nauk SSSR, Ser. Fiz. **38**, 444 (1974)
5.135 V.A. Fomichev, R.L. Barinskii: Zh. Strukt. Khim. **11**, 875 (1970)
5.136 A.P. Sadovskii, L.I. Nasonova: Zh. Strukt. Khim. **18**, 673 (1977)
      L.N. Mazalov, V.V. Volkov, S.Ya. Dvurechenskaya, L.I. Nasonova: Zh. Strukt. Khim. **23**, 1860 (1978)
      A.P. Sadovskii, E.A. Kravtsova: Koord. Khim. **5**, 197, 890 (1979)
5.137 R. Szargan, E. Hartmann, H. Sommer, A. Meisel: In [Ref.1.64, p.717]
5.138 W.H.E. Schwarz, L. Mensching, K.-H. Hallmeier, R. Szargan: Chem. Phys. **82**, 57 (1983)
5.139 R.D. Deslattes: Phys. Rev. A **133**, 390, 395, 399 (1964)
5.140 R.C. Ehlert, R.A. Mattson: J. Chem. Phys. **48**, 5471 (1968)
5.141 F.A. Gianturco, C. Guidotti, V. Lamanna: J. Chem. Phys. **57**, 840 (1972)
      F.A. Gianturco: In [Ref.1.60, Vol.I, p.321]
5.142 M.J. Hanus, E. Gilberg: In [Ref.1.60, Vol.I, p.338]; J. Phys. B **9**, 137 (1976)
5.143 F. Hopfgarten, R. Manne: J. Electron Spectrosc. **2**, 13 (1973)
      A. Stogard: Chem. Phys. Lett. **36**, 357 (1975)
5.144 R.E. LaVilla, R.D. Deslattes: J. Chem. Phys. **45**, 3446 (1966); Appl. Optics **6**, 39 (1967)
5.145 R.C.C. Perera, J. Barth, R.E. LaVilla, R.D. Deslattes, A. Henins: Phys. Rev. A **32**, 1489 (1985)
5.146 R.C.C. Perera, R.E. LaVilla, G.V. Gibbs: J. Chem. Phys. **86**, 4824 (1987)
5.147 R.C.C. Perera, R.E. LaVilla, P.L. Cowan, T. Jach, B. Karlin: Phys. Scr. **36**, 132 (1987)
5.148 R. Manne: Chem. Phys. Lett. **5**, 125 (1970)
5.149 J.A. Stephens, D. Dill, J.L. Dehmer: J. Chem. Phys. **84**, 3638 (1986)
5.150 J. Nordgren, L. Selander, L. Pettersson, R. Brammer, M. Bäckström, C. Nordling: Phys. Scr. **27**, 169 (1983)
5.151 L. Pettersson, M. Bäckström, J. Nordgren, C. Nordling: Chem. Phys. Lett. **106**, 425 (1984)
5.152 R. Brammer, J.-E. Rubensson, N. Wassdahl, J. Nordgren: Phys. Scr. **36**, 262 (1987)
5.153 V.N. Akimov, A.S. Vinogradov, A.A. Pavlychev, V.N. Sivkov: Opt. Spektrosk. **59**, 342 (1985)
5.154 T.M. Zimkina, A.S. Vinogradov: Izv. Akad. Nauk SSSR, Ser. Fiz. **36**, 248 (1972)
      A.S. Vinogradov, A.Yu. Dukhnyakov, T.M. Zimkina, V.M. Ipatov, I.V. Kavunina, D.E. Onopko, A.A. Pavlychev, S.A. Titov, E.O. Filatova: Fiz. Tverd. Tela **22**, 2602 (1980)
5.155 A.A. Pavlychev, A.S. Vinogradov, T.M. Zimkina, D.E. Onopko, S.A. Titov: Opt. Spektrosk. **52**, 506 (1982)
5.156 S. Bodeur, I. Nenner, P. Millie: Phys. Rev. A **34**, 2986 (1986)
5.157 P.E. Best: J. Chem. Phys. **47**, 4002 (1967)
5.158 D. Blechschmidt, R. Haensel, E.E. Koch, U. Nielsen, T. Sagawa: Chem. Phys. Lett. **14**, 33 (1972)

5.159 J.W.D. Conolly, K.H. Johnson: Chem. Phys. Lett. **10**, 616 (1971)

5.160 J.L. Dehmer: J. Chem. Phys. **56**, 4496 (1972)

5.161 F.A. Gianturco, C. Guidotti: Chem. Phys. Lett. **9**, 539 (1971)
F.A. Gianturco, C. Guidotti, V. Lamanna, R. Moccia: Chem. Phys. Lett. **10**, 269 (1971)

5.162 R.E. LaVilla: J. Chem. Phys. **57**, 899 (1972)

5.163 V.I. Nefedov, V.A. Fomichev: Zh. Strukt. Khim. **9**, 126, 217, 268, 279 (1968)

5.164 V.I. Nefedov: Zh. Strukt. Khim. **11**, 292, 299 (1970)

5.165 V.P. Sachenko: Izv. Akad. Nauk SSSR, Ser. Fiz. **36**, 226, 232 (1972)

5.166 R.V. Vedrinskii, A.P. Kovtun, V.V. Kolesnikov, Yu.F. Migal, E.V. Polozhent-sev, V.P. Sachenko: Izv. Akad. Nauk SSSR, Ser. Fiz. **38**, 434 (1974)
R.V. Vedrinskii, V.L. Kraizman: Zh. Eksp. Teor. Fiz. **74**, 1215 (1978)

5.167 R.V. Vedrinskii. V.L. Kraizman: Zh. Neorg. Khim. **25**, 2858 (1980)

5.168 T.M. Zimkina, V.A. Fomichev: Dokl. Akad. Nauk SSSR **169**, 1309 (1966)

5.169 A.A. Pavlychev, A.S. Vinogradov: Fiz. Tverd. Tela **23**, 3564 (1981)

5.170 S. Bodeur, A.P. Hitchcock: Chem. Phys. **111**, 467 (1987)

5.171 J. Hormes, R. Chauvistre, U. Kuetgens, U. Fischer, I. Ruppert: J. Physique **48**, C9-1113 (1987)

5.172 S. Bodeur, J.L. Ferrer, I. Nenner, P. Millie, M. Benfatto, C.R. Natoli: J. Physique **48**, C9-1117 (1987)

5.173 S. Bodeur, I. Nenner, P. Millie: Phys. Rev. A **34**, 2986 (1986)

5.174 O. Edqvist, E. Lindholm, L. Asbrink: Ark. Fysik **40**, 439 (1970)

5.175 P. Venkateswarlu: Canad. J. Phys. **48**, 1055 (1970)

5.176 A.W. Potts, W.C. Price: Proc. Roy. Soc. A **326**, 181 (1972)

5.177 F.P. Boer, W.N. Lipscomb: J. Chem. Phys. **50**, 989 (1969)

5.178 I.H. Hillier, V.R. Saunders: Trans. Faraday Soc. **66**, 1544 (1970); Mol. Phys. **22**, 193, 1025 (1971)

5.179 A.D. McLean: J. Chem. Phys. **32**, 1595 (1960)

5.180 W.W. Palke, W.N. Lipscomb: J. Am. Chem. Soc. **88**, 2384 (1966)

5.181 G. Wiech: In [Ref.1.57, p.343]

5.182 A.V. Potts, H.-J. Lempka, D.G. Street, W.C. Price: Phil. Trans. Roy. Soc. A **268**, 59 (1970)

5.183 J.A. Bearden: *X-Ray Wavelengths* (US Atomic Energy Commission, Oak Ridge 1964)
J.A. Bearden, A.F. Burr: *Atomic Energy Levels* (US Atomic Energy Commission, Oak Ridge 1965)

5.184 T.D. Thomas: J. Am. Chem. Soc. **92**, 4184 (1970)

5.185 R.E. LaVilla: Bull. Am. Phys. Soc. **11**, 389 (1966)

5.186 U. Gelius, B. Roos, P. Siegbahn: Theor. Chim. Acta **23**, 59 (1971)

5.187 E.L. Rosenberg, M.E. Dyatkina: Zh. Strukt. Khim. **11**, 299, 323 (1970); ibid. **12**, 548 (1971)

5.188 V.I. Nefedov, Yu.A. Buslaev, N.P. Sergushin, L. Beyer, Yu.V. Kokunov, A.A. Kuznetsova: Izv. Akad. Nauk SSSR, Ser. Fiz. **38**, 448 (1974)
V.I. Nefedov, Yu.A. Buslaev, N.P. Sergushin, Yu.V. Kokunov, V.V. Kovalev: J. Electron. Spectrosc. **6**, 221 (1975)

5.189 M.O. Krause, T.A. Carlson, R.D. Dismukes: Phys. Rev. **170**, 37 (1968)

5.190 H.W. Schnopper: Phys. Rev. **131**, 2558 (1963)

5.191 B.Yu. Khelmer, V.I. Nefedov, L.N. Mazalov: Izv. Akad. Nauk SSSR, Ser. Fiz. **40**, 329 (1976)
Yu.A. Zhdanov, B.Yu. Khelmer, L.N. Mazalov, A.T. Shuvaev, P.I. Vadash, O.E. Shelepin: Zh. Strukt. Khim. **18**, 677 (1977)

5.192 E. Tegeler, N. Kosuch, G. Wiech, A. Faessler: Phys. Stat. Sol. B 84, 561 (1977); ibid. 91, 223 (1979)
5.193 R. Horn, D. Urch: J. Physique 48, C9-1009 (1987)
5.194 R.C.C. Perera, R.E. LaVilla: J. Chem. Phys. 84, 4228 (1986)
5.195 L.N. Mazalov, V.D. Yumatov, G.N. Dolenko: Zh. Strukt. Khim. 21, 21 (1980)
5.196 A.T. Shuvaev, A.P. Semlyanov, Yu.V. Kolodyazhnyi, O.A. Osipov, M.N. Tatevosyan, V.N. Eliseev, M.M. Morgunova: Izv. Akad. Nauk SSSR, Ser. Fiz. 38, 541 (1974)
5.197 G.L. Bendazzoli: Theoret. Chim. Acta 36, 77 (1974)
5.198 G.A.D. Collins, D.W.J. Cruickshank, A. Breeze: J. Chem. Soc., Faraday Trans. II 68, 1189 (1972)
5.199 S.P. Dolin, M.E. Dyatkina: Zh. Strukt. Khim. 13, 901 (1972)
5.200 Y. Gohshi: Adv. X-Ray Anal. 12, 518 (1969); in [Ref.1.60, Vol.2, p.250]
     Y. Gohshi, H. Kamada: Proc. Jpn. Acad., Ser. B 56, 167 (1980)
5.201 K.H. Johnson, F.C. Smith: Chem. Phys. Lett. 7, 541 (1970); ibid. 10, 219 (1971)
     K.H. Johnson: In *Advances in Quantum Chemistry*, ed. by P.-O. Löwdin, Vol.7 (Academic, New York 1973) p.143
5.202 G. Karlsson, R. Manne: Phys. Scr. 4, 119 (1971)
5.203 R. Szargan, H.-J. Köhler, A. Meisel: Spectrochim. Acta B 27, 43 (1972)
5.204 S. Aksela, M. Karras: Chem. Phys. Lett. 20, 356 (1973)
5.205 R. Manne, E. Suoninen: Chem. Phys. Lett. 15, 34 (1972)
     E.-K. Kortela, M. Karras: Spectrochim. Acta A 29, 1293 (1973)
5.206 A. Meisel, I. Steuer, R. Szargan: Spectrochim. Acta B 23, 527 (1968)
5.207 J. Merritt, E.J. Agazzi: Analyt. Chem. 38, 1954 (1966)
5.208 A.P. Sadovskii, L.N. Mazalov, G.N. Dolenko, A.A. Krasnoperova, V.D. Yumatov: Zh. Strukt. Khim. 14, 1048 (1973)
5.209 C. Sugiura, Y. Fujino, S. Kiyono: Techn. Rep. Tohoku Univ. 34, 107, 307 (1969)
5.210 A. Calabrese, R.G. Hayes: J. Am. Chem. Soc. 96, 5054 (1974); J. Electron Spectrosc. 6, 1 (1975)
5.211 R. Prins, T. Novakov: Chem. Phys. Lett. 9, 593 (1971); ibid. 16, 86 (1972)
5.212 R. Prins: J. Chem. Phys. 61, 2580 (1974);
     R. Prins, P. Biloen: Chem. Phys. Lett. 30, 340 (1975)
5.213 N. Kosuch, E. Tegeler, G. Wiech, A. Faessler: In [Ref.1.62, p.60]
     N. Kosuch, G. Wiech, A. Faessler: J. Electron Spectrosc. 20, 11 (1980)
5.214 A.V. Nikolaev, L.N. Mazalov, A.P. Sadovskii, E.A. Galtsova, V.V. Murakhtanov, T.I. Gushavina: Izv. Sib. Otd. Akad. Nauk SSSR, Ser. Khim. Nauk (1970) Nr.12, H.5, p.3
     V.V. Murakhtanov, T.I. Guzhavina, L.N. Mazalov, A.P. Sadovskii: Izv. Sib. Otd. Akad. Nauk SSSR, Ser. Khim. Nauk (1971) Nr.7, H.3, p.8
5.215 A. Faessler: Z. Physik 72, 734 (1931); Proc. 10th Coll. Spectrosc. Int'l Washington (1963) p.307
5.216 B.J. Lindberg, K. Hamrin, G. Johansson, U. Gelius, A. Fahlman, C. Nordling, K. Siegbahn: Phys. Scr. 1, 286 (1970)
5.217 H. Mendel: Proc. Kon. Ned. Akad. Wetensch. B 70, 276 (1967)
5.218 D. Urch: J. Phys. C 3, 1275 (1970)
5.219 J.A. Connor, M.B. Hall, I.H. Hillier, W.N.E. Meredith, M. Barber, Qu. Herd: J. Chem. Soc., Faraday Trans. II 69, 1677 (1973)
5.220 J.A. Connor, L.M.R. Derrick, M.B. Hall, I.H. Hillier, M.V. Guest, B.R. Higginson, D.R. Lloyd: Mol. Phys. 28, 1193 (1974)
     J.A. Connor, L.M.R. Derrick, I.H. Hillier: J. Chem. Soc., Faraday Trans. II 70, 941 (1974)

5.221 J.A. Connor, I.M.R. Derrick, I.H. Hillier, M.F. Guest, D.R. Lloyd: Mol. Phys. 31, 23 (1976)

5.222 M.-M. Coutiere, J. Demuynck, A. Veillard: Theor. Chim. Acta 27, 281 (1972)
J. Demuynck, A. Veillard, U. Wahlgren: J. Am. Chem. Soc. 95, 5563 (1973)

5.223 S. Evans, M.F. Guest, I.H. Hillier, A.F. Orchard: J. Chem. Soc., Faraday Trans. II 70, 417 (1974)

5.224 M.F. Guest, B.R. Higginson, D.R. Lloyd, I.H. Hillier: J. Chem. Soc., Faraday Trans. II 71, 902 (1975)

5.225 M.F. Guest, I.H. Hillier, B.R. Higginson, D.R. Lloyd: Mol. Phys. 29, 113 (1975)

5.226 B.R. Higginson, D.R. Lloyd, P. Burroughs, D.M. Gibson, A.F. Orchard: J. Chem. Soc., Faraday Trans. II 69, 1659 (1973)
B.R. Higginson, D.R. Lloyd, J.A. Connor, I.H. Hillier: J. Chem. Soc., Faraday Trans. II 70, 1418 (1974)

5.227 I.H. Hillier, M.F. Guest, B.R. Higginson, D.R. Lloyd: Mol. Phys. 27, 215 (1974)

5.228 A.T. Shuvaev, S.A. Prosandeev, I.A. Zarubin: Izv. Akad. Nauk SSSR, Ser. Fiz. 46, 753 (1982)

5.229 P.E. Best: J. Chem. Phys. 44, 3248 (1966)

5.230 D. Heintz: Z. Angew. Physik 27, 98 (1969)

5.231 A.P. Sadovskii, E.A. Kravtsova, L.N. Mazalov, S.V. Zemskov, Yu.I. Nikonorov: Izv. Sib. Otd. Akad. Nauk SSSR, Ser. Khim. Nauk Nr.9, H.4, 30 (1975)

5.232 A.P. Sadovskii, L.N. Mazalov, E.A. Kravtsova, L.I. Nasonova: Izv. Sib. Akad. Nauk, SSSR, Ser. Khim. Nauk Nr.9, H.4, 113 (1975)
A.P. Sadovskii, E.A. Kravtsova: Koord. Khim. 4, 418 (1978)

5.233 I.A. Zarubin, A.T. Shuvaev, V.N. Uvarov, N.E. Kolobova: Izv. Sib. Otd. Akad. Nauk SSSR, Ser. Khim. Nauk Nr.9, H.4, 37 (1975)
I.A. Zarubin, A.T. Shuvaev, V.V. Tyazhkorov, V.M. Danyushin: Izv. Akad. Nauk SSSR, Ser. Fiz. 40, 340 (1976)
A.T. Shuvaev, I.A. Zarubin, V.N. Uvarov, B.Yu. Khelmer, S.P. Gubin, N.E. Kolobova, V.M. Danyushin: Izv. Akad. Nauk SSSR, Ser. Fiz. 40, 333 (1976)

5.234 R. Szargan, E. Hartmann, H. Sommer, A. Meisel: In [Ref.1.64, p.717]

5.235 K.-H. Hallmeier, R. Szargan, K. Fritsche, A. Meisel: Phys. Scr. 35, 827 (1987)

5.236 A. Meisel, R. Szargan, K.-H. Hallmeier, I. Uhlig, J.-A., Momand: J. Physique 48, C9-981 (1987)

5.237 S. Evans, M.L.H. Green, B. Jewitt, A.F. Orchard, C.F. Pygall: J. Chem. Soc., Faraday Trans. II 68, 249, 1847 (1972)

5.238 J.V. Rabalais, L.O. Werme, T. Bergmark, L. Karlsson, M. Hussain, K. Siegbahn: J. Chem. Phys. 57, 1185 (1972)

5.239 N. Rösch, K.H. Johnson: Chem. Phys. Lett. 24, 179 (1974)

5.240 C.J. Ballhausen: Introduction to Ligand Field Theory (McGraw-Hill, New York 1962)

5.241 C.S. Fadley, D.A. Shirley: Phys. Rev. Lett. 23, 1397 (1969)

5.242 D.W. Fischer: Appl. Spectrosc. 25, 263 (1971)

5.243 D.W. Fischer: J. Phys. Chem. Solids 32, 2455 (1971)

5.244 H. Sommer, V.F. Volkov: Izv. Vyssh. Uchebn. Zaved., Fiz. 16, 153 (1973)

5.245 D.W. Clack: J. Chem. Soc., Faraday Trans. II 68, 1672 (1972)

5.246 I.H. Hillier, V.R. Saunders: Proc. Roy. Soc. A 320, 161 (1970); Chem. Phys. Lett. 9, 219 (1971)

5.247 L.N. Mazalov: Zh. Strukt. Khim. 18, 607 (1977)

5.248 E.A. Galtsova, L.N. Mazalov, V.D. Yumatov, A.P. Sadovskii, V.G. Torgov, S.S. Shatskaya, G.K. Parygina, G.N. Dolenko, S.V. Larionov: Izv. Sib. Otd. Akad. Nauk SSSR, Ser. Khim. Nauk Nr.9, H.4, 41 (1975)

G.K. Parygina, L.N. Mazalov, A.P. Sadovskii, V.I. Minkin, O.A. Osipov, V.P. Kurbatov: Zh. Strukt. Khim. 16, 355 (1975)

G.K. Parygina, L.N. Mazalov, Yu.N. Kukushkin, S.V. Larionov, G.N. Dolenko: 13 Vses. Chugaev Soveshch. po Khimii Kompleks. Soedin. 307 (1978)

5.249 V.D. Yumatov, G.N. Dolenko, E.A. Galtsova, L.N. Mazalov, A.P. Sadovskii: Izv. Sib. Otd. Akad. Nauk SSSR, Ser. Khim. Nauk Nr.9, H.4, 124 (1975)

5.250 V.I. Nefedov, A.P. Sadovskii, L.N. Mazalov, Ya.V. Salyn, E.A. Kravtsova, L. Beyer, N.P. Sergushin: Koord. Khim. 1, 950 (1975)

5.251 A.T. Shuvaev, V.A. Kondakov, V.N. Uvarov, K. Hallmeier, N.D. Lapkina, V.A. Postnikov, Yu.N. Novikov, M.E. Volpin: Zh. Strukt. Khim. 20, 736 (1979)

A.V. Nefedev, R.A. Stukan, V.A. Makarov, V.A. Kondakov, A.T. Shuvaev, N.D. Lapkina, Yu.N. Novikin, M.E. Volpin: Zh. Strukt. Khim. 21, 68 (1980)

5.252 J. Wong, F.W. Lytle, R.P. Messmer, D.H. Maylotte: Phys. Rev. B 30, 5596 (1984)

5.253 K.-H. Frank, E.-E. Koch, H.-W. Biester: J. Physique 47, C8-653 (1986)

5.254 K.-H. Hallmeier, R. Szargan, G. Werner, R. Meier, M.A. Sheromov: Spectrochim. Acta A 42, 841 (1986)

5.255 M. Verdaguer, C. Cartier, M. Momenteau, E. Dartyge, A. Fontaine, G. Tourillon, A. Michalowicz: J. Physique 47, C8-563, 623, 649 (1986)

5.256 M.F. Ruiz-Lopez, D. Rinaldi, C. Esselin, J. Goulon, J.-L. Poncet, R. Guilard: J. Physique 47, C8-637 (1986)

5.257 J. Garcia, M. Benfatto, C.R. Natoli, A. Bianconi, I. Davoli, A. Marcelli: Solid State Commun. 58, 595 (1986)

5.258 M. Benfatto, C.R. Natoli: J. Physique 48, C9-1077 (1987)

5.259 M. Benfatto, C.R. Natoli, A. Bianconi, J. Garcia, A. Marcelli, M. Fanfoni, I. Davoli: Phys. Rev. B 34, 5774 (1986)

5.260 N. Binsted, S.L. Cook, J. Evans, G. Greaves, R.J. Price: J. Am. Chem. Soc. 109, 3669 (1987)

5.261 G. Dräger, R. Frahm, G. Materlik, O. Brümmer: Phys. Stat. Sol. B 146, 287 (1988)

5.262 W. Gädeke, W.F. Koch, G. Dräger, R. Frahm, V. Saile: Chem. Phys. 124, 113 (1988)

5.263 B.I. Peshchevitskii, S.V. Zemskov, A.P. Sadovskii, E.A. Kravtsova, V.N. Mitkin: Koord. Khim. 5, 1838 (1979)

A.P. Sadovskii, S.V. Tsemskov, E.A. Kravtsova, V.N. Mitkin: Koord. Khim. 6, 1727 (1980)

5.264 V.I. Nefedov, A.P. Sadovskii, Ya.V. Salyn: Koord. Khim. 5, 1204 (1979)

5.265 D. Haycock, D.S. Urch, C.D. Garner, I.H. Hillier: J. Electron Spectrosc. 17, 345 (1979)

5.266 P. Biloen, R. Prins: Chem. Phys. Lett. 16, 611 (1972)

5.267 V.I. Nefedov, A.P. Sadovskii, L.N. Mazalov, A.V. Belaev, E.S. Gluskin: Zh. Strukt. Khim. 12, 681 (1971)

5.268 L.N. Mazalov, A.P. Sadovskii, A.V. Belaev, L.I. Chernyavski, E.S. Gluskin, L.F. Berkhoer: Izv. Sib. Otd. Akad. Nauk SSSR, Ser. Khim. Nauk (1971) Nr.2, H.1, p.51

5.269 H. Basch, H.B. Gray: Inorg. Chem. 6, 365 (1967)

5.270 P. Ros, G.C.A. Schuit: Theoret. Chim. Acta 4, 1 (1966)

5.271 W.T.A.M. van der Lugt: Chem. Phys. Lett. 10, 117 (1971)

5.272 R. Szargan, E. Suoninen, M. Lähdeniemi, M. Pessa: Spectrochim. Acta A 33, 129 (1977)

5.273 S. Foster, S. Felps: J. Am. Chem. Soc. 95, 5521, 6578 (1973)

5.274 R. Kebabcioglu, A. Muller: Chem. Phys. Lett. **8**, 59 (1971)
5.275 L.N. Mazalov, V.I. Baranovskii, A.P. Sadovskii, Yu.N. Kukushkin, G.K. Parygina, N.S. Panina: Zh. Strukt. Khim. **15**, 51 (1974)
5.276 V.I. Nefedov, Ya.V. Salyn, I.I. Moiseev, A.P. Sadovskii, A.S. Berenblyum, A.G. Knizhnik, S.L. Mund: Inorg. Chim. Acta **35**, L343 (1979)
5.277 R.L. Barinskii, I.M. Kulikova, N.P. Lapatova: Zh. Strukt. Khim. **13**, 1089 (1972)
    R.L. Barinskii, I.M. Kulikova, I.B. Barskaya, G.M. Tuptygina: Izv. Akad. Nauk SSSR, Ser. Fiz. **38**, 516 (1974)
    I.M. Kulikova, R.L. Barinskii, V.B. Aleksandrov, E.G. Proshchenko: Dokl. Akad. Nauk SSSR **210**, 1423 (1973)
5.278 B.Yu. Khelmer, V.F. Volkov, A.I. Platkov: Izv. Sib. Otd. Akad. Nauk SSSR, Ser. Khim. Nauk (1975) Nr.9, H.4, p.26
5.279 T.K. Sham: Phys. Rev. B **31**, 1903 (1985)
    T.K. Sham, B.S. Brunschwig: In [Ref.1.82, p.168]
5.280 H. Johansen: Theor. Chim. Acta **32**, 273 (1974)

Chapter 6

6.1   C. Kittel: *Introduction to Solid State Physics* (Wiley, New York 1966)
6.2   W.E. Spicer: In [Ref.1.101, p.139]
    M. Cardona, L. Ley (eds.): *Photoemission in Solids I, ,II*, Topics Appl. Phys., Vols.26, 27 (Springer, Berlin, Heidelberg 1978/79)
    S. Hüfner: *Introduction to Photoemission Spectroscopy*, Springer Ser. Solid-State Sci., Vol.82 (Springer, Berlin, Heidelberg 1989) in print
6.3   K. Siegbahn, C. Nordling, A. Fahlman, R. Nordberg, K. Hamrin, J. Hedman, G. Johansson, T. Bergmark, S.-E. Karlsson, I. Lindgren, B. Lindberg: *ESCA (Atomic, Molecular and Solid State Structure Studied by Means of Electron Spectroscopy)* (North-Holland, Amsterdam 1967)
6.4   G.F. Amelio: Surf. Sci. **22**, 301 (1970)
6.5   H.D. Hagstrum: In [Ref.1.101, p.349]
6.6   J.C. Phillips: Rev. Mod. Phys. **42**, 317 (1970); *Band and Bonds in Semiconductors* (Academic, New York 1973)
6.7   G. Klein: In [Ref.1.60, Vol.II, p.362]
6.8   G. Leonhardt, I. Topol, K. Unger, A. Meisel: Ann. Physik **28**, 245 (1972)
6.9   G.A. Rooke: J. Phys. Chem. **1**, 776 (1968); in [Ref.1.78, p.185]
6.10   C.N. Berglund, W.E. Spicer: Phys. Rev. A **136**, 1030, 1044 (1964)
6.11   D. Brust: Phys. Rev. A **139**, 489 (1965)
6.12   R.G. Cavell, S.P. Kowalczyk, L. Ley, R.A. Pollak, B. Mills, D.A. Shirley, W. Perry: Phys. Rev. B **7**, 5313 (1973)
6.13   G. Leonhardt: J. Electron Spectrosc. **5**, 603 (1974)
6.14   C.S. Fadley, D.A. Shirley: Phys. Rev. A **2**, 1109 (1970); in [Ref.1.101, p.163]
6.15   W.C. Price, A.W. Potts, D.G. Streets: In *Electron Spectroscopy*, ed. by D.A. Shirley (North Holland, Amsterdam, 1972) p.187
6.16   M. Klasson, J. Hedman, A. Berndtsson, R. Nilsson, C. Nordling: Phys. Scr. **5**, 93 (1972)
6.17   A. Faessler: In [Ref.1.78, p.93]
6.18   J.E. Holliday: In [Ref.1.78, p.101]
6.19   D.S. Fischer, W.L. Baun: J. Appl. Phys. **38**, 4830 (1967)
6.20   R.I. Liefeld: In [Ref.1.78, p.133]

6.21 A.P. Lukirskii, E.P. Savinov, O.A. Ershov, I.I. Zhukova, V.A. Fomichev: Opt. Spektrosk. 19, 425 (1965)

6.22 G. Wiech: Untersuchungen an der $L_{II,III}$-Röntgenemissionsbande von Aluminium mit einem neuen Konkavgitterspektrographen. Dissertation (Universität München 1964)

6.23 H. Neddermeyer: Beiträge zur Spektroskopie der ultraweichen Röntgenstrahlung. Dissertation (Universität München 1969)

6.24 L. Hedin, S. Lundqvist: Solid State Physics 23, 2 (Academic, New York 1969)
L. Hedin, B.I. Lunqvist, S. Lundqvist: In [Ref.1.101, p.233]

6.25 M.A. Blokhin: Methods of X-Ray Spectroscopic Research (Pergamon, Oxford 1965)

6.26 K.D. Sevier: Low Energy Electron Spectrometry (Wiley, New York 1972)

6.27 V.G. Aleshin, V.P. Smirnov: In [Ref.1.58, Vol.I, p.314]
V.G. Aleshin, E.P. Moiseenko, V.V. Nemoshkalenko, N.N. Sirota, V.T. Sharai: Metallofizika 66, 57 (1976)

6.28 J. Drahokoupil: J. Phys. C 5, 2259 (1972)

6.29 J. Klima: J. Phys. C 3, 70 (1970)

6.30 I.Ya. Nikiforov: In [Ref.1.57, p.241]; in [Ref.1.58, Vol.I, p.147]

6.31 G.A. Rooke: In [Ref.1.78, p.3]; in [Ref.1.101, p.287]

6.32 I. Topol, E. Hess: Phys. Stat. Sol. B 56, K9 (1973); Ann. Physik 30, 211 (1973)

6.33 I.A. Topol, G. Leonhardt, K. Unger, E. Hess: Phys. Stat. Sol. B 61, 285 (1974)
G. Leonhardt: Fiz. Tverd. Tela 17, 3 (1975)

6.34 R. Eibler, J. Redinger, A. Neckel: J. Phys. F 17, 1533 (1987)

6.35 V.I. Anisimov, A.L. Ivanovskii, V.A. Gubanov, E.Z. Kurmaev: Izv. Akad. Nauk SSSR, Neorg. Mater. 23, 685 (1987)

6.36 J. Redinger, J. Yu, A.J. Freeman, P. Weinberger: Phys. Lett. A 124, 463 (1987)
J. Redinger, A.J. Freeman, J. Yu, S. Massidda: Phys. Lett. A 124, 469 (1987)

6.37 M.A. Khan, C. Koenig: J. Physique 48, C9-1067 (1987)

6.38 I.Ya. Nikiforov: X-ray spectroscopic investigation of the electronic distributions in the 3d metals (in Russian). Dissertation (University of Rostov/Don 1966)

6.39 J.R. Cuthill, A.J. McAlister, M.L. Williams, R.C. Dobbyn: In [Ref.1.78, p.151]

6.40 I.I. Geguzin, I.A. Topol, I.Ya. Nikiforov, G. Leonhardt: Ann. Physik 28, 341 (1973)

6.41 J.B. Conklin, F.W. Averill, T.M. Mattox: J. Physique 33, C3-213 (1972)

6.42 D.W. Fischer: In [Ref.1.79, p.669]

6.43 H.W.B. Skinner: Phil. Trans. Roy. Soc. A 239, 95 (1940)

6.44 G. Wiech, E. Zöpf: In [Ref.1.101, p.335]

6.45 G. Leonhardt: Informationsgehalt der Röntgen- und Photoelektronenspektren binärer Festkörper. Dissertation (Karl-Marx-Universität, Leipzig 1976)

6.46 A. Wenger, G. Burri, S. Steinemann: Solid State Commun. 9, 1125 (1971)

6.47 J. Drahokoupil, A. Simunek: J. Phys. C 7, 610 (1974); Czech. J. Phys. B 25, 542 (1975)
A. Simunek: Czech. J. Phys. (Engl. Transl.) B 26, 239 (1976)

6.48 C.G. Dodd, G.L. Glen: J. Appl. Phys. 39, 5377 (1968); ibid. 40, 2361 (1969)

6.49 E.-K. Kortela, R. Manne: In [Ref.1.60, Vol.II, p.41]; J. Phys. C 7, 1749 (1974)

6.50 D.J. Nagel: Adv. X-Ray Anal. 13, 182 (1970)

6.51 V.I. Nefedov: Zh. Strukt. Khim. 8, 686, 1037 (1967)

6.52 D. Urch: J. Phys. C 3, 1275 (1970)

6.53 T.L. Gilbert, W.J. Stevens, H. Schrenk, M. Yoshimine, P.S. Bagus: Phys. Rev. B **8**, 5977 (1973)

6.54 H.-U. Chun, G. Klein: Z. Naturforsch. **24a**, 930 (1969)
G. Klein, H.-U. Chun: Phys. Stat. Sol. B **49**, 167 (1972)

6.55 C. Sugiura, Y. Hayasi: Jpn. J. Appl. Phys. **11**, 327, 598 (1972)

6.56 G. Wiech, E. Zöpf: In [Ref.1.79, p.629]

6.57 D.W. Fischer, W.L. Baun: Phys. Rev. A **138**, 1047 (1965)

6.58 R.H. Willens: In [Ref.1.101, p.281]

6.59 G.A. Burdick: Phys. Rev. **129**, 138 (1963)

6.60 A. Mansour, S.E. Schnatterly: Phys. Rev. B **36**, 9234 (1987)

6.61 I.B. Borovskii: J. Physique **32**, C 4-207 (1971)

6.62 O. Brümmer, G. Dräger, V.A. Fomichev, A.S. Shulakov: In [Ref.1.60, Vol.I, p.78]
U. Berg, G. Dräger, O. Brümmer: Phys. Stat. Sol. B **74**, 341 (1976)
G. Dräger, O. Brümmer: Phys. Stat. Sol. B **78**, 729 (1976)

6.63 J. Müller, K. Feser, G. Wiech, A. Faessler: Phys. Lett. A **44**, 263 (1973)
C. Beyreuther, G. Wiech: Phys. Fenn. **9**, S 1-176 (1974)
C. Beyreuther, R. Hierl, G. Wiech: Ber. Bunsenges. Phys. Chemie **79**, 1081 (1975)

6.64 E. Tegeler, N. Kosuch, G. Wiech, A. Faessler: Phys. Stat. Sol. B **84**, 561 (1977); ibid. **91**, 223 (1979)

6.65 L.V. Azaroff: Phys. Fenn. **9**, S 1-57 (1974)

6.66 D.J. Nagel, D.A. Papaconstantopoulos, J.W. McCaffrey, J.W. Criss: In [Ref.1.60, Vol.II, p.51]

6.67 J. Jortner, E.E. Koch, N. Schwenter: In *Photophysics and Photochemistry in the Vacuum Ultraviolet*, ed. by G. Findley, S. McGlynn, R. Huebner (Reidel, Dordrecht 1985) p.515

6.68 A.B. Kunz, D.J. Mickish, T.C. Collins: Phys. Rev. Lett. **31**, 756 (1973)

6.69 W.P. Menzel, C.C. Lin, F.F. Fouquet, E.E. Lafan, R.C. Chaney: Phys. Rev. Lett. **3**, 1313 (1973)

6.70 W. Gudat, C. Kunz, H. Petersen: Phys. Rev. Lett. **32**, 1370 (1974)

6.71 R. Haensel, C. Kunz, B. Sonntag: Phys. Rev. Lett. **20**, 262 (1968)
R. Haensel: DESY F 41-70/1 (1970)

6.72 A.S. Vinogradov, T.M. Zimkina: Fiz. Tverd. Tela **12**, 1492 (1970); Opt. Spektrosk. **32**, 33 (1972)

6.73 S.T. Pantelides, R.M. Martin, P.N. Sen: In [Ref.1.86, p.387]

6.74 F.C. Brown, C. Gähwiller, H. Fujita, A.B. Kunz, W. Scheifley, N. Carrera: Phys. Rev. B **2**, 2126 (1970)

6.75 M. Elango, A. Maiste: Phys. Fenn. **9**, S 1-86 (1974)

6.76 V.I. Nefedov: Zh. Strukt. Khim. **11**, 292, 299 (1970)

6.77 T. Åberg, J.L. Dehmer: J. Phys. C **6**, 1450 (1973); ibid. C **7**, L 278 (1974)

6.78 A.A. Pavlychev, I.V. Kondrateva: Fiz. Tverd. Tela **28**, 837 (1986)

6.79 J.L. Dehmer, A.F. Starace, U. Fano, J. Sugar, J.W. Cooper: Phys. Rev. Lett. **26**, 1521 (1971)

6.80 V.A. Fomichev, T.M. Zimkina, S.A. Gribovskii, I.I. Zhukova: Fiz. Tverd. Tela **9**, 1163 (1967)

6.81 P.S. Bagus, A.J. Freeman, F. Sasaki: Phys. Rev. Lett. **30**, 850 (1973)

6.82 G. Dräger, R. Frahm, G. Materlik, O. Brümmer: Phys. Stat. Sol. B **146**, 287 (1988)

6.83 K. Ulmer: In [Ref.1.79, p.521]

6.84 A. Kohm, H. Merz: In [Ref.1.60, Vol.II, p.394]

6.85 I.B. Borovskii, E.Ya. Komarov: Izv. Akad. Nauk SSSR, Ser. Fiz. **38**, 478 (1974)

6.86 E. Tegeler, N. Kosuch, G. Wiech, A. Faessler: Jpn. J. Appl. Phys. **17**, S 2-97 (1978)
6.87 E. Tegeler, M. Iwan, E.E. Koch: J. Electron Spectrosc. **22**, 297 (1981)
E. Tegeler, G. Wiech, A. Faessler: J. Phys. B **14**, 1273 (1981)
6.88 L.G. Parratt: Rev. Mod. Phys. **31**, 616 (1959)
6.89 B. Bergersen, F. Brouers: J. Phys. C **2**, 651 (1969); in [Ref.1.101, p.273]
6.90 P.H. Citrin, G.K. Wertheim, M. Schluter: Phys. Rev. B **20**, 3067 (1979)
6.91 R.A. Ferrell: Rev. Mod. Phys. **28**, 308 (1956)
6.92 S. Foster, S. Felps: J. Am. Chem. Soc. **95**, 5521, 6578 (1973)
6.93 A.J. Glick, P. Longe, S.M. Bose: In [Ref.1.78, p.319]
P. Longe: Phys. Rev. B **8**, 2572 (1973)
S.M. Bose, A.J. Glick: Phys. Rev. B **10**, 2733 (1974); ibid B **17**, 2073 (1978)
6.94 L. Hedin, R. Sjöström: In [Ref.1.101, p.269]
6.95 J.J. Hopfield: Comm. Solid State Phys. **2**, 40 (1969)
6.96 G.D. Mahan: Phys. Rev. **163**, 612 (1967); in [Ref.1.101, p.253]
6.97 T. McMullen, B. Bergersen: Canad. J. Phys. **50**, 1002 (1970)
6.98 P. Nozieres, C.T. Dominicis: Phys. Rev. **178**, 1097 (1969)
6.99 M.J. Stott, N.H. March: In [Ref.1.78, pp.283, 303]
6.100 C. Kunz, R. Haensel, G. Keitel, P. Schreiber, B. Sonntag: In [Ref.1.101, p.275]
6.101 J.D. Dow, J.E. Robinson, T.R. Carrer: Phys. Rev. Lett. **31**, 759 (1973)
6.102 I. Topol, J. Tilgner, G. Leonhardt, A. Meisel: J. Phys. Chem. Solids **35**, 1657 (1974)
6.103 I. Nagakura, V. Aita, K. Ichikawa, S. Suzuki, S. Kono, T. Ishii, T. Sagawa: Proc. 3rd Int'l Conf. VUV Rad. Phys. Tokyo 1971
6.104 M. Cardona, R. Haensel, D.W. Lynch, B. Sonntag: Phys. Rev. B **2**, 1117 (1970)
6.105 H. Winick: Nucl. Instr. Meth. A **261**, 9 (1987)
6.106 J.-E. Rubensson, N. Wassdahl, G. Bray, J. Rindstedt, R. Nyholm, S. Cramm, N. Martensson, J. Nordgren: Phys. Rev. Lett. **60**, 1759 (1988)
6.107 M.A. Blokhin, V.P. Sachenko: Izv. Akad. Nauk SSSR, Ser. Fiz. **24**, 397 (1960)
6.108 P.T. Landsberg: Proc. Phys. Soc. A **62**, 806 (1949)
6.109 P. Longe, A.J. Glick: Phys. Rev. **177**, 526 (1961)
F. Brouers: Phys. Stat. Sol. **22**, 213 (1967)
6.110 K. Läuger: Über den Einfluß der Bindungsart und der Kristallstruktur auf das K-Röntgenemissionsspektrum von Aluminium und Silizium. Dissertation (Universität München 1968)
6.111 V.V. Nemoshkalenko, V.V. Gorskii: Ukr. Fiz. Zh. **12**, 819 (1967); Phys. Stat. Sol. **28**, K 15 (1968); in [Ref.1.95, pp.169, 177]
V.V. Nemoshkalenko, L.S. Voskrekasenko, V.P. Krivitskii, L.I. Nikolaev: Fiz. Met. Metalloved. **43**, 191 (1977)
6.112 B.K. Agarwal, R.K. John: Phys. Stat. Sol. B **88**, 309 (1978)
R.K. Johri, B.K. Agarwal: J. Phys. F **8**, 555 (1978)
6.113 R.S. Crisp: Phil. Mag. **25**, 167 (1972); ibid. **36**, 609 (1977); J. Phys. F **10**, 511 (1980)
6.114 O. Aita, T. Sagawa: J. Phys. Soc. Jpn. **27**, 164 (1969)
6.115 T. Sagawa, Y. Iguchi, M. Sasanuma, A. Ejiri, S. Fujiwara, M. Yokota, S. Yamaguchi: J. Phys. Soc. Jpn. **21**, 2602 (1966)
6.116 H. Neddermeyer: Phys. Lett. A **38**, 329 (1972); in [Ref.1.79, p.153]; Phys. Fenn. **9**, S 1-195 (1974); in [Ref.1.86, p.665]; Phys. Rev. B **13**, 2411 (1976); Phys. Stat. Sol. B **78**, 609 (1976); ibid. **80**, 611 (1977)
6.117 C. Senemaud, C. Hague: J. Physique **32**, C 4-193 (1971)
C. Senemaud: J. Physique **32**, 89 (1971); Phys. Rev. B **18**, 3929 (1978)
6.118 I. Nagakura: Sci. Rep. Tohoku Univ. **149**, 1, 166 (1965)

6.119 A.N. Gusatinskii, S.A. Nemnonov: Fiz. Tverd. Tela 11, 1528 (1969)
6.120 M.A. Blokhin: *Physik der Röntgenstrahlen* (Verlag Technik, Berlin 1957)
6.121 L.M. Watson, R.K. Dimond, D.J. Fabian: In [Ref.1.78, p.45]; in [Ref.1.58, Vol.II, p.56]
6.122 H.M. O'Bryan, H.W.B. Skinner: Phys. Rev. 45, 370 (1934); Proc. Roy. Soc. A 176, 229 (1940)
6.123 J.A. Bearden: X-Ray Wavelengths, US Atomic Energy Commission, Oak Ridge, TN (1964)
     J.A. Bearden, A.F. Burr: Atomic Energy Levels, US Atomic Energy Commission, Oak Ridge, TN (1965)
6.124 D.H. Tomboulian, W.M. Cady: Phys. Rev. 59, 422 (1941)
6.125 F.S. Ham: Phys. Rev. 128, 82, 2524 (1962)
6.126 D.A. Goodings: Proc. Phys. Soc. 86, 75 (1965)
6.127 F.K. Allotey: Phys. Rev. 157, 467 (1967); Solid State Commun. 9, 91 (1971)
6.128 J.A. Tagle, E.T. Arakawa, T.A. Callcott: Phys. Rev. B 21, 4552 (1980)
6.129 H.-W. Wolff, K. Radler, B. Sonntag, R. Haensel: Z. Physik 257, 353 (1972)
6.130 T.L. Loucks, P.H. Cutter: Phys. Rev. A 133, 819 (1964)
6.131 V.A. Fomichev, I.I. Zhukova: Fiz. Tverd. Tela 10, 3073, 3753 (1968)
6.132 C. Kunz: J. Physique 32, C4-180 (1971)
6.133 I.B. Borovskii, V.A. Batyrev: Izv. Akad. Nauk SSSR, Ser. Fiz. 24, 441 (1960)
6.134 J. Clift, C. Curry, B.J. Thompson: Phil. Mag. 8, 593, 639 (1963)
6.135 I.Ya. Dekhtyar, V.V. Nemoshkalenko: *Electronic Structure and Electronic Properties of Transition Metals and Their Alloys* (Naukova Dumka, Kiev 1971)
6.136 L.D. Finkelshtein, S.A. Nemnonov: Fiz. Met. Metalloved. 22, 843 (1966)
     I.V. Gribov, V.I. Minin, L.D. Finkelshtein: Fiz. Met. Metalloved. 47, 949 (1979)
6.137 D.W. Fischer, W.L. Baun: J. Appl. Phys. 39, 4757 (1968)
     D.W. Fischer: J. Appl. Phys. 41, 3561 (1970)
6.138 D.W. Fischer: J. Appl. Phys. 40, 4151 (1969); ibid. 41, 3922 (1970)
6.139 D.W. Fischer: J. Phys. Chem. Solids 32, 2455 (1971)
6.140 S. Hanzely, R.J. Liefeld: In [Ref.1.101, p.319]
6.141 S.A. Nemnonov, E.Z. Kurmaev, V.A. Fomichev, B.Kh. Ishmukhametov, V.P. Belash, A.V. Rudnev: Izv. Akad. Nauk SSSR, Ser. Fiz. 36, 317 (1972)
     A.Z. Menshikov, E.Z. Kurmaev: Fiz. Met. Metalloved. 41, 748 (1976)
     E.Z. Kurmaev, V.P. Belash, S.A. Nemnonov: Fiz. Met. Metalloved. 43, 443 (1977)
     E.Z. Kurmaev, Yu.M. Yarmoshenko, V.E. Dolgikh, S.A. Nemnonov, F. Werfel, O. Brümmer: Solid State Commun. 29, 59 (1979)
6.142 A.S. Shulakov, I.I. Lyakhovskaya, V.A. Fomichev: Fiz. Tverd. Tela 15, 2246 (1973)
     I.I. Lyakhovskaya, T.M. Zimkina, V.A. Fomichev: Fiz. Tverd. Tela 17, 1417 (1975)
6.143 A.S. Shulakov, T.M. Zimkina, V.A. Fomichev, V.Ya. Nagornyi: Fiz. Tverd. Tela 15, 3598 (1973)
6.144 H. Sommer: X-ray L spectra of some elements of the 4th period (in Russian). Dissertation (University of Rostov/Don 1971)
6.145 B. Sonntag, R. Haensel, C. Kunz: Solid State Commun. 7, 597 (1969)
6.146 W.W. Beeman, H.F. Friedman: Phys. Rev. 56, 392 (1939)
6.147 W. Blau: In [Ref.1.58, Vol.II, p.188]
6.148 J.P. Irkhin: Fiz. Met. Metalloved. 11, 10 (1961)
6.149 M. Lähdeniemi, E. Suoninen, J. Bremer: Jpn. J. Appl. Phys. 17, S 2-129 (1978)

6.150 V.V. Nemoshkalenko, M.A. Mindlina, B.P. Mamko: Phys. Stat. Sol. B **30**, 703 (1968)

6.151 S.A. Nemnonov, E.Z. Kurmaev, K.M. Kolobova, A.Z. Menshikov: Fiz. Met. Metalloved. **25**, 1064 (1968)
S.A. Nemnonov, I.A. Brytov: Fiz. Met. Metalloved. **26**, 45 (1968)

6.152 U. Fano, J.W. Cooper: Rev. Mod. Phys. **40**, 441 (1968); ibid. **41**, 724 (1969)

6.153 H. Merz, K. Ulmer: Z. Physik **210**, 92 (1968)

6.154 H. Merz, K. Ulmer: Z. Physik **212**, 435 (1968)
K. Ulmer: In [Ref.1.62, p.92]

6.155 M.A. Blokhin, V.P. Sachenko, I.Ya. Nikiforov: J. Physique **32**, C 4-211 (1971)
M.A. Blokhin, E.G. Orlova, I.G. Shveitser: Fiz. Tverd. Tela **14**, 2470 (1972)

6.156 E.C. Snow, T.J. Waber: Phys. Rev. **157**, 570 (1967)

6.157 V.V. Nemoshkalenko: In [Ref.1.58, Vol.I, p.77]

6.158 C. Bonnelle: In [Ref.1.78, p.163]

6.159 G.A. Burdick: Phys. Rev. **129**, 138 (1963)

6.160 D.A. Papaconstantopoulos, D.J. Nagel, C. Jones-Björklund: Int'l J. Quantum Chem. **12**, 497 (1978)

6.161 I.I. Geguzin, G.I. Alperovich, I.Ya. Nikiforov: Izv. Akad. Nauk SSSR, Ser. Fiz. **36**, 305 (1972)

6.162 M.A. Blokhin: In [Ref.1.58, Vol.I, p.6]

6.163 M.A. Blokhin, I.Ya. Nikiforov, H. Sommer, I.I. Geguzin: In [Ref.1.79, p.321]

6.164 I.G. Shveitser, M.A. Blokhin: Izv. Akad. Nauk SSSR, Ser. Fiz. **31**, 947 (1967)
I.G. Shveitser, V.P. Sachenko, I.Ya. Nikiforov: Izv. Akad. Nauk SSSR, Ser. Fiz. **31**, 949 (1967)

6.165 M.I. Korsunskii, Ya.E. Genkin, M.M. Omarov, V.G. Lifshits: In [Ref.1.58, Vol.I, p.123]
M.I. Korsunskii, Ya.E. Genkin, V.G. Lifshits, M.M. Omarov: Izv. Akad. Nauk Kaz. SSR, Ser. Fiz.-Mat. Nauk **8**, 20 (1970)
M.I. Korsunskii, Ya.E. Genkin, V.G. Lifshits, V.I. Andryushin, A.N. Noerenchuk: Izv. Akad. Nauk Kaz. SSR, Ser. Fiz.-Mat. Nauk **14**, 68 (1976)

6.166 J.E. Holliday: In *The Electron Microprobe*, ed. by T.D. McKinley, K.F.V. Heinrich, D.B. Wittry (Wiley, New York 1966) p.3

6.167 A.P. Lukirskii, T.M. Zimkina: Izv. Akad. Nauk SSSR, Ser. Fiz. **27**, 330 (1963)

6.168 V.A. Fomichev, A.V. Rudnev, S.A. Nemnonov: Izv. Akad. Nauk SSSR, Ser. Fiz. **36**, 291 (1972)

6.169 V.I. Minin, L.D. Finkelshtein, S.A. Nemnonov, N.N. Yefremova, M.L. Kusminych: Izv. Akad. Nauk SSSR, Ser. Fiz. **36**, 424 (1972)

6.170 Y. Baer, P.F. Hedén, J. Hedman, M. Classon, C. Nordling, K. Siegbahn: Phys. Scr. **1**, 55 (1970)

6.171 I.Ya. Nikiforov, I.I. Geguzin, G.I. Alperovich: In [Ref.1.60, Vol.II, p.122]

6.172 A.I. Kozlenkov, A.I. Shulgin, A.V. Postnikov, A.I. Ivanovskii, V.A. Gubanov: J. Phys. C **18**, 3581 (1985)

6.173 P.A. Bruhwiler, Y. Shen, S.E. Schnatterly, S.J. Poon: Phys. Rev. B **36**, 7347 (1987)
P.A. Bruhwiler, J.L. Wagner, B.D. Biggs, Y. Shen, K.M. Wong, S.E. Schnatterly, S.J. Poon: Phys. Rev. B **37**, 6529 (1988)

6.174 W. Franz, P. Lamparter, S. Steeb: Z. Naturforsch. **42a**, 1385 (1987)
W. Franz, S. Steeb, H. Ebert, H. Winter, J. Voigtländer: Z. Physik B **69**, 257 (1987)

6.175 N.Z. Negm, L.M. Watson, A. Szasz: J. Physique **48**, C9-1033 (1987)
N.Z. Negm, L.M. Watson, P.R. Norris, A. Szasz: J. Phys. F **17**, 2295 (1987)

6.176 T.A. Callcott, K.-L. Tsang, C.H. Zhang, D.L. Ederer, E.T. Arakawa: J. Physique **48**, C9-1053 (1987)

6.177 D.J. Fabian: J. Physique 32, C 4-317 (1971)

6.178 V.V. Nemoshkalenko, V.G. Aleshin: *Theory of X-Ray Emission Spectroscopy* (in Russian) (Naukova Dumka, Kiev 1974)

6.179 V.V. Nemoshkalenko, ed.: *Electron Structure of Transition Metals, Their Alloys and Intermetallic Compounds* (Naukova Dumka, Kiev 1979)

6.180 L.M. Watson: In [Ref.1.79, p.125]

6.181 C. Curry: In [Ref.1.78, p.173]

6.182 L.V. Azaroff, B.N. Das: Phys. Rev. A 134, 747 (1964)
R.S. Brown, L.V. Azaroff, R.J. Donahue: In [Ref.1.79, p.491]

6.183 R.J. Donahue, L.V. Azaroff: J. Appl. Phys. 38, 2813 (1967)

6.184 J. Eggs, K. Ulmer: Z. Physik 213, 293 (1968)

6.185 S.A. Nemnonov, V.G. Syryanov, V.I. Minin, M.F. Sorokina: Phys. Stat. Sol. B 43, 319 (1971); ibid. 46, 77 (1971)

6.186 K. Kudrnovsky, L. Smrcka, B. Velicky: In [Ref.1.60, .Vol.II, p.94]

6.187 V.A. Fomichev: Fiz. Tverd. Tela 8, 2892 (1966); Izv. Akad. Nauk SSSR, Ser. Fiz. 31, 957 (1967)

6.188 J.E. Holliday: Norelco Report 14, 84 (1967)

6.189 L. Hoffmann, G. Wiech, E. Zöpf: Z. Physik 229, 131 (1969)

6.190 R.S. Crisp, S.E. Williams: Phil. Mag. 5, 1205 (1960); ibid. 6, 365 (1961)

6.191 D.W. Fischer, W.L. Baun: J. Appl. Phys. 37, 768 (1966); ibid. 38, 229 (1967)

6.192 T. Sagawa: J. Phys. Soc. Jpn. 21, 49 (1966)

6.193 F.C. Chalklin: Proc. Roy. Soc. A 194, 42 (1948)

6.194 M. Umeno, G. Wiech: Phys. Stat. Sol. B 59, 145 (1973)
G. Wiech: In [Ref.1.62, p.195]

6.195 E.A. Zhurakovskii: Dokl. Akad. Nauk SSSR 184, 1317 (1969); ibid. 194, 312 (1970)

6.196 C.A. Coulson, R. Taylor: Proc. Phys. Soc. A 65, 815 (1952)

6.197 V.I. Nefedov, V.I. Matyskin, I.B. Borovskii: Zh. Strukt. Khim. 12, 893 (1971)

6.198 O. Aita, I. Nagakura, T. Sagawa: J. Phys. Soc. Jpn. 30, 516, 1414 (1971)

6.199 V.P. Sachenko: Izv. Akad. Nauk SSSR, Ser. Fiz. 36, 226, 232 (1972)

6.200 C. Beyreuther, G. Wiech: In [Ref.1.86, p.517]
G. Wiech: In [Ref.1.62, p.195]

6.201 L. Ley, S. Kowalczyk, R. Pollak, D.A. Shirley: Phys. Rev. Lett. 29, 1088 (1972)
L. Ley, R.A. Pollak, S.P. Kowalczyk, F.R. McFeely, .D.A. Shirley: Phys. Rev. B 8, 641 (1973)

6.202 F.C. Brown, Om P. Rustgi: Phys. Rev. Lett. 28, 497 (1972)

6.203 R.S. Crisp, D. Haneman, V. Chacorn: J. Physique 48, C9-931 (1987)

6.204 A. Bianconi, R. Del Sole, A. Selloni, P. Chiaradia, M. Fanfoni, I. Davoli: Solid State Commun. 64, 1313 (1987)

6.205 R.D. Deslattes: Phys. Rev. 172, 625 (1968)

6.206 V.A. Fomichev: Fiz. Tverd. Tela 12, 2639 (1970); ibid. 13, 907 (1971)

6.207 V.A. Fomichev: Fiz. Tverd. Tela 9, 3034, 3167 (1967)

6.208 G. Wiech: Z. Physik 216, 472 (1968)

6.209 C. Sugiura: J. Phys. Soc. Jpn. 30, 1766 (1971); ibid. 32, 404, 571, 49K (1972)

6.210 B.F. Levine: J. Chem. Phys. 59, 1463 (1973); Phys. Rev. B 7, 2591 (1973)

6.211 V.A. Fomichev, M.A. Rumsh: J. Phys. Chem. Solids 29, 1015 (1968)
V.A. Fomichev, I.I. Zhukova, I.K. Polushina: J. Phys. Chem. Solids 29, 1025 (1968)

6.212 W.D. Grobman, D.E. Eastman, M.L. Cohen: Phys. Lett. A 43, 49 (1973)

6.213 N.O. Lipari, A.B. Kunz: Phys. Rev. B 3, 491 (1971)

6.214 G. Wiech: Z. Physik **193**, 490 (1966)
G. Wiech, E. Zöpf: J. Physique **32**, C4-200 (1971)
6.215 A.V. Ivanov, A.M. Tyutikov: Izv. Akad. Nauk SSSR, Ser. Fiz. **36**, 267 (1972)
A.V. Ivanov, L.D. Timakova, V.N. Kupriyanov: Izv. Akad. Nauk SSSR, Ser. Fiz. **36**, 272 (1972)
6.216 R. Szargan: Zur Interpretation röntgenspektroskopischer Untersuchungen der Valenzelektronenstruktur in Schwefelverbindungen auf der Grundlage von Molekül- und Festkörperrechnungen. Dissertation (Karl-Marx-Universität, Leipzig 1970)
R. Szargan, A. Meisel: Wiss. Z. Karl-Marx-Univ. Leipzig, Math.-Naturwiss. R. **20**, 41 (1971)
6.217 R.A. Pollak, L. Ley, S. Kowalczyk, D.A. Shirley, J.D. Joannopoulos, D.J. Chadi, M.L. Cohen: Phys. Rev. Lett. **29**, 1103 (1972)
6.218 C.J. Vesely, D.W. Langer, R.L. Hengehold: In *Electron Spectroscopy*, ed. by D.A. Shirley (North Holland, Amsterdam 1972) p.535
6.219 J. Drahokoupil, H. Klokocnikova, A. Simunek: Phys. Fenn. **9**, S 1-197 (1974); in [Ref.1.62, p.154]; J. Phys. C **9**, 2667 (1976)
6.220 C. Sugiura, Y. Hayasi, H. Konuma, S. Kiyono: J. Phys. Soc. Jpn. **31**, 1784 (1971)
6.221 D.J. Stukel, R.N. Euwema, T.C. Collins, F. Herman, R.L. Kortum: Phys. Rev. **179**, 740 (1969)
6.222 T.C. Collins, G.G. Wepfer, R.N. Euwema: Int'l J. Quantum Chem. **5**, 451 (1971)
6.223 C. Sugiura, Y. Hayasi, H. Konuma, C. Sato, M. Watanabe: J. Phys. Soc. Jpn. **29**, 1645 (1970)
6.224 G. Wiech: In [Ref.1.78, p.59]
6.225 I.I. Zhukova, V.A. Fomichev, A.S. Vinogradov, T.M. Zimkina: Fiz. Tverd. Tela **10**, 1383 (1968)
6.226 V.A. Fomichev: Fiz. Tverd. Tela **10**, 763 (1968)
6.227 I.A. Brytov, E.Z. Kurmaev: Fiz. Met. Metalloved. **32**, 520 (1971)
6.228 E. Tegeler, N. Kosuch, G. Wiech, A. Faessler: Phys. Stat. Sol. B **84**, 561 (1977); ibid. **91**, 223 (1979)
6.229 R.N. Euwema, T.C. Collins, D.G. Shankland, J.S. Dewitt: Phys. Rev. **162**, 710 (1967)
6.230 H.-U. Chun, D. Hendel: Z. Naturforsch. **22a**, 1401 (1967)
6.231 Y. Cauchois: In [Ref.1.78, p.71]
6.232 D.W. Fischer: J. Appl. Phys. **36**, 2048 (1965); Adv. X-Ray Anal. **13**, 159 (1970)
6.233 H. Mendel: Proc. Kon. Ned. Akad. Wetensch. B **70**, 276 (1967)
6.234 P.F. Walch, D.E. Ellis: Phys. Rev. B **8**, 5920 (1973)
6.235 C. Sugiura, S. Kiyono: Techn. Rep. Tohoku Univ. **35**, 243 (1970); J. Phys. Soc. Jpn. **32**, 494 (1972); ibid. **33**, 455 (1972); Phys. Rev. B **6**, 1709 (1972); ibid. B **8**, 823 (1973); ibid B **9**, 2679 (1974); J. Chem. Phys. **58**, 3527 (1973)
6.236 R.D. Deslattes: Phys. Rev. A **133**, 390, 395, 399 (1964)
6.237 L.N. Mazalov, E.E. Vainshtein, V.G. Syryanov: Dokl. Akad. Nauk SSSR **164**, 545 (1965); Zh. Strukt. Khim. **7**, 475 (1966)
6.238 J. Valasek: Phys. Rev. **52**, 250 (1937); ibid. **53**, 274 (1938)
6.239 G.J. Lapeyre, J. Anderson, J.A. Knopp, P.L. Gobby: In [Ref.1.86, p.380]
J. Anderson, G.J. Lapeyre: In [Ref.1.86, p.404]
6.240 V.A. Fomichev: Izv. Akad. Nauk SSSR, Ser. Fiz. **38**, 533 (1974)
6.241 P.E. Best: Phys. Rev. B **3**, 4377 (1971)
6.242 T Åberg, G. Graeffe, J. Utriainen, M. Linkoaho: J. Phys. C **3**, 1112 (1970)

6.243 N. Kosuch, J. Müller, G. Wiech, A. Faessler: Phys. Fenn. **9**, 51–189 (1974)
6.244 P.E. Best: J. Chem. Phys. **54**, 1512 (1971)
6.245 T.I. Liberberg-Kukher: Zh. Eksp. Teor. Fiz. **30**, 724 (1956)
6.246 P.H. Citrin, R.W. Shaw, A. Packer, T.D. Thomas: In *Electron Spectroscopy*, ed. by D.A. Shirley (North Holland, Amsterdam 1972) p.691
6.247 V.V. Nemoshkalenko, A.I. Senkevich, V.G. Aleshin: In [Ref.1.79, p.107]
6.248 R.T. Poole, J. Liesegang, R.C.G. Leckey, J.G. Jenkin: Chem. Phys. Lett. **23**, 194 (1973); ibid. **26**, 514 (1974); ibid. **31**, 308 (1975)·
6.249 A.P. Lukirskii, T.M. Zimkina: Izv. Akad. Nauk SSSR, Ser. Fiz. **28**, 765 (1964) A.S. Vinogradov, T.M. Zimkina, Yu.F. Maltsev: Fiz. Tverd. Tela **11**, 3354 (1969)
6.250 V. Saile, M. Skibowski: Phys. Stat. Sol. B **50**, 661 (1972)
6.251 C. Satoko, S. Sugano: J. Phys. Soc. Jpn. **34**, 701 (1973); In [Ref.1.86, p.368] C. Satoko: Solid State Commun. **13**, 1851 (1973)
6.252 S. Nakai, T. Sagawa: J. Phys. Soc. Jpn. **26**, 1427 (1969)
6.253 F.C. Brown, C. Gähwiller, H. Fujita, A.B. Kunz, W. Scheitley, N.J. Carrera: Phys. Rev. B **2**, 2126 (1970)
6.254 T.M. Zimkina, V.A. Fomichev: Fiz. Tverd. Tela **10**, 1392 (1968)·
6.255 F.C. Brown, C. Gähwiller, A.B. Kunz, N.O. Lipari: Phys. Rev. Lett. **25**, 927 (1970)
6.256 C. Sugiura, S. Kiyono: Techn. Rep. Tohoku Univ. **35**, 61, 249 (1970)
6.257 Y. Onodera: J. Phys. Soc. Jpn. **25**, 469 (1968)
6.258 K. Unger, H. Neumann: Phys. Stat. Sol. B **64**, 117 (1974)
6.259 F.R. McFeely, S. Kowalczyk, L. Ley, R.A. Pollak, D.A. Shirley: Phys. Rev. B **7**, 5228 (1973)
6.260 P.J. Lin, L. Kleinman: Phys. Rev. **142**, 478 (1966)
6.261 M.H. Reilly: J. Phys. Chem. Solids **31**, 1041 (1970)
6.262 I.A. Brytov, V.S. Neshpor, Yu.N. Romashchenko: Izv. Sib. Otd. Akad. Nauk SSSR, Ser. Khim. Nauk (1975) Nr.9, H.4, p.142
I.A. Brytov, L.E. Mstibovskaya, E.A. Obolenskii, L.G. Rabinovich, Yu.N. Romashchenko: Izv. Sib. Otd. Akad. Nauk, Ser. Khim. Nauk (1975) Nr.9, H.4, p.148
I.A. Brytov, Yu.N. Romashchenko: App. Met. Rentgen. Anal. **19**, 168 (1977); Fiz. Tverd. Tela **20**, 664 (1978)
I.A. Brytov, Yu.N. Romashchenko, B.F. Shchegolev: Zh. Strukt. Khim. **20**, 277 (1979)
I.A. Brytov, K.I. Konashenok, Yu.N. Romashchenko: Geokhimiya 1979, 261
L.E. Mstibovskaya, Yu.N. Romashchenko, I.A. Brytov: Fiz. Tverd. Tela **20**, 2526 (1978)
6.263 A. Breeze, P.G. Perkins: J. Chem. Soc., Faraday Trans. II **69**, 1237 (1973)
6.264 A.R. Ruffa: J. Appl. Phys. **43**, 4263 (1972)
6.265 Y. Iguchi: Phys. Fenn. **9**, S 1–162 (1974); In Proc. 7th Int'l Conf. X-Ray Optics Microanal., Moscow 1974, pp. 123, 158; in [Ref.1.86, p.506]; Sci. Light, Tokyo **26**, 161 (1977)
6.266 F.J. Arlinghaus, W.A. Albers: In [Ref.1.101. p.111]
6.267 V.V. Nemoshkalenko, V.G. Aleshin, I.A. Brytov, K.K. Sidorin, Yu.N. Romashchenko: Izv. Akad. Nauk SSSR, Ser. Fiz. **38**, 626 (1974)
K.K. Sidorin, V.V. Nemoshkalenko, V.G. Aleshin, I.A. Brytov, Yu.N. Romashchenko, A.I. Senkevich: Metallofizika **60**, 48 (1975)
6.268 N.V. Starostin, V.A. Ganin: Fiz. Tverd. Tela **15**, 3404 (1973); ibid. **16**, 572 (1974)
6.269 P. Rabe, B. Sonntag, T. Sagawa, R. Haensel: Phys. Stat. Sol. B **50**, 559 (1972)
6.270 G. Leonhardt, H. Neumann, A. Kosakov, T. Götze, M. Petke: Phys. Scr. **16**, 448 (1977)

6.271 G. Leonhardt, A. Kosakov, H. Sommer, M. Petke: In [Ref.1.62, p.57]
A. Kosakov, H. Neumann, G. Leonhardt: Phys. Lett. A 61, 57 (1977); ibid. A 62, 95 (1977)

6.272 J. Chen: Phys. Rev. B 2, 1053 (1970); ibid. B 8, 1440 (1973)

6.273 C. Beyreuther, G. Wiech: Phys. Fenn. 9, S 1-168 (1974)

6.274 M.A. Blokhin, L.M. Monastyrskii, I.G. Shveitser: Fiz. Met. Metalloved. 37, 640 (1974)

6.275 I.V. Kavich: Ukr. Fiz. Zh. 20, 156 (1975)
G.I. Ilkiv, I.V. Kavich, R.A. Antonyuk, Ya.N. Sinitski: Metallofizika 66, 62 (1976)
G.V. Samsonov, Ya.I. Dutchak, I.V. Kavich, P.I. Shevchuk: Dopov. Akad. Nauk Ukr. SSR A (1976) 446
Ya.I. Dutchak, I.V. Kavich, P.I. Shevchuk, V.G. Sinyushko: Izv. Akad. Nauk SSSR, Neorg. Mater. 12, 589 (1976)

6.276 E.A. Zhurakovskii, I.N. Frantsevich, V.V. Shvaiko, I.L. Kungurov: Ukr. Fiz. Zh. 20, 1324 (1975)

6.277 E.E. Vainshtein, I.B. Stary, M.N. Bril: Izv. Akad. Nauk SSSR, Ser. Fiz. 20, 784 (1956)

6.278 S.A. Nemnonov, K.M. Kolobova: Fiz. Met. Metalloved. 14, 874 (1962); ibid. 22, 680 (1966)

6.279 V. Ern, A.C. Switendick: Phys. Rev. A 137, 1927 (1965)

6.280 S.A. Nemnonov, A.Z. Menshikov, K.M. Kolobova, E.Z. Kurmaev, V.A. Trapeznikov: Trans. Met. Soc. AIME 245, 1191 (1969)
N.P. Sergushin, I.N. Shabanova, K.M. Kolobova, V.A. Trapeznikov, V.I. Nefedov: Fiz. Met. Metalloved. 35, 947 (1973)
V.A. Trofimova, S.A. Nemnonov: Fiz. Met. Metalloved. 39, 215 (1975)
V.A. Trofimova, K.M. Kolobova, S.A. Nemnonov, O.I. Klyushnikov, V.A. Trapeznikov: Fiz. Met. Metalloved. 40, 524 (1975)
V.A. Trofimova, K.M. Kolobova, S.A. Nemnonov: In [Ref.1.100, p.178]

6.281 J. Klima: J. Phys. C 15, 689 (1982)

6.282 E. Beauprez, C.F. Hague, J.-M. Mariot, F. Teyssandier: J. Microsc. Spectrosc. Electron. 11, 251 (1986)

6.283 E. Beauprez, C.F. Hague, J.-M. Mariot, F. Teyssandier, J. Redinger, P. Marksteiner, P. Weinberger: Phys. Rev. B 34, 886 (1986)

6.284 A. Neckel, P. Rastl, P. Weinberger, R. Mechtler: Theor. Chim. Acta 24, 170 (1972)
A. Neckel, P. Rastl, K. Schwarz, R. Eibler-Mechtler: Z. Naturforsch. 29a, 107 (1974)
A. Neckel, K. Schwarz, R. Eibler, P. Rastl, P. Weinberger: Mikrochim. Acta Suppl. 6, 257 (1975)
K. Schwarz, A. Neckel: Ber. Bunsen-Ges. Phys. Chemie 79, 1071 (1975)
P. Weinberger: Theor. Chim. Acta 41, 169 (1976)
K. Schwarz: J. Phys. C 10, 195 (1977)
P. Weinberger, F. Rosicky: Theor. Chim. Acta 48, 349 (1978)

6.285 V.I. Potorocha, V.A. Tskhai, P.V. Geld, E.Z. Kurmaev: Fiz. Met. Metalloved. 33, 666, 960 (1972); Dokl. Akad. Nauk SSSR 203, 1118 (1972)

6.286 K. Schwarz: Monatsh. Chemie 102, 1400 (1971)

6.287 P. Weinberger: Phys. Stat. Sol. B 97, 565 (1980); ibid. B 98, 207, 591 (1980)

6.288 D. Adler: Solid State Phys. 21, 1 (1968)

6.289 N.F. Mott, K.W.H. Stevens, J.B. Goodenough: Phil. Mag. 2, 1364 (1957)
F.J. Marin: Bell System Techn. 37, 1047 (1958)

6.290 A.Z. Menshikov, I.A. Brytov, E.Z. Kurmaev: In [Ref.1.58, Vol.I, p.115]

6.291 S.A. Nemnonov, S.S. Mikhailova: Izv. Akad. Nauk SSSR, Ser. Fiz. **38**, 493 (1974); Fiz. Met. Metalloved. **39**, 1178 (1975)

     K.M. Kolobova, S.A. Nemnonov: Izv. Sib. Otd. Akad. Nauk SSSR, Ser. Khim. Nauk (1975) Nr.9, H.4, p.34

     S.S. Mikhailova, S.A. Nemnonov, V.I. Minin, F.B. Maksyutov, V.R. Galakhov: Izv. Akad. Nauk SSSR, Ser. Fiz. **40**, 439 (1976)

     K.M. Kolobova, O.I. Klyushnikov, S.A. Nemnonov, V.A. Trapeznikov: Fiz. Met. Metalloved. **41**, 1201 (1976)

6.292 C. Sugiura. Jpn. J. Appl. Phys. **10**, 1120 (1971)

     C. Sugiura, S. Nakai, T. Matsukawa, M. Obashi, J. Kashiwakura, Y. Gohshi: In [Ref.1.62, p.168]

     T. Matsukawa, M. Obashi, S. Nakai, C. Sugiura: Jpn. J. Appl. Phys. **17**, S 2-184 (1978)

6.293 C. Sugiura, Y. Gohshi, I. Suzuki: Jpn. J. Appl. Phys. **11**, 911 (1972)

6.294 J.G. Bednorz, K.A. Müller: Z. Physik B **64**, 189 (1986)

6.295 M.K. Wu, J.R. Ashburn, C.J. Torng, P.H. Hor, R.C. Meng, L. Gao, Z.H. Huang, Y.Q. Wang, C.W. Chu: Phys. Rev. Lett. **58**, 908 (1987)

6.296 K.L. Tsang, C.H. Zhang, T.A. Callcott, L.R. Canfield, D.L. Ederer, J.E. Blendell, C.W. Clark, N. Wassdahl, J.E. Rubensson, G. Bray, N. Mortensson, J. Nordgren, R. Nyholm, S. Cramm: J. Physique **48**, C9-1193 (1987); Phys. Rev. B **37**, 2293 (1988)

6.297 R.C.C. Perera, B.L. Henke, P.J. Batson, J.A. Kerner, D. Berkeland: J. Physique **48**, C9-1185 (1987)

6.298 J.-M. Mariot, V. Barnole, C.F. Hague, V. Geiser, H.-J. Güntherodt: J. Physique **48**, C9-1203 (1987)

6.299 J. Drahokoupil, M. Polcik, E. Pollert: Solid State Commun. **66**, 405 (1988)

6.300 V.I. Glazyrina, N.N. Efremova, M.A. Korotin, E.Z. Kurmaev, L.D. Finkelshtein, Yu.M. Yarmoshenko: Fiz. Met. Metalloved. **65**, 399 (1988)

6.301 V.I. Anisimov, V.R. Galakhov, E.Z. Kurmaev, M.A. Korotin, V.L. Kozhevnikov, G.V. Bazuev: Fiz. Met. Metalloved. **65**, 207 (1988)

6.302 A. Bianconi, A. Clozza, A. Congiu Castellano, S. Della Longa, M. de Santis, A. di Cicco, K. Garg, P. Delogu, A. Gargano, R. Giorgi, P. Lagarde, A.M. Flank, A. Marcelli: J. Physique **48**, C9-1179 (1987)

     A. Bianconi, A. Congiu Castellano, M. de Santis, P. Rudolf, P. Lagarde, A.M. Flank, A. Marcelli: Solid State Commun. **63**, 1009 (1987)

6.303 M. Sacchi, F. Corni, G.M. Antonini, C. Calandra, F.C. Matacotta, R. Frahm: Z. Physik B **72**, 335 (1988)

6.304 F. Baudelet, G. Collin, E. Dartyge, A. Fontaine, J.P. Kappler, G. Krill, J.P. Itie, J. Jegoudez, M. Maurer, Ph. Monod, A. Revcolevschi, H. Tolentino, G. Tourillon, M. Verdaguer: Z. Physik B **69**, 141 (1987)

6.305 A. Bianconi, J. Budnick, A.M. Flank, A. Fontaine, P. Lagarde, A. Marcelli, H. Tolentino, B. Chamberland, C. Michel, B. Raveau, G. Demazeau: Phys. Lett. A **127**, 285 (1988)

6.306 F.W. Lytle, R.B. Greegor, A.J. Panson: Phys. Rev. B **37**, 1150 (1988)

6.307 S. Horn, K. Cai, T. Guo, C.L. Chang, M.L. Denboer: AIP Conf. Proc. **165**, 311 (1988)

6.308 Y. Suzuki, H. Sekiyama, Y. Hirai, K. Hayakawa, H. Kajiyama, T. Hirano, R. Usami, K. Takagi: Jpn. J. Appl. Phys. **27**, L149 (1988)

6.309 H. Oyanagi, H. Ihara, T. Matsubara, M. Tokumoto, T. Matsushita, M. Hirabayashi, K. Murata, N. Terada, T. Yao, H. Iwasaki, Y. Kimura: Jpn. J. Appl. Phys. **26**, L1561 (1987)

6.310 M. Grioni, J.C. Fuggle, P.J.M. Weijs, J.B. Goedkoop, G. Rossi, F. Schaefers, J. Fink, N. Nücker: J. Physique 48, C9-1189 (1987)

6.311 J.C. Fuggle, P.J.W. Weijs, R. Schoorl, G.A. Sawatzky, J. Fink. N. Nücker, P.J. Durham, W.M. Temmerman: Phys. Rev. B 37, 123 (1988)

6.312 H.M. Meyer III, Y. Gao, T.J. Wagener, D.M. Hill, J.H. Weaver, B.K. Flandermeyer, D.W. Capone II: AIP Conf. Proc. 165, 254 (1988)

6.313 P. Steiner, S. Hüfner, V. Kinsinger, I. Sander, B. Siegwart, H. Schmitt, R. Schulz, S. Junk, G. Schwitzgebel, A. Gold, C. Politis, H.P. Müller, R. Hoppe, S. Kemmler-Sack, C. Kunz: Z. Physik B 69, 449 (1988)

6.314 L.F. Mattheiss: Phys. Rev. Lett. 58, 1028 (1987)
L.F. Mattheiss, D.R. Hamann: Solid State Commun. 63 395 (1987)

6.315 J. Yu, A.J. Freeman, J.-H. Xu: Phys. Rev. Lett. 58, 1035 (1987)

6.316 G. Wendin: J. Physique 48, C9-1157 (1987)

6.317 L.F. Mattheiss: Phys. Rev. B 5, 290, 306, 315 (1972)

6.318 K. Hübner, G. Leonhardt: Phys. Stat. Sol. B 68, K175 (1975)

6.319 S. Aksela, M. Karras, E. Kortela, E. Suoninen: In [Ref.1.60, Vol.I, p.352]

6.320 A. Goldmann, J. Tejeda, N.J. Shevchik, M. Cardona: Phys. Rev. B 10, 4388 (1974); Solid State Commun. 15, 1093 (1974)

Appendix: Survey of X-Ray-Spectra Investigations

3 Lithium

K emission
- *band*
- - metal A.54; B.81; C.2,20,63,98,99; D.95,
  G.107; K.146; M.94; N.65; O.8; R.68; S.20;
  T.2; Y.22
- - alloys C.4
- - compounds A.28; F.66; M.13; T.70
- - theory
- - - metal A.51,52,84; B.218; G.46; H.53
- - - alloys S.250
- - - fluoride A.4; K.187; P.13
- *satellites* A.78; R.43; S.221
K absorption
- metal A.54; B.81; C.2,63; D.95; H.4;
  K.188,190; M.94; O.8; Y.22
- compounds M.13
- halides H.4; K.188,190; R.10; S.211; V.36; W.9
- theory F.50; H.48; M.12; R.49

4 Beryllium

K emission
- *band*
- - metal A.27; B.231; C.3,100; F.11; H.42,65;
  S.20; T.1; W.41,44
- - alloys E.6; K.158; O.7; S.183
- - compounds B.144; E.6; I.8; T.88
- - oxide F.60; O.7; S.183
- - theory A.51; G.46; R.38
- *satellites* S.221
K absorption
- metal C.3,100; F.60; H.8; L.89; R.17; S.287
- compounds B.144; I.8
- oxide F.60; L.89; S.287

5 Boron

K emission
- element F.11; M.176
- alloys T.33
- compounds A.47; D.90; E.6; F.36,37,74; G.112;
  H.42,65; L.94; M.166; O.7; S.17,78; T.88;
  Z.18,29
- borides F.36,37,74; G.112; H.42,76; M.96;
  N.61; S.123; T.88; V.44; Z.18

- phosphide F.37,62; N.85
- metallic glasses T.12
- theory B.194; G.24,95; T.71
K absorption
- element S.200
- compounds B.46; F.55; J.2; L.94; Z.29,31
- borides O.25
- fluoride B.40; G.32; H.41; M.83; S.94
- theory T.57

6 Carbon

K emission
- element B.112,193,217,218,263; C.26; D.99;
  E.9; F.12,72; K.21,76,201; M.49,50,102,176;
  N.41; S.71,103,169,184; U.7; Z.22,23
- compounds B.211,217; C.26; D.82; F.12; G.41;
  I.24; K.71,139; L.27; M.74,93,165;
  N.47,54,158,159; P.37; R.65; S.17,164,169,184;
  T.22,24; W.35; Y.24
- carbides A.80; B.238,247; C.39; F.34,37;
  G.87,96; H.42,80; I.26; K.131; N.61,112; S.147;
  Z.16,20
- theory
- - element K.132; L.87
- - compounds C.80; F.49; G.24; H.87; I.25,29;
  K.96,103; L.15; M.43,125; N.28,39; P.61,70;
  S.171; W.25
K absorption
- element F.72; K.80; M.11,101; R.72; Z.22,23
- compounds A.33; H.19; K.112; M.101; O.46;
  S.118,249; T.63
- oxides B.222; C.54,55; K.50,111; N.18; S.182;
  V.18
- adsorbed molecules A.86; B.7,9; C.15; F.30,86;
  M.153; O.45; S.243
- theory
- - element K.91; L.87
- - compounds B.13; K.123; S.94,275; T.56
Isochromats K.75

7 Nitrogen

K emission
- element and gaseous compounds L.24; M.84;
  P.57; W.34,35
- solid compounds B.238,245,246,248; D.58;

345

24 Chromium

K emission
- $\alpha_{1,2}$ *lines*
- -- metal K.173; O.34; S.111
- -- alloys N.81,84
- -- compounds B.84; G.53; K.143; L.58–60; M.112; N.2; S.155,175; T.74; Z.21
- α *satellites* A.24; K.63; S.138; Z.7
- $\beta_{1,3}$ *lines*
- -- metal N.139
- -- alloys N.81,84
- -- compounds C.85; K.135,143; L.58–60; M.112,173; N.36; S.155; T.74
- -- theory B.26
- β_5 *band*
- -- metal B.139,151; G.60; K.47,169; L.6; N.139,140; R.62
- -- alloys B.65,66; F.19; K.195; N.81,84; S.286
- -- compounds B.147; C.85; F.75; I.3,26; K.135; M.88; N.36,37,54,67,98; R.50; S.155,164; Z.2,12,17,19
- -- nitride R.62
- -- silicides A.98; K.195; W.31
- -- theory A.57; B.64; J.9; W.24
- β *satellites* D.56; S.178,226

K absorption
- *XANES*
- -- metal B.151,229; K.47; R.62; S.232
- -- alloys D.76; F.19,76; K.154; L.82; M.180
- -- compounds B.35,36; C.34; G.9; K.70,84; L.72; M.51,68,97,185; N.37,67; R.50; S.96,225; V.3,21
- -- oxides B.229; K.47; S.232; W.37
- -- hydrides B.229; L.50; S.232
- -- nitrides R.62
- -- chromates B.89; I.3; K.179,205; M.88; W.37
- -- catalysts L.98,99; S.276
- *EXAFS* C.34; L.65; M.152; S.112

L emission
- metal B.170,174; H.14; L.91; N.36,63,77; R.62; S.185; U.2
- alloys F.75; H.79,84; L.35; N.117; P.34; S.286
- compounds D.87; K.141; N.54; S.205; U.2
- oxides B.246; F.42–44; H.14; L.91
- carbides H.79,84; I.26
- nitrides R.62
- silicides A.98; Z.19,21
- chromates F.42–44; G.89; M.88; S.14,17
- theory C.85; G.19; N.140; W.24

L absorption
- metal B.170,174; R.62
- compounds F.42–44; H.21; R.62
- theory D.35

M emission
- compounds F.59,64,67
- theory G.19; N.140

M absorption
- element M.48
- compounds L.95; N.11

- theory D.35
Isochromats
- metal M.130
- alloys P.34
- compounds B.94; N.135; S.175

25 Manganese

K emission
- $\alpha_{1,2}$ *lines*
- -- metal S.85
- -- alloys N.88
- -- compounds B.29; G.53; M.109,117; P.44,46; R.66; S.66; V.5; W.65; Z.21
- -- fluoride G.52,54; P.47; Y.17
- -- glasses K.87
- -- theory D.37; N.32; P.48
- α *satellites* B.214; D.39; K.63; Z.7
- $\beta_{1,3}$ *lines*
- -- metal E.16; K.135; S.85
- -- alloys N.88
- -- compounds B.29,89; C.85; D.34; G.53; K.135; M.23,173; N.14; P.47; R.66; S.66; T.7; W.65
- -- oxides K.140; V.5
- -- fluoride G.52,54; P.47; Y.17
- -- theory B.26
- β_5 *band*
- -- alloys A.8; F.19; N.88
- -- compounds B.29; E.29; F.19; G.106; I.3; L.54; M.33,88; N.99,134; S.163; T.76; V.5; W.65; Z.2
- -- borides A.81; S.122,124
- -- carbides A.83
- -- nitrides M.169
- -- ferrites K.82
- -- silicides M.142; Z.21
- β *satellites* G.51; S.141,176,226; T.73; W.10

K absorption
- *XANES*
- -- metal A.77; W.37
- -- alloys D.75; F.19; K.154; L.52; P.30
- -- compounds A.87; B.223; K.70,151,179; L.54; M.33,36; N.14,134; O.49; P.63; R.25; S.263; Y.6
- -- oxides A.87; K.84,86; P.6; S.60; T.7; W.37
- -- carbides A.83
- -- ferrites K.84,86; P.6; S.60
- -- permanganates B.89; G.9; I.3; M.154; R.5; S.113; T.7
- -- solutions B.78
- -- theory G.44; L.65; S.96; T.84
- *EXAFS*
- -- alloys B.25,207; D.7; H.57; M.1; S.5,241
- -- inorganic compounds A.77; B.73,223; C.69,103; J.7; R.5; S.239
- -- organic compounds B.73; J.7
- -- solutions B.77

L emission
- metal S.185
- alloys G.2; N.14; S.188; W.29

S.299; T.61; W.4
- - alloys A.100,101; B.101; D.8,73,75,76,92,120; K.20,35,119,154; P.30,32
- - compounds B.143; F.4; G.98; K.84,148,151; L.55; M.33,51; N.134; O.1; P.66; S.41,225,256; T.75; U.8; V.17
- - organic compounds B.36; E.25; M.108; P.2
- - oxides K.84; N.36; S.251
- - catalysts L.67
- - theory E.7; N.8
- *EXAFS*
- - metal A.20,74; C.79; E.26; H.69; M.60; S.116,231
- - alloys C.69,72; L.81; O.22; Z.35
- - organic compounds K.78
- - oxide and hydride L.50; T.48
- - catalysts G.76; J.30
- - solutions B.56; C.89; L.68; S.44
- - metallic glasses H.11; M.25,54; S.3,6; T.31; W.62

L emission
- metal B.134,173; C.45; D.119; F.6; H.14,31; L.70; R.47; S.290; U.6
- alloys B.127; D.117,120; F.34,51; G.97; K.7,10,73; L.7; N.73,86,117; S.25,127; T.79; V.48; W.29; Y.16
- compounds B.134,143,173; F.80; H.14; N.73; S.124
- complex compounds N.54; S.13,17; T.22
- oxides B.248; D.12; M.127; N.36; T.80,82; W.65
- phosphides D.87; S.297
- metallic glasses B.72; F.5; M.56; N.51; T.12
- theory D.36; E.7; H.92; N.149

L absorption
- metal B.95,134,173; F.6; H.14,31; K.94; S.290,299
- alloys B.127; K.10; M.181; N.117; P.30,32; R.17; V.48
- compounds B.95,134,143,173; H.14,21; O.15; V.15
- oxides B.171,248; N.36
- theory N.8

M emission
- metal and alloys A.75,76; C.70,108,109; D.119; K.37; L.34; M.66; S.25,89,145,290
- compounds B.143; I.28; N.96; S.297; T.78
- zeolites B.182
- glasses H.18
- theory B.163

M absorption
- element B.235; O.28; S.290
- alloys G.91; H.7,29
- compounds I.28; N.11; P.62; T.78
- oxide B.248
- zeolites B.182
- theory N.8

Isochromats
- metal A.95; R.47
- alloys C.58; F.51; K.95; L.33; U.5
- compounds H.96

29 Copper

K emission
- $\alpha_{1,2}$ *lines*
- - metal D.9,58; S.43; W.28
- - alloys P.45
- - compounds M.115; T.74
- $\beta_{1,3}$ *lines*
- - metal N.136,137
- - compounds L.54
- - theory B.29
- β *satellites* L.31; S.178
- β_5 *band*
- - metal A.34; B.11; G.10; L.8; N.76; T.79
- - alloys H.56; N.68
- - compounds A.34,66; D.107,108,110; S.207

K absorption
- *XANES*
- - metal A.34,42,94; B.11,21,196; D.80; E.26; F.97; H.89; I.14; K.179; L.44,97,99; M.35; O.30; S.38,70,120,235; T.61; W.4,20,58
- - alloys A.100,101; D.8,72,75,92; H.56; J.17; K.154,157; L.79,80; P.30,76
- - compounds A.11,12; B.17; C.33; D.48,50,74; H.62; K.48,148; L.54; M.186; N.121; P.66; R.24,75; S.225,229; T.65
- - complex compounds A.34,45,85; B.98,104; K.168,182; M.108; N.132; O.2; P.77,78; S.75,197,229; T.72
- - oxides A.14,34; H.89; M.140; O.17; S.231
- - catalysts C.68; V.42
- - superconductors A.70; B.120; J.16; O.57
- - solutions E.10; S.11; T.59
- - theory B.18; R.34
- *EXAFS*
- - metal B.155,162; D.26; I.13; M.59,60,152; S.239
- - alloys C.66; F.69; L.48,81; M.2,3; R.28; Y.21
- - compounds A.39,71; B.162; D.49; I.13; J.31; M.59,60,68,133; S.239; T.86
- - organic compounds A.41; B.99,156,225; C.28; G.5; K.78; P.75; S.152
- - oxides A.19,70; B.162; F.71; M.59,60
- - iodides B.204; L.73
- - catalysts A.74; G.75; N.155; S.60
- - superconductors O.57
- - metallic glasses C.30; H.11; K.110; M.54; S.3; T.43
- - solutions D.26; G.63; L.2; S.48,112
- - theory F.87; K.104; L.39; R.33; S.79; V.28

L emission
- metal B.136; C.5; D.12; G.65; H.16,31,92; L.70; N.65; W.65
- alloys B.53; C.5; D.117; F.5; H.56,83; N.30,60,98; W.23,29
- compounds D.87,98; F.96; G.3; K.138; R.40; T.33
- oxides B.173; D.12; N.65; T.82; W.65
- surfaces N.153

L absorption
- metal A.42; H.31,59; K.94

M emission
- metal H.80; K.162; M.150
- theory N.101; S.91
N emission
- metal F.61; R.45
- theory N.101

40 Zirconium

K emission
- $\alpha_{1,2}$ *lines*
- - oxide M.19; P.55
- - theory A.44; K.68
- α *satellites* S.39
- $\beta_{1,3}$ *lines*
- - compounds M.114
- - oxide M.19; P.55
- - theory K.68
K absorption
- *XANES*
- - alloys C.67; L.52; P.53
- - compounds A.15; B.97,111; C.52; D.13,41;
  G.98; P.36,53
- - theory M.177
- *EXAFS*
- - compounds L.50; T.83
- - glasses H.11; R.20; S.3,6
L emission
- $\alpha_{1,2}\beta_1$ *lines*
- - metal B.141; D.12; L.77; N.3,100
- - alloys N.151
- - oxide H.12; N.3
- - carbide N.3
- - nitrides L.77
- - chloride K.144
- α *satellites* J.35; W.69
- $\beta_2\gamma_1$ *bands*
- - metal L.77; N.80; R.22; S.286
- - metallic glasses H.18; M.56
- - oxide K.72
- - alloys M.107; N.80,91; S.286; T.12
- - hydrides N.91,151; T.10; Z.17
- - carbides K.129; R.22; Z.17
- - zirconates B.131
- - nitrides K.129; L.77
- - theory K.161; N.107; P.70; S.92
L absorption
- metal M.7
- complex compounds B.39,41
- zirconates B.131
M emission
- metal and oxide Z.25
- alloys D.125
- compounds B.169; H.42; N.75,91; S.145-147
- carbides H.76,77; Z.17
- theory P.70; S.92
M absorption
- metal Z.8
N emission
- metal and compounds F.59,61,64; S.145

- theory S.92
Isochromats
- metal M.132

41 Niobium

K emission
- $\alpha_{1,2}$ *lines*
- - metal and oxide S.277
- - compounds B.17,108; M.19
- $\beta_{1,3}$ *lines*
- - compounds B.17,108; M.19
K absorption
- metal B.196
- alloys C.14,67; L.52
- compounds B.108,183; D.13; S.63; T.43
- complex compounds N.126
- hydride L.50
L emission
- $\alpha_{1,2}\beta_1$ *lines*
- - metal B.141; K.9; N.2,4,100; P.80
- - alloys N.86; W.6
- - compounds B.16; K.144,178
- - oxide N.2,4
- - carbide D.44; E.15; K.9
- - nitride K.9
- $\beta_2\gamma_1$ *bands*
- - metal B.66; E.24; K.9,198
- - alloys C.40; D.125; G.23; H.13; K.129-131;
  N.75,80,83
- - oxides B.66,232; E.24; G.90,96; H.13;
  K.72,198; R.37
- - hydrides G.38; N.75,83
- - boride K.129-131
- - carbide B.66; C.40; D.44; E.24; G.23,96;
  K.9,129-131,198; R.22,37; Z.17
- - nitride B.66; E.24; G.96; K.9,129-131,198
- - silicides N.75,83
- - niobate K.39
- - theory A.57; B.138; C.80; N.28,101,149; P.70
- γ *satellites* S.213
L absorption
- compounds M.6; O.14
- complex compounds and minerals B.39,41
- theory B.138
M emission
- metal A.57; W.41
- alloys B.66; D.84; N.83
- hydride N.83
- carbide H.80,81; S.146,147
- nitride B.66; D.84; S.146,147
- silicide N.83; S.146,147
- theory C.80; N.28,101,149; P.70
M absorption
- metal Z.8
N emission
- metal A.57; F.61,64; S.146
- silicide F.61,64; S.146
Isochromats
- metal A.96; M.130

42 Molybdenum

K emission
- $\alpha_{1,2}$ *lines*
- - metal P.55; S.278; T.41; W.28
- - compounds F.22,23,28; M.19,51
- - oxide P.55; S.278
- α *satellites* N.129; S.36; Z.5
- $\beta_{1,3}$ *lines*
- - metal and oxide P.55; S.278
- - compounds F.22,23,28; M.19,51,172
- - theory C.92

K absorption
- *XANES*
- - metal M.177
- - compounds B.110;K.205
- - glasses R.27
- - theory M.51
- *EXAFS*
- - alloys T.31
- - compounds C.43; K.206; S.45; T.26
- - organic compounds B.185; C.95; S.152; T.85
- - clusters T.30
- - salt solutions C.95; S.152
- - theory H.54

L emission
- $\alpha_{1,2}\beta_1$ *lines*
- - metal B.141; D.12; K.130; N.100; P.80
- - alloys B.191
- - compounds F.24-26; K.130,141; N.2
- - theory C.92
- $\beta_2\gamma_1$ *bands*
- - metal H.80; K.130; N.95
- - alloys B.191; N.64,83,92; S.107; Y.12
- - compounds B.34,134; K.130; N.54,83; S.14,295
- - silicides N.95
- - sulfides D.105; H.46,47; K.199
- - theory A.57; B.138; K.44
- *satellites* D.30; K.150

L absorption
- *XANES*
- - metal K.43,46,128; S.230
- - compounds B.134; K.70
- - oxide and sulfide K.43,46; O.14; S.230
- - theory B.138; K.44
- *EXAFS*
- - theory H.54

M emission
- metal D.12,F.61,64; H.80; N.95; S.146; W.41; Z.19
- alloys N.83,92
- carbide and nitride N.83
- silicides F.61,64; N.95; S.146

M absorption
- metal Z.8

N emission
- metal and silicides F.61,64; H.80; S.146; Z.19
- theory A.57

Isochromats
- metal A.95; M.106,130

43 Technetium

K emission M.172
L emission K.100
- theory N.106,107
M and N emission
- theory N.106,107

44 Ruthenium

K emission
- $\alpha_{1,2}\beta_{1,3}$ *lines*
- - compounds L.42; M.19

K absorption
- *XANES*
- - compounds B.109
- *EXAFS*
- - alloys H.57
- - catalysts A.92; G.75; L.101
- - polysterene complexes B.118
- - theory J.33

L emission
- $\alpha_{1,2}\beta_1$ *lines*
- - metal D.12, P.80
- $\beta_2\gamma_1$ *bands*
- - metal B.138; S.168
- - alloys N.91
- - complex compounds M.183; P.38,41; S.18
- - theory N.101,106,107

L absorption
- metal S.168
- complex compounds P.41; S.114,266,268
- theory B.138

M emission
- metal D.12
- alloys N.91
- theory N.101,106,107

N emission
- metal F.61,64
- theory N.101,106,107

45 Rhodium

K absorption
- compounds and catalysts G.75; R.31; V.22

L emission
- $\alpha_{1,2}\beta_1$ *lines*
- - metal B.141; D.12; N.100; P.80
- α *satellites* J.34; S.136
- $\beta_2\gamma_1$ *bands*
- - metal E.14; H.56; K.43,131; M.56; S.168
- - complex compounds F.29; M.80; N.50
- - metallic glasses M.56
- - theory A.56; K.44; N.101; P.56

L absorption
- *XANES*
- - metal E.14; K.43,131; S.117,168
- - complex compounds M.80; N.50; S.266,268; V.47

- - oxide K.43
- - theory A.56; B.138; K.44
- *EXAFS*
- - compounds B.30
- - catalysts K.125
M emission
- metal D.12
- theory A.56
N emission
- metal F.61,64
Isochromats
- alloys F.52; K.97; M.106,132; U.4,5

46 Palladium

K absorption
- *XANES*
- - alloys M.30
- *EXAFS*
- - alloys L.51
L emission
- $\alpha_{1,2}\beta_1$ *lines*
- - metal B.141; N.100
- β_1 *line*
- - metal and compounds G.25; P.80
- $\gamma_{2,3}$ *lines*
- - metal O.10; P.80
- $\beta_2\gamma_1$ *bands*
- - metal B.174; G.39; M.56; N.80,91; P.65
- - alloys B.1,153,238; H.56; K.196; M.30;
  N.60,62,63,67,70,73,78,91
- - compounds K.165; M.93; N.40,49; S.262
- - silicides T.12
- - metallic glasses M.56
- - theory A.57; K.44; N.53
L absorption
- *XANES*
- - metal B.76; D.22; G.39; S.117,168
- - alloys B.1, C.87; H.18,56; M.30,52; S.51,117
- - complex compounds N.40,49; S.265,267,269
- - oxide B.76; S.262
- - silicides D.21; R.73; S.51
- - chloride S.117,262
- - theory B.138; K.44; N.53
- *EXAFS*
- - clusters D.22; S.247
M emission
- theory A.57
M absorption
- metal Z.8
- theory A.57; M.178
N emission
- metal F.61,64
Isochromats
- alloys F.52; K.97; M.106,132; S.51; U.4,5

47 Silver

K emission
- metal B.125; K.66; T.41

- compounds M.17; P.55
- *satellites* S.139
K absorption
- metal and alloys K.157
- theory D.118
L emission
- metal B.1,125,141; D.12; H.56; K.43,48;
  N.3,78,80,100; P.80
- alloys B.1; D.122; H.56; N.78,80
- compounds D.87,88; N.3
- theory K.44; M.103
- *satellites* C.37; D.43; H.17; N.127; S.139
L absorption
- *XANES*
- - metal K.43,48,157; S.117
- - alloys C.87; K.157
- - compounds K.43,48
- - theory K.44
- *EXAFS*
- - alloys D.7
- - iodides B.203; D.3
- - borate glasses B.86; D.6
- - clusters G.75; M.152; S.247
- - impurities W.21
M emission
- metal D.12
M absorption
- metal Z.8
N absorption
- metal H.3
Isochromats E.5; R.35

48 Cadmium

K emission
- α_1 *line*
- - compounds M.16,19,20
K absorption
- metal D.4; H.73; I.31; T.44; W.19,20
L emission
- metal B.141; D.12,85,87; G.19; N.80,100; P.80
- compounds D.85,87; S.262
- *satellites* R.14
L absorption
- *XANES*
- - metal D.85,87; K.43; N.161; S.120
- - compounds D.85,87; S.262
- - telluride N.161
- *EXAFS*
- - tellurides B.25
- - bromide solutions S.4
M emission
- metal D.12
M absorption
- metal C.81
N absorption
- telluride C.11

49 Indium

K emission
- α_1 *line*
- – compounds M.21
- – oxide P.54
K absorption
- metal S.117,120
- solid solutions M.138
- superlattices C.7
L emission
- metal D.12,52; N.80; P.79,81
- compounds G.106; N.141
- antimonide D.52
L absorption
- metal and alloys G.105
- compounds G.106
- selenide A.68
M emission
- metal D.12
N absorption
- compounds C.10

50 Tin

K emission
- metal N.56
- compounds D.11; M.18,19; S.129
- oxide G.58; P.54; S.278,279
- theory A.44; C.92; K.170
L emission
- metal D.12; L.30; N.80; P.80; W.6
- alloys W.6; Z.1
- compounds G.53; W.6; Z.1
- theory F.25; K.170; N.149
L absorption
- metal B.15; O.39; S.117; T.69
- alloy G.94
- oxide B.15; T.69
- aqueous solutions Y.4
M emission
- metal D.12; J.15
- theory K.170
N absorption
- metal H.3
N and O spectra S.143; Z.27

51 Antimony

K emission
- α_1 *line*
- – metal R.80
- – oxides P.54
L emission
- metal D.12; N.80; P.80
- compounds D.85; G.20; J.11
- theory N.149
L absorption
- compounds D.5; G.103,105; S.19

M emission
- metal D.12
N absorption
- metal E.13; S.19
- alloys E.13
- compounds C.10

52 Tellurium

K emission
- α_1 *line*
- – compounds M.18; P.54; S.278
K absorption
- metal S.278
- oxide S.61,278
- tellurides S.61
L emission
- metal L.30
- tellurides M.139
L absorption
- metal and compounds T.13
- alloys B.25; F.47; R.39
M emission
- metal D.12; S.106
N emission
- tellurides T.34,36
N absorption
- metal and alloys C.13; S.209,210; Z.27

53 Iodine

K emission
- compounds M.18,147; R.56
K absorption
- element and compounds C.59; R.4
- solid solution B.206
L emission
- compounds L.30; R.56
L absorption
- iodides and adsorbed iodine C.61; F.93; J.27,28; K.172
- SbSI D.5
M emission
- iodide D.12
M absorption
- iodide E.17
N emission
- iodide I.23; Z.28
N absorption
- element B.220; C.12; P.50
- iodides B.220; C.12; I.6,23; P.50; Z.28

54 Xenon

K emission
- compounds M.148
- theory M.19

M emission
- oxide E.32
M absorption
- oxide K.26
N absorption
- metal G.83

66 Dysprosium

L emission D.42
L absorption
- compounds G.11; I.17
M emission
- metal B.176
- oxide D.30
M absorption
- metal B.176
N emission
- intermetallics S.189

68 Erbium

L emission D.2; N.122,125
L absorption K.207
N absorption G.43

69 Thulium

K emission
- $\alpha_{1,2}$ *lines*
- - metal B.184
- - theory S.278
L absorption
- compounds H.2; L.18; V.45
- intermetallics P.76
M absorption
- compounds K.5
N emission G.56
N absorption G.43

70 Ytterbium

K emission
- oxide and fluoride L.42
- theory S.278
L emission
- compounds B.149,150; T.69; V.46
- *satellites* N.122,125
L absorption
- metal N.122,125; R.73; S.289
- alloys F.73; M.27; P.22
- compounds B.149,150; D.46; F.20;: H.89
- oxides N.122,125; P.22
- silicides R.73
- intermediate-valence compounds B.50,186;
  K.29; R.23
- solutions L.46

M emission
- metal B.180
- oxide L.28
M absorption
- metal and compounds C.77
- dilute mixed-valence materials K.2
N emission
- metal M.141
N absorption
- theory G.43

71 Lutetium

K emission
- α_1 *line*
- - oxide S.277
L emission N.122,125; W.70
N and O emission F.61; M.141; N.104; S.147
Iosochromats M.131

72 Hafnium

K emission
- α_1 *line*
- - oxide S.277
L absorption
- metal J.18
- selenide D.13
M emission B.247
N emission F.61; M.141; N.104; S.147
O emission N.104

73 Tantalum

K emission
- α_1 *line*
- - oxide S.277
- $\beta_{1,3}$ *lines*
- - metal N.56
L emission
- metal R.13,22
- compounds Z.11
- carbides R.22
- gallides D.125
L absorption
- metal L.100
- compounds G.79; P.72; S.97
- sulfide H.51; S.236
M emission
- metal D.12
N emission F.61; M.141; S.147
O emission F.61; S.147
O absorption H.5

74 Tungsten

K emission
- α_1 *line*

- - metal K.65
- - oxide S.277
- - chlorides K.142
L emission
- metal G.62
- oxide and sulfide G.18
- *satellites* N.124
L absorption
- metal J.18; P.4; R.16
- clusters T.30
- oxide G.18; H.91; K.70; T.14
- sulfide G.18
- selenide H.51
- carbonyle K.70
M emission
- metal D.12
- *satellites* K.176
N emission F.61; M.141
O emission F.61
O absorption H.5
Isochromats B.161; K.97; M.132; N.150; U.5

75 Rhenium

L absorption
- metal P.4; R.16
- oxide and alloys I.17
- compounds C.74; M.38,39; P.5,63; S.59; T.69
- perrhenates F.7
- catalysts Y.18
- clusters G.75
- solution S.115
N emission
- metal F.61; S.143
- theory N.105
O emission
- metal F.61; S.143
O absorption
- element H.5

76 Osmium

L absorption
- metal P.4; R.16
- compounds N.1; R.6
- clusters B.88; D.116; G.75; J.32; L.100,102
N and O emission
- metal F.61; S.143

77 Iridium

K emission
- α_1 line
- - metal N.56
L emission
- metal C.18; G.61
L absorption
- metal P.4; R.16; S.47

- compounds H.90; T.60
- oxide R.7
- catalysts L.100,102
- clusters G.75
N emission
- metal F.61
- theory N.105
O emission
- metal F.61
Isochromats
- metal E.3; R.36

78 Platinum

K emission
- metal A.60
- theory A.44
L emission
- compounds F.27; M.179; N.53
L absorption
- metal B.208; J.18; M.5; P.4; R.33; S.47,239
- alloys B.208; C.66; L.52
- compounds C.8; F.27; H.90; L.100,102; M.179; P.7,23
- complex compounds C.74; D.41; O.21; S.58,117; T.17
- organic compounds T.28,60
- catalysts B.58; K.125; L.100,102; M.57
- zeolites G.6; L.67; M.159
- metallic glasses P.32
- clusters G.75; K.52,53; M.57
- interfaces O.54; R.73,74
- theory B.123
M emission
- metal D.12
- catalysts F.83
M absorption
- silicides C.8
N emission
- metal D.69; F.61; P.1
- theory A.43; N.105; W.26
O emission
- metal F.61
O absorption
- metal H.5
Isochromats E.3; R.36

79 Gold

K emission
- metal V.52
- theory A.44
L emission
- metal B.125; C.18
L absorption
- metal B.20,208; J.18; L.97,100,102; P.4;R.3,4
- alloys A.99; B.208; C.88; M.30,75; P.22; Y.21
- compounds D.41; T.18,60
- clusters B.19,20,48; C.71; G.75; T.61; Z.4

M emission
- metal D.12
M absorption
- metal C.5; K.45
- alloy C.5
N emission
- metal M.95
- alloys C.21; F.64,88; K.16; M.95,175
- theory A.43; N.105; W.26
N absorption
- metal G.93; H.3
O emission
- metal F.61
O absorption
- metal G.93; H.3
Isochromats
- metal E.3; R.36

80 Mercury

L emission
- metal and sulfide K.34
L absorption
- alloys B.191
- compounds A.11,12; D.60
M emission
- metal D.12

81 Thallium

K emission
- metal A.60
L absorption
- compounds A.11,12; D.60
M emission
- metal D.12
M absorption
- metal K.45

82 Lead

K emission
- $\alpha_{1,2}$ *lines*
- - metal C.6; L.41
L emission
- α_1 *line*
- - metal T.4
L absorption
- compounds D.60; R.26; V.40,41
- organic complex compounds S.142,224,229
- oxides B.14; W.37
M emission
- metal D.12
M absorption
- metal K.45
- chloride S.260
- theory K.48

N emission
- oxide D.90
- sulfide S.259
O absorption
- compounds C.11,13

83 Bismuth

L absorption
- alloys L.52
- compounds D.60; K.12; V.40,41
- oxide B.23; K.152
- halides S.57
- glasses E.20
M emission
- metal D.12
M absorption
- metal K.45
- theory K.48
N and O emission
- oxide D.90
N and O absorption
- metal H.3; S.19
- alloys E.13

88 Radium

K emission
- metal B.44 (until 98 Californium)

90 Thorium

K emission
- $\alpha_{1,2}$ *lines*
- - metal B.184
L emission
- metal A.59
- compounds V.23
L absorption
- alloys L.52
- fluoride glasses R.26
M emission
- metal D.12
M absorption
- metal and oxide C.24; P.51
- oxide and nitrate B.31
O emission L.96; S.143
O absorption C.104; L.96; S.143; W.18

91 Protactinium

M absorption
- metal and oxide C.24

92 Uranium

K emission
- metal B.184
- compounds L.45
K absorption
- compounds L.45
L emission
- compounds C.18; V.23
L absorption
- compounds C.31; K.6; O.27; T.42
- oxides B.122,190; J.23; M.41; P.51
- intermetallics M.136
- solutions M.41
M emission
- metal B.179,181; D.12; K.60,100; O.12
- oxides B.179,181; N.22
M absorption
- metal B.181; C.24
- oxides B.181,190; C.24; P.51
- intermetallics K.6; M.136
N absorption
- compounds K.6
O emission
- oxides L.96
O absorption
- metal C.104
- compounds K.6; L.96

93 Neptunium

K emission
- metal N.55
L emission
- metal D.25
- compounds V.23
M emission
- metal D.25; K.100
Isochromats
- compounds B.12

94 Plutonium

K emission
- metal B.184; N.55
L spectra
- oxide C.22
M emission
- metal B.181; K.100
M absorption
- metal and oxide B.181; C.24

95 Americium

K emission
- metal N.55
M emission
- metal K.100

Literature

A.1   T. Åberg, J. Utriainen: Phys. Rev. Lett. 22, 1346 (1969)
A.2   T. Åberg, G. Graeffe, J. Utriainen, M. Linkoaho: J. Phys. C 3, 1112 (1970)
A.3   T. Åberg, J. Utriainen: J. Physique 32, C 4-295 (1971)
A.4   T. Åberg, J.L. Dehmer: J. Phys. C 6, 1450 (1973); ibid. 7, L 278 (1974)
A.5   H. Adachi, K. Taniguchi: Technol. Rep. Osaka Univ. 30, 365 (1980); J. Phys.
       Soc. Jpn. 49, 1944 (1980)
A.6   M.Y. Adam et al.: Jpn. J. Appl. Phys. 17, S 2-170 (1978)
A.7   E. Adelson, A.E. Austin: Adv. X-Ray Anal. 12, 507 (1969)
A.8   E. Adelson, A.E. Austin: J. Solid State Commun. 7, 1819 (1969)
       A.E. Austin, E. Adelson: Solid State Chem. 1, 229 (1970)
A.9   S.V. Adhyapak, A.S. Nigavekar: J. Phys. Chem. Solids 37, 1037 (1976); ibid.
       39, 171 (1978)
A.10   B.K. Agarwal, M.P. Givens: J. Phys. Chem. Solids 6, 178 (1958)
A.11   B.K. Agarwal: J. Phys. C 1, 1658 (1968); ibid. 3, 535 (1970)
A.12   B.K. Agarwal, L.P. Verma: J. Phys. C 1, 208 (1968); ibid. 2, 104 (1969)
       A.K. Dey, B.K. Agarwal: Lett. Nuovo Cim. 2, 947 (1971)
A.13   B.K. Agarwal, R.B. Singh: Lett. Nuovo Cim. 4, 765 (1972)
A.14   B.K. Agarwal, C.B. Bhargava, A.N. Vishnoi, V.P. Seth: J. Phys. Chem. Solids
       37, 725 (1976)
A.15   B.K. Agarwal, R.K. John: Phys. Stat. Sol. (b) 88, 309 (1978)
       R.K. Johri, B.K. Agarwal: J. Phys. F 8, 555 (1978)
A.16   B.K. Agarwal, V. Balakrishnan: J. Phys. F 12, 1519 (1982)
A.17   B.R.K. Agarwal, L.P. Verma, B.K. Agarwal: Lett. Nuovo Cim. 2, 581 (1971)
A.18   R.M. Agarwal, A.N. Nigam: Proc. Indian Acad. Sci. 63, 200 (1966)
A.19   O.P. Aggarwal, N.N. Saxena: Phys. Stat. Sol. (b) 122, 669 (1984)
A.20   A.K. Agnihotri, A.N. Nigram: Indian J. Phys. 61A, 236 (1987)
A.21   H. Ågren, S. Svensson, U.I. Wahlgren: Chem. Phys. Lett. 35, 336 (1975)
A.22   H. Ågren, J. Nordgren, L. Selander, C. Nordling, K. Siegbahn: J. Electron
       Spectrosc. 14, 27 (1978)
       H. Ågren, J. Muller: J. Chem. Phys. 72, 4078 (1980)
A.23   M. Ahonen, G. Graeffe: Phys. Fenn. 9, S 1-421 (1974)
A.24   J. Ahopelto, E. Rantavuori, O. Keski-Rahkonen: Phys. Scr. 20, 71 (1979)
A.25   O. Aita, T. Sagawa: J. Phys. Soc. Jpn. 27, 164 (1969)
A.26   O. Aita, I. Nagakura, T. Sagawa: J. Phys. Soc. Jpn. 33, 750 (1972); ibid. 34,
       1112 (1973)
A.27   O. Aita, K. Ichikawa, H. Nakamura, Y. Iwasaki, K. Tsutsumi: Jpn. J. Appl.
       Phys. 17, 595 (1978)
A.28   O. Aita, K. Tsutsumi. K. Ichikawa: Phys. Rev. B23, 5676 (1981)
A.29   C. Aita, T. Watanabe, Y. Fujimoto, K. Tsutsumi: J. Phys. Soc. Jpn. 51, 483
       (1982)
A.30   O. Aita, K. Ichikawa, M. Okusawa, K. Tsutsumi: Phys. Rev. B 34, 8230
       (1986)

A.31 O. Aita et al.: J. Phys. Soc. Jpn. **56**, 649 (1987)
A.32 K. Akimoto: Appl. Phys. Lett. **41**, 49 (1982)
A.33 V.N. Akimov, A.S. Vinogradov, A.A. Pavlychev, V.N. Sivkov: Opt. Spektrosk. **59**, 342 (1985)
A.34 R.G. Akopdzhanov: Fiz. Met. Metalloved. **24**, 245 (1967); Kin. Katalis **8**, 91, (1967); Fiz. Tverd. Tela **12**, 1393 (1970); Deposited Doc. VINITI 2087 (1978)
A.35 R.G. Akopdzhanov: Fiz. Tverd. Tela **18**, 882 (1976)
A.36 S. Aksela, M. Karras: Chem. Phys. Lett. **20**, 356 (1973)
A.37 S. Aksela, M. Karras, E. Kortela, E. Suoninen: In [Ref.1.60, Vol.I, p.352]
A.38 L. Alagna, A. Bianconi, A. Desideri, A.G. Tomlinson: Ital. J. Biochem. **29**, 73 (1980)
A.39 L. Alagna, A.A.G. Tomlinson: J. Chem. Soc., Faraday Trans. I **78**, 3009 (1982)
     L. Alagna et al.: J. Chem. Soc., Faraday Trans. I **79**, 1039 (1983)
A.40 L. Alagna, T. Prosperi, A.G. Tomlinson, R. Rizzo: J. Phys. Chem. **90**, 6853 (1986)
A.41 L. Alagna, R.W. Strange, P. Durham, S.S. Hasnain: Preprint DL/SCI/P 509 E, Daresbury 1986
A.42 R.C. Albers, A.K. McMahon, J.E. Muller: Phys. Rev. B **31**, 3435 (1985)
A.43 R.C. Albers, A.M. Boring, P. Weinberger, N.E. Christensen: Phys. Rev. B **32**, 7571 (1985)
A.44 K. Alder, G. Baur, U. Raff: Helv. Phys. Acta **45**, 765 (1972)
A.45 I.P. Aleksandrova, D.N. Shigorin, A.P. Skoldinov: Zh. Fiz. Khim. **38**, 1203 (1964)
A.46 V.G. Aleshin, V.P. Smirnov, M.S. Nakhmanson: Fiz. Tverd. Tela **10**, 1585 (1968); ibid. **11**, 2010 (1969)
A.47 V.G. Aleshin, V.P. Smirnov: In [Ref.1.58, Vol.I, p.314]
     V.G. Aleshin et al.: Metallofizika **66**, 57 (1976)
A.48 V.G. Aleshin: Dopov. Akad. Nauk Ukr. SSR, Ser. A 1976, 647
A.49 M.A.M. Al-Kadier, C.J. Nicholls, D.S. Urch: Iraqi J. Sci. **24**, 109 (1983)
A.50 F.K. Allotey: Phys. Rev. **157**, 467 (1967); Solid State Commun. **9**, 91 (1971)
A.51 F.K. Allotey: In [Ref.1.79, p.361]
A.52 C.O. Almbladh, U. von Barth: J. Phys. C **8**, 4117 (1975); Phys. Rev. B **13**, 3307 (1976)
     U. von Barth, G. Grossmann: Solid State Commun. **32**, 645 (1979)
     U. von Barth, G. Grossmann: Phys. Scr. **28**, 107 (1983)
A.53 C.O. Almbladh, U. von Barth: In [Ref.1.87, p.15]
A.54 C.O. Almbladh: Phys. Rev. B **16**, 4343 (1977)
A.55 M. Alouani, J.M. Koch, M.A. Khan: Solid State Commun. **60**, 657 (1986)
A.56 M. Alouani, M.A. Khan: J. Phys. F **17**, 519 (1987)
A.57 G.I. Alperovich et al.: Phys. Stat. Sol. (b) **54**, K 127 (1972)
     I.I. Geguzin et al.: Izv. Vyssh, Uchebn. Zaved, Fiz. **16**, 103 (1973)
     I.Ya. Nikiforov, I.I. Geguzin: Phys. Fenn. **9**, S 1-186 (1974)
     G.I. Alperovich et al.: Izv. Akad. Nauk SSSR, Ser. Fiz. **40**, 251 (1976)
A.58 G.I. Alperovich et al.: Akad. Nauk SSSR. Ser. Fiz. **41**, 224 (1977)
     A.N. Gusatinskii, G.I. Alperovich, A.V. Soldatov: Phys. Stat. Sol. (b) **112**, 599 (1982)
A.59 P. Amorim, L. Salgueiro, F. Parente, J.G. Ferreira: Nucl. Instrum. Methods Phys. Res. A**255**, 56 (1987)
A.60 P. Amorim, F. Parente, L. Salgueiro, J.G. Ferreira: J. Phys. B **20**, L 295 (1987)
A.61 G. Andermann, H.C. Whitehead: Adv. X-Ray Anal. **14**, 453 (1971)

F. Fujiwara, G. Andermann: Spectrosc. Lett. **13**, 211 (1980)

A.62   G. Andermann, R.A. Lynch, H.C. Whitehead, K. Myers: In [Ref.1.60, Vol.I, p.283]

R.A. Lynch, G. Andermann: Phys. Fenn. **9**, S 1-138 (1974)

D.R. Phillips, G. Andermann: Phys. Fenn. **9**, S 1-141 (1974)

A.63   G. Andermann, B. Henke, D.S. Urch, G. Wiech: Jpn. J. Appl. Phys. **17**, S 2-428 (1978)

A.64   G. Andermann, L. Bergknut, M. Karras, G. Grieshaber: Spectrosc. Lett. **11**, 571 (1978)

G. Andermann et al.: Rev. Sci. Instrum. **51**, 814 (1980)

A.65   N.S. Andrushchenko, Yu.P. Kostikov: Mikrochim. Acta 1974, 783

V.N. Pak, Yu.P. Kostikov et al.: Kin. Katalis **15**, 1358 (1974)

A.G. Avramenko et al.: Zh. Fiz. Khim. **49**, 1339 (1975)

A.66   V.I. Anisimov, V.A. Gubanov, E.Z. Kurmaev: Zh. Strukt. Khim. **21**, 46 (1980)

A.67   V.I. Anisimov, A.V. Postnikov, E.Z. Kurmaev, G. Wiech: Fiz. Met. Metalloved. **62**, 730 (1986)

A.68   F. Antonangeli et al.: Physica B **105**, 25 (1981)

A.69   F. Antonangeli et al.: Phys. Rev. B **32**, 6644 (1985)

A.70   G.M. Antonini et al.: Europhys. Lett. **4**, 851 (1987)

A.71   G. Antonioli et al.: In Proc. 7th Int'l Conf. Ternary and Multinary Compounds. Mater. Res. Soc. Pittsburgh, ed. by S.K. Deb, A. Zunger, (1987) p.149

A.72   R.A. Antonyuk, I.V. Kavich: Ukr. Fiz. Zh. **20**, 1343 (1975)

R.A. Antonyuk: Ukr. Fiz. Zh. **30**, 294 (1985)

A.73   A.I. Antoshchuk et al.: Fiz. Tverd. Tela **22**, 2826 (1980)

A.74   G. Apai, J.F. Hamilton, J. Stöhr, A. Thompson: Phys. Rev. Lett **43**, 165 (1979)

A.75   A. Appleton: Contemp. Phys. **6**, 50 (1964); Phil. Mag. **12**, 245 (1965)

A.76   A. Appleton, C. Curry: Phil. Mag. **16**, 1031 (1967)

A.77   M.Y. Apte, C. Mande: J. Phys. Chem. Solids **41**, 307 (1980)

M.Y. Apte, C. Mande: J. Phys. C **15**, 607 (1982)

A.78   E.T. Arakawa, M.W. Williams: Phys. Rev. B **8**, 4075 (1973); Phys. Rev. Lett. **36**, 333 (1976)

A.79   M.P. Arbuzov, E.E. Vainshtein, B.I. Kotlyar, V.V. Krasnova: Fiz. Met. Metalloved. **21**, 464 (1966)

B.I. Kotlyar et al.: Metallofizika **64**, 29 (1976)

M.P. Arbuzov, L.G. Dobrovolskaya, B.I. Kotlyar: Metallofizika **67**, 87 (1977)

A.80   M.P. Arbuzov, E.A. Zhurakovskii et al.: Izv. Akad. Nauk SSSR, Neorg. Mater. **7**, 2183 (1971)

A.81   M.P. Arbuzov, B.I. Kotlyar et al.: Izv. Vyssh. Uchebn. Zaved., Fiz. **19**, 145 (1976)

M.P. Arbuzov, P.P. Kirichok, M.M. Yatsura, O.R. Yatsura: Izv. Vyssh. Uchebn. Zaved., Fiz. **23**, 128 (1980)

A.82   M.P. Arbuzov, B.I. Kotlyar et al.: Izv. Vyssh. Uchebn. Zaved., Fiz. **19**, 152 (1976)

A.83   M.P. Arbuzov, B.I. Kotlyar, O.Kh. Tadeush, L.G. Dobrovolskaya: In [Ref.1.100, p.127]

A.84   K. Arisawa: J. Phys. Soc. Jpn. **42**, 1783 (1977)

A.85   M.V. Artemko et al.: Zh. Neorg. Khim. **17**. 1006, 1009 (1972)

A.86   D. Arvanitis et al.: Surf. Sci. **178**, 686 (1986)

D. Arvanitis, K. Baberschke, L. Wenzel, U. Döbler: Phys. Rev. Lett. **57**, 3175 (1986)

A.87 E. Asada: Chem. Lett. 1974, 1467; Jpn. J. Appl. Phys. 12, 1946 (1973); ibid. 15, 1417 (1976)
      E. Asada, T. Takiguchi: Jpn. J. Appl. Phys. 17, 687 (1978)
A.88 E. Asada, T. Takiguchi, Y. Suzuki: X-Ray Spectrom. 4, 186 (1975)
A.89 S. Asada, C. Satoko, S. Sugano: J. Phys. Soc. Jpn. 38, 855 (1975)
A.90 T. Asada, T. Hoshino, M. Kataoka: J. Phys. F 15, 1497 (1985)
A.91 H. Asahina, T. Morita, Y. Nakajima: Sharp Tech. J. 36, 47 (1986)
A.92 K. Asakura, N. Kosugi, Y. Iwasawa, H. Kuroda: In [Ref.1.83, p.190]
A.93 S. Asbrink, G.N. Greaves, P.D. Hatton, K. Garg: J. Appl. Crystallogr. 19, 331 (1986)
A.94 C.A. Ashley, S. Doniach: Phys. Rev. B11, 1279 (1975)
A.95 J. Auleytner: In [Ref.1.60, Vol.II, p.383]
A.96 J. Auleytner, K. Lawniczak, E. Sobczak: Ann. Physik 32, 476 (1975)
A.97 M.I. Auslender, L.D. Finkelshtein, N.N. Efremova, M.I. Simonova: Fiz. Tverd. Tela 28, 552 (1986)
A.98 A.O. Avetisyan, Yu.M. Goryachev, B.A. Kobenskaya, T.M. Yarmola: In [Ref.1.100, p.94]
A.99 V.A. Avetisyan: Izv. Vyssh. Uchebn. Zaved., Fiz. (1964) 46
A.100 L.V. Azaroff: Mater. Res. Bull. 2, 137 (1967)
A.101 L.V. Azaroff, R.J. Donahue: J. Physique 32, C 4-312 (1971)
      D.M. Pease, F. Szmulowicz, L.V. Azaroff: Phys. Lett. A 101, 38 (1984)
A.102 L.V. Azaroff: Phys. Fenn. 9, S 1-57 (1974)
A.103 M. Azizan et al.: J. Physique 48, 81 (1987)

B.1   Yu.A. Babanov, O.B. Sokolov, M.F. Sorokina: Phys. Stat. Sol. (b) 52, 155 (1972)
B.2   Yu.A. Babanov et al.: J. Non-Cryst. Solids 79, 1 (1986)
      Yu.A. Babanov, N.V. Ershov, A.V. Serikov, V.R. Shvetsov: Fiz. Met. Metalloved. 61, 779 (1986)
B.3   G.R. Babu et al.: Nuovo Cimento 97A, Ser. 2, 124 (1987)
B.4   R.Z. Bachrach et al.: Nucl. Instrum. Methods 152, 53, 135 (1978)
B.5   J. Backovsky: Czech. J. Phys. 4, 118 (1954)
B.6   M. Bäckström, L. Petterson, N. Wassdahl, J. Nordgren: UUIP 1135 (1985)
B.7   M. Bader et al.: Phys. Rev. Lett. 56, 1921 (1986)
B.8   M. Bader, A. Puschmann, C. Ocal, J. Haase: Phys. Rev. Lett. 57, 3273 (1986)
B.9   M. Bader et al.: Phys. Rev. B 35, 5900 (1987)
B.10 R. Bader, A. Perez, C. Desuzinges, M. Romand: Appl. Surf. Sci. 24, 173 (1985)
B.11 A.R. Badzian, A. Klokocki: Phys. Stat. Sol. (b) 96, 529 (1979)
B.12 Y. Baer: Physica B+C 102, 104 (1980)
B.13 P.S. Bagus, M. Krauss, R.E. LaVilla: Chem. Phys. Lett. 23, 13 (1973)
B.14 P.S. Bagus, A.J. Freeman, F. Sasaki: Phys. Rev. Lett. 30, 850 (1973)
B.15 A.A. Bahgat, K. Das Gupta: Adv. X-Ray Anal. 23, 203 (1980)
B.16 M.K. Bahl, A.S. Koster: J. Phys. C 7, 4537 (1974); ibid. 8, 1776 (1975)
B.17 M.K. Bahl: J. Phys. C 8, 389, 4107 (1975)
B.18 R.A. Bair, W.A. Goddard: Phys. Rev. B 22, 2767 (1980)
B.19 A. Balerna et al.: Phys. Rev. B 31, 5058 (1985)
B.20 A. Balerna, S. Mobilio: Phys. Rev. B 34, 2293 (1986)
B.21 M.M. Ballal, C. Mande, J. Phys. Chem. Solids 38, 843, 1383 (1977); J. Phys. C 11, 837 (1978)
B.22 D. Bally, L. Muller: Studii Fizica 3, 437 (1962)
B.23 A. Balzarotti et al.: Solid State Commun. 35, 145 (1980)

A. Balzarotti, F. Antonangeli, R. Girlanda, G. Martin: Solid State Commun. **44**, 275 (1982)

A. Balzarotti, A.P. Menushenkov, N. Motta, J. Purans: Solid State Commun. **49**, 887 (1984)

B.24   A. Balzarotti, M. De Crescenzi: Phys. Rev. B **25**, 6349 (1982)

B.25   A. Balzarotti et al.: Phys. Rev. B **30**, 2295 (1984)

A. Balzarotti et al.: Phys. Rev. B **31**, 7526 (1985)

N. Motta et al.: Solid State Commun. **53**, 509 (1985)

B.26   I.M. Band, A.P. Kovtun, M.A. Listengarten, M.B. Trzhaskovskaya: J. Electron Spectrosc. **36**, 59 (1985)

B.27   V.I. Baranovskii, M.S. Nakhmanson: Izv. Akad. Nauk SSSR, Ser. Fiz. **36**, 389 (1972)

B.28   V.I. Baranovskii, Yu.N. Kukushkin, N.S. Panina: Koord. Khim. **3**, 166 (1977)

B.29   M. Barber, J.D. Clark, A. Hinchliffe, D.S. Urch: Preprint 1978

B.30   A.V. Barhate, A.V. Pendharkar, V.B. Sapre, C. Mande: Solid State Commun. **36**, 473 (1980)

B.31   R.L. Barinskii: Izv. Akad. Nauk SSSR, Ser. Fiz. **20**, 133 (1956)

B.32   R.L. Barinskii, E.E. Vainshtein: Dokl. Akad. Nauk SSSR **82**, 355 (1952)

B.33   R.L. Barinskii, E.E. Vainshtein, K.I. Narbutt: Izv. Akad. Nauk SSSR, Ser. Fiz. **21**, 1351 (1957)

K.I. Narbutt, I.S. Smirnova: Izv. Akad. Nauk SSSR, Ser. Fiz. **27**, 340 (1963)

L.N. Mazalov, E.E. Vainshtein, L.N. Trushnikova: Izv. Sibir. Otd. Akad. Nauk SSSR, Ser. Khim. Nauk 1966, 42

I.S. Smirnova, K.I. Narbutt: Izv. Akad. Nauk SSSR, Ser. Fiz. **31**, 927 (1967)

B.34   R.L. Barinskii, E.E. Vainshtein: Izv. Akad. Nauk SSSR, Ser. Fiz. **21**, 1387 (1957)

B.35   R.L. Barinskii, B.A. Malyukov: Izv. Akad. Nauk SSSR, Ser. Fiz. **26**, 418 (1962); ibid. **27**, 351 (1963)

B.36   R.L. Barinskii, A.E. Shvelashvili, B.A. Malyukov: Zh. Strukt. Khim. **5**, 497, 608 (1964)

B.37   R.L. Barinskii, B.A. Malyukov: Izv. Akad. Nauk SSSR, Ser. Fiz. **28**, 805 (1964)

B.38   R.L. Barinskii: In [Ref.158, Vol.II, p.222]

B.39   R.L. Barinskii et al.: Zh. Strukt. Khim. **13**, 1089 (1972); Izv. Akad. Nauk SSSR, Ser. Fiz. **38**, 516 (1974)

I.M. Kulikova et al.: Dokl. Akad. Nauk SSSR **210**, 1423 (1973)

B.40   R.L. Barinskii, I.M. Kulikova: Zh. Strukt. Khim. **14**, 372 (1973); Izv. Akad. Nauk SSSR, Ser. Fiz. **38**, 444 (1974)

B.41   R.L. Barinskii, I.M. Kulikova: Izv. Akad. Nauk SSSR, Ser. Fiz. **40**, 279 (1976); Phys. Chem. Miner. **1**, 325 (1977)

B.42   R.L. Barinskii, I.M. Kulikova, E.P. Shevchenko: Izv. Akad. Nauk SSSR, Ser. Fiz. **46**, 724 (1982)

B.43   R.L. Barinskii, E.P. Shevchenko, I.M. Kulikova, A.I. Likhtenshtein: Dokl. Akad. Nauk SSSR **274**, 562 (1984)

B.44   G. Barreau, H.G. Borner, T. von Egidy, R.W. Hoff: Z. Physik A **308**, 209 (1982)

B.45   D.M. Barrus, R.L. Blake, A.J. Byrek, K.C. Chambers, A.L. Pregenter: Phys. Rev. A **20**, 1045 (1979)

B.46   J. Barth, C. Kunz, T.M. Zimkina: Solid State Commun. **36**, 453 (1980)

B.47   R.D. Bartosevich, N.D. Savchenko, V.P. Tsvetkov: Ukr. Fiz. Zh. **11**, 424 (1966)

B.48   I.W. Bassi, F.W. Lytle, G. Parravano: J. Catal. **42**, 139 (1976)

B.49   V.A. Batyrev, A.V. Shatunova: Izv. Akad. Nauk SSSR, Ser. Fiz. **31**, 883 (1967)
I.B. Borovskii et al.: Izv. Akad. Nauk SSSR, Ser. Fiz. **31**, 887 (1967)

B.50   K.R. Bauchspiess et al.: In *Valence Fluctuations in Solids*, ed. by L.M. Falicov, W. Hanke, M.B. Maple (North-Holland, Amsterdam 1981) p.417

B.51   W.L. Baun, D.W. Fischer: Nature **204**, 642 (1964); Spectrochim. Acta **21**, 1471 (1965)

B.52   W.L. Baun, D.W. Fischer: Adv. X-Ray Anal. **8**, 371 (1965)
D.W. Fischer, W.L. Baun: J. Appl. Phys. **36**, 534 (1965); ibid. **38**, 2404 (1967)

B.53   W.L. Baun, D.W. Fischer: J. Appl. Phys. **38**, 2092 (1967)

B.54   W.L. Baun: J. Appl. Phys. **40**, 4210 (1969); Adv. X-Ray. **13**, 49 (1970)

B.55   W.L. Baun, T.J. Wild, J.S. Solomon: J. Electrochem. Soc. **123**, 72 (1976)

B.56   N.V. Bausk, L.N. Mazalov, S.V. Larionov: Zh. Strukt. Khim. **27**, No.3, 163 (1986)

B.57   D.V. Baxter: J. Non-Cryst. Solids **79**, 41 (1986)

B.58   D. Bazin, H. Dexpert, P. Lagarde, J.P. Bournonville: In [Ref.1.83, p.195]

B.59   E. Beauprez et al.: Phys. Rev. **B34**, 886 (1986)

B.60   E. Beaurepaire, J.P. Kappler, G. Krill: Solid State Commun. **57**, 145 (1986)

B.61   D.E. Bedo, D.H. Tomboulian: Phys. Rev. **95**, 621 (1954)

B.62   M.J. Bedzyk, G. Materlik: Phys. Rev. B **32**, 4228 (1985)

B.63   V.L. Bekenev, A.A. Lisenko, A.A. Ostroukhov: Fiz. Met. Metalloved. **41**, 301 (1976)

B.64   V.L. Bekenev, A.A. Lisenko, E.A. Zhurakovskii: Izv. Vuz. Fiz. **28**, 72 (1985)

B.65   V.P. Belash, S.A. Nemnonov, E.Z. Kurmaev: Fiz. Met. Metalloved. **33**, 1324 (1972)

B.66   V.P. Belash, E.Z. Kurmaev, A.S. Menshikov, S.A. Nemnonov: Izv. Akad. Nauk SSSR, Ser. Fiz. **38**, 658 (1974)
E.Z. Kurmaev, V.P. Belash et al.: Solid State Commun. **16**, 1139 (1975)
S.A. Nemnonov, K.M. Kolobova, V.A. Trofimova, I.N. Shabanova: Fiz. Met. Metalloved. **54**, 412 (1982)

B.67   E. Belin, C. Senemaud, C. Bonnelle: J. Non-Cryst. Solids **27**, 119 (1978)

B.68   E. Belin et al.: J. Physique C 8-427 (1980)

B.69   E. Belin, C. Bonnelle, S. Zuckerman, F. Machizaud: J. Phys. F **14**, 625 (1984)

B.70   E. Belin, C. Cardinaud, C. Senemaud, J. Dixmier: J. Non-Cryst. Solids **77-78**, 331 (1985)

B.71   E. Belin, M. Gupta, P. Zolliker, K. Yvon: J. Less-Common Met. **130**, 267 (1987)

B.72   E. Belin: J. Phys. F **17**, 1913 (1987)

B.73   M. Belli et al.: Ital. J. Biochem. **29**, 77 (1980); Solid State Commun. **35**, 355 (1980)

B.74   G.L. Bendazzoli, P. Palmieri, C. Zauli: Boll. Sci. Fac. Chim. Ind. Bologna **22**, 97 (1964)

B.75   G.L. Bendazzoli: Theoret. Chim. Acta **36**, 77 (1974)

B.76   M. Benfatto et al.: Solid State Commun. **46**, 367 (1983)
I. Davoli et al.: Solid State Commun. **48**, 475 (1983)

B.77   M. Benfatto et al.: Phys. Rev. B **34**, 5774 (1986)

B.78   M. Benfatto et al.: In *Proceedings of the NATO Advanced Study Institute Amorphous and Liquid Materials*, ed. by E. Lüscher, G. Fritsch, G. Jacucci (Nijhoff, Dordrecht 1987) p.142

B.79   G. Beni, P.M. Platzman: Phys. Rev. B **14**, 1514 (1976)

B.80   O. Benka, R.L. Watson: Phys. Lett. A **94**, 143 (1983)
O. Benka, M. Uda: Phys. Rev. Lett. **56**, 54 (1986)

B.81   B. Bergersen, P. Jena, T. McMullen: J. Phys. F **4**, L219 (1974)

B.82   S. Bergwall: Z. Physik 187, 495 (1965)
B.83   S. Bergwall, M. Elango: Phys. Lett. 24 A, 230 (1967)
B.84   S. Bergwall, A.S. Nigavekar: Phys. Rev. 175, 33 (1968)
       A.S. Nigavekar, S. Bergwall: J. Phys. B 2, 507 (1969)
B.85   S. Bergwall, A.S. Nigavekar: Physik Kondens. Materie 10, 107 (1969)
B.86   E. Bernieri et al.: Solid State Commun. 48, 421 (1983)
B.87   E. Bernieri, E. Burattini: Phys. Rev. A 35, 3322 (1987)
B.88   B. Besson, B. Moraweck et al.: J. Chem. Soc. Chem. Commun. 1980, 569
B.89   P.E. Best: J. Chem. Phys. 44, 3248 (1966)
B.90   P.E. Best: J. Chem. Phys. 47, 4002 (1967)
B.91   P.E. Best: J. Chem. Phys. 49, 2797 (1968)
B.92   P.E. Best: J. Chem. Phys. 54, 1512 (1971)
B.93   P.E. Best: Phys. Rev. B3, 4377 (1971)
B.94   P.E. Best, C.C. Chu: Jpn. J. Appl. Phys. 17, S 2-317 (1978)
B.95   M.G. Betti et al.: Phys. Scr. 36, 153 (1987)
B.96   C. Beyreuther, G. Wiech: Phys. Fenn. 9, S 1-168 (1974)
B.97   C.B. Bhargava, A.N. Vishnoi, U.C. Srivastava: Indian J. Pure Appl. Phys. 14,
       945 (1976)
B.98   C.B. Bhargava, A.N. Vishnoi, U.C. Srivastava: Indian J. Chem. 16 A, 774
       (1978)
B.99   O.K. Bhaskare, S.Y. Kulkarni: Proc. Indian Acad. Sci. Chem. Sci. 97, 25
       (1986)
B.100 N.V. Bhat: Spectrochim. Acta B 28, 257 (1973)
B.101 N.V. Bhat, P.H. Umadikar, D.H. Gohil, S.V. Salvi: Proc. Nucl. Phys. Solid
       State Phys. Symp. 17 C, 147 (1974)
B.102 N.V. Bhat, A. Syamal: J. Mol. Struct. 30, 161 (1976)
B.103 N.V. Bhat, A. Syamal, S.V. Salvi, P.H. Umadikar: Spectrochim. Acta 35 B,
       489 (1980)
B.104 P. Bhattacharya, A.R. Chetal: Phys. Stat. Sol. (b) 119, 179 (1983); J. Phys.
       Chem. Solids 45, 519 (1984)
       P. Bhattacharya, A.R. Chetal: Physica 133 B+C, 87 (1985)
B.105 V.G. Bhide, N.V. Bhat: J. Chem. Phys. 48, 3103 (1968); ibid. 50, 42 (1969)
       N.V. Bhat: Acta Crystallogr. A 27, 71 (1971)
B.106 V.G. Bhide, N.V. Bhat, K.R. Rambhad: J. Appl. Phys. 39, 4744 (1968)
       V.G. Bhide, M.K. Bahl: J. Phys. Chem. Solids 32 1001 (1971)
B.107 V.G. Bhide, N.V. Bhat: J. Appl. Phys. 41, 3159 (1970)
B.108 V.G. Bhide, M.K. Bahl: J. Phys. Chem. Solids 33, 285, 1669 (1972); ibid. 34,
       667 (1973); J. Phys. C 6, 2214 (1973)
B.109 V.G. Bhide, S.K. Kaicker: J. Phys. Chem. Solids 35, 695 (1974)
B.110 V.G. Bhide, S.K. Kaicker, M.K. Bahl: J. Phys. Chem. Solids 35, 901 (1974)
B.111 V.G. Bhide, B.A. Patki: Pramana 2, 290 (1974)
B.112 A. Bianconi, S.B.M. Hagström, R.Z. Bachrach: Phys. Rev. B 16, 5543 (1977)
B.113 A. Bianconi, D. Jackson, K. Monahan: Phys. Rev. B 17, 2021 (1978)
       A. Bianconi, S. Doniach, D. Lublin: Chem. Phys. Lett. 59, 121 (1978)
B.114 A. Bianconi, C.R. Natoli: Solid State Commun. 27, 1177 (1978)
       A. Bianconi: Phys. Rev. B 26, 2741 (1982)
B.115 A. Bianconi, R.Z. Bachrach: Phys. Rev. Lett. 42, 104 (1979)
       A. Bianconi, R.Z. Bachrach, S.B.M. Hagström, S.A. Flodström: Phys. Rev. B
       19, 2837 (1979)
       A. Bianconi, R.Z. Bachrach, S.A. Flodström: Phys. Rev. B 19, 3879 (1979)
B.116 A. Bianconi: Surf. Sci. 89, 41 (1979)
       A. Bianconi, R.S. Bauer: Surf. Sci. 99, 76 (1980)

B.117 A. Bianconi, M. Campagna, S. Stizza, I. Davoli: Phys. Rev. B 24, 6139 (1981)
A. Bianconi, M. Campagna, S. Stizza: Phys. Rev. B 25, 2477 (1982)
A. Bianconi, A. Marcelli, M. Tomellini, I. Davoli: J. Magn. Magn. Mater.
47-48, 209 (1985)

B.118 A. Bianconi et al.: Chem. Phys. Lett. 90, 257 (1982)

B.119 A. Bianconi, M. Dell'Ariccia, P.J. Durham. J.B. Pendry: Phys. Rev. B 26,
6502 (1982)

B.120 A. Bianconi et al.: Z. Physik B 67, 307 (1987)

B.121 A. Bianconi et al.: Solid State Commun. 63, 1009 (1987)

B.122 A. Bianconi et al.: Phys. Rev. B 35, 806 (1987)

B.123 V.A. Biebesheimer et al.: J. Chem. Phys. 81, 2599 (1984)

B.124 N. Binsted et al.: J. Am. Chem. Soc. 109, 3669 (1987)

B.125 G. Bissinger, P.H. Nettles, S.M. Shafroth, A.W. Waltner: Phys. Rev. A 10,
1932 (1974)

B.126 W. Blau: In [Ref.1.58, Vol.II, p.188]

B.127 W. Blau, J. Weisbach, G. Merz, K. Kleinstück: Phys. Stat. Sol. (b) 93, 713
(1979)

B.128 D. Blechschmidt, R. Haensel et al.: Phys. Stat. Sol. (b) 44, 787 (1971)

B.129 D. Blechschmidt, V. Saile, M. Skibowski, W. Steinmann: Phys. Lett. 35A, 221
(1971)
V. Saile, M. Skibowski: Phys. Stat. Sol. (b) 50, 661 (1972)

B.130 D. Blechschmidt, R. Haensel et al.: Chem. Phys. Lett. 14, 33 (1972)

B.131 M.A. Blokhin, A.T. Shuvaev: Izv. Akad. Nauk SSSR, Ser. Fiz, 22, 1453 (1958)

B.132 M.A. Blokhin, A.T. Shuvaev: Izv. Akad. Nauk. SSSR, Ser. Fiz. 26, 429 (1962)

B.133 M.A. Blokhin, A.T. Shuvaev, V.V. Gorski: Izv. Akad. Nauk SSSR, Ser. Fiz.
28, 801 (1964)

B.134 M.A. Blokhin, V.F. Demekhin: Izv. Akad. Nauk SSSR, Ser. Fiz. 28, 830, 834
(1964)

B.135 M.A. Blokhin, V.F. Demekhin, L.S. Gorbatenko: Izv. Vyssh. Uchebn. Zaved.,
Fiz. 1966, 138
M.A. Blokhin, A.T. Shuvaev et al.: Izv. Akad. Nauk SSSR, Ser. Fiz. 38, 544
(1974)
A.T. Shuvaev, M.A. Blokhin et al.: Izv. Sib. Otd. Akad. Nauk SSSR, Ser.
Khim. Nauk 1975, Nr. 9, H. 4, 86

B.136 M.A. Blokhin, V.F. Volkov: Dokl. Akad. Nauk SSSR 183, 64 (1968)

B.137 M.A. Blokhin, H. Sommer et al.: Fiz. Tverd. Tela 11, 17 (1969)

B.138 M.A. Blokhin, I.G. Shveitser: In [Ref.1.96, Vol.II, p.3]

B.139 M.A. Blokhin, I.Ya. Nikiforov. H. Sommer, I.I. Geguzin: In [Ref.1.79, p.321]

B.140 M.A. Blokhin, L.M. Monastyrskii, I.G. Shveitser: Fiz. Met. Metalloved. 35,
213 (1973)

B.141 M.A. Blokhin, I.Ya. Nikiforov et al.: Fiz. Met. Metalloved. 35, 743 (1973)

B.142 M.A. Blokhin, L.M. Monastyrskii, I.G. Shveitser: Fiz. Met. Metalloved. 37,
640 (1974)

B.143 M.A. Blokhin, Ya.I. Dutchak et al.: Izv. Akad. Nauk SSSR, Ser. Fiz. 38, 631
(1974); Metallofizika 1974, Nr. 53, 116
I.G. Shveitser, L.M. Monastyrskii: Deposited Doc. 1975 (VINITI 2000 - 75)
M.A. Blokhin, L.M. Monastyrskii, I.G. Shveitser: Izv. Akad. Nauk SSSR, Ser.
Fiz. 40, 321 (1976)

B.144 M.A. Blokhin, E.G. Orlova, I.G. Shveitser: Izv. Akad. Nauk SSSR, Ser. Fiz.
38, 652 (1974)
M.A. Blokhin, E.G. Orlova, I.G. Shveitser, G. Kokh: Izv. Akad. Nauk SSSR,
Ser. Fiz. 40, 323 (1976)
M.A. Blokhin, E.G. Orlova, I.G. Shveitser: Zh. Strukt. Khim. 18, 687 (1977)

        M.A. Blokhin, E.G. Orlova, I.G. Shveitser: Phys. Stat. Sol. (b) 103, 63 (1981)
B.145   M.A. Blokhin et al.: In [Ref.98, p.206]
B.146   M.A. Blokhin, A.T. Shuvaev et al.: Izv. Akad. Nauk SSSR, Ser. Fiz. 38, 544
        (1974)
        A.T. Shuvaev, M.A. Blokhin et al.: Izv. Sib. Otd. Akad. Nauk SSSR, Ser.
        Khim. Nauk 1975, Nr. 9, H. 4, 86
B.147   M.A. Blokhin, A.T. Shuvaev: In Proc. XX Colloq. Spectrosc. Int., Invited
        Lect., Vol.1, ed. by I. Rubesca, F. Mostecky, Z. Ksandr (Statni Pedagogicke
        Nakladatelstvi, Prague 1977) p.173
B.148   M.A. Blokhin, S.M. Blokhin, A.A. Polyakov, S.A. Prosandeev: In [Ref.1.66,
        p.287]
B.149   S.M. Blokhin, E.E. Vainshtein: Dokl. Akad. Nauk SSSR 158, 694 (1964)
B.150   S.M. Blokhin: Thesis Rostov/Don 1966
B.151   S.M. Blokhin, V.M. Bertenev, V.I. Chirkov: In [Ref.1.58, Vol.II, p.294]
B.152   S.M. Blokhin, A.P. Sadovskii et al.: Zh. Strukt. Khim. 10, 833 (1969)
B.153   S.M. Blokhin, V.I. Chirkov et al.: In [Ref.1.96, Vol.I, p.98]
B.154   J. Blomqvist, B. Roos, M. Sundbom: Chem. Phys. Lett. 9, 160 (1971)
B.155   L. Blum et al.: J. Chem. Phys. 85, 6732 (1986)
B.156   W.E. Blumberg, J. Peisach, P. Eisenberger, J.A. Fee: Biochemistry 17, 1842
        (1978)
B.157   S. Bodeur, C. Senemaud, C. Bonnelle, J.P. Connerade: In [Ref.1.86, p.94]
B.158   S. Bodeur, J.M. Esteva: Chem. Phys. 100, 415 (1985)
B.159   S. Bodeur, I. Nenner, P. Millie: Phys. Rev. A 34, 2986 (1986)
B.160   S. Bodeur, A.P. Hitchcock: Chem. Phys. 111, 467 (1987)
B.161   G. Böhm, K. Ulmer, Z. Angew. Physik 29, 287 (1970); J. Physique 32, C
        4-241 (1971)
B.162   W. Böhmer, P. Rabe: J. Phys. C 12, 2465 (1979)
B.163   T.K. Boletskaya, V.E. Egorushkin, E.M. Savistskii, V.P. Fadin: Dokl. Akad.
        Nauk SSSR 252, 87 (1980)
B.164   T. Bolze, J. Peisl: Z. Physik B 62, 9 (1985)
B.165   T. Bolze, J. Peisl: Mater. Sci. Forum 15, 575 (1987)
B.166   T.N. Bondarenko, V.P. Dseganovskii, E.A. Zhurakovskii: Ukr. Fiz. Zh. 18,
        683, 1121 (1973)
        I.N. Frantsevich, E.A. Zhurakovskii, V.V. Shvaiko: Ukr. Fiz. Zh. 20, 1177
        (1975)
        E.A. Zhurakovskii et al.: Ukr. Fiz. Zh. 20, 1324 (1975)
        I.N. Frantsevich, E.A. Zhurakovskii, V.V. Shvaiko: Visn. Akad. Nauk Ukr.
        SSR 1976, 9; Ukr. Fiz. Zh. 21, 1339 (1976)
B.167   T.N. Bondarenko, E.A. Zhurakovskii, V.P. Dseganovskii: Izv. Akad. Nauk
        SSSR, Neorg. Mater. 11, 2015 (1975)
B.168   T.N. Bondarenko, L.M. Monastyrskii, E.A. Zhurakovskii: Ukr. Fiz. Zh. 21,
        1209 (1976)
B.169   T.N. Bondarenko, Yu.A. Teterin, A.S. Baev, V.P. Dzeganovskii: Izv. Akad.
        Nauk SSSR, Ser. Fiz. 49, 1550 (1985)
B.170   C. Bonnelle: C.R. Acad. Sci. 254, 2313 (1962)
B.171   C. Bonnelle, C.K. Jörgensen: J. Chim. Physique Physico-Chim. Biol. 61, 826
        (1964)
B.172   C. Bonnelle, F. Wuilleumier, C. Senemaud: In [Ref.1.57, p.20]
B.173   C. Bonnelle: Ann. Physique 1, 439 (1966); J. Physique 28, C 3-65 (1967)
        C. Bonnelle, E. Belin, C. Senemaud: Jpn. J. Appl. Phys. 17, S 2-125 (1978)
B.174   C. Bonnelle: In [Ref.1.78, p.163]
B.175   C. Bonnelle, C. Senemaud: C.R. Acad. Sci. 268 B 65 (1969)

B.176   C. Bonnelle, R.C. Karnatak: J. Physique 32, C 4-230 (1971)
        J.M. Mariot, R.C. Karnatak: Phys. Fenn. 9, S 1-96 (1974)
B.177   C. Bonnelle, C. Hague: In [Ref.1.98, p.78]
B.178   C. Bonnelle, R.C. Karnatak, J. Sugar: Phys. Rev. A 9, 1920 (1974)
B.179   C. Bonnelle, G. Lachere: J. Physique 35, 295 (1974)
B.180   C. Bonnelle, F. Vergand: J. Phys. Chem. Solids 36, 575 (1975)
B.181   C. Bonnelle: Struct. Bonding 31, 23 (Springer, Berlin, Heidelberg 1976)
        C. Lachere, C. Bonnelle: J. Physique C 5-15 (1980)
B.182   C. Bonnelle: In [Ref.1.66, p.317]
        P. Motais, E. Belin, C. Bonnelle: Phys. Rev. B 30, 4399 (1984)
B.183   U. Bonse, A. Henning: Nucl. Instrum. Methods A 246, 814 (1986)
B.184   G.L. Borchert: Z. Naturforsch. 31a, 102 (1976)
B.185   J. Bordas et al.: J. Inorg. Biochem. 11, 181 (1979)
        J. Bordas et al.: J. Biochem. 191, 499 (1980)
B.186   P. Bordet et al.: J. Magn. Magn. Mater. 63, 524 (1987)
B.187   N.D. Borisov, V.V. Nemoshkalenko: Vopr. Fiz. Met. Metalloved. 202 (1964)
B.188   N.A. Borovoi, A.Z. Zhmudskii, V.F. Surzhko, V.I. Shiyanovskii: Ukr. Fiz. Zh. 30, 1701 (1985)
        N.A. Borovoi, V.I. Shiyanovskii: Ukr. Fiz. Zh. 31, 401 (1986)
B.189   N.A. Borovoi, A.S. Erniyazov, A.Z. Zhmudskii, V.I. Shiyanovskii: Opt. Spektrosk. 59, 1324 (1985)
B.190   I.B. Borovskii, R.L. Barinskii: Dokl. Akad. Nauk SSSR 72, 31 (1950); Izv. Akad. Nauk SSSR, Ser. Fiz. 15, 225 (1951)
B.191   I.B. Borovskii, K.P. Gurov: Fiz. Met. Metalloved. 7, 225 (1959)
B.192   I.B. Borovskii, V.I. Matyskin: Dokl. Akad. Nauk SSSR 192, 63, (1970); ibid. 195, 1072 (1970)
B.193   I.B. Borovskii: J. Physique 32, C 4-207 (1971)
B.194   I.B. Borovskii, P.N. Semochkin, M. Rzaeva: Fiz. Met. Metalloved. 40, 537 (1975)
        P.N. Semochkin, I.B. Borovskii: Fiz. Met. Metalloved. 39, 497 (1975)
B.195   S. Bose, P. Longe: Phys. Rev. B 18, 3921 (1978)
B.196   T.A. Boster, J.E. Edwards: Phys. Rev. 170, 12 (1968)
B.197   J. Bouat, D. Bonnin, L. Facchini: Synth. Met. 7, 233 (1983)
B.198   M. Boudart, R. Dalla Betta, K. Foger, D.G. Loffler: SSRL Report No. 84/06, p.26 (1984)
B.199   C.E. Bouldin, E.A. Stern, B. von Roerdern, M. Azoulay: Phys. Rev. B 30, 4462 (1984)
B.200   C.E. Bouldin et al.: Proc. SPIE 690, 45 (1986)
B.201   C.E. Bouldin, R.A. Forman, M.I. Bell: Phys. Rev. B 35, 1429 (1987)
B.202   A.J. Bourdillon, S.J. Bull, P.J. Burnett, T.F. Page: J. Mater. Sci. 21, 1547 (1986)
B.203   J.B. Boyce, T.M. Hayes, W. Stutius, J.C. Mikkelsen, Jr.: Phys. Rev. Lett. 38, 1362 (1977)
        W. Stutius, J.B. Boyce, J.C. Mikkelsen, Jr.: Solid State Commun. 31, 539 (1979)
B.204   J.B. Boyce, T.M. Hayes: In Physics of Superionic Conductors, ed. by M.B. Salamon, Topics Curr. Phys., Vol.15 (Springer, Berlin, Heidelberg 1979) p.5
        J.B. Boyce, T.M. Hayes, C.J. Mikkelsen, W. Stutius: Solid State Commun. 33, 183 (1980)
B.205   J.B. Boyce, R.M. Martin, J.W. Allen: In [Ref.1.82, p.187]
B.206   J.B. Boyce, J.C. Mikkelsen: Phys. Rev. B 31, 6903 (1985)
B.207   J.B. Boyce, J.C. Mikkelsen, F. Bridges, T. Egami: Phys. Rev. B 33, 7314 (1986)

B.208 Boyun Qi et al.: Phys. Rev. B. 36, 2972 (1987)

B.209 A.P. Braiko et al.: Fiz. Tverd. Tela 26, 1068 (1984)

B.210 R. Brammer: J. Chem. Phys. 87, 1153 (1987)

B.211 R. Brammer, J.-E. Rubensson, N. Wassdahl, J. Nordgren: Phys. Scr. 36, 262 (1987)

B.212 W. Braun et al.: Surf. Interface Anal. 10, 250 (1987)

B.213 J.P. Briand et al.: J. Phys. B 9, 1055 (1976)

B.214 J.P. Briand et al.: In [Ref.1.62, p.335]

B.215 V. Brijunas, A. Kupliaukiene, Z. Kupliauskis: Izv. Vyssh. Uchebn. Zaved., Fiz. 23, 108 (1980)

B.216 G. Brogren: Ark. Mat. Astr. Fysik 31 A, 1 (1945)

B.217 H. Broili, R. Glocker, H. Kiessig: Z. Physik 92, 27 (1934)

B.218 H. Bross: Phys. Cons. Matter 17, 55 (1973)

B.219 F. Brouers, P. Longe, B. Bergersen: Solid State Commun. 8, 1423 (1970)

B.220 F.C. Brown, C. Gähwiller et al.: Phys. Rev. B 2, 2126 (1970)

B.221 F.C. Brown, R.Z. Bachrach, M. Skibowski: Phys. Rev. B 15, 4781 (1977)

B.222 F.C. Brown, R.Z. Bachrach, A. Bianconi: In [Ref.1.87, p.17]; Chem. Phys. Lett. 54, 425 (1978)

R.Z. Bachrach, A. Bianconi, F.C. Brown: Nucl. Instrum. Methods 152, 53 (1978)

A. Bianconi, H. Petersen, F.C. Brown, R.Z. Bachrach: Phys. Rev. A 17, 1907 (1978)

B.223 F.C. Brown et al.: Phys. Rev. B 34, 7698 (1986)

B.224 G.S. Brown, P. Eisenberger, P. Schmidt: Solid State Commun. 24, 201 (1977)

B.225 J.M. Brown et al.: J. Am. Chem. Soc. 102, 4210 (1980)

B.226 O. Brümmer, G. Dräger: Phys. Stat. Sol. 14, K 175 (1966)

B.227 O. Brümmer, G. Dräger: Phys. Stat. Sol. 27, 513 (1968)

B.228 O. Brümmer, G. Dräger, K. Machlitt: In [Ref.1.58, Vol.II, p.300]

B.229 O. Brümmer, G. Dräger, W. Starke: Ann. Physik 24, 200 (1970)

B.230 O. Brümmer, G. Dräger, W. Starke: J. Physique 32, C 4-169 (1971)

B.231 O. Brümmer, G. Dräger, V.A. Fomichev, A.S. Shulakov: In [Ref.1.60, Vol.I, p.78]

U. Berg, G. Dräger, O. Brümmer: Phys. Stat. Sol. (b) 74, 341 (1976)

G. Dräger, O. Brümmer: Phys. Stat. Sol. (b) 78, 729 (1976)

B.232 O. Brümmer, G. Dräger, F. Werfel: Microchim. Acta Suppl. 7, 279 (1977)

F. Werfel, G. Dräger, U. Berg: Cryst. Res. Technol. 16, 119 (1981)

F. Werfel, D. Rohländer: Cryst. Res. Technol. 20, 1657 (1985)

B.233 O. Brümmer et al.: Mikrochim. Acta Suppl. 9, 205 (1981)

G. Dräger et al.: Phys. Stat. Sol. (b) 131, 193 (1985)

B.234 R. Bruhn, B. Sonntag, H.W. Wolff: Phys. Lett. 69 A, 9 (1978)

B.235 R. Bruhn, B. Sonntag, H.W. Wolff: J. Phys. B 12, 203 (1979)

B.236 G. Brunner, M. Nagel, E. Hartmann, E. Arndt: J. Phys. B 15, 4517 (1982)

B.237 G.W. Bryant: J. Phys. F 10, 321 (1980)

B.238 I.A. Brytov, E.Z. Kurmaev, S.A. Nemnonov: Fiz. Met. Metalloved. 26, 366 (1968)

B.239 I.A. Brytov, M.A. Rumsch, A.S. Parobets: Fiz. Tverd. Tela 10, 794 (1968)

B.240 I.A. Brytov, M.A. Rumsh: In [Ref.1.95, p.141]

Yu.N. Romashchenko et al.: Fiz. Khim. Stekla 7, 391 (1981)

B.241 I.A. Brytov, N.I. Komyak, A.P. Lukirskii: In [Ref.1.58, Vol.I, p.284]

B.242 I.A. Brytov: App. Met. Rentgen. Anal. 4, 201 (1969)

B.243 I.A. Brytov, E.Z. Kurmaev: Fiz. Met. Metalloved. 32, 520 (1971)

B.244 I.A. Brytov, E.Z. Kurmaev et al.: Izv. Akad. Nauk SSSR, Neorg. Mater. 9, 137 (1973)

B.245 I.A. Brytov, V.S. Neshpor et al.: Fiz. Met. Metalloved. **37**, 905 (1974)

B.246 I.A. Brytov, Yu.N. Romashchenko: In [Ref.1.98, p.122]

B.247 I.A. Brytov, N.I. Komyak et al.: In [Ref.1.98, p.227]; Izv. Akad. Nauk SSSR, Ser. Fiz. **38**, 572 (1974)
I.A. Brytov, L.E. Mstibovskaya: Opt. Spektrosk. **39**, 404 (1975)

B.248 I.A. Brytov et al.: Izv. Sib. Otd. Akad. Nauk SSSR, Ser. Khim. Nauk 1975 (4), 142, 148; App. Met. Rentgen. Anal. **19**, 168 (1977); Fiz. Tverd. Tela **20**, 664 (1978); Zh. Strukt. Khim. **20**, 277 (1979); Geokhimiya 1979, 261
L.E. Mstibovskaya, Yu.N. Romashchenko, I.A. Brytov: Fiz. Tverd. Tela **20**, 2526 (1978)

B.249 I.A. Brytov, Yu.P. Dikov, Yu.N. Romashchenko: Izv. Akad. Nauk SSSR, Ser. Fiz. **40**, 413 (1976)

B.250 I.A. Brytov, Yu.N. Romashchenko: Geokhimiya 1977, 1722

B.251 I.A. Brytov et al.: Dokl. Akad. Nauk SSSR **240**, 1111 (1978)

B.252 I.A. Brytov, L.E. Mstibovskaya, E.A. Obolenskii, L.G. Rabinovich: Fiz. Tverd. Tela **25**, 3461 (1983)

B.253 I.A. Brytov, V.A. Gritsenko, Yu.N. Romashchenko: Zh. Eksp. Teor. Fiz. **89**, 562 (1985)

B.254 I.A. Brytov, L.E. Mstibovskaya, L.G. Rabinovich: Izv. Akad. Nauk SSSR, Ser. Fiz. **49**, 1490 (1985)

B.255 J.I. Budnick et al.: Proc. SPIE **690**, 58 (1986)

B.256 B. Buffat et al.: Solid State Commun. **59**, 17 (1986)

B.257 B. Buffat et al.: J. Phys. Chem. Solids **47**, 491 (1986)

B.258 L.A. Bugaev, R. V. Vedrinskii: Phys. Stat. Sol. (b) **132**, 459 (1985)
L.A. Bugaev et al.: Phys. Stat. Sol. (b) **133**, 195 (1986)
V.N. Datsyuk, I.I. Geguzin, R.V. Vedrinskii: Phys. Stat. Sol. (b) **134**, 175 (1986)
I.I. Geguzin et al.: Phys. Stat. Sol. (b) **134**, 641 (1986)

B.259 P.G. Burkhalter, A.R. Knudson, D.J. Nagel: Phys. Rev. A **6**, 2093 (1972)
A.R. Knudson, D.J. Nagel, P.G. Burkhalter: Phys. Lett. **42 A**, 69 (1972)

B.260 P.G. Burkhalter, A.R. Knudson, D.J. Nagel: Phys. Rev. A **7**, 1936 (1973)

B.261 W. Butscher, R.J. Buenker, S.D. Peyerimhoff: Chem. Phys. Lett. **52**, 449 (1977)
R.J. Buenker, S.D. Peyerimhoff, W. Butscher: Mol. Phys. **35**, 771 (1978)

B.262 W. Butscher, W.H.E. Schwarz, H. Friedrich, B. Sonntag, P. Rabe: Chem. Phys. Lett. **64**, 360 (1979)

B.263 M.P. Bytzman et al.: Carbon **20**, 293 (1982)

C.1 G. Calas, J. Petiau: Solid State Commun. **48**, 625 (1983); in *Structure of Non-crystalline Materials*, ed. by P.H. Gaskell (Taylor and Francis, London 1983) p.18; J. Physique **43**, C9-47 (1982); ibid. **46**, C8-41 (1985)
A. Bianconi, E. Fritsch, G. Calas, J. Petiau: Phys. Rev. B **32**, 4292 (1985)

C.2 T.A. Callcott, E.T. Arakawa: Phys. Rev. Lett. **38**, 442 (1977)
T.A. Callcott, E.T. Arakawa, D.L. Ederer: Phys. Rev. B **16**, 185 (1977)

C.3 T.A. Callcott, E.T. Arakawa, D.L. Ederer: Jpn. J. Appl. Phys. **17**, S 2-149 (1978); Phys. Rev. B **18**, 6622 (1978)

C.4 T.A. Callcott, J.A. Tagle, E.T. Arakawa, G.M. Stocks: Appl. Opt. **19**, 4035 (1980)

C.5 P. Callon: C. R. Acad. Sci. **248**, 1985 (1959); Thesis Paris 1965

C.6 J.L. Campell, G.W. Schulte: J. Phys. B **15**, 47 (1982)

C.7 E. Canova et al.: Phys. Rev. B **31**, 8308 (1985)

C.8 C. Carbone et al.: Thin Solid Films **140**, 105 (1986)

C.9	C. Cardinaud, C. Senemaud, G. Villela: J. Non-Cryst. Solids **88**, 55 (1986)
C.10	M. Cardona, W. Gudat, B. Sonntag, P.Y. Yu: DESY F 41-70/6 (1970)
C.11	M. Cardona, R. Haensel: Phys. Rev. B **1**, 2605 (1970)
C.12	M. Cardona, R. Haensel, D.W. Lynch, B. Sonntag: Phys. Rev. B **2**, 1117 (1970)
C.13	M. Cardona, C.M. Penchina, E.E. Koch, P.Y. Yu: Phys. Stat. Sol. (b) **53**, 327 (1972)
C.14	G.S. Cargill, R.F. Boehme, W. Weber: Phys. Rev. Lett. **50**, 1391 (1983)
C.15	R.G. Carr, T.K. Sham, W.E. Eberhardt: Chem. Phys. Lett. **113**, 63 (1985)
C.16	R.D. Carson, S.E. Schnatterly: Phys. Rev. B **33**, 2432 (1986)
C.17	D.E. Carter, M.P. Givens: Phys. Rev. **101**, 1469 (1956)
C.18	M.L. Carvalho, F. Parente, L. Salgueiro: J. Phys. B **20**, 935 (1987)
C.19	C.R.A. Catlow et al.: Solid State Ionics 9-10, 1107 (1983)
C.20	J.A. Catterall, J. Trotter: Phil. Mag. **4**, 1164 (1959)
C.21	J.A. Catterall, J. Trotter: Proc. Phys. Soc. **79**, 691 (1962); ibid. **81**, 1043 (1963)
C.22	Y. Cauchois, I. Manescu, F. LeBerquier: C. R. Acad. Sci. **239**, 1780 (1954)
C.23	Y. Cauchois, C. Bonnelle, C. Senemaud: C. R. Acad. Sci. **257**, 1051 (1963)
C.24	Y. Cauchois, C. Bonnelle, I. Manescu: C. R. Acad. Sci. 267 B 817 (1968) Y. Cauchois, C. Bonnelle, C. Senemaud, R.C. Karnatak: In [Ref.1.58, Vol.I, p.43]
C.25	V.D. Chafekar, S.C. Sen: In [Ref.1.83, p.92]
C.26	F.C. Chalkin: Proc. Roy. Soc. A **194**, 42 (1948)
C.27	M.B. Chamberlain, A.F. Burr, R.J. Liefeld: Phys. Rev. A **9**, 663 (1974) M.B. Chamberlain, A.F. Burr, R.J. Liefeld: J. Vac. Sci. Technol. A **1**, 1169 (1983) M.B. Chamberlain, A.F. Burr, R.J. Liefeld: J. Vac. Sci. Technol. **2**, 973 (1984)
C.28	S.I. Chan et al.: J. Mol. Struct. **45**, 239 (1978) S.I. Chan, R.C. Gamble: Methods Enzym. **54**, 323 (1980)
C.29	D. Chandesris, P. Roubing, G. Rossi, J. Lecante: Surf. Sci. **169**, 57 (1986)
C.30	Chang Longcun et al.: Acta. Metall. Sin. **22**, 8105 (1986)
C.31	P. Charpin et al.: J. Chim. Phys., Phys.-Chim. Biol. **82**, 925 (1985)
C.32	S. Chattopadhyay, V.B. Sapre, C. Mande: X-Ray Spectrom. **13**, 153 (1984) S. Chattopadhyay, C. Mande: Phys. Stat. Sol. (b) **126**, 53 (1984)
C.33	T. Chattopadhyay, A.R. Chetal: Phys. Stat. Sol. (b) **122**, 193 (1984); J. Phys. Chem. Solids **46**, 427 (1985); J. Phys. C **18**, 5373 (1985) K. Hemachandran, A.R. Chetal: Phys. Stat. Sol. (b) **132**, 503 (1985)
C.34	S. Chattopadhyay, A.P. Deshpande: J. Phys. Soc. Jpn. **55**, 2320 (1986)
C.35	F.Y. Chen, H.W. Huang: Chin. J. Phys. **21**, 45 (1983)
C.36	H. Chen: J. Phys. Chem. Solids **41**, 641 (1980) J.W. Park, H. Chen: Phys. Chem. Glasses **23**, 107 (1981)
C.37	M.H. Chen, B. Crasemann, M. Aoyagi, H. Mark: Phys. Rev. A **15**, 2312 (1977)
C.38	V.M. Cherkashenko et al.: J. Solid State Chem. **32**, 377 (1980) V.M. Cherkashenko, E.Z,. Kurmaev, V.L. Volkov: J. Electron Spectrosc. **28**, 1 (1982) V.M. Cherkashenko et al.: Solid State Commun. **54**, 933 (1985)
C.39	V.M. Cherkashenko et al.: Zh. Strukt. Khim. **25**, Nr.5, 55 (1984)
C.40	V.M. Cherkashenko, A.A. Rempel, A.I. Gusev: Fiz. Tverd. Tela **27**, 1387 (1985)
C.41	V.M. Cherkashenko, V.R. Galakhov, E.Z. Kurmaev: Zh. Strukt. Khim. **27**, Nr.4, 185 (1986)
C.42	P. Chevallier, M. Tavernier, J.P. Briand: J. Phys. B **11**, L 171 (1978)
C.43	N.S. Chiu, S.H. Bauer: Acta Crystallogr. C **40**, 1646 (1984)

C.44   T.H. Chiu, D. Gibbs, J.E. Cunningham, C.P. Flynn: J. Phys. F **13**, L 23 (1983)
D. Gibbs, T.H. Chiu, J.E. Cunningham, C.P. Flynn: Phys. Rev. Lett. **49**, 815 (1982)

C.45   D. Chopra: Phys. Rev. A **1**, 230 (1970)

C.46   D. Chopra, F. Keyvan: Proc. Symp. Nondestr. Eval., 10th San Antonio (1975) 230

C.47   D.R. Chopra: J. Less-Common Met. **127**, 373 (1987)

C.48   S.-H. Chou, F.W. Kutzler, D.E. Ellis, P.-L. Cao: Surf. Sci. **164**, 85 (1985)

C.49   S.-H. Chou, J. Guo, D.E. Ellis: Phys. Rev. B **34**, 12 (1986)

C.50   S.-H. Chou, J.J. Rehr, E.A. Stern, E.R. Davidson: Phys. Rev. B **35**, 2604 (1987)

C.51   B.K. Chougule, S.A. Patil: Sci. J. Shivaji Univ. **15**, 177 (1975)

C.52   B.K. Chougule, R.N. Patil: J. Phys. Chem. Solids **39**, 1141 (1978)

C.53   A.R. Chourasia, V.D. Chafekar, S.D. Deshpande, C. Mande: Pramana **24**, 787 (1985)
A.R. Chourasia, V.D. Chafekar, C. Mande: Pramana **24**, 867 (1985)

C.54   H.-U. Chun, Ha Tae-Kju, R. Mann: Z. Phys. Chemie (Frankfurt) **47**, 313 (1965)

C.55   H.-U. Chun, H. Gebelein: Z. Naturforsch. **22a**, 1813 (1967)
H.-U. Chun: Phys. Lett. **30 A**, 445 (1969)

C.56   H.-U. Chun, G. Klein: Z. Naturforsch. **24a**, 930 (1969)
G. Klein, H.-U. Chun: Phys. Stat. Sol. (b) **49**, 167 (1972)

C.57   H.-U. Chun: In [Ref.160, Vol.I, p.426]; Phys. Fenn. **9**, S 1-144 (1974)

C.58   H.-U. Chun, H. Klein: Jpn. J. Appl. Phys. **17**, S 2-141 (1978)

C.59   B. Cioffari: Phys. Rev. **51**, 630 (1937)

C.60   P.H. Citrin, P. Eisenberger, B.M. Kincaid: Phys. Rev. Lett. **36**, 1346 (1976)

C.61   P.H. Citrin, P. Eisenberger, R.C. Hewitt: Phys. Rev. Lett. **41**, 309 (1978); ibid. **45**, 1948 (1980)
P.H. Citrin: Phys. Rev. B **31**, 700 (1985)

C.62   P.H. Citrin, G.K. Wertheim, M. Schluter: Solid State Commun. **32**, 429 (1979)

C.63   P.H. Citrin, G.K. Wertheim, M. Schluter: Phys. Rev. B **20**, 3067 (1979)

C.64   P.H. Citrin, J.E. Rowe, P. Eisenberger: Phys. Rev. B **28**, 2299 (1983)

C.65   T. Cleason, J.B. Boyce, T.H. Geballe: Phys. Rev. B **25**, 6666 (1982)

C.66   T. Claeson, J.B. Boyce: Phys. Rev. B **29**, 1551 (1984)

C.67   T. Claeson, J.B. Boyce, W.P. Lowe, T.H. Geballe: Phys. Rev. B **29**, 4969 (1984)

C.68   B.S. Clausen, B. Lengeler, B.S. Rasmussen: J. Phys. Chem. **89**, 2319 (1985)

C.69   R. Clement, A. Michalowicz: Rev. Chim. Miner. **21**, 426 (1984)

C.70   J. Clift, C. Curry, B.J. Thompson: Phil Mag. **8**, 593, 639 (1963)

C.71   G. Cocco et al.: J. Phys. Chem. **83**, 2527 (1979)

C.72   G. Cocco, S. Enzo, L. Incoccia, S. Mobilio: Z. Naturforsch. A **38**, 1391 (1983)

C.73   P.I. Cohen et al.: Applic. Surf. Sci. **1**, 538 (1978)

C.74   V. Collet: Thesis Paris (1959); C. R. Acad. Sci. **248**, 1314 (1959)

C.75   G.A.D. Collins, D.W.J. Cruickshank, A. Breeze: J. Chem. Soc., Faraday Trans. II **68**, 1189 (1972)

C.76   F. Combet-Farnoux: Phys. Fenn. **9**, S 1-80 (1974)

C.77   F.H. Combley, E.A. Stewardson, J.E. Wilson: J. Phys. B **1**, 120 (1968)

C.78   F.J. Comes, R. Haensel, U. Nielsen, W.H.E. Schwarz: J. Chem. Phys. **58**, 516 (1973)

C.79   F. Comin, J.E. Rowe, P.H. Citrin: Phys. Rev. Lett. **51**, 2402 (1983)

C.80   J.B. Conklin, F.W. Averill, T.M. Mattox: J. Physique **33**, C 3-213 (1972)

C.81   J.P. Connerade, M.W.D. Mansfield, I. Pollard, K. Thimm: Phys. Rev. A **6**,

1955 (1972); Proc. Roy. Soc. A **339**, 533 (1974)

C.82 J.P. Connerade et al.: In [Ref.1.86, p.99]

C.83 J.P. Connerade, M.W.D. Mansfield: Proc. Roy. Soc. A **343**, 415 (1975)

C.84 J.P. Connerade, M.W.D. Mansfield: Proc. Roy. Soc. A **348**, 539 (1976)

C.85 J.A. Connor, I.H. Hillier et al.: Mol. Phys. **23**, 81 (1972); ibid. **24**, 497 (1972)

C.86 J.A. Connor, I.H. Hillier, M.H. Wood, M. Barber: J. Chem. Soc., Faraday Trans. II **70**, 1040 (1974)

C.87 B. Cordts, D. Pease, L.V. Azaroff: Phys. Rev. B **24**, 538 (1981)

C.88 F. Corni, G.M. Antonini: Phys. Stat. Sol. (a) **97**, K21 (1986)

C.89 A. Corrias, A. Musinu, G. Pinna: Chem. Phys. Lett. **120**, 295 (1985)

C.90 F.A. Cotton, C.J. Ballhausen: J. Chem. Phys. **25**, 617 (1956)

F.A. Cotton, H.P. Hanson: J. Chem. Phys. **25**, 619 (1956); ibid. **26**, 1758 (1957); ibid. **28**, 83 (1958)

C.91 C.A. Coulson, C. Zauli: Mol. Phys. **6**, 525 (1963)

C.92 M.A. Coulthard: J. Phys. B **7**, 440 (1974)

C.93 A.D. Cox, J.H. Beaumont: Phil. Mag. B **42**, 115 (1980)

C.94 S.P. Cramer et al.: J. Am. Chem. Soc. **98**, 8059 (1976)

C.95 S.P. Cramer et al.: J. Am. Chem. Soc. **100**, 2748, 3398 (1978); ibid. **101**, 2770 (1979); Progr. Inorg. Chem. **25**, 1 (1979)

C.96 M.D. Crapper et al.: Europhys. Lett. **2**, 857 (1986)

C.97 M.D. Crapper, D.P. Woodruff, M. Bader, J. Haase: Surf. Sci. **182**, L241 (1987)

M.D. Crapper, C.E. Riley, D.P. Woodruff: Surf. Sci. **184**, 184 (1987)

C.98 R.S. Crisp, S.E. Williams: Phil. Mag. **5**, 1205 (1960); ibid **6**, 365 (1961)

R.S. Crisp: J. Phys. F **12**, 1529 (1982)

C.99 R.S. Crisp: Phil. Mag. **25**, 167 (1972); ibid. **36**, 609 (1977); J. Phys. F. **10**, 511 (1980); ibid. F **11**, 1705 (1981); ibid. F **13**, 1317 (1983)

C.100 R.S. Crisp: Phil. Mag. **36**, 609 (1977); J. Phys. F **10**, 511, 2313 (1980); ibid. F **11**, 219 (1981)

C.101 E.D. Crozier, F.W. Lytle, D.E. Sayers, E.A. Stern: Can. J. Chem. **55**, 1968 (1977)

C.102 E.D. Crozier, A.J. Seary: Can. J. Phys. **58**, 1388 (1980)

C.103 E.D. Crozier, N. Alberding, B.R. Sundheim: Acta Crystallogr. C **39**, 808 (1983); J. Chem. Phys. **79**, 939 (1983)

C.104 M. Cukier et al.: In [Ref.1.87, p.29]

C.105 I.M. Curelaru, G. Wendin: Phys. Scr. **22**, 513 (1980)

C.106 C. Curry: In [Ref.1.78, p.173]

C.107 C. Curry, R. Harrison: Phil. Mag. **32**, 659 (1970)

C.108 J.R. Cuthill, A.J. McAlister, M.L. Williams: Phys. Rev. Lett. **16**, 993 (1966); Phys. Rev. **164**, 1006 (1967); J. Appl. Phys. **39**, 2204 (1968)

C.109 J.R. Cuthill, A.J. McAlister, M.L. Williams, R.C. Dobbyn: In [Ref.1.78, p.151]

C.110 J.R. Cuthill, R.C. Dobbyn, A.J. McAlister, M.L. Williams: In [Ref.1.60, Vol.II, p.208]

D.1 A.A. Dakhel: Phys. Stat. Sol. (a) **74**, K 73 (1982); Jpn. J. Appl. Phys. I **21**, 1521 (1982); ibid. **21**, 1101 (1982); Acta Phys. Pol. A **63**, 89 (1983)

D.2 A.A. Dakhel: Acta Phys. Pol. A **70**, 791 (1986)

D.3 G. Dalba et al.: Solid State Ionics **9-10**, 597 (1983)

G. Dalba, P. Fornasini, F. Rocca: J. Physique **46**, C8-101 (1985)

D.4 G. Dalba, P. Fornasini, E. Burattini: J. Phys. C **16**, L 165 (1983)

D.5 G. Dalba, P. Fornasini, E. Burattini: J. Phys. C **16**, L 1091 (1983); in [Ref.1.83, p.461]

E. Burattini, G. Dalba, P. Fornasini: Nuovo Cimento D **7**, 293 (1986)

D.6 G. Dalba et al.: J. Non-Cryst. Solids **91**, 153 (1987)

D.7 E. Dartyge, A. Fontaine: J. Phys. F **14**, 721 (1984)

D.8 B.N. Das, L.V. Azaroff: Acta Metallurg. **13**, 827 (1965)
L.V. Azaroff: J. Appl. Phys. **38**, 2809 (1967)
H.C. Yeh, L.V. Azaroff: J. Appl. Phys. **38**, 4034 (1967)

D.9 K. Das Gupta: Phys. Lett. **46** A, 179 (1973)

D.10 L.M. Dautov et al.: In [Ref.1.98, p.320]

D.11 L.M. Dautov, B.D. Zhurkabaev, E.R. Ishkenov, N.K. Ergard: Fiz. Tverd. Tela **27**, 3150 (1985)
L.M. Dautov, B.D. Zhurkabaev, N.K. Ergard: J. Phys. Chem. Solids **47**, 175 (1986)

D.12 F.D. Davidson, R.W.G. Wyckoff: Adv. X-Ray Anal. **9**, 344 (1966)

D.13 B.M. Davies, F.C. Brown: Phys. Rev. B **25**, 2997 (1982)

D.14 L.C. Davis, L.A. Feldkamp: Phys. Rev. A **17**, 2012 (1978)

D.15 I. Davoli, L. Palladino, S. Stizza, A. Bianconi: Solid State Commun. **44**, 1585 (1982)

D.16 I. Davoli et al.: Phys. Rev. B **33**, 2979 (1986)

D.17 D.E. Day: Nature **200**, 649 (1963)

D.18 G. Deconninck, S. Van den Broek: J. Phys. C **13**, 3329 (1980)

D.19 G. Deconninck, A. Lefebvre: Mater. Sci. Eng. **90**, 167 (1987)

D.20 M. De Crescenzi, G. Chiarello, E. Colavita, R. Memeo: Phys. Rev. B **29**, 3730 (1984)

D.21 M. De Crescenzi et al.: Phys. Rev. B **32**, 612 (1985)

D.22 M. De Crescenzi, M. Diociaiuti, P. Picozzi, S. Santucci: Phys. Rev. B **34**, 4334 (1986)

D.23 A. Defrain et al.: J. Non-Cryst. Solids **61–62**, 439 (1984)

D.24 J.L. Dehmer: J. Chem. Phys. **56**, 4496 (1972)

D.25 F. De Keroulas, D. Calais, G. Lachere: J. Less-Common Met. **46**, 39 (1976)

D.26 J.A. Del Cueto, N.J. Shevchik: J. Phys. F **7**, L215 (1977); ibid. C **11**, L833 (1980); ibid. E **11**, 616 (1978)
N.J. Shevchik, D.A. Fischer: Rev. Sci. Instrum. **50**, 577 (1979)
D.A. Fischer, G.G. Cohen, N.J. Shevchik: J. Phys. F **10**, L139 (1980)

D.27 J.A. Del Cueto, N.J. Shevchik: J. Phys. C **11**, L829 (1978)

D.28 V.F. Demekhin, M.A. Blokhin: Izv. Akad. Nauk SSSR, Ser. Fiz. **28**, 825 (1964); Izv. Vyssh. Uchebn. Zaved., Fiz. 1966, 154

D.29 V.F. Demekhin, V.P. Sachenko: Izv. Akad. Nauk SSSR, Ser. Fiz. **31**, 900, 907 (1967); Bull. Acad. Sci. USSR **31**, 921 (1967)

D.30 V.F. Demekhin et al.: Izv. Akad. Nauk SSSR, Ser. Fiz. **36**, 264, 350 (1972)

D.31 V.F. Demekhim et al.: Izv. Akad. Nauk SSSR, Ser. Fiz. **36**, 352 (1972); Zh. Eksp. Teor. Fiz. **62**, 49 (1972); Zh. Strukt. Khim. **18**, 644 (1977)

D.32 V.F. Demekhin, T.M. Poltinnikova: Opt. Spektrosk. **36**, 663 (1974)

D.33 V.F. Demekhin, T.V. Shelkovich: Fiz. Tverd. Tela **16**, 1020 (1974)

D.34 V.F. Demekhin et al.: Izv. Akad. Nauk SSSR, Ser. Fiz. **38**, 587, 593 (1974); Izv. Sib. Otd. Akad. Nauk SSSR, Ser. Khim. Nauk 1975, Nr. 9, H. 4, 53, 68; Izv. Akad. Nauk SSSR, Ser. Fiz. **40**, 255 (1976)

D.35 V.F. Demekhin, T.V. Shelkovich: Izv. Sib. Otd. Akad. Nauk SSSR, Ser. Khim. Nauk 1975, Nr. 9, H. 4, 57; Fiz. Tverd. Tela **17**, 1806 (1975)

D.36 V.F. Demekhin et al.: Izv. Akad. Nauk SSSR, Ser. Fiz. **40**, 255, 263 (1976)

D.37 V.F. Demekhin et al.: Zh. Strukt. Khim. **20**, 38 (1979)

D.38 L.A. Demekhina, V.L. Sukhorukov, V.F. Demekhin, V.A. Yarna: Opt. Spektrosk. **49**, 861 (1980)

D.39 A.V. Demyanchuk et al.: Fiz. Met. Metalloved. **43**, 1307 (1977)

D.40   D. Denley, R.S. Williams, P. Perfetti, D.A. Shirley, J. Stöhr: Phys. Rev. B **19**, 1762 (1979)

D.41   G.B. Deodhar: In [Ref.1.57, p.65]

D.42   G.B. Deodhar, P.P. Varma: Physica **43**, 209 (1969)

D.43   G.B. Deodhar, S. Rai: J. Phys. B **4**, 1119 (1971)

D.44   G.L. De Poorter, E. Storms: High Temp. Sci. **1**, 294 (1969)

D.45   R.C. Der, R.J. Fortner, T.M. Kavanagh: Phys. Lett. **42 A**, 337 (1973)

D.46   P. Deshmukh, C. Mande: Pramana 2, 138 (1974); ibid. **6**, 305 (1976)

D.47   P. Deshmukh, C. Mande: J. Phys. C **10**, 3421 (1977)

D.48   A.P. Deshpande, V.B. Sapre, C. Mande: J. Phys. C **16**, L 433 (1983); Phys. Stat. Sol. (b) **118**, K 39 (1983)

D.49   A. Deshpande, C. Mande: Proc. Natl. Acad. Sci. India A **56**, 77 (1986)

D.50   A.P. Deshpande, P.C. Deshmukh: In *Proc. 7th Int'l Conf. Ternary and Multinary Compounds*, ed. by S.K. Deb., A. Zunger (Mater. Res. Soc. Pittsburgh, 1987) p.105
        A.P. Deshpande, C. Mande: In *Proc. 7th Int'l Conf. Ternary and Multinary Compounds*, ed. by S.K. Deb, A. Zunger (Mater. Res. Soc. Pittsburgh, 1987) p.145

D.51   R.D. Deslattes, H.S. De Ben: Phys. Rev. **115**, 71 (1959)

D.52   R.D. Deslattes: Phys. Rev. **129**, 1511 (1963)

D.53   R.D. Deslattes: Phys. Rev. A **133**, 390, 395, 399 (1964)

D.54   R.D. Deslattes: Phys. Rev. **172**, 625 (1968)

D.55   R.D. Deslattes: Phys. Rev. Lett. **20**, 483 (1968)

D.56   R.D. Deslattes, R.E. La Villa, A. Henins: Nucl. Instrum. Methods **152**, 179 (1978)

D.57   R.D. Deslattes, R.E. La Villa, P.L. Cowan, A. Henins: Phys. Rev. A **27**, 923 (1983)
        R.D. Deslattes, P.L. Cowan, R.E. La Villa: In [Ref.1.65, p.100]

D.58   M. Deutsch, M. Hart: Phys. Rev. B **26**, 5558 (1982)

D.59   H. Dexpert et al.: Phys. Rev. B **36**, 1750 (1987)

D.60   A.K. Dey, B.K. Agarwal: Lett. Nuovo Cim. **1**, 803 (1971)

D.61   A.K. Dey, B.K. Agarwal: J. Chem. Phys. **59**, 1397 (1973)

D.62   A. Di Cicco, A. Bianconi, N.V. Pavel: Solid State Commun. **61**, 635 (1987)

D.63   E.R. Dietz: J. Phys. B **16**, 4593 (1983)

D.64   Yu.P. Dikov, Yu.N. Romashchenko, I.A. Brytov, V.V. Nemoshkalenko: Dokl. Akad. Nauk SSSR **224** 88 (1975)
        Yu.P Dikov et al.: Phys. Chem. Miner. **1**, 27 (1977)

D.65   D. Dill, J.L. Dehmer: J. Chem. Phys. **61**, 692 (1974)
        J.L. Dehmer, D. Dill: Phys. Rev. Lett. **35**, 213 (1975); J. Chem. Phys. **65**, 5327 (1976)

D.66   R.K. Dimond: Phil. Mag. **15**, 631 (1967)

D.67   Y.S. Ding, D.J. Yarusso, H.K.D. Pan, S.L. Cooper: J. Appl. Phys. **56**, 2396 (1984)

D.68   R.C. Dobbyn, M.L. Williams, J.R. Cuthill: Phys. Rev. B **2**, 1563 (1970)

D.69   R.C. Dobbyn, A.J. McAlister, J.R. Cuthill, N.E. Erickson: Phys. Lett. **47 A**, 251 (1974)

D.70   U. Döbler, K. Baberschke, J. Haase, A. Puschmann: Phys. Rev. Lett. **52**, 1437 (1984)
        U. Döbler, K. Baberschke, J. Stöhr, D.A. Outka: Phys. Rev. B **31**, 2532 (1985)
        U. Döbler, K. Baberschke, J. Haase, A. Puschmann: Surf. Sci. **152-153**, 569 (1985)
        K. Baberschke et al.: Phys. Rev. B **33**, 5910 (1986)
        A. Puschmann, J. Haase: Surf. Sci. **144**, 559 (1984)

D.71 U. Döbler, K. Baberschke, D.D. Vvedensky, J.B. Pendry: Surf. Sci. 178, 679 (1986)

D.72 V.D. Dobrovolskii et al.: Ukr. Fiz. Zh. 13, 928 (1968); ibid. 15, 1206, 1300, 1304 (1970)

D.73 V.D. Dobrovolskii et al.: Ukr. Fiz. Zh. 13, 1127 (1968); ibid. 14, 1284 (1969); ibid. 15, 849 (1970)

D.74 V.D. Dobrovolskii et al.: Izv. Vyssh. Uchebn. Zaved., Khim. Tekhnol. 13, 1395 (1970); Ukr. Khim. Zh. 37, 650 (1971)

D.75 V.D. Dobrovolskii, S.M. Karalnik, S.A. Nemnonov: Izv. Vyssh. Uchebn. Zaved., Fiz. 1972, Nr. 11, 128

D.76 V.D. Dobrovolskii, S.M. Karalnik, A.V. Koval: Metallofizika 1972, Nr. 41, 68; Ukr. Fiz. Zh. 18, 323 (1973)
V.P. Maiboroda, V.D. Dobrovolskii, A.K. Sinelnichenko: Fiz. Met. Metalloved. 59, 895 (1985)

D.77 C.G. Dodd, G.L. Glen: J. Appl. Phys. 39, 5377 (1968); ibid. 40, 2361 (1969)

D.78 C.G. Dodd, G.L. Glen: Am. Mineral. 54, 1299 (1969); J. Am. Ceram. Soc. 53, 322 (1970)

D.79 C.G. Dodd, P.H. Ribbe: Phys. Chem. Miner. 3, 145 (1978)

D.80 E. Döring: Thesis, Moscow (1966); in [Ref.1.57, p.80]

D.81 G.N. Dolenko, A.P. Sadovskii et al.: Izv. Akad. Nauk SSSR, Ser. Fiz. 38, 603 (1974)
A.V. Nikolaev, L.M. Mazalov et al.: Dokl. Akad. Nauk SSSR 217, 865 (1974)

D.82 G.N. Dolenko, G.K. Parygina et al.: Zh. Strukt. Khim. 17, 539 (1976)
L.N. Mazalov, E.A. Kravtsova, S.V. Tsemkov, Yu.I. Nikonorov: Zh. Strukt. Khim. 18, 565 (1977)

D.83 G.N. Dolenko et al.: Zh. Strukt. Khim. 17, 435 (1976); ibid. 20, 334 (1979); Izv. Akad. Nauk SSSR, Ser. Khim. 1979, 343; Izv. Sib. Otd. Akad. Nauk SSSR, Ser. Khim. Nauk 1979, 32, 1980, 81
M.G. Voronkov et al.: Dokl. Akad. Nauk SSSR 248, 897 (1979)
A.V. Zibarev et al.: Izv. Sib. Otd. Akad. Nauk SSSR, Ser. Khim. Nauk 1980, 73
G.N. Dolenko, A.V. Zibarev, S.A. Krupoder, G.G. Furin: Izv. Sib. Otd. Akad. Nauk SSSR, Ser. Khim. Nauk 1980, 81
G.N. Dolenko, A.A. Voityuk, T.N. Dolenko, L.N. Mazalov: Zh. Strukt. Khim. 23, Nr.2, 34 (1982)

D.84 V.E. Dolgikh, E.Z. Kurmaev et al.: Fiz. Met. Metalloved. 40, 664 (1975)

D.85 E.P. Domashevskaya: Thesis. Woronezh 1966

D.86 E.P. Domashevskaya, L.N. Marshakova, Ya.A. Ugai, J.A. Yurakov: Izv. Vyssh. Uchebn. Zaved., Fiz. 17, 142 (1974)

D.87 E.P. Domashevskaya, L.N. Marshakova, V.A. Terekhov, Ya.A. Ugai: Izv. Akad. Nauk SSSR, Ser. Fiz. 38, 562, 567 (1974)
V.I. Nefedov, Ya.V. Salyn, E.P. Domashevskaya et al.: J. Electron Spectrosc. 6, 231 (1975)
V.A. Terekhov, E.P. Domashevskaya, L.N. Marshakova, K.B. Aleinikova: Izv. Akad. Nauk SSSR, Ser. Fiz. 40, 385 (1976)
Ya.A. Uagi et al.: Zh. Neorg. Khim. 22, 490 (1977)
E.P. Domashevskaya et al. Fiz. Tverd. Tela 19, 3610 (1977); ibid. 20, 2675, 3719 (1978); J. Electron Spectrosc. 16, 441 (1979)
E.P. Domashevskaya, V.A. Terekhov: Phys. Stat. Sol. (b) 105, 121 (1981)
V.A. Terekhov et al.: Fiz. Tverd. Tela 24, 283 (1982)

D.88 E.P. Domashevskaya et al.: Izv. Akad. Nauk SSSR, Ser. Fiz. 40, 389 (1976)
E.P. Domashevskaya et al.: Phys. Stat. Sol. (b) 106, 429 (1981)

D.89   E.P. Domashevskaya et al.: Izv. Akad. Nauk SSSR, Ser. Fiz. **40**, 2431 (1976);
        Metallofizika **2**, 24 (1980)

D.90   E.P. Domashevskaya et al.: Izv. Akad. Nauk SSSR, Neorg. Mater. **14**, 1183
        (1978)

D.91   E.P. Domashevskaya et al.: J. Non-Cryst. Solids **90**, 127 (1987)
        E.P. Domeshavskaya, O.A. Golikova, V.A. Terekhov, S.N. Trostyanskii: J.
        Non-Cryst. Solids **90**, 135 (1987)

D.92   R.J. Donahue, L.V. Azaroff: J. Appl. Phys. **38**, 2813 (1967)

D.93   M. Dorst: Wiss. Z. Karl-Marx-Univ. Leipzig, math.-naturwiss. R. **14**, 903
        (1965)

D.94   J.D. Dow: Phys. Rev. B **9**, 4165 (1974)

D.95   J.D. Dow, L.N. Watson, D.J. Fabian: J. Phys. F **4**, L 76 (1974)
        J.D. Dow, J.E. Robinson, J.H. Slowik, B.F. Sonntag: Phys. Rev. B **10**, 432
        (1974); in [Ref.1.86, p.649]
        R.P. Gupta, A.J. Freeman, J.D. Dow: Phys. Lett. A **59**, 226 (1976)
        J.D. Dow, D.L. Smith, D.R. Franceschetti, J.E. Robinson, T.R. Carver: Phys.
        Rev. B **16**, 4707 (1977)
        C.A. Swarts, J.D. Dow, C.P. Flynn: Phys. Rev. Lett. **43**, 158 (1979)

D.96   J.D. Dow, D.L. Smith, B.F. Sonntag: Phys. Rev. B **10**, 3092 (1974)

D.97   L.G. Dowell, J.M. Bennett, D.E. Passoja: Adv. X-Ray Anal. **20**, 471 (1977)
        R.L. Patton, E.M. Flanigen, L.G. Dowell, D.E. Passoja: ACS Symp. Ser. **40**, 64
        (1977)

D.98   H. Drack, M. Grasserbauer: Mikrochim. Acta 1977, S 7-289
        H. Drack, S. Kosina, M. Grasserbauer: Fresenius Z. Anal. Chem. **295**, 30
        (1979)
        M. Grasserbauer, H. Drack, H. Malissa, Jr.: Mikrochim. Acta **1**, 29 (1979)
        M. Grasserbauer: Angew. Chemie **12**, 1059 (1981)

D.99   G. Dräger, W. Beier, O. Brümmer, P. Krönert: Krist. Technik **9**, 1291 (1974)
        O. Brümmer, G. Dräger: Mikrochim. Acta. S6-321 (1975)
        O. Brümmer et al.: Wiss. Z. Martin-Luther-Univ. Halle, math.-naturwiss. R.
        24, 5 (1975)

D.100   G. Dräger, O. Brümmer: Izv. Akad. Nauk SSSR, Ser. Fiz. **40**, 2437 (1976)
        U. Berg, G. Dräger, K. Mosebach, O. Brümmer: Phys. Stat. Sol. (b) **75**, K 89
        (1976)

D.101   G. Dräger, O. Brümmer, J. Bonitz: Phys. Stat. Sol. (b) **94**, K 111 (1979)

D.102   G. Dräger, O. Brümmer: Phys. Stat. Sol. (b) **98**, K 33 (1980)

D.103   G. Dräger et al.: Phys. Stat. Sol. (b) **104**, 219 (1981)

D.104   G. Dräger, F. Werfel, O. Brümmer: Phys. Stat. Sol. (b) **113**, K 15 (1982)
        G. Dräger, R. Wippermann, O. Brümmer, A. Simunek: Phys. Stat. Sol. (b)
        118, K 113 (1983)

D.105   G. Dräger, O. Brümmer: In [Ref.1.66, p.337]
        G. Dräger, W. Czolbe, A. Simunek, F. Levy: Cryst. Res. Technol. **20**, 1451
        (1985)

D.106   J. Drahokoupil, J. Urban, P. Vilim: Czech. J. Phys. B **18**, 1034 (1968)

D.107   J. Drahokoupil: J. Phys. C **5**, 2259 (1972)

D.108   J. Drahokoupil, A. Simunek: In [Ref.1.60, Vol.II, p.456]

D.109   J. Drahokoupil, H. Klokocnikova, A. Simunek: Phys. Fenn. **9**, S 1-197 (1974);
        in [Ref.1.62, p.154]; J. Phys. C **9**, 2667 (1976)

D.110   J. Drahokoupil, A. Simunek: J. Phys. C **7**, 610 (1974); Czech. J. Phys. B **25**,
        542 (1975)
        A. Simunek: Czech. J. Phys. (Engl. Transl.) B **26**, 239 (1976)

D.111   J. Drahokoupil, A. Simunek: Czech. J. Phys. B **36**, 702 (1986)

D.112   J. Drahokoupil et al.: J. Non-Cryst. Solids **86**, 43 (1986)

D.113   B. Drevillon et al.: Philos. Mag. B **54**, 335 (1986)
D.114   V.P. Dseganovskii; E.A. Zhurakovskii, T.N. Bondarenko, I.N. Frantsevich: Dopov. Akad. Nauk. Ukr. SSR A **34**, 161 (1972)
D.115   V.S. Dubey, B.D. Shrivastava: Phys. Stat. Sol. (b) **53**, K 51 (1972)
        M.S. Kushwaha, B.D. Shrivastava, V.S. Dubey: J. Phys. F. **5**, 597 (1975)
D.116   F.B.M. Duivenvoorden, D.C. Koningsberger, Y.S. Uh, B.C. Gates: J. Am. Chem. Soc. **108**, 6254 (1986)
D.117   P.J. Durham et al.: J. Phys. F **9**, 1719 (1979)
D.118   P.J. Durham, J.B. Pendry: Solid State Commun. **38**, 159 (1981)
D.119   P.J. Durham: J. Phys. F **12**, 1539 (1982)
D.120   Ya.I. Dutchak, I.V. Kavich et al.: Izv. Akad. Nauk SSSR, Neorg. Mat. **9**, 1729 (1973)
        G.I. Ilkiv, I.V. Kavich: Ukr. Fiz. Zh. **16**, 153 (1974)
        Ya.I. Dutchak et al.: Ukr. Fiz. Zh. **24**, 1556 (1979)
D.121   Ya.I. Dutchak, I.V. Kavich et al.: Ukr. Fiz. Zh. **20**, 1194 (1975)
        Ya.I. Dutchak, I.V. Kavich, R.A. Antonyuk: Ukr. Fiz. Zh. **21**, 1476 (1976)
        I.V. Kavich, R.A. Antonyuk: Metallofizika **71**, 45 (1978); Ukr. Fiz. Zh. **23**, 1465 (1978)
D.122   Ya.I. Dutchak, I.V. Kavich, P.O. Shevchuk: Ukr. Fiz. Zh. **22**, 822 (1977)
D.123   Ya.I. Dutchak et al.: Ukr. Fiz. Zh. **23**, 1692 (1978)
D.124   Ya. I. Dutchak, I.V. Kavich, V.G. Sinyushko, B.N. Yatsyk: In [Ref.1.100, p.131]
D.125   Ya.I. Dutchak, I.E. Shcherba, B.N. Yatsyk: Ukr. Fiz. Zh. **28**, 559 (1983)
        Ya.I. Dutchak et al.: Izv. Akad. Nauk SSSR, Ser. Fiz. **46**, 818 (1982)
D.126   Ya.I. Dutchak, I.V. Kavich, P.I. Shevchuk: Sci. Sintering **16**, 41 (1984)
D.127   K.G. Dyall, F.P. Larkins: J. Phys. B **15**, 1811 (1982)
        K.G. Dyall, I.P. Grant: J. Phys. B **17**, 1281 (1984)
D.128   K.G. Dyall, R.E. LaVilla: Phys. Rev. A **34**, 5123 (1986)

E.1   R.W. Eason, D.K. Bradley, P.J. Dobson, J.D. Hares: Appl. Phys. Lett. **47**, 442 (1985)
E.2   W. Eberbeck: Z. Physik **149**, 412 (1957)
E.3   F. Edelmann, J. Eggs, K. Ulmer: In [Ref.1.58, Vol.I, p.13]
E.4   N.N. Efremova, L.D. Finkelshtein, N.D. Samsonov, S.A. Nemnonov: Izv. Akad. Nauk SSSR, Ser. Fiz. **40**, 420 (1976)
E.5   J. Eggs, K. Ulmer: Phys. Lett. **26** A 246 (1968)
E.6   R.C. Ehlert, R.A. Mattson: Adv. X-Ray Anal. **9**, 456 (1966)
E.7   R. Eibler, J. Redinger, A. Neckel: J. Phys. F **17**, 1533 (1987)
E.8   G. Eichhoff: Ann. Physik **1**, 55 (1958)
E.9   R. Eisberg, P. Josuks, G. Wiech, R. Schlogl: Solid State Commun. **60**, 827 (1986)
E.10   P. Eisenberger, B.M. Kincaid: Chem. Phys. Lett. **36**, 134 (1975)
E.11   P. Eisenberger, R.G. Shulman, G.S. Brown, S. Ogawa: Proc. Natl. Acad. Sci. USA **73**, 491 (1976)
        R.G. Shulman, Y. Yafet, P. Eisenberger, W.E. Blumberg: Proc. Natl. Acad. Sci. USA **73**, 1384 (1976)
        P. Eisenberger et al.: Nature (London) **274**, 30 (1978)
E.12   A. Ejiri, S. Yamaguchi et al.: Opt. Commun. **1**, 349 (1970)
E.13   A. Ejiri, F. Sugawara, H. Onuki: Jpn. J. Appl. Phys. **17**, S 2-204 (1978)
E.14   B. Ekstig: Ark. Fysik **37**, 107 (1968)
E.15   B. Ekstig, E. Källne, E. Noreland, R. Manne: Phys. Scr. **2**, 38 (1970)
E.16   B. Ekstig, E. Källne, E. Noreland, R. Manne: J. Physique **32**, C 4-214 (1971)

E.17   M. Elango, A. Maiste, R. Ruus: Phys. Lett. 72 A, 16 (1979)
E.18   D.C. Eldridge: Thesis. 1974
E.19   V.P. Elin, V.P. Feshin, M.G. Voronkov: J. Mol. Struct. 83, 329 (1982)
E.20   S.R. Elliott, A.T. Steel: Phys. Rev. Lett. 57, 1316 (1986)
E.21   M. Emili et al.: J. Non-Cryst. Solids 74, 129 (1985)
E.22   H. Endo, M. Uda, K. Maeda: Phys. Rev. A 22, 1436 (1980)
E.23   A. Erbil, W. Weber, G.S. Cargill, R.F. Boehme: Phys. Rev. B 34, 1392 (1986)
E.24   M. Erbudak, V.A. Gubanov, E.Z. Kurmaev: J. Phys. Chem. Solids 39, 1157
        (1978)
        E.Z. Kurmaev, V.M. Cherkashenko, V.A. Gubanov: J. Electron Spectrosc. 16,
        455 (1979)
E.25   S.B. Erenburg, I.I. Kukushkina, I.A. Ovsyannikova, E.N. Yurchenko: Zh. Fiz.
        Khim. 47, 1878 (1973)
        S.B. Erenburg, I.A. Ovsyannikova: React. Kin. Catal. Lett. 1, 507 (1974)
        I.A. Ovsyannikova et al.: Izv. Akad. Nauk SSSR, Ser. Fiz. 40, 230 (1976);
        Kinet. Katal. 17, 1072 (1976)
        B.N. Kusnetsov et al.: Kinet. Katal. 19, 749 (1978)
E.26   N.V. Ershov, Yu.A. Babanov, V.R. Galakhov: Phys. Stat. Sol. (b) 117, 749
        (1983)
E.27   O.A. Ershov, D.A. Gogonov, A.P. Lukirskii: Fiz. Tverd. Tela 7, 2355 (1965)
        V.I. Baranovskii, M.S. Nakhmanson, Yu.M. Zaitsev: Zh. Strukt. Khim. 13, 848
        (1972)
E.28   O.A. Ershov, A.P. Lukirskii: Fiz. Tverd. Tela 8, 1699, 2137 (1966)
E.29   E.I. Esmail, C. Nicholls, D.S. Urch, P. Wood: In [Ref.1.60, Vol.I, p.415]
E.30   J.M. Esteva, B. Gauthe, P. Dhez, R.C. Karnatak: J. Phys. B 16, L 263 (1983)
E.31   J.M. Esteva, R.C. Karnatak, J.C. Fuggle, G.A. Sawatzky: Phys. Rev. Lett. 50,
        910 (1983)
E.32   J.-M. Esteva, R.C. Karnatak: In *Giant Resonances in Atoms, Molecules and
        Solids*, ed. by J.P. Connerade, J.-M Esteva, R.C. Karnatak (Plenum, 1987)
        p.361
E.33   F. Evangelisti et al.: Solid State Commun. 37, 413 (1981)
        M.G. Proietti et al.: In [Ref.1.83, p.26]

F.1   G.W. Fabel, W.B. White, E.W. White, R. Rou: Proc. 3rd Lunar Sci. Conf. MIT
        Cambridge, Mass. 1972, p.939
F.2   A. Faessler, E.D. Schmid: Z. Physik 138, 71 (1954)
F.3   A. Faessler, M. Goehring: Z. Physik 142, 558 (1955)
        A. Faessler, P. Mecke: Z. Elektrochemie 64, 587 (1960)
F.4   A. Faessler, R. Schmid: Z. Physik 190, 10 (1966)
F.5   S. Falch, P. Lamparter, S. Steeb: Z. Naturforsch. A 39, 1175 (1984)
F.6   D. Fargues, F. Vergand, E. Belin, C. Bonnelle: Surf. Sci. 106, 239 (1981)
        D. Fargues, F. Vergand, C. Bonnelle: J. Phys. F 14, 1897 (1984); Surf. Sci.
        156, 590 (1985); ibid. 163, 489 (1985)
        B. Iraqi, F. Vergand, D. Fargues, C. Bonnelle: Surf. Sci. 162, 871 (1985)
F.7   S.S.P. Farkin, J.V. Acrivos: J. Physique 44, C3-1011 (1983)
F.8   J.L. Feldman et al.: Solid State Commun. 49, 1023 (1984)
F.9   J.L. Feldman et al.: Phys. Rev. B 33, 7961 (1986)
F.10   D.E. Fenton, C.J. Nicholls, D.S. Urch: Chem. Phys. Lett. 23, 211 (1973)
        C.J. Nicholls, D.S. Urch: J. Chem. Soc., Dalton Trans. 1975, 2143
        A.J.C.L. Hogarth, D.S. Urch: J. Chem. Soc., Dalton Trans. 1976, 794
        M. Al-Kadier, C. Tolon, D.S. Urch: J. Chem. Soc., Faraday Trans. II 80, 669
        (1984)

F.11   K. Feser: Phys. Rev. Lett. **28**, 1013 (1972); ibid. **29**, 901 (1972)
A. Faessler: In [Ref.1.86, p.801]

F.12   K. Feser, J. Müller, A. Faessler, G. Wiech: In [Ref.1.60, Vol.II, p.304]

F.13   B. Feuerbacher, R.P. Godwin, T. Sasaki, M. Skibowski: J. Opt. Soc. Am. **58**, 1434 (1968)
M. Cardona, W. Gudat et al.: Phys. Rev. Lett. **25**, 659 (1970)

F.14   M. Fichter: Thesis. München 1966; in [Ref.1.57, p.112]; Spectrochim. Acta **30** B, 417 (1975)

F.15   E.O. Filatova, A.S. Vinogradov, T.M. Zimkina: Fiz. Tverd. Tela **27**, 997 (1985)

F.16   J.M. Fine et al.: J. Vac. Sci. Technol. A **1**, 1036 (1983)

F.17   J. Fink et al.: Phys. Rev. B **32**, 4899 (1985) ·
J. Zaanen et al.: Phys. Rev. B **32**, 4905 (1985)

F.18   L.D. Finkelshtein, S.A. Nemnonov: Fiz. Met. Metalloved. **22**, 843 (1966)
I.V. Gribov, V.I. Minin, L.D. Finkelshtein: Fiz. Met. Metalloved. **47**, 949 (1979)

F.19   L.D. Finkelshtein, S.A. Nemnonov: Fiz. Met. Metalloved. **26**, 481 (1968); ibid. **32**, 662 (1971)

F.20   L.D. Finkelshtein, S.A. Nemnonov et al.: Izv. Akad. Nauk SSSR, Ser. Fiz. **38**, 654 (1974)
N.N. Efremova, S.A. Nemnonov, L.D. Finkelshtein: Fiz. Met. Metalloved. **38**, 868 (1974)

F.21   L.D. Finkelshtein, N.D. Samsonova, G.V. Batsuev: Zh. Neorg. Khim. **25**, 2029 (1980)
L.D. Finkelshtein et al.: Zh. Strukt. Khim. **23**, Nr.1, 87 (1982)
L.D. Finkelshtein, N.D. Samsonova: Fiz. Met. Metalloved. **53**, 718 (1982)
N.D. Samsonova, L.D. Finkelshtein, E.M. Levin: Fiz. Tverd. Tela **24**, 3711 (1983)
N.D. Samsonova, R.V. Skolozdra, L.D. Finkelshtein: Fiz. Tverd. Tela **27**, 2203 (1985)
L.D. Finkelshtein, N.D. Samsonova: Fiz. Met. Metalloved. **56**, 466 (1983)
L.D. Finkelshtein, N.D. Samsonova: Fiz. Tverd. Tela **25**, 3167 (1983)
L.D. Finkelshtein et al.: Fiz. Tverd. Tela **26**, 3442 (1984)
L.D. Finkelshtein, N.I. Lobachevskaya, N.N. Efremova, B.A. Gizhevskii: Fiz. Tverd. Tela **27**, 204 (1985)
L.D. Finkelshtein, N.D. Samsonova, E.M. Levin: Fiz. Met. Metalloved. **59**, 1036 (1985)
L.D. Finkelshtein, N.I. Lobachevskaya, N.N. Efremova: Fiz. Tverd. Tela **27**, 3689 (1985)

F.22 · J. Finster, A. Meisel: In [Ref.1.58, Vol. II, p.350]

F.23   J. Finster, G. Leonhardt, A. Meisel: J. Physique **32**, C 4-218 (1971)

F.24   J. Finster et al.: Z. Chemie **13**, 146 (1973)

F.25   J. Finster, P. Müller, N. Meusel, A. Meisel: In [Ref.1.60, Vol. II, p.467]

F.26   J. Finster, N. Meusel, A. Meisel: Phys. Fenn. **9**, S 1-425 (1974)
N. Meusel: Thesis, Leipzig 1976

F.27   J. Finster, P. Müller, F. Thiel, A. Meisel: In [Ref.1.62, p.214]

F.28   J. Finster, A. Meisel: In [Ref.1.100, p.28]

F.29   M.N. Firsov, V.I. Nefedov, E.P. Domashevskaya, I.S. Shaplygin: Zh. Strukt. Khim. **20**, 49 (1979)
M.N. Firsov, V.I. Nefedov, I.S. Shaplygin: Zh. Strukt. Khim. **23**, 72 (1982)
V.I. Nefedov, M.N. Firsov, I.S. Shaplygin: J. Electron Spectrosc. **26**, 65 (1982)
M.N. Firsov, V.I. Nefedov, I.S. Shaplygin: Izv. Akad. Nauk SSSR, Neorg. Mater. **19**, 660 (1983)

F.30   D.A. Fischer et al.: Surf. Sci. **177**, 114 (1986)
F.31   D.W. Fischer, W.L. Baun: Adv. X-Ray Anal. **7**, 489 (1963); Spectrochim. Acta **21**, 443 (1965)
F.32   D.W. Fischer: J. Chem. Phys. **42**, 3814 (1965)
F.33   D.W. Fischer, W.L. Baun: Analyt. Chem. **37**, 902 (1965)
F.34   D.W. Fischer, W.L. Baun: J. Chem. Phys. **43**, 2075 (1965); Phys. Rev. **145**, 555 (1966)
F.35   D.W. Fischer: J. Appl. Phys. **36**, 2048 (1965); Adv. X-Ray Anal. **13**, 159 (1970)
F.36   D.W. Fischer, W.L. Baun: Adv. X-Ray Anal. **9**, 329 (1966)
F.37   D.W. Fischer, W.L. Baun: J. Appl. Phys. **37**, 768 (1966); ibid. **38**, 229 (1967)
F.38   D.W. Fischer, W.L. Baun: J. Appl. Phys. **38**, 4830 (1967)
F.39   D.W. Fischer, W.L. Baun: J. Appl. Phys. **39**, 4757 (1968)
       D.W. Fischer: J. Appl. Phys. **41**, 3561 (1970)
F.40   D.W. Fischer: J. Appl. Phys. **40**, 4151 (1969); ibid. **41**, 3922 (1970)
F.41   D.W. Fischer: Appl. Spectrosc. **25**, 263 (1971)
F.42   D.W. Fischer: J. Phys. Chem. Solids **32**, 2455 (1971)
F.43   D.W. Fischer: Phys. Rev. B **4**, 1778 (1971); ibid. **5**, 4219 (1972)
F.44   D.W. Fischer: In [Ref.1.79, p.669]
F.45   D.W. Fischer: Phys. Rev. B **8**, 3576 (1973)
F.46   E.O. Fischer, G. Joos, E. Vogg: Z. Phys. Chemie (Frankfurt) **18**, 80 (1958)
       A. Schneider: Z. Phys. Chemie (Frankfurt) **31**, 249 (1962)
       G. Fiedler: Z. Phys. Chemie (Frankfurt) **37**, 79 (1963)
F.47   A.M. Flank et al.: J. Non-Cryst. Solids **91**, 306 (1987)
F.48   M. Fliyou, M.A. Khan, J. Ringeissen: Opt. Commun. **49**, 135 (1984)
F.49   A. Flores-Riveros et al.: J. Chem. Phys. **83**, 2053 (1985)
F.50   C.P. Flynn, N.O. Lipari: Phys. Rev. B **7**, 2215 (1973)
F.51   H. Föll: Z. Physik B **26**, 329 (1977)
F.52   H. Föll, K. Ulmer: Phys. Stat. Sol. (a) **41**, 113 (1977)
F.53   V.A. Fomichev: Fiz. Tverd. Tela **8**, 2892; Izv. Akad. Nauk SSSR, Ser. Fiz. **31**, 957 (1967)
F.54   V.A. Fomichev, A.P. Lukirskii: Opt. Spektrosk. **22**, 796 (1967)
F.55   V.A. Fomichev, M.A. Rumsh: J. Phys. Chem. Solids **29**, 1015 (1968)
       V.A. Fomichev, I.I. Zhukova, I.K. Polushina: J. Phys. Chem. Solids **29**, 1025 (1968)
F.56   V.A. Fomichev: Fiz. Tverd. Tela **10**, 763 (1968)
F.57   V.A. Fomichev, I.I. Zhukova: Fiz. Tverd. Tela **10**, 3073, 3753 (1968)
F.58   V.A. Fomichev, R.L. Barinskii: Zh. Strukt. Khim. **11**, 875 (1970)
F.59   V.A. Fomichev, A.V. Rudnev, T.M. Zimkina: In [Ref.1.96, Vol.II, p.175]
       A.S. Shulakov, T.M. Zimkina, V.A. Fomichev, E.Z. Kurmaev: Fiz. Tverd. Tela **15**, 2092 (1973)
F.60   V.A. Fomichev: Fiz. Tverd. Tela **12**, 2639 (1970); ibid. **13**, 907 (1971)
F.61   V.A. Fomichev, A.V. Rudnev, S.A. Nemnonov: Izv. Akad. Nauk SSSR, Ser. Fiz. **36**, 291 (1972)
F.62   V.A. Fomichev: Fiz. Tverd. Tela **15**, 1286 (1973)
F.63   V.A. Fomichev, S.A. Gribovskii, T.M. Zimkina: Fiz. Tverd. Tela **15**, 2817 (1973)
F.64   V.A. Fomichev, T.M. Zimkina, A.V. Rudnev: In [Ref.1.79, p.259]
F.65   V.A. Fomichev, S.A. Gribovskii, T.M. Zimkina: Fiz. Tverd. Tela **15**, 201 (1973)
       A. Karosiene et al.: Izv. Akad. Nauk SSSR, Ser. Fiz. **38**, 426 (1974)
F.66   V.A. Fomichev: Izv. Akad. Nauk SSSR, Ser. Fiz. **38**, 533 (1974)
F.67   V.A. Fomichev et al.: In [Ref.1.98, p.173]

F.68 A. Fontaine, P. Lagarde, D. Raoux, J.M. Esteva: J. Phys. F 9, 2143 (1979)
F.69 A. Fontaine et al.: Phil. Mag. B 40, 17 (1979)
F.70 M.P. Fontana et al.: Solid State Commun. 43, 561 (1982)
F.71 R. Fox, S.J. Gurman: J. Phys. C 13, L249 (1980)
F.72 C.P. Franck et al.: Phys. Rev. B 31, 5366 (1985)
F.73 K.H. Frank, B. Reihl, Z. Fisk, G. Schmiester: J. Less-Common Met. 111, 251 (1985)
F.74 I.N. Frantsevich, E.A. Zhurakovskii, A.V. Kurdyumov, N.N. Vasilenko: Dokl. Akad. Nauk SSSR 203, 87 (1972)
F.75 I.N. Frantsevich, E.A. Zhurakovskii, V.V. Shvaiko, Ukr. Fiz. Zh. 20, 1177 (1975); Visn. Akad. Nauk Ukr. SSR (1976) 9
F.76 I.N. Frantsevich, E.A. Zhurakovskii, V.V. Shvaiko: Ukr. Fiz. Zh. 21, 1339 (1976)
F.77 S.P. Freidman, V.A. Gubanov, E.Z. Kurmaev: Zh. Strukt. Khim. 21, No.4, 37 (1980)
      S.P. Freidman et al.: Zh. Strukt. Khim. 25, No.1, 42 (1984)
F.78 S.P. Freidman et al.: Izv. Akad. Nauk SSSR, Ser. Fiz. 46, 779 (1982)
F.79 S.P. Freidman et al.: Zh. Strukt. Khim. 27, No.2, 70 (1986)
F.80 V.S. Frenchko et al.: Izv. Akad. Nauk SSSR, Ser. Fiz. 49, 1535 (1985)
F.81 F. Freund, M. Hamich: Z. Anorg. Allg. Chemie 385, 209 (1971); Fortschr. Mineral. 48, 243 (1971)
F.82 F. Freund: Phys. Stat. Sol. (b) 66, 271 (1974)
F.83 F. Freund: J. Catalysis 32, 159 (1974)
F.84 F. Freund: Bull. Soc. Chim. Belg. 84, 101 (1975)
F.85 H. Friedrich et al.: Chem. Phys. Lett. 64, 360 (1979); J. Phys. B 13, 25 (1980)
F.86 C.M. Friend et al.: J. Chem. Phys. 87, 1847 (1987)
F.87 V. Fritzsche, P. Rennert: Phys. Stat. Sol. (b) 142, 15 (1987)
F.88 J.C. Fuggle, E. Källne, L.M. Watson, D.J. Fabian: Phys. Rev. B 16, 750 (1977)
F.89 J.C. Fuggle, F.U. Hillebrecht, J.M. Esteva, R.C. Karnatak: Phys. Rev. B 27, 4637 (1983)
      J.C. Fuggle: Physica B+C 130, 56 (1985)
F.90 J.C. Fuggle: Phys. Scr. T 17, 64 (1987)
F.91 T. Fujikawa: J. Electron Spectrosc. 26, 79 (1982)
F.92 T. Fujikawa: J. Phys. Soc. Jpn. 52, 4001 (1983)
F.93 T. Fujikawa et al.: J. Phys. Soc. Jpn. 55, 4074 (1986)
F.94 T. Fujikawa et al.: J. Phys. Soc. Jpn. 55, 4090 (1986)
F.95 H. Fujimori, H. Iba, H. Tsuchiya: Jpn. J. Appl. Phys. 22, 1025 (1983)
F.96 K. Fujimori: Sci. Rep. Tohoku Univ. I 47, 1, 37, 50 (1963)
F.97 H. Fujimoto: Sci. Rep. Tohoku Univ. I 49, 28, 32, 37, 43 (1965)
      H. Sugawara, H. Fujimoto: J. Phys. Soc. Jpn. 31, 1527 (1971)
F.98 F. Fujiwara, R. Chong, G. Andermann: Spectrosc. Lett. 13, 521 (1980)
F.99 Y. Fukai, S. Kazama, K. Tanaka, M. Matsumoto: Solid State Commun. 19, 507 (1976)
      Y. Sato, K. Tanaka, M. Yasuda: J. Less-Common Met. 38, 251 (1982)
F.100 S. Fukushima et al.: J. Am. Ceram. Soc. 68, 490 (1985)
F.101 S. Furukawa, M. Seki, S. Maeyama: Phys. Rev. Lett. 57, 2029 (1986)

G.1   V.R. Galakhov, E.Z. Kurmaev, V.M. Cherkashenko: Izv. Akad. Nauk SSSR, Ser. Fiz. 49, 1513 (1985)
G.2   V.R. Galakhov, E.Z. Kurmaev, A.V. Postnikov: Solid State Commun. 58, 143 (1986)
G.3   V.R. Galakhov et al.: Fiz. Met. Metalloved. 64, 186 (1987)

G.4   B. Gale, J.A. Catterall, J. Trotter: Phil. Mag. 20, 79 (1969)

G.5   D. Galland et al.: Polymer 27, 883 (1985)

G.6   P. Gallezot, R. Weber, R.A. Dalla Betta, M. Bondart: Z. Naturforsch. A 34, 40 (1979)

G.7   E.A. Galtsova, L.N. Mazalov et al.: Izv. Sib. Otd. Akad. Nauk SSSR, Ser. Khim. Nauk (1975) Nr. 9, H.4, 41
        G.K. Parygina, L.N. Mazalov et al.: Zh. Strukt. Khim. 16, 355 (1975)
        G.K. Parygina et al.: 13 Vses. Chugaev Soveshch. po Khimii Kompleks. Soedin. (1978) 307

G.8   T.M. Galushka, A.Z. Zhmudskii, L.A. Musatenko, V.I. Shiyanovskii: Ukr. Fiz. Zh. 22, 840 (1977)

G.9   J. Garcia et al.: Solid State Commun. 58, 595 (1986); Nuovo Cimento D 7, 493 (1986)

G.10   K.B. Garg, E. Källne, B. Ekstig. E. Noreland: In [Ref.1.98, p.155]
        K.B. Garg, E. Källne: Phys. Stat. Sol. (b) 70, K 121 (1975)

G.11   K.B. Garg, B.K. Sharma, D.C. Jain, A.I.P. Sinha: Phys. Stat. Sol. (b) 102, K 37 (1980)
        K.B. Garg, J. Leiro, M. Heinonen, M. Richardson: Phys. Stat. Sol. (a) 83, 279 (1984)

G.12   K.B. Garg, M. Heinonen, J. Leiro: J. Phys. F 12, 1547 (1982)

G.13   K.B. Garg, D. Normal, R.B. Beeken: Phys. Stat. Sol. (a) 96, 301 (1986)

G.14   K.B. Garg et al.: Solid State Commun. 62, 575 (1987)

G.15   E. Garstein, M. Radler, T.O. Mason, J.B. Cohen, In *Characterization of Defects in Materials*, ed. by R.W. Siegel, J.R. Weertman, R. Sinclair (Mater. Res. Soc. Pittsburgh, 1987) p.65

G.16   M. Gasgnier et al.: J. Less-Common Met. 127, 367 (1987)

G.17   P.H. Gaskell et al.: J. Phys. C 15, L 597 (1982)

G.18   Gy. Gati, A. Meisel: Z. Chemie 4, 171 (1964)

G.19   I.I. Geguzin, G.I. Alperovich, I.Ya. Nikiforov: Izv. Akad. Nauk SSSR, Ser. Fiz. 36, 305 (1972)
        I.I. Geguzin: Thesis. Rostov/Don 1973

G.20   I.I. Geguzin, I.A. Topol, I.Ya Nikiforov, G. Leonhardt: Ann. Physik 28, 341 (1973)

G.21   I.I. Geguzin, A.P. Kovtun, V.P. Sachenko, I.A. Topol: Izv. Akad. Nauk SSSR, Ser. Fiz. 40, 288 (1976)

G.22   I.I. Geguzin, V.N. Datsyuk, R.V. Vedrinskii: Phys. Stat. Sol. (b) 109, 563 (1982)
        R.V. Vedrinskii et al.: Solid State Commun. 44, 1401 (1982); Phys. Stat. Sol. (b) 111, 433 (1982)
        R.V. Vedrinskii, L.A. Bugaev: Fiz. Tverd. Tela 27, 280 (1985)

G.23   P.V. Gel, E.A. Zhurakovskii, S.G. Avdeev: In [Ref.1.100, p.149]

G.24   F.Kh. Gelmukhanov, L.N. Mazalov, A.V. Kondratenko: Chem. Phys. Lett. 46, 133 (1977)
        A.V. Kondratenko et al.: Zh. Strukt. Khim. 18, 622 (1977)

G.25   G. Genchev: Dokl. Bolg. Akad. Nauk 30, 997, 1113 (1977)

G.26   H. Grenz, W. Low, A. Richter, W. Schäfer: J. Magn. Magn. Mater. 45, 309 (1984)

G.27   H.C. Gerritsen et al.: J. Appl. Phys. 60, 1774 (1986)

G.28   M.N. Ghatikar, B.D. Padalia: J. Phys. C 11, 1941 (1978)

G.29   F.A. Gianturco: J. Phys. B 1, 614 (1968)

G.30   F.A. Gianturco: Chem. Phys. Lett. 17, 127 (1972)

G.31   F.A. Gianturco, C. Guidotti, V. Lamanna: J. Chem. Phys. 57, 840 (1972); F.A. Gianturco: In [Ref.1.60, Vol.I, p.321]

G.32   F.A. Gianturco, E. Semprini, F. Stefani: Nuovo Cimento D **2**, 687 (1983)

G.33   P.D. Gigle, G.A. Savanick, E.W. White: J. Electrochem. Soc. **117**, 15 (1970)

       H.B. Krause, G.A. Savanick, E.W. White: J. Electrochem. Soc. **117**, 557 (1970)

G.34   D. Gignoux et al.: J. Physique **43**, 173 (1982)

G.35   E. Gilberg: Thesis München 1969

G.36   E. Gilberg: In [Ref.1.58, Vol.I, p.277]; Z. Physik **236**, 21 (1970)

G.37   E. Gilberg: In [Ref.1.60, Vol.I, p.373]

G.38   E. Gilberg: In [Ref.1.60, Vol.II, p.486]; Phys. Stat. Sol. (b) **69**, 477 (1975)

G.39   E. Gilberg: In [Ref.1.62, p.229]

G.40   E. Gilberg, W. Schätzl, H.W. Schrenk: Chem. Phys. **13**, 115 (1976)

       E. Gilberg: Mol. Phys. **44**, 871 (1981)

G.41   E. Gilberg, M.J. Hanus, B. Föltz: Jpn. J. Appl. Phys. **17**, S2-101 (1978); Rev. Sci. Instrum. **52**, 7 (1981)

       E. Gilberg: Proc. SPIE Int. Soc. Opt. **447**, 68 (1984)

G.42   T.L. Gilbert, W.J. Stevens, H. Schrenk, M. Yoshimine, P.S. Bagus: Phys. Rev. B **8**, 5977 (1973)

G.43   I.I. Glembocki et al.: Liet. Fis. Rinkinys **12**, 25, 235 (1972)

G.44   G.L. Glen, C.G. Dodd: J. Appl. Phys. **39**, 5372 (1968)

G.45   G.L. Glen, C.V. Hurst: Develop. Appl. Spectrosc. **9**, 307 (1971)

G.46   A.J. Glick, P. Longe, S.M. Bose: In [Ref.1.78, p.319]

       P. Longe: Phys. Rev. B **8**, 2572 (1973)

       S.M. Bose, A.J. Glick: Phys. Rev. B **10**, 2733 (1974); Phys. Rev. B **17**, 2073 (1978)

G.47   E.S. Gluskin, A.P. Sadovskii, L.N. Mazalov: Zh. Strukt. Khim. **14**, 739 (1973)

       E.S. Gluskin: Thesis Novosibirsk (1973)

       E.S. Gluskin, L.N. Mazalov, A.P. Sadovskii, D.A. Zhogolev: Zh. Strukt. Khim. **16**, 1061 (1975)

G.48   E.S. Gluskin, L.N. Mazalov et al.: Izv. Sib. Otd. Akad. Nauk SSSR, Ser. Khim. Nauk (1974) 3; Zh. Strukt. Khim. **15**, 304 (1974)

G.49   E.S. Gluskin et al.: Izv. Akad. Nauk SSSR, Ser. Fiz. **40**, 226 (1976)

G.50   E.S. Gluskin, A.A. Krasnoperava, L.N. Mazalov: Zh. Strukt. Khim. **18**, 185, 665 (1977)

G.51   J.P. Goel, O.K. Harsh, K.S. Srivastava: Physica **142** B&C, 55 (1986)

G.52   Y. Gohshi: Adv. X-Ray Anal. **12**, 518 (1969); in [Ref.1.60, Vol.II, p.250]

       Y. Gohshi, H. Kamada: Proc. Jpn. Acad., Ser. B **56**, 167 (1980)

G.53   Y. Gohshi, K. Yanagase: J. Fuel Soc. Jpn. **52**, 926 (1973)

       Y. Gohshi, A. Ohtsuka: Spectrochim. Acta B **28**, 179 (1973)

       Y. Gohshi, O. Hirao, I. Suzuki: Adv. X-Ray Anal. **18**, 406 (1975)

       I. Nakamura, Y. Gohshi: Chem. Lett. 1171 (1975)

G.54   Y. Gohshi, J. Kashiwakura: Phys. Fenn. **9**, S 1-330 (1974); Spectrochim. Acta B **30**, 471 (1975)

G.55   Y. Gohshi, O. Hirao: X-Ray Spectrom. **4**, 74 (1975)

G.56   Y. Gohshi, I. Nakamura, M. Yoshimura: X-Ray Spectrom. **4**, 117 (1975)

G.57   Y. Gohshi, H. Miyamoto, M. Kudo, H. Kamada: Jpn. J. Appl. Phys. **17**, S 2-412 (1978)

G.58   B.G. Gokhale, R.B. Chester, F. Boehm: Phys. Rev. Lett. **18**, 957 (1967)

G.59   B.G. Gokhale, S.N. Shukla: J. Phys. B **3**, 438, 1392 (1970)

G.60   B.G. Gokhale, S. Rai, S.D. Rai: J. Phys. F **7**, 299 (1977)

G.61   B.G. Gokhale, U.D. Misra: J. Phys. B **13**, 1317 (1980)

G.62   B.G. Gokhale, S.N. Shukla, R.N. Srivastava: Phys. Rev. A **28**, 858 (1983); Phys. Rev. A **31**, 2715 (1985)

G.63   A.I. Goldman et al.: Appl. Phys. Lett. **43**, 836 (1983)

G.64   A.I. Goldman, F. Canova, Y.H. Kao: Solid State Commun. **58**, 277 (1986)

G.65   D.A. Goodings, R. Harris: Proc. Phys. Soc. (Solid State Phys.) 2, 1808 (1969)

G.66   V.V. Gorbachev, A.S. Bystrikov, S.K. Vasilev, L.V. Bogomolova: Fiz. Khim. Stekla 9, 648 (1983)

G.67   Yu.K. Gorelenko et al.: Ukr. Fiz. Zh. 30, 301 (1985)
Ya.I. Dutchak, I.D. Shcherba, Yu.K. Gorelenko: Fiz. Tverd. Tela 27, 2166 (1985)

G.68   L.E. Gorsh, G.N. Dolenko: Izv. Akad. Nauk SSSR, Neorg. Mater. 15, 1521 (1979)

G.69   G. Graeffe, H. Juslén, E.K. Viinikka: In [Ref.1.62, p.157]

G.70   G. Graeffe, H. Juslén, M. Karras: J. Phys. B 10, 3219 (1977)

G.71   P.M. Grant et al.: Phys. Rev. B 30, 6973 (1984)

G.72   M. Grasserbauer: Z. Anal. Chemie 273, 401 (1975); Mikrochim. Acta (1975) Vol.I, 145, 563, 597
S. Kosina, M. Grasserbauer, H. Drack: Talanta 26, 765 (2979)

G.73   G.N. Greaves: J. Physique 42, C4-225 (1981)

G.74   G.N. Greaves, X.L. Jiang, S.R. Elliott, T.G. Fowler: J. Non-Cryst. Solids 77–78, 1165 (1985)

G.75   R.B. Greegor, F.W. Lytle: Phys. Rev. B 20, 4902 (1979); J. Catal. 63, 476 (1980)
J.H. Sinfelt, G.H. Via, F.W. Lytle: J. Chem. Phys. 72, 4832 (1980)
J.H. Sinfelt, G.H. Via, F.W. Lytle, R.B. Greegor: J. Chem. Phys. 75, 5527 (1981)
G. Meitzner, G.H. Via, F.W. Lytle, J.H. Sinfelt: J. Chem. Phys. 78, 882 (1983); ibid. 79, 1527 (1983); ibid. 83, 353 (1985); ibid. 83, 4793 (1985)

G.76   R.B. Greegor, F.W. Lytle, R.L. Chin, D.M. Hercules: J. Phys. Chem. 85, 1232 (1981)

G.77   R.B. Greegor et al.: J. Non-Cryst. Solids 55, 27 (1983)
R.B. Greegor, F.W. Lytle, R.C. Ewing, R.F. Haaker: Nucl. Instrum. B 229, 587 (1984)
C.A. Yarker et al.: J. Non-Cryst. Solids 79, 117 (1986)

G.78   R.B. Greegor, F.W. Lytle, J. Kortright, A. Fischer-Colbrie: J. Non-Cryst. Solids 89, 311 (1987)

G.79   R.B. Greegor et al.: In *Scientific Basis for Nuclear Waste Management*, ed. by J.K. Bates, W.B. Seefeldt (Mater. Res. Soc. Pittsburgh, 1987) p.645

G.80   I.V. Gribov, V.I. Minin, L.D. Finkelshtein: Fiz. Met. Metalloved. 38, 216 (1974); ibid. 39, 1295 (1975)

G.81   I.V. Gribov et al.: Fiz. Met. Metalloved. 43, 1308 (1977); ibid. 44, 643 (1977)

G.82   I.V. Gribov, V.I. Minin, L.D. Finkelshtein: Fiz. Met. Mettaloved. 47, 949 (1979)

G.83   S.A. Gribovski, T.M. Zimkina: Fiz. Tverd. Tela 15, 300, 1312 (1973)

G.84   V.V. Grigoreva, S.M. Karalnik, A.V. Koval, N.G. Bondarenko: Visnik Kiiv. Univ., Ser. Khim. (1976) 26

G.85   L.A. Grunes: Phys. Rev. B 27, 2111 (1983)

G.86   Yu.S. Grushko et al.: Phys. Stat. Sol. (b) 128, 591 (1985)

G.87   V.A. Gubanov, N.I. Lazukova, E.Z. Kurmaev: Izv. Sib. Otd. Akad. Nauk SSSR, Ser. Khim. Nauk (1975) Nr. 9, H.4, 18
V.A. Gubanov, E.Z. Kurmaev: Zh. Strukt. Khim. 16, 731 (1975)
V.A. Gubanov, B.G. Kasimov, E.Z. Kurmaev: J. Phys. Chem. Solids 36, 861 (1975)
V.A. Gubanov, E.Z. Kurmaev, G.P. Shveikin: J. Phys. Chem. Solids 38, 201 (1977)

G.88   V.A. Gubanov, J. Weber, J.W.D. Connolly: Chem. Phys. 11, 319 (1975)
V.A. Gubanov: J. Electron Spectrosc. 9, 85 (1976)

G.89   V.A. Gubanov, E.Z. Kurmaev: Tr. Instr. Khim. Ural. Nauchn. Tsentr. Akad. Nauk SSSR (1976) 56
V.A. Gubanov, N.I. Lazukova, E.Z. Kurmaev: J. Solid State Chem. 19, 1 (1976)
V.M. Cherkashenko et al.: J. Solid State Chem. 22, 217 (1977)
E.Z. Kurmaev, V.M. Cherkashenko, V.A. Gubanov: J. Electron Spectrosc. 16, 455 (1979)

G.90   V.A. Gubanov, M. Erbudak. E.Z. Kurmaev: Inorg. Nucl. Chem. Lett. 14, 75 (1978)
V.E. Dolgikh, V.M. Cherkashenko, E.Z. Kurmaev: Prob. Tekh. Eksp. 26, 60 (1983)

G.91   W. Gudat, C. Kunz: Phys. Stat. Sol. (b) 52, 433 (1972)

G.92   W. Gudat, C. Kunz: Phys. Rev. Lett. 29, 169 (1972)
W. Gudat: Thesis Hamburg 1974

G.93   W. Gudat, J. Karlau, C. Kunz: In [Ref.1.60, Vol.I, p.205]

G.94   D. Guenzburger et al.: Phys. Rev. B 32, 4398 (1985)

G.95   A. Gupta, J.A. Tossell: Phys. Chem. Miner. 7, 159 (1981)

G.96   M. Gupta, V.A. Gubanov, D.E. Ellis: J. Phys. Chem. Solids 38, 499 (1977)

G.97   M. Gupta, E. Belin, L. Schlapbach: J. Less-Common Met. 103, 389 (1984)

G.98   M.K. Gupta, A.K. Nigam: J. Phys. F 2, 1174 (1972); ibid. 4, 947 (1974)

G.99   M.K. Gupta, A.K. Nigam: J. Phys. B 5, 1790 (1972)
N.N. Saxena, S.N. Gupta, R.G. Anikhindi: J. Phys. Chem. Solids 35, 1451 (1974)

G.100 M.K. Gupta: J. Phys. F 5, 359 (1975)

G.101 R.P. Gupta, A.J. Freeman: Phys. Rev. Lett. 36, 1194 (1976)

G.102 S.N. Gupta et al.: Phys. Stat. Sol. (b) 82, 603 (1977)

G.103 A.N. Gusatinskii, S.A. Nemnonov: Izv. Akad. Nauk SSSR, Ser. Fiz. 28, 922 (1964)

G.104 A.N. Gusatinskii, K.M. Kolobova, N.S. Mikhailov, S.A. Nemnonov: Izv. Akad. Nauk SSSR, Neorg. Mater. 1, 877 (1965)

G.105 A.N. Gusatinskii, S.A. Nemnonov: Izv. Akad. Nauk SSSR, Neorg. Mater. 1, 838 (1965); in [Ref.1.57, p.124]
A.N. Gusatinskii, S.A. Ishchenko: Izv. Akad. Nauk SSSR, Ser. Fiz. 31, 1002 (1967)

G.106 A.N. Gusatinskii: In [Ref.1.58, Vol.I, p.328]; in [Ref.1.62, p.269]; Izv. Akad. Nauk SSSR, Ser. Fiz. 40, 368, 2415 (1976); Fiz. Tekh. Poluprovodn. 11, 379 (1977); Dokl. Akad. Nauk SSSR 233, 342 (1977)
A.N. Gusatinskii, G.I. Alperovich, I.I. Geguzin, L.M. Monastyrskii: Fiz. Tverd. Tela 22, 3091 (1980)
A.N. Gusatinskii et al.: Phys. Stat. Sol. (b) 108, 405 (1981)

G.107 A.N. Gusatinskii, S.A. Nemnonov: Fiz. Tverd. Tela 11, 1528 (1969)

G.108 A.N. Gusatinskii et al.: Phys. Stat. Sol. (b) 100, 739 (1980)
M.A. Bunin et al.: Fiz. Tekh. Poluprovodn. 15, 1617 (1981)
A.N. Gusatinskii et al.: Izv. Akad. Nauk SSSR, Neorg. Mater. 17, 565 (1981)

G.109 A.N. Gusatinskii, A.A. Lavrentev, M.A. Blokhin, V.Yu. Slivka: Phys. Stat. Sol. (b) 131, K 139 (1985)

G.110 A.N. Gusatinskii, A.A. Lavrentev, M.A. Blokhin, V.Yu. Slivka: Solid State Commun. 57, 389 (1986)
A.V. Soldatov et al.: Izv. Akad. Nauk SSSR, Neorg. Mater. 19, 1052 (1983)
A.V. Soldatov, A.N. Gusatinskii, O.A. Sadovskaya: Izv. Akad. Nauk SSSR, Neorg. Mater. 19, 1056 (1983)

G.111 T.I. Guzhavina, L.N. Mazalov, V.V. Murakhtanov: Izv. Sib. Otd. Akad. Nauk SSSR, Ser. Khim. Nauk (1975) Nr.9, H.4, 14

L.N. Mazalov et al.: Theor. Chim. Acta **44**, 257 (1977)

G.112 E. Gwinner, H. Kiessig: Z. Physik **107**, 449 (1937)
E. Gwinner: Z. Physik **108**, 523 (1938)

G.113 B. Gyorffy: In [Ref.1.60, Vol.II, p.29]

H.1   J. Haase: Appl. Phys. A **38**, 181 (1985)
M.D. Crapper et al.: Surf. Sci. **171**, 1 (1986)
M. Bader, A. Puschmann, J. Haase: Phys. Rev. B **33**, 7336 (1986)

H.2   P. Haen et al.: J. Magn. Magn. Mater. **47–48**, 490 (1985)

H.3   R. Haensel, C. Kunz, T. Sasaki, B. Sonntag: Appl. Optics **7**, 301 (1968)

H.4   R. Haensel, C. Kunz, B. Sonntag: Phys. Rev. Lett. **20**, 262 (1968)
R. Haensel: Habilitationsschrift, Universität Hamburg (1970); DESY F 41-70/1 (1970)

H.5   R. Haensel, K. Radler, B. Sonntag, C. Kunz: Solid State Commun. **7**, 1495 (1969)

H.6   R. Haensel, G.Keitel, P. Schreiber, C. Kunz: Phys. Rev. **188**, 1375 (1969)

H.7   R. Haensel, R. Sonntag, C. Kunz, T. Sasaki: J. Appl. Phys. **40**, 3046 (1969)
H.-J. Hagemann, W. Gudat, C. Kunz: Solid State Commun. **15**, 655 (1974); Phys. Stat. Sol. (b) **74**, 507 (1976)

H.8   R. Haensel, G. Keitel et al.: Phys. Stat. Sol. (a) **2**, 85 (1970)

H.9   R. Haensel, P. Rabe, B. Sonntag: Solid State Commun. **8**, 1845 (1970)

H.10 R. Haensel et al.: Phys. Rev. Lett. **25**, 208, 1281 (1970); J. Physique **32**, C4-236 (1971); Phys. Rev. B **7**, 1577 (1973)

H.11 R. Haensel, P. Rabe, G. Tolkiehn, A. Werner: DESY SR-80/06 (1980)
W. Thulke, R. Haensel, P. Rabe: Phys. Stat. Sol. (a) **78**, 539 (1983)
R. Frahm, R. Haensel, P. Rabe: J. Phys. F **14**, 1029, 1333 (1984)

H.12 P. Haglund: Ark. Mat. Astr. Fysik **28A** (1941) Nr. 8

H.13 C.F. Hague, C. Bonnelle: In [Ref.1.79, p.251]

H.14 C.F. Hague, R.C. Karnatak: In [Ref.1.60, Vol.I, p.53]
J.M. Mariot, R.C. Karnatak, C. Bonnelle: J. Phys. Chem. Solids **35**, 657 (1974); Solid State Commun. **16**, 611 (1975)

H.15 C.F. Hague: Inst. Phys. Conf. Ser. **30**, 360 (1976)
C.F. Hague et al.: J. Phys. F **6**, 899 (1976)

H.16 C.F. Hague, J.-M. Mariot, H. Ostrowiecki: Jpn. J. Appl. Phys. **17**, S2-105 (1978); Phys. Lett. A **67**, 121 (1978)

H.17 C.F. Hague, J.-M. Mariot, G. Dufour: Phys. Lett. A **78**, 328 (1980)
J.M. Mariot, C.F. Hague, G. Dufour: In [Ref.1.64, p.513]

H.18 C.F. Hague et al.: J. Phys. F **11**, L 95 (1981)
C.F. Hague, J.-M. Mariot, E. Belin, E. Beauprez: In *Amorphous Metals and Non-Equilibrium Processing*, ed. by M. von Allmen (Les Ulis 1984) p.331

H.19 C.F. Hague et al.: Solid State Commun. **48**, 1 (1983)

H.20 F.G. Halaka, J.J. Boland, J.D. Baldeschweiler: J. Am. Chem. Soc. **106**, 5408 (1984)

H.21 K.-H. Hallmeier, R. Szargan, A. Meisel, A.Yu. Dukhnyakov: Spectrochim. Acta A **38**, 1333 (1982)

H.22 K.-H. Hallmeier et al.: Spectrochim. Acta A **42**, 841 (1986)

H.23 K.-H. Hallmeier, R. Szargan, K. Fritsche, A. Meisel: Phys. Scr. **35**, 827 (1987)

H.24 Y. Hammoud, J.C. Parlebas, F. Gautier: J. Phys. F **17**, 503 (1987)

H.25 T. Hanada, T. Aikawa, N. Soga: J. Non-Cryst. Solids **50**, 397 (1982)
T. Hanada, N. Soga: J. Am. Ceram. Soc. **65**, C 84 (1982)
T. Hanada, T. Aikawa, N. Soga: J. Am. Ceram. Soc. **67**, 52 (1984)

H.26 T. Hanada, N. Soga, M. Ohkawa: J. Non-Cryst. Solids **88**, 236 (1986)

H.27 W. Hansch, W. Ekardt: Phys. Rev. B **25**, 7815 (1982)

H.28 M.J. Hanus, E. Gilberg: In [Ref.1.60, Vol.I, p.338]; J. Phys. B **9**, 137 (1976)

H.29 T. Hanyu, S. Yamaguchi, H. Koike, S. Sato: J. Phys. Soc. Jpn. **46**, 599 (1979)

H.30 T. Hanyu et al.: Solid State Commun. **56**, 381 (1985)

H.31 S. Hanzely, R.J. Liefield: In [Ref.1.101, p.319]

H.32 J.E. Harries, D.W.L. Hukins: J. Phys. C **19**, 6859 (1986)

H.33 R. Harrison: Phil. Mag. **22**, 131 (1970)

H.34 E. Hartmann: J. Phys. B **19**, 1899 (1986)

H.35 H. Hartmann, H.-U. Chun: Theoret. Chim. Acta **2**, 1 (1964)

H.36 T.K. Hatwar, S.K. Malik, M.N. Chatikar, B.D. Padalia: Phys. Stat. Sol. (b) **95**, 621 (1979)

H.37 T.K. Hatwar et al.: Solid State Commun. **34**, 617 (1980)

H.38 T.K. Hatwar, D.R. Chopra: J. Electron Spectrosc. **36**, 319 (1985)

H.39 T. Hayasi: Sci. Rep. Tohoku Univ. **49**, 13 (1965)

H.40 T. Hayasi, S. Kiyono: Proc. 3rd Int. Conf. VUV Rad. Phys. (1971) Tokyo

H.41 T. Hayasi, Y. Hayasi: In [Ref.1.93, p.136]; Phys. Fenn. **9**, S 1-89 (1974)

H.42 Y. Hayasi: Sci. Rep. Tohoku Univ. **51**, 1, 43 (1968); ibid. **52**, 135 (1969)

H.43 Y. Hayasi: Z. Physik B **32**, 67 (1978)

H.44 Y. Hayasi et al.: Phys. Rev. B **30**, 1891 (1984)

H.45 D.E. Haycock, D.S. Urch: X-Ray Spectrom. **7**, 206 (1978)
       D.E. Haycock, D.S. Urch, G. Wiech: J. Chem. Soc., Faraday Trans. II **79**, 1692 (1979)

H.46 D.E. Haycock et al.: Jpn. J. Appl. Phys. **17**, S2-138 (1978); J. Chem. Soc., Chem. Commun. 262 (1978)

H.47 D. Haycock, D.S. Urch, C.D. Garner, I.H. Hillier: J. Electron Spectrosc. **17**, 345 (1979)

H.48 T.M. Hayes, P.N. Sen: Phys. Rev. Lett. **34**, 956 (1975)

H.49 T.M. Hayes, P.N. Sen, S.H. Hunter: J. Phys. C **9**, 4357 (1976)
       T.M. Hayes: J. Non-Cryst. Solids **31**, 57 (1978)
       G. Lucovsky, T.M. Hayes: In *Amorphous Semiconductors*, 2nd. ed., ed. by H.H. Brodsky, Topics Appl. Phys., Vol.36 (Springer, Berlin, Heidelberg 1985) p.275

H.50 W. Hayes: Contemp. Phys. **13**, 441 (1972); Phys. Rev. A **6**, 21 (1972)

H.51 S.M. Heald, E.A. Stern: Phys. Rev. B **16**, 5549 (1977)

H.52 S.M. Heald et al.: J. Am. Chem. Soc. **101**, 67 (1979)

H.53 L. Hedin, A. Rosengren: J. Phys. F **7**, 1339 (1977)

H.54 B. Hedman, J.F. Penner-Hahn, K.O. Hodgson: In [Ref.1.83, p.64]

H.55 B. Hedman et al.: Nucl. Instrum. Methods A **246**, 797 (1986)

H.56 J. Hedman, M. Klasson, R. Nilsson, C. Nordling, M.F. Sorokina, O.I. Kluy-zhnikov, S.A. Nemnonov, V.A. Trapeznikov, V.G. Zyryanov: Phys. Scr. **4**, 1 (1971)
       M.F. Sorokina, S.A. Nemnonov, V.G. Zyryanov: Izv. Akad. Nauk SSSR, Ser. Fiz. **36**, 429 (1972)

H.57 P.A. Heiney, P.A. Bancel, A.I. Goldman, P.W. Stephens: Phys. Rev. B **34**, 6746 (1986)

H.58 W. Heinle, A. Faessler: Phys. Lett. A **28**, 783 (1969)

H.59 M.H. Heinonen, J.A. Leiro: Phil. Mag. **46**, 669 (1982); Phys. Lett. A **98**, 385 (1983); Phil. Mag. B **49**, L 43 (1984)

H.60 D. Heintz: Dissertation, Universität München 1969; Z. Angew. Physik **27**, 98 (1969)

H.61 R. Hellmann, K. Ulmer: Z. Physik B **58**, 259 (1985)

H.62 K. Hemachandran, A.R. Chetal: Phys. Stat. Sol. (b) **136**, 181 (1986)
       K. Hemachandran, A.R. Chetal, G. Joshi: Phys. Stat. Sol. (b) **141**, 441 (1987)

398

H.63 K. Hemachandran, A.R. Chetal: Phys. Stat. Sol. (b) **138**, 229 (1986)
H.64 B.L. Henke: Adv. X-Ray Anal. **9**, 430 (1966)
H.65 B.L. Henke, E.N. Smith: J. Appl. Phys. **37**, 922 (1966)
H.66 B.L. Henke, K. Tanigushi: J. Appl. Phys. **47**, 1027 (1976)
K. Tanigushi, B.L. Henke: J. Chem. Phys. **64**, 3021 (1976)
H.67 B.L. Henke, R.C.C. Perera, E.M. Gullikson, M.L. Schattenburg: J. Appl. Phys. **49**, 480 (1978)
B.L. Henke, R.C.C. Perera, D.S. Urch: J. Chem. Phys. **68**, 3692 (1978)
B.L. Henke: Nucl. Instrum. Methods **177**, 161 (1980)
H.68 M. Hida et al.: J. Non-Cryst. Solids **61**, 415 (1984)
H.69 M. Hida et al.: Jpn. J. Appl. Phys. **24**, L 3 (1985)
H.70 V. Hietschold, G. Seifert: Phys. Stat. Sol. (b) **129**, K 163 (1985)
H.71 F.J. Himpsel et al.: Phys. Rev. Lett. **56**, 1497 (1986)
H.72 A.P. Hitchcock, C.E. Brion: J. Electron Spectrosc. **14**, 417 (1978)
H.73 A.P. Hitchcock, T. Tyliszczak: J. Phys. C **20**, 981 (1987)
H.74 A.P. Hitchcock, S. Bodeur, M. Tronc: Chem. Phys. **115**, 93 (1987)
H.75 W. Hodge, K. Roberts et al.: J. Phys. B **8**, L355 (1975)
H.76 J.E. Holliday: Adv. X-Ray Anal. **9**, 38, 365 (1966)
H.77 J.E. Holliday: Develop. Appl. Spectrosc. **5**, 77 (1966)
H.78 J.E. Holliday: Norelco Report **14**, 84 (1967)
H.79 J.E. Holliday: J. Appl. Phys. **38**, 4720 (1967)
H.80 J.E. Holliday: In [Ref.1.78, p.101]
H.81 J.E. Holliday: In [Ref.1.92, p.177]
H.82 J.E. Holliday: J. Phys. Chem. Solids **32**, 1825 (1971)
H.83 J.E. Holliday et al.: J. Electrochem. Soc. **119**, 1190 (1972); ibid. **120**, 470 (1973)
H.84 J.E. Holliday: In [Ref.1.79, p.713]
H.85 J.E. Holliday: Adv. X-Ray Anal. **16**, 53 (1973); Surf. Sci. **48**, 137 (1975)
H.86 D.J. Holmes et al.: J. Vac. Sci. Technol. A **5**, 703 (1987)
H.87 F. Hopfgarten, R. Manne: J. Electron Spectrosc. **2**, 13 (1973)
A. Stogard: Chem. Phys. Lett. **36**, 357 (1975)
H.88 Y. Horikawa, H. Soezima: Clay Sci. **5**, 97 (1977)
H.89 S. Horn et al.: Phys. Rev. B **36**, 3895 (1987)
H.90 J.A. Horsley: J. Chem. Phys. **76**, 1451 (1982)
H.91 J.A. Horsley et al.: J. Phys. Chem. **91**, 4014 (1987)
H.92 D. House, P.V. Smith, B.L. Gyorffy: J. Phys. F **3**, 745 (1973)
H.93 H.W. Huang, S.H. Hunter, W.K. Warburton, S.C. Moss: Science **204**, 191 (1979)
C.M. Dutta, H.W. Huang: Phys. Rev. Lett. **44**, 643 (1980)
H.94 K. Hübner: Physik Halbleiteroberfläche **8**, 95, 108 (1977)
H.95 S. Hüfner, G.K. Wertheim: Phys. Lett. A **44**, 47 (1973)
H.96 S. Hüfner, T. Riesterer: Phys. Rev. B **33**, 7267 (1986)
H.97 S.L. Hulbert, B.A. Bunker, F.C. Brown: Phys. Rev. B **30**, 2120 (1984)
H.98 R.G. Hurley, E.W. White: Anal. Chem. **46**, 2234 (1974)

I.1 K. Ichikawa: J. Phys. Soc. Jpn. **37**, 377 (1974)
I.2 K. Ichikawa, O. Teraskai, T. Sagawa: J. Phys. Soc. Jpn. **36**, 706 (1974)
I.3 K. Ichikawa et al.: Jpn. J. Appl. Phys. **17**, S2-157 (1978)
I.4 K. Ichikawa, A. Nisawa, K. Tsutsumi: Phys. Rev. B **34**, 6690 (1986)
I.5 Y. Iguchi, T. Sagawa et al.: Solid State Commun. **6**, 575 (1968)
I.6 Y. Iguchi: Sci. Light **19**, 1 (1970); Proc. 6th Int'l Conf. Optics Microanal., Univ. of Tokyo 1972, p.573

I.7 Y. Igushi: Phys. Fenn. **9**, S1-162 (1974); in *Proc. 7th Int'l Conf. X-Ray Optics Microanal.* (Nauka, Moscow 1974) p.123; in [Ref.1.86, p.506]; Sci. Light, Tokyo **26**, 161 (1977)

I.8 Y. Iguchi: In [Ref.1.62, p.130]

I.9 G.I. Ilkiv, I.V. Kavich, R.A. Antonyuk, Ya.N. Sinitskii: Metallofizika **66**, 62 (1976)

I.10 L.I. Ilyashev et al.: In [Ref.1.58, Vol.II, p.276]; in [Ref.1.94, p.101]
A.T. Shuvaev et al.: In [Ref.1.98, p.137]; Izv. Sib. Otd. Akad. Nauk SSSR, Ser. Khim. Nauk 1975, Nr. 9, H.4, 7.

I.11 L. Incoccia et al.: Phys. Rev. B **31**, 1028 (1985)

I.12 R. Ingalls, G.A. Garcia, E.A. Stern: Phys. Rev. Lett. **40**, 334 (1978)

I.13 R. Ingalls et al.: J. Appl. Phys. **51**, 3158 (1980)
J.M. Tranquada, R. Ingalls: Phys. Rev. B **28**, 3520 (1983)

I.14 R. Ingalls, E.D. Crozier, A.J. Seary: Physica B+C **139-140**, 505 (1986)

I.15 M.S. Ioffe, V.I. Nefedov: Zh. Strukt. Khim. **15**, 424 (1974)

I.16 T. Iseki, H. Tagai: J. Am. Ceram. Soc. **53**, 582 (1970)

I.17 M.S. Islam, C. Mande: Phys. Stat. Sol. (a) **81**, 197 (1984)
M.S. Islam et al.: Indian J. Phys. A **59**, 194 (1985)
M.S. Islam: Phys. Stat. Sol. (b) **132**, 471 (1985); J. Phys. C **19**, 6673 (1986)
M.L. Jungfleisch, M.S. Island, V.B. Sapre, C. Mande: Indian J. Phys. A **57**, 250 (1983)

I.18 M. Ito, T. Kawamura: Phil. Mag. A **49**, L 9 (1984)
M. Ito et al.: J. Phys. Soc. Jpn. **54**, 1843 (1985)

I.19 F. Itoh, T. Kitano, K. Suzuki: Sci. Rep. Res. Inst. Tohoku Univ. A **33**, 15 (1986)

I.20 A.V. Ivanov: Izv. Akad. Nauk SSSR, Ser. Fiz. **27**, 359 (1963)

I.21 A.V. Ivanov et al.: Opt. Spektrosk. **29**, 805 (1970); ibid. **31**, 317 (1971); Izv. Akad. Nauk SSSR, Ser. Fiz. **36**, 277 (1972)

I.22 A.V. Ivanov: Fiz. Tverd. Tela **13**, 1818 (1971)

I.23 A.V. Ivanov et al.: Izv. Akad. Nauk SSSR, Ser. Fiz. **36**, 267, 272 (1972)

I.24 A.L. Ivanovskii, V.A. Gubanov, G.P. Shveikin: Zh. Neorg. Khim. **24**, 629 (1979)
A.L. Ivanovskii et al.: Zh. Strukt. Khim. **23**, Nr.6, 42 (1982)

I.25 A.L. Ivanovskii, V.A. Gubanov, V.P. Zhukov, G.P. Shveikin: Phys. Stat. Sol. (b) **98**, 79 (1980)

I.26 A.L. Ivanovskii et al.: Izv. Akad. Nauk SSSR, Neorg. Mater. **21**, 1149 (1985)

I.27 N. Iwamoto, N. Umesaki, T. Atsumi: J. Mater. Sci. Lett. **6**, 271 (1987)

I.28 Y. Iwasaki et al.: Jpn. J. Appl. Phys. **17**, S2-164 (1978)

I.29 S. Iwata: Nippon Kessho Gakhaishi **11**, 102 (1969)
S. Iwata, N. Kosugi, O. Nomura: Jpn. J. Appl. Phys. **17**, S2-109 (1978)

I.30 M. Iwatsuki, T. Fukusawa: Nippon Kagaku Kaishi 1976, 444; ibid. 1978, 470

I.31 L.K. Izraileva: In [Ref.1.58, Vol.II, p.211]

J.1 W.B. Jackson, S.-J. Oh, C.C. Tsai, J.W. Allen: AIP Conf. Proc. No. 120, 341 (1984)

J.2 L. Jacob, R. Noble, H. Yee, B. Fraenkel: In [Ref.1.78, p.81]

J.3 R.L. Jacobs: Phys. Lett. A **30**, 523 (1969); J. Phys. F **3**, L166 (1973)

J.4 D. Jain, K.B. Garg, B.K. Sharma: Proc. Nucl. Phys. Solid State Phys. Symp. C **21**, 42 (1978)

J.5 D.C. Jain, U. Chandra, K.B. Garg, B.K. Sharma: J. Phys. D **13**, 1113 (1980)

J.6 D.C. Jain et al.: Key Eng. Mater. **13-15**, 148 (1987)

J.7 J.J. Jaklevic et al.: Solid State Commun. **23**, 679 (1977)

J.8   K.A. Jamison, C.W. Woods, R.L. Kauffman, P. Richards: Phys. Rev. A 11, 505 (1975)
J.9   T. Jarlborg, P.O. Nilsson: J. Phys. C 12, 265 (1979)
J.10   St. Jasienska, J. Janowski, D. Tomkowicz: Acta Phys. Pol. A 62, 41 (1982)
        St. Jasienska, D. Tomkowicz: J. Physique 45, C2-617 (1984)
J.11   G. Jasiolek, J. Raczynska, Z. Furmanik, J. Makowski: Acta Phys. Pol. A 69, 469 (1986)
J.12   G. Jasiolek, M. Berkowski, W. Piekarczyk: Chemtronics 1, 121 (1986)
J.13   G. Jasiolek, J. Raczynska, J. Gorecka: J. Cryst. Growth 78, 105 (1986)
J.14   G. Jasiolek, H.A. Dabkowska: J. Cryst. Growth 79, 534 (1986)
J.15   R. Jenkins: X-Ray Spectrom. 2, 207 (1973)
J.16   Y. Jeon et al.: Phys. Rev. B 36, 3891 (1987)
J.17   K.S. Jerath, K.B. Garg: Key Eng. Mater. 13-15, 99 (1987)
J.18   W. Jitschin, U. Werner, G. Materlik, G.D. Doolen: Phys. Rev. A 35, 5038 (1987)
J.19   T. Jo, A. Kotani: Solid State Commun. 54, 451 (1985)
J.20   T. Jo, A. Kotani: Phys. Scr. 35, 570 (1987)
J.21   B. Johnson, M. Senglaub, P. Richard, C.F. Moore: Z. Physik 261, 413 (1973)
J.22   R.W. Johnson et al.: J. Non-Cryst. Solids 83, 251 (1986)
J.23   D.J. Jones, J. Roziere, G.C. Allen, P.A. Tempest: J. Chem. Phys. 84, 6075 (1986)
J.24   J.B. Jones, D.S. Urch: J. Chem. Soc., Dalton Trans. 1975, 1885
J.25   J.B. Jones, M. Kasrai, D.S. Urch: In [Ref.1.62, p.136]
J.26   R.G. Jones, D.P. Woodruff: Surf. Sci. 114, 38 (1982)
        S.M. El-Mashri, R.G. Jones, A.J. Forty: Phil. Mag. 48, 665 (1983)
J.27   R.G. Jones et al.: Surf. Sci. 152-153, 443 (1985)
J.28   R.G. Jones et al.: Surf. Sci. 179, 425 and 442 (1987)
J.29   J.A. Jope: J. Phys. C 3, 21 (1970)
J.30   R.W. Joyner: Daresbury Lab. Rep. R 13, 114 (1979)
J.31   R.W. Joyner: Chem. Phys. Lett. 72, 162 (1980)
J.32   R.W. Joyner: J. Chem. Soc., Faraday Trans. II 76, 357 (1980)
J.33   R.W. Joyner, K.J. Martin, P. Meeham: J. Phys. C 20, 4005 (1987)
J.34   H. Juslen, M. Pessa, G. Graeffe: Jpn. J. Appl. Phys. 17, S2-120 (1978)
J.35   H. Juslen, M. Pessa, G. Graeffe: Phys. Rev. A 19, 196 (1979)

K.1   N.M. Kabachnik, I.P. Sazhina: Phys. Fenn. 9, S1-415 (1974); Fiz. Tverd. Tela 17, 397 (1975)
K.2   G. Kaindl, W.D. Brewer, G. Kalkowski, F. Holtzberg: Phys. Rev. Lett. 51, 2056 (1983)
        W.D. Brewer, G. Kalkowski, G. Kaindl, F. Holtzberg: Phys. Rev. B 32, 3676 (1985)
        G. Kaindl et al.: J. Appl. Phys. 55, 1910 (1984)
        R. Suryanarayanan et al.: J. Magn. Magn. Mater. 47-48, 487 (1985)
K.3   G. Kaindl, G.K. Wertheim, G. Schmiester, E.V. Sampathkumaran: Phys. Rev. Lett. 58, 606 (1987)
K.4   A.J. Kalb, E.A. Stern, S.M. Heald: J. Mol. Biol. 135, 501 (1979)
K.5   G. Kalkowski et al.: Phys. Rev. B 32, 2717 (1985)
        G. Kalkowski, W.D. Brewer, G. Kaindl, F. Holtzberg: J. Magn. Magn. Mater. 47-48, 215 (1985)
K.6   G. Kalkowski, G. Kaindl, W.D. Brewer, W. Krone: Phys. Rev. B 35, 2667 (1987)
K.7   E. Källne, K.B. Garg: In [Ref.1.60, Vol.I, p.62]

K.8   E. Källne: J. Phys. F **4**, 167 (1974)
K.9   E. Källne, M. Pessa: J. Phys. C **8**, 1985 (1975)
K.10   A.V. Kalenichenko, R.D. Bartosevich, E.T. Kovalenko, N.W. Soloveva: In [Ref.1.98, p.179]
K.11   N. Kamijo et al.: J. Phys. Soc. Jpn. **53**, 4210 (1984)
K.12   N. Kamijo et al.: J. Phys. Soc. Jpn. **55**, 2217 (1986)
K.13   D. Kantelhardt, W. Waidelich: Z. Angew. Physik **26**, 239 (1969)
K.14   Y.H. Kao et al.: Surf. Sci. **174**, 567 (1986)
K.15   Q.S. Kapoor, K.B. Garg, A.N. Nigam: Phys. Lett. A **30**, 228 (1969); J. Phys. B **3**, 1180 (1970)
       K.B. Garg et al.: J. Phys. B **5**, 2152 (1972)
K.16   Q.S. Kapoor, L.M. Watson, D. Hart, D.J. Fabian: In [Ref.1.79, p.215]; Solid State Commun. **11**, 503 (1972)
       L.M. Watson, Q.S. Kapoor, D.J. Fabian: In [Ref.1.98, p.213]
       C.M. Sayers, N.H. March, L.M. Watson, D.J. Fabian: J. Phys. F **7**, L77 (1977)
K.17   S.M. Karalnik, A.P. Nesenyuk, V.D. Dobrovolskii: Ukr. Fiz. Zh. **10**, 668 (1965)
K.18   S.M. Karalnik, E.P. Domashevskaya: In [Ref.1.92, p.161]
K.19   S.M. Karalnik, A.V. Koval, M.A. Masin: Zh. Fiz. Khim. **17**, 683 (1972)
K.20   S.M. Karalnik, A.V. Koval, M.A. Masin: Ukr. Fiz. Zh. **17**, 1853, 1857 (1972)
K.21   V.Yu. Karasov et al.: Fiz. Tverd. Tela **25**, 1964 (1983)
K.22   R.I. Karazia, D. Grabauskas, A.A. Kiselev: Liet. Fis. Rinkinys **14**, 235, 249 (1974)
       R.I. Karazia, J.J. Grudzinskas, S.A. Kucas, A.V. Karosene: Litov. Fiz. Sb. **22**, 125 (1982)
       R.I. Karazia et al.: Opt. Spektrosk. **57**, 395 (1984)
K.23   G. Karlsson, R. Manne: Phys. Scr. **4**, 119 (1971)
K.24   R.C. Karnatak, P. Sakellaridis: J. Chim. Physique **62**, 883 (1965)
K.25   R.C. Karnatak, P. Motais deNarbonne: Jpn. J. Appl. Phys. **17**, S2-127 (1978)
K.26   R.C. Karnatak et al.: Phys. Rev. B **36**, 1745 (1987)
K.27   M. Karras, R. Juslen, G. Graeffe: Phys. Fenn. **9**, S1-153 (1974); Appl. Phys. **6**, 185 (1975)
K.28   M. Karras, M. Lindroos, G. Graeffe: Jpn. J. Appl. Phys. **17**, S2-181 (1978)
K.29   M. Kasaya et al.: J. Magn. Magn. Mater. **31-34**, 389 and 437 (1983)
       A. Ochiai et al.: J. Magn. Magn. Mater. **47-48**, 570 (1985)
K.30   J. Kashiwakura, Y. Gohshi: Spectrochim. Acta B **36**, 625 (1981)
K.31   G.G. Kasimov, N.D. Samsonova, M.F. Sorokina, L.D. Finkelshtein: Izv. Akad. Nauk SSSR, Neorg. Mater. **14**, 301 (1978)
K.32   M. Kasrai, D.S. Urch: Jpn. J. Appl. Phys. **17**, S2-131 (1978); J. Chem. Soc., Faraday Trans. II **75**, 1522 (1979)
       P.M. Montague, D.S. Urch: Nature (London) **272**, 804 (1978)
       L. Bergknut et al.: J. Chem. Soc., Faraday Trans. II **77**, 1879 (1981)
K.33   R.L. Kauffman, J.H. McGuire, P. Richard, C.F. Moore: Phys. Rev. A **8**, 1233 (1973)
       J. McWherter, J. Bolger, C.F. Moore, P. Richard: Z. Physik **263**, 283 (1973)
K.34   I.V. Kavich, G.M. Ilkiv: Ukr. Fiz. Zh. **8**, 1267 (1963)
K.35   I.V. Kavich et al.: Zh. Strukt. Khim. **6**, 318 (1965); Fiz. Met. Metalloved. **21**, 632 (1966); Ukr. Fiz. Zh. **18**, 2050 (1973)
       Ya.I. Dutchak et al.: Dopov. Akad. Nauk Ukr. SSR, Ser. A 1977, 61
K.36   I.V. Kavich: Ukr. Fiz. Zh. **20**, 156 (1975)
       G.I. Ilkiv, I.V. Kavich, R.A. Antonyuk, Ya.I. Sinitski: Metallofizika **66**, 62 (1976)
       G.V. Samsonov, Ya.I. Dutchak, I.V. Kavich, P.I. Shevchuk: Dopov. Akad.

Nauk Ukr. SSR A (1976) 446

Ya.I. Dutchak et al.: Izv. Akad. Nauk SSSR, Neorg. Mater. **12**, 589 (1976)

K.37 I.V. Kavich, L.I. Nikolaev, L.S. Voskrekasenko, B.N. Yatsyk: Ukr. Fiz. Zh. **22**, 2051 (1977)

K.38 I.V. Kavich, L.P. Shevchuk: Ukr. Fiz. Zh. **23**, 624 (1978)

I.V. Kavich, R.A. Antonyuk, Ya.I. Sinitskii, I.D. Shcherba: In [Ref.1.100, p.159]

I.V. Kavich, I.D. Shcherba: Fiz. Met. Metalloved. **51**, 962 (1981)

K.39 I.V. Kavich, L.I. Nikolaev, Ya.I. Mikhailishin: Ukr. Fiz. Zh. **27**, 1708 (1982)

K.40 J. Kawai, Y. Gohshi: Spectrochim. Acta B **41**, 265 (1986)

K.41 J. Kawai, C. Satoko, K. Fujisawa, Y. Gohshi: Phys. Rev. Lett. **57**, 988 (1986); Spectrochim. Acta B **42**, 729 (1987)

J. Kawai, C. Satoko, Y. Gohshi: Spectrochim. Acta B **42**, 745, 1125 (1987)

K.42 J. Kawai, C. Satoko, Y. Gohshi: J. Phys. C **20**, 69 (1987)

K.43 S. Kawata, K. Maeda: J. Phys. Soc. Jpn. **32**, 778 (1972); J. Phys. F **3**, 167 (1973)

S. Kawata: J. Phys. F **5**, 324 (1975)

K.44 S. Kawata: J. Phys. F **3**, 1511 (1973)

K.45 S. Kawata, K. Maeda: Phys. Lett. A **50**, 71 (1974); J. Phys. Chem. Solids **38**, 1289 (1977)

K.46 S. Kawata: J. Phys. F **5**, 324 (1975)

K.47 S. Kawata, K. Maeda: J. Phys. C **11**, 2391 (1978)

K.48 S. Kawata, K. Maeda: J. Phys. C **13**, 3983 (1980)

K.49 H. Kawazoe: J. Non-Cryst. Solids **42**, 281 (1980)

H. Kawazoe, H. Kokumai, T. Kanazawa: J. Phys. Chem. Solids **42**, 579 (1981)

K.50 R.B. Kay, Ph.E. Van der Leeuw, M.J. Van der Wiel: J. Phys. B **10**, 2513 (1977)

K.51 R.O. Keeling: J. Chem. Phys. **31**, 279 (1959); Develop. Appl. Spectrosc. **2**, 263 (1963)

K.52 H. Keilacker, A. Meisel: Wiss. Z. Karl-Marx-Univ., Math.-Naturwiss. R. **22**, 585 (1973)

K.53 H. Keilacker, A. Meisel: In [Ref.1.60, Vol.I, p.509]; Ann. Physik **30**, 236 (1973); Kristallografiya **20**, 245 (1975)

K.54 E. Kellner, E.A. Stern: In [Ref.1.83, p.507]

K.55 B. Kern: Thesis Freiburg 1956; Z. Physik **159**, 178 (1960)

K.56 N. Kerr del Grande, A.J. Oliver: In [Ref.1.60, Vol.I, p.183]

K.57 L. Kertesz, J. Kojnok, A. Szasz: Cryst. Lattice Defects **9**, 219 (1982)

L. Kertesz, M.A. Ahmed: J. Phys. F **13**, 2235 (1983)

K.58 L. Kertesz, A. Szasz, M.I. Zakharova, A.G. Khundzhua: Phys. Stat. Sol. (a) **98**, 107 (1986)

K.59 O. Keski-Rahkonen: Phys. Scr. **7**, 173 (1973)

K.60 O. Keski-Rahkonen, M.O. Krause: Phys. Fenn. **9**, S1-261 (1974); Phys. Rev. A **15**, 959 (1977)

K.61 O. Keski-Rahkonen: J. Utriainen: J. Phys. B **7**, 55 (1974)

K.62 O. Keski-Rahkonen: J. Phys. C **8**, 541 (1975)

K.63 O. Keski-Rahkonen, J. Saijanmaa, M. Suvanen, A. Servomaa: Phys. Scr. **16**, 105 (1977)

B. Johansson, H. Ludvigsen, O. Keski-Rahkonen: Phys. Rev. B **28**, 3622 (1983)

O. Keski-Rahkonen, K. Reinikainen, E. Mikkola: Phys. Scr. **28**, 179 (1983)

O. Keski-Rahkonen, E. Mikkola, K. Reinikainen, M. Lehkonen: J. Phys. C **18**, 2961 (1985)

K.64 O. Keski-Rahkonen, E. Mikkola: J. Phys. C **16**, L 505 (1983)

K.65 E.G. Kessler, R.D. Deslattes, A. Henins: Phys. Rev. A 19, 215 (1979)

K.66 E.G. Kessler et al.: NBS Report Washington DC 1982

K.67 P.V. Khadikar, S.P. Pandharkar: Jpn. J. Appl. Phys. Part 1 26, 1146 (1987)

K.68 P.L. Khare, P. Deshmukh, C. Mande: J. Phys. B 10, 2531 (1977)

K.69 B.V. Khasbardar, A.S. Vaingankar, R.N. Patil: J. Phys. F 10, 1879 (1980)

K.70 B.Yu. Khelmer, V.F. Volkov, A.I. Platkov: Izv. Sib. Otd. Akad. Nauk SSSR, Ser. Khim. Nauk (1975) Nr. 9, H.4, 26

B.Yu. Khelmer: Izv. Sev.-Kavk. Nauchn. Tsentra Vyssh. Shk., Ser. Estestv. Nauk 4, 86 (1976)

K.71 B.Yu. Khelmer, V.I. Nefedov, L.N. Mazalov: Izv. Akad. Nauk SSSR, Ser. Fiz. 40, 329 (1976)

Yu.A. Zhdanov et al.: Zh. Strukt. Khim. 18, 677 (1977)

K.72 M.Ya. Khodos et al.: Izv. Akad. Nauk SSSR, Neorg. Mater. 21, 2059 (1985)

K.73 L.G. Khokha, A.S. Shulakov: Fiz. Met. Metalloved. 63, 205 (1987)

K.74 M.B. Khusidman: Fiz. Tverd. Tela 14, 3287 (1972)

K.75 J. Kieser: In [Ref.1.79, p.557]

K.76 J. Kieser: Phys. Fenn. 9, S1-173 (1974); Z. Physik B 26, 1 (1977)

K.77 N. Kikuchi, T. Maekawa, T. Yokokawa: Bull. Chem. Soc. Jpn. 52, 1260 (1979)

K.78 B.M. Kincaid, P. Eisenberger, K.O. Hodgson, S. Doniach: Proc. Natl. Acad. Sci. USA 72, 2340 (1975)

K.79 B.M. Kincaid, P. Eisenberger: Phys. Rev. Lett. 34, 1361 (1975)

B.-K. Teo et al.: J. Am. Chem. Soc. 99, 3854 (1977)

K.80 B.M. Kincaid, A.E. Meixner, P.M. Platzman: Phys. Rev. Lett. 40, 1296 (1978)

K.81 M.M. Kindrat, Yu.N. Kucherenko, L.M. Sheludchenko: Izv. Akad. Nauk SSSR, Ser. Fiz. 46, 807 (1982)

K.82 P.P. Kirichok, S.M. Karalnik: Ukr. Fiz. Zh. 13, 95 (1968)

K.83 P.P. Kirichok et al.: Ukr. Fiz. Zh. 13, 1627 (1968)

V.F. Belov et al.: Metallofizika 37, 42 (1971)

G.S. Podvalnykh et al.: Izv. Vyssh. Uchebn. Zaved., Fiz. 15, 153 (1972)

K.84 P.P. Kirichok. G.S. Podvalnykh et al.: Izv. Vyssh. Uchebn. Zaved., Fiz. 14, 129 (1971); Izv. Akad. Nauk SSSR, Ser. Fiz. 36, 397 (1972); in [Ref.1.94, p.131]

K.85 P.P. Kirichok, G.S. Podvalnykh et al.: Zh. Fiz. Khim. 46, 1550 (1972); Izv. Akad. Nauk SSSR, Ser. Fiz. 36, 402 (1972)

K.86 P.P. Kirichok, L.G. Kustovskii et al.: Izv. Akad. Nauk SSSR, Ser. Fiz. 38, 617 (1974)

P.P. Kirichok, A.I. Antonshchuk: Izv. Akad. Nauk SSSR, Ser. Fiz. 40, 435 (1976)

P.P. Kirichok, P.P. Zarin, G.S. Podvalnykh: Izv. Vyssh. Uchebn. Zaved., Fiz. 20, 141 (1977)

D.E. Bondarev, P.P. Kirichok: Dokl. Akad. Nauk SSSR 234, 1055 (1977)

P.P. Kirichok, A.V. Kopaev, B.N. Budivskaya, P.M. Bugai: Izv. Akad. Nauk SSSR, Neorg. Mater. 20, 134 (1984)

P.P. Kirichok et al.: Izv. Akad. Nauk SSSR, Neorg. Mater. 20, 659 (1984)

K.87 P.P. Kirichok, T.O. Kostur, N.I. Minko, M.M. Kindrat: Issled. po Molekulyar. Fiz. i Fiz. Tverd. Tela (1976) 99

K.88 P.P. Kirichok, N.I. Minko, T.O. Kostur: Izv. Vyssh. Uchebn. Zaved., Fiz. 19, 30 (1976)

K.89 M. Kitamura, S. Muramatsu, C. Sugiura: Phys. Rev. B 33, 5294 (1986)

K.90 M. Kitamura, S. Muramatsu, C. Sugiura: Phys. Stat. Sol. (b) 142, 191 (1987)

K.91 M. Kitamura, C. Sugiura, S. Muramatsu: Solid State Commun. 62, 663 (1987)

K.92 S. Kiyono: Sci. Rep. Tohoku Univ. I 37, 249 (1953)

C. Sugiura: Sci. Rep. Tohoku Univ. I **46**, 1 (1962)

S. Kiyono, C. Sugiura: Techn. Rep. Tohoku Univ. **31**, 1, 85 (1966)

C. Sugiura, S. Muramatsu: Phys. Stat. Sol. (b) **132**, K 111 (1985)

K.93 S. Kiyono: Sci. Rep. Tohoku Univ. I **39**, 129 (1956); ibid. **40**, 1 (1956)

K.94 S. Kiyono et al.: Jpn. J. Appl. Phys. **17**, S2-212 (1978)

K.95 S. Kiyono, Y. Hayasi, T. Muranaka: In [Ref.1.62, p.241]

S. Kiyono, T. Muranaka, K. Aota: Tech. Rep. Tohoku Univ. **43**, 423 (1978)

S. Kiyono, T. Muranaka, T. Watanabe: Jpn. J. Appl. Phys. **18**, 1865 (1979); ibid. **20**, 1939 (1981)

K.96 M. Klasson, R. Manne: In *Electron Spectroscopy*, ed. by D.A. Shirley (North Holland, Amsterdam 1972) p.471

K.97 R. Kleber: Z. Physik **264**, 309 (1973)

K.98 B.M. Klein, L.L. Boyer, D.A. Papaconstantopoulos, L.F. Mattheiss: Phys. Rev. B **18**, 6411 (1978)

K.99 G. Klein: In [Ref.1.60, Vol. II, p.362]

K.100 H. Kleykamp. Z. Naturforsch. **36** A, 1388 (1981); ibid. **41** A, 681 (1986)

K.101 J. Klima: J. Phys. C **3**, 70 (1970)

K.102 J. Klima: Czech. J. Phys. B **30**, 905 (1980)

K.103 J. Klima: J. Phys. C **15**, 689 (1982)

K.104 G.S. Knapp, H. Chen. T.E. Klippert: Rev. Sci. Instrum. **49**, 1658 (1978)

K.105 G.S. Knapp, B.W. Veal, H.K. Pan, T. Klippert: Solid State Commun. **44**, 1343 (1982)

K.106 G.S. Knapp, M.V. Nevitt, A.T. Aldred, T.K. Klippert: J. Phys. Chem. Solids **46**, 1321 (1985)

K.107 J.C. Knights, T.M. Hayes, J.C. Mikkelsen, Jr.: Phys. Rev. Lett. **39**, 712 (1977)

K.108 M.L. Knotek, J.E. Houston: Phys. Rev. B **15**, 4580 (1977)

K.109 H. Kobayashi, A. Kobayashi, Y. Sasaki: Mol. Cryst. Liq. Cryst. **118**, 427 (1985)

K.110 S. Kobayashi, S. Takeuchi: J. Phys. F **12**, 1273 (1982)

K.111 E.E. Koch et al.: Chem. Phys. Lett. **9**, 429 (1971); ibid. **12**, 476 (1972); ibid. **16**, 131 (1972); ibid. **21**, 501 (1973)

W. Eberhardt et al.: Chem. Phys. Lett. **40**, 180 (1976)

K.112 E.E. Koch, Y. Jugnet, F.J. Himpsel: Chem. Phys. Lett. **116**, 7 (1985)

K.113 A.N. Kocharyan: J. Magn. Magn. Mater. **63–64**, 499 (1987)

K.114 B.N. Kodess, I.Ya. Nikiforov, S.V. Stolbov: Solid State Commun. **31**, 1011 (1979)

K.115 D.M. Koffmann, S.H. Moll: Adv. X-Ray Anal. **9**, 323 (1966)

K.116 A. Kohm, H.Merz: In [Ref.1.60, Vol.II, p.394]

K.117 K.M. Kolobova, S.A. Nemnonov: Fiz. Met. Metalloved. **25**, 267, 634, 1010 (1968)

K.118 K.M. Kolobova, S.A. Nemnonov, E.V. Agapova: Fiz. Tverd. Tela **10**, 729 (1968)

K.119 K.M. Kolobova, V.A. Trofimova: In [Ref.1.58, Vol.I, p.172]

K.120 K.M. Kolobova, S.A. Nemnonov, F.A. Sidorenko: Fiz. Met. Metalloved. **30**, 309 (1970)

K.121 K.M. Kolobova, I.N. Shabanova, V.R. Galakhov: Fiz. Met. Metalloved. **61**, 278 (1986)

K.122 E.Ya. Komarov, I.B. Borovskii: Fiz. Met. Metalloved. **54**, 806 (1982)

K.123 A.V. Kondratenko, L.N. Mazalov et al.: Zh. Strukt. Khim. **18**, 546 (1977); ibid. **20**, 203, 919 (1979); Opt. Spektrosk. **48**, 1072 (1980); ibid. **49**, 488 (1980)

I.A. Topol et al.: Izv. Akad. Nauk SSSR, Ser. Fiz. **46**, 770 (1982)

V.D. Yumatov et al.: Zh. Strukt. Khim. **24**, Nr.5, 20 (1984)

V.D. Yumatov, L.N. Mazalov, A.V. Okotrub, I.A. Topol: Zh. Strukt. Khim. **25**, Nr.4, 43 (1984); ibid. **25**, Nr.5, 63 (1984)

K.124 A.V. Kondratenko, L.N. Mazalov et al.: Zh. Strukt. Khim. **18**, 622 (1977); Theor. Chim. Acta **52**, 311 (1979)

K.125 D.C. Koningsberger, D.E. Sayers: Solid State Ionics **16**, 23 (1985)

D.C. Koningsberger et al.: J. Phys. Chem. **89**, 4075 (1985)

K.126 H. Konuma, Y. Hayasi, S. Kiyono: Technol. Rep. Tohoku Univ. **42**, 235 (1977)

K.127 H. Konuma, Kagaku Keisatsu Kenkyusho Hokoku **31**, 170 (1979)

K.128 M.I. Korsunkii, Ya.E. Genkin, M.M. Omarov: Zavodskaya Lab. **32**, 381 (1966); in [Ref.1.95, p.16]

K.129 M.I. Korsunkii, Ya.E. Genkin: Izv. Akad. Nauk Kaz. SSR, Ser. Fiz.-Mat. Nauk **5**, 77 (1967); Poroshk. Metallurg. **7**, 82 (1967)

K.130 M.I. Korsunskii et al.: In [Ref.1.58, Vol.I, p.123]; Izv. Akad. Nauk Kaz. SSR, Ser. Fiz.-Mat. Nauk **8**, 20 (1970); ibid. **14**, 68 (1976)

K.131 M.I. Korsunskii, Ya.E. Genkin, E.A. Zhurakovskii, V.G. Lifshits: Izv. Akad. Nauk Kaz. SSR, Ser. Fiz.-Mat. **4**, 68 (1972)

M.I. Korsunskii, Ya.E. Genkin, V.G. Lifshits: Izv. Akad. Nauk Kaz. SSR, Ser. Fiz.-Mat. **13**, 61 (1975)

M.I. Korsunskii et al.: Izv. Akad. Nauk SSSR, Ser. Fiz. **40**, 352 (1976); Izv. Akad. Nauk Kaz. SSR, Ser. Fiz.-Mat. **14**, 1 (1976)

V.I. Andryushin, M.I. Korsunskii: Fiz. Met. Metalloved. **49**, 737 (1980)

K.132 E.-K. Kortela, R. Manne: In [Ref.1.60, Vol.II, p.41]; J. Phys. C **7**, 1749 (1974)

K.133 S. Korsina, J. Kristin: Geol. ZB (Bratislava) **24**, 197 (1973)

K.134 A.I. Kostarev: Fiz. Met. Metalloved. **20**, 26 (1965)

K.135 A.S. Koster, G.D. Rieck, H. Mendel: J. Phys. Chem. Solids **31**, 2505, 2511, 2523 (1970)

K.136 A.S. Koster: Proc. Kon. Ned. Akad. Wetensch. **74**, 332 (1971)

K.137 A.S. Koster: Appl. Phys. Lett. **18**, 170 (1971); J. Phys. Chem. Solids **32**, 2685 (1971)

K.138 A.S. Koster: Mol. Phys. **26**, 625 (1973)

K.139 A.S. Koster: Chem. Phys. Lett. **23**, 18 (1973)

K.140 Yu.P. Kostikov et al.: Mikrochim. Acta **2**, 95 (1976)

K.141 G.P. Kostikova: Sovrem. Probl. Khim. 1973, 11

K.142 G.P. Kòstikova, D.V. Korolkov, Yu.P. Kostikov: Teor. Eksp. Khim. **9**, 87 (1973)

K.143 G.P. Kostikova, M.P. Morosova, Yu.P. Kostikov, D.V. Korolkov: Teor. Eksp. Khim. **10**, 69 (1974)

K.144 G.P. Kostikova, Yu.P. Kostikov, S.I. Troyanov, D.V. Korolkov: Inorg. Chem. **17**, 2279 (1978)

K.145 V.O. Kostroun, R.W. Fairchild, C.A. Kukkoner, J.W. Wilkins: Phys. Rev. B **13**, 3268 (1976)

K.146 N. Kosuch, J. Müller, G. Wiech, A. Faessler: Phys. Fenn. **9**, S1-189 (1974)

N. Kosuch, E. Tegeler, G. Wiech, A. Faessler: J. Electron Spectrosc. **13**, 263 (1978)

K.147 N. Kosuch, E. Tegeler, G. Wiech, A. Faessler: In [Ref.1.62, p.60]

N. Kosuch: Thesis München 1977

N. Kosuch, G. Wiech, A. Faessler: J. Electron Spectrosc. **20**, 11 (1980)

K.148 K. Kosugi, T. Yokoyama, K. Asakura, H. Kuroda: Chem. Phys. **91**, 249 (1984)

T. Yokoyama, N. Kosugi, H. Kuroda: Chem. Phys. **103**, 101 (1986); ibid. **104**, 449 (1986)

K.149 A. Kotani et al.: J. Phys. Soc. Jpn. 56, 798 (1987)
       A. Kotani, M. Okada, T. Jo: Phys. Scr. 35, 566 (1987)
K.150 V.G. Kotetishvili: Tr. Gruz. Politekh. Inst. 1968, 73
K.151 B.I. Kotlyar, T.B. Shashkina et al.: In [Ref.1.58, Vol.I, p.163]
       P.M. Ovrutskaya: In [Ref.1.58, Vol.I, p.294]
       T.B. Shashkina, B.I. Kotlyar, E.A. Zhurakovskii: Izv. Vyssh. Uchebn. Zaved.,
       Fiz. (1970) 16; Teor. Eksp. Khim. 7, 686 (1971)
K.152 K. Koto, H. Mori. Y. Ito: Solid State Ionics 18-19, 720 (1986)
K.153 P.N. Koul, B.D. Padalia, M.N. Ghatikar: J. Phys. Soc. Jpn. 50, 246 (1981)
K.154 A.V. Koval, S.M. Karalnik et al.: Metallofizika 49, 99 (1973); ibid. 59, 65
       (1975); ibid. 75, 94 (1979); Ukr. Fiz. Zh. 23, 689 (1978)
K.155 A.V. Koval, S.M. Karalnik: Izv. Vyssh. Uchebn. Zaved., Fiz. 21, 17 (1978)
K.156 A.V. Kozinkin et al.: Fiz. Tverd. Tela 26, 697 (1984)
K.157 A.I. Kozlenkov, A.V. Anikin, I.B. Borovskii: In [Ref.1.58, Vol.I, p.33]
K.158 A.I. Kozlenkov et al.: J. Phys. C 18, 3581 (1985)
K.159 H. Kramer: Ann. Physik 4, 263 (1959); Thesis München 1960
K.160 A.A. Krasnoperova, E.S. Gluskin, L.N. Mazalov, V.A. Kochubei: Zh. Strukt.
       Khim. 17, 1113 (1976)
K.161 M.O. Krause, F. Wuilleumier, C.W. Nestor: Phys. Rev. A 6, 871 (1972)
K.162 M.O. Krause: Adv. X-Ray Anal. 16, 74 (1973); Phys. Lett. A 74, 303 (1979)
K.163 M.O. Krause, J.G. Ferreira: J. Phys. B 8, 2007 (1975)
K.164 M.O. Krause, J.H. Oliver: J. Phys. Chem. Ref. Data 8, 329 (1979)
K.165 E.A. Kravtsova, I.A. Sakharova, L.N. Mazalov, A.P. Sadovskii: Izv. Sib. Otd.
       Akad. Nauk SSSR, Ser. Khim. 1973, Nr. 2, 138
       A.V. Ablov, L.N. Mazalov et al.: Zh. Neorg. Khim. 18, 553 (1973)
K.166 G. Krill et al.: Solid State Commun. 35, 547 (1980)
       C. Godart, L.C. Gupta, M.F. Ravet-Krill: J. Less-Common Met. 94, 187
       (1983)
       C. Godart, J.C. Achard, G. Krill, M.F. Ravet-Krill: J. Less-Common Met. 94,
       177 (1983)
K.167 V. Krishna, J. Prasad, H.L. Nigam: Inorg. Chim. Acta 20, 193 (1976)
K.168 V. Krishna, J. Prasad, H.L. Nigam: Indian J. Pure Appl. Phys. 17, 95 (1979)
       H.L. Nigam, K.M. Kanth, V. Krishna: Indian J. Chem. 19 A, 804 (1980)
K.169 V.V. Krivitskii, I.Ya Nikiforov: Izv. Akad. Nauk SSSR, Ser. Fiz. 31, 970
       (1967)
K.170 V.V. Krivitskii, A.T. Shuvaev: Izv. Akad. Nauk SSSR, Ser. Fiz. 38, 522 (1974)
       V.V. Krivitskii et al.: Izv. Akad. Nauk SSSR, Ser. Fiz. 40, 284 (1976)
K.171 W. Krone et al.: Solid State Commun. 52, 253 (1984)
K.172 W. Krone et al.: Synth. Met. 17, 479 (1987)
K.173 E. Krousky, J. Hrdy: Jpn. J. Appl. Phys. 17, S2-433 (1978)
K.174 J. Krueger, G.G. Long, M. Kuriyama, A.I. Goldman: In Passivity of Metals
       and Semiconductors, ed. by M. Froment (Elsevier, Amsterdam 1983) p.163
K.175 V.I. Kruglov, T.M. Zimkina: Fiz. Tverd. Tela 10, 226 (1968)
K.176 I.Ya. Kudryavtsev, V.D. Demekhin, A.T. Kozakov: Vopr. Obshch. Prikl. Fiz.,
       Tr. Respubl. Konf. 1969, 57
K.177 G.H. Kuehl: J. Phys. Chem. Solids 38, 1259 (1977)
K.178 I.M. Kulikova, R.L. Barinskii: Izv. Akad. Nauk SSSR, Ser. Fiz. 49, 1476
       (1985)
K.179 D.K. Kulkarni, C. Mande: Indian J. Pure Appl. Phys. 12, 60 (1974)
K.180 A. Kumar, A.N. Nigam: X-Ray Spectrom. 8, 135 (1979)
K.181 A. Kumar, A.N. Nigam, U. Agarwala: Jpn. J. Appl. Phys. 18, 2299 (1979)
       A. Kumar, A.N. Nigam, B.D. Shrivastava: X-Ray Spectrom. 9, 77 (1980)
K.182 A. Kumar, A.N. Nigam, B.D Shrivastava: J. Phys. C 13, 3523 (1980)

S. Patil, A. Kumar, B.D. Padalia, S.V. Deshpande: Spectrochim. Acta A 41, 495 (1985)

K.183 A. Kumar, A.N. Nigam, U. Agarwala: Z. Phys. Chem. 262, 34,908 (1981)
A. Kumar, A.N. Nigam, B.D. Srivastava: X-Ray Spectrom. 10, 25 (1981)
A. Kumar, A.N. Nigam: Indian J. Pure Appl. Phys. 19, 1127 (1981)

K.184 A. Kumar, S.I. Salem: X-Ray Spectrom. 11, 164 (1982)

K.185 V. Kumar, A.R. Chetal, K.S. Srivastava: Indian J. Phys. 56 A, 373 (1982)

K.186 E.M. Kunoff, M.S. Dresselhaus, Y.H. Kao: Phys. Rev. B 34, 8460 (1986)

K.187 A.B. Kunz, D.J. Mickish, T.C. Collins: Phys. Rev. Lett. 31, 756 (1973)
A.B. Kunz, J.C. Boisvert, T.O. Woodruff: Phys. Rev. B 30, 2158 (1984)

K.188 C. Kunz: J. Physique 32, C4-180 (1971)

K.189 C. Kunz, R. Haensel, G. Keitel, P. Schreiber, B. Sonntag: In [Ref.1.101, p.275]

K.190 C. Kunz, H. Petersen, D.V. Hynch: Phys. Rev. Lett. 33, 1556 (1974)
H. Petersen: Phys. Rev. Lett. 35, 1363 (1975); J. Phys. F 7, 2495 (1977)

K.191 E.Z. Kurmaev, S.A. Nemnonov et al.: Izv. Akad. Nauk SSSR, Ser. Fiz. 31, 996 (1967)
V.M. Cherkashenko, E.Z. Kurmaev et al.: Fiz. Tverd. Tela 17, 280 (1975)
V.A. Gubanov, N.I. Lazukova, E.Z. Kurmaev: Izv. Sib. Otd. Akad. Nauk SSSR, Ser. Khim. Nauk 1975, Nr. 9, H.4, p.18

K.192 E.Z. Kurmaev: Phys. Stat. Sol. (b) 43, K49 (1971)
V.I. Anisimov et al.: Solid State Commun. 29, 185 (1979)
Yu.M. Yarmoshenko, E.Z. Kurmaev: Fiz. Met. Metalloved. 51, 302 (1981)

K.193 E.Z. Kurmaev, G.P. Shveikin, S.A. Nemnonov: Phys. Stat. Sol. (b) 60, K65 (1973)
V.A. Gubanov, E.Z. Kurmaev, G.P. Shveikin: J. Phys. Chem. Solids 38, 201 (1977)
V.A. Gubanov, E.Z. Kurmaev, D.E. Ellis: J. Phys. C 14, 5567 (1981)

K.194 E.Z. Kurmaev, V.G. Syryanov, I.Ya Gusman, T.N. Sabruskova: Izv. Akad. Nauk SSSR, Neorg. Mater. 9, 867 (1973)

K.195 E.Z. Kurmaev, V.P. Belash, S.A. Nemnonov, A.S. Shulakov: Phys. Stat. Sol. (b) 61, 365 (1974)

K.196 E.Z. Kurmaev, M.F. Sorokina: Phys. Stat. Sol. (b) 62, K99 (1974)

K.197 E.Z. Kurmaev, I.A. Brytov, V.M. Cherkashenko, Yu.N. Romashchenko: App. Met. Rentgen. Anal. 17, 112 (1975)

K.198 E.Z. Kurmaev, V.E. Dolgikh, Yu.M. Yarmoshenko, S.A. Nemnonov: Fiz. Met. Metalloved. 48, 1302 (1979)

K.199 E.Z. Kurmaev et al.: Solid State Commun. 37, 647 (1981)
Yu.M. Yarmoshenko, V.M. Cherkashenko, E.Z. Kurmaev: J. Electron Spectrosc. 32, 103 (1983)

K.200 E.Z. Kurmaev, V.R. Galakhov, A.V. Postnikov: J. Phys. F 15, 2041 (1985)

K.201 E.Z. Kurmaev, S.N. Shamin, K.M. Kolobova, S.V. Shulepov: Carbon 24, 249 (1986)

K.202 H. Kuroda et al.: Solid State Commun. 46, 235 (1983)

K.203 C. Kurylenko: Cah. Physique 17, 163 (1963); ibid. 20, 333 (1966); ibid. 21, 206, 407 (1967)

K.204 M.S. Kushwaha, B.D. Shrivastava, V.S. Dubey: Phys. Stat. Sol. (b) 64, 65 (1974); Nuovo Cim. B 19, 169 (1974); indian J. Pure Appl. Phys. 15, 290 (1977)

K.205 F.W. Kutzler et al.: J. Chem. Phys. 73, 3274 (1980)
F.W. Kutzler, K.O. Hodgson, S. Doniach: Phys. Rev. A 26, 3020 (1982)

K.206 F.W. Kutzler, R.A. Scott, J.M. Berg. K.O. Hodgson: J. Am. Chem. Soc. 103, 6083 (1981)

408

K.207 F.W. Kutzler, K.O. Hodgson, D.K. Misemer, S. Doniach: Chem. Phys. Lett. **92**, 626 (1982)
K.208 F.W. Kutzler et al.: Solid State Commun. **46**, 803 (1983)
      S.H. Chou et al.: Phys. Rev. B **31**, 1069 (1985)
K.209 F.W. Kutzler, D.E. Ellis: Phys. Rev. B **29**, 6890 (1984)
K.210 V.G. Kuznetsov, I.I. Tupitsyn: Vestn. Leningr. Univ., Fiz. Khim. **22**, 33 (1976)

L.1   A. Labhardt, C. Yuen: Nature (London) **277**, 150 (1979)
L.2   S. Laderman, A. Bienenstock, K.S. Liang: Sol. Energy Mater. **8**, 15 (1982)
L.3   P. Lagarde: Phys. Rev. B **14**, 741 (1976)
L.4   P. Lagarde, A.M. Flank: J. Physique **47**, 1389 (1986)
L.5   M. Lähdeniemi, E. Suoninen: In [Ref.1.62, p.142]
L.6   M. Lähdeniemi, E. Suoninen, J. Bremer: Jpn. J. Appl. Phys. **17**, S2-129 (1978)
L.7   M. Lähdeniemi, E. Suoninen, J. Vanhatalo: In [Ref.1.110, p.175]
L.8   M. Lähdeniemi, E. Ojala, E. Suoninen, I. Terakura: J. Phys. F **11**, 1531 (1981)
L.9   M. Lähdeniemi, E. Ojala, I. Terakura, K. Terakura: J. Phys. F **13**, 521 (1983)
      M. Lähdeniemi, E. Ojala, M. Okochi: J. Phys. F **13**, 513 (1983)
L.10 G.M. Lamble et al.: Phys. Rev. B **34**, 2975 (1986)
      G.M. Lamble, R.S. Brooks: Preprint 1986
L.11 G.M. Lamble et al.: Phys. Rev. B **34**, 2975 (1986); ibid. B **36**, 1796 (1987)
L.12 J. Lang, W.S. Watson, J. Phys. B **8**, L339 (1975)
L.13 C. Lapeyre, J. Petiau, G. Calas: In *Structure of Non-Crystalline Materials 1982*, ed. by P.H. Gaskell, J.M. Parker, E.A. Davis (Taylor and Francis, London 1983) p.42
L.14 I.P. Laputina, K.I. Narbutt: Izv. Akad. Nauk SSSR, Ser. Fiz. **31**, 912 (1967)
      I.P. Laputina: In [Ref.1.58, Vol.II, p.281]
      K.I. Narbutt: Izv. Akad. Nauk SSSR, Ser. Fiz. **40**, 407 (1976)
L.15 F.P. Larkins, T.W. Rowlands: J. Phys. B **19**, 591 (1986)
L.16 K. Läuger: Über den Einfluß der Bindungsart und der Kristallstruktur auf das K-Röntgenemissionsspektrum von Aluminium und Silizium. Dissertation, Universität München (1968)
L.17 K. Läuger: J. Phys. Chem. Solids **32**, 609 (1971); ibid. **33**, 1343 (1972)
L.18 H. Launois et al.: Phys. Rev. Lett. **44**, 1271 (1980)
L.19 R.E. LaVilla: Bull. Am. Soc. **11**, 389 (1966)
L.20 R.E. LaVilla, R.D. Deslattes: J. Chem. Phys. **44**, 4399 (1966)
L.21 R.E. LaVilla, R.D. Deslattes: J. Chem. Phys. **45**, 3466 (1966)
      R.D. Deslattes, R.E. LaVilla: Appl. Optics **6**, 39 (1967)
L.22 R.E. LaVilla: Phys. Rev. A **4**, 476 (1971)
L.23 R.E. LaVilla, R.D. Deslattes: J. Physique **32**, C4-160 (1971)
L.24 R.E. LaVilla: J. Chem. Phys. **56**, 2345 (1972)
L.25 R.E. LaVilla: J. Chem. Phys. **57**, 899 (1972)
L.26 R.E. LaVilla: Phys. Rev. A **8**, 1143 (1973)
L.27 R.E. LaVilla: J. Chem. Phys. **58**, 3841 (1973); ibid. **63**, 2733 (1975)
L.28 R.E. LaVilla: Phys. Rev. A **9**, 1801 (1974)
L.29 R.E. LaVilla: J. Chem. Phys. **62**, 2209 (1975)
      R.C.C. Perera, R.E. LaVilla: J. Chem. Phys. **81**, 3375 (1984); ibid. **84**, 4228 (1986)
L.30 R.E. LaVilla: Phys. Rev. A **17**, 1018 (1978); ibid. B **18**, 644 (1978)
L.31 R.E. LaVilla: Phys. Rev. A **19**, 717 (1979)
L.32 R.E. LaVilla: Phys. Rev. A **19**, 1999 (1979)
L.33 K. Lawniczak, J. Auleytner: In [Ref.1.100, p.168]

L.34   K. Lawniczak-Jablonska, J. Auleytner: J. Phys. F 12, 2729 (1982)
L.35   K. Lawniczak-Jablonska et al.: Phys. Stat. Sol. (b) 123, 627 (1984)
L.36   D.F. Lawrence, D.S. Urch: Spectrochim. Acta 25 B, 305 (1970)
        M. Kasrai, D.S. Urch: J. Chem. Soc., Faraday Trans. II 75, 1522 (1979)
        D.S. Urch: J. Chem. Soc., Chem. Commun. 10, 526 (1982)
        E.I. Esmail, D.S. Urch: Spectrochim. Acta A 39, 573 (1983)
L.37   N.I. Lazukova et al.: Zh. Strukt. Khim. 21, 41 (1980)
L.38   V.I. Lebedev: Mineral. Geokhim. 5, 70 (1975)
L.39   P.A. Lee, G. Beni: Phys. Rev. B 15, 2862 (1977)
L.40   P.L. Lee, E.C. Seltzer, F. Boehm: Phys. Lett. A 38, 29 (1972)
L.41   P.L. Lee, F. Boehm: Phys. Rev. C 8, 819 (1973)
L.42   P.L. Lee, F. Boehm, P. Vogel: Phys. Rev. A 9, 614 (1974)
L.43   P.L. Lee, S.I. Salem: Phys. Rev. A 10, 2027 (1974)
L.44   P.L. Lee, J.B. Pendry: Phys. Rev. B 11, 2795 (1975)
L.45   P.L. Lee, F. Boehm, P. Vogel: Phys. Lett. A 63, 251 (1977)
L.46   J.P. Lelieur, J. Goulon, R. Cortes, P. Friant: J. Phys. Chem. 88, 3730 (1984)
L.47   G.F. Lemeshko: Multiplet structure of the $K\alpha_{1,2}$ and $K\beta_1\beta'$ spectra of elements with partly filled 3d shells. Dissertation, University of Rostov/Don (1974).
L.48   B. Lengeler, P. Eisenberger: Phys. Rev. B 21, 4507 (1980)
        P. Eisenberger, B. Lengeler: Phys. Rev. B 22, 3551 (1980)
L.49   B. Lengeler, G. Materlik, J.E. Muller: Phys. Rev. B 28, 2276 (1983)
L.50   B. Lengeler: Phys. Rev. Lett. 53, 74 (1984)
        B. Lengeler, R. Zeller: J. Less-Common Met. 103, 337 (1984)
        B. Lengeler, R. Zeller: Solid State Commun. 51, 889 (1984)
        B. Lengeler, M.A. Pick: Solid State Commun. 53, 297 (1985)
L.51   B. Lengeler: Solid State Commun. 55, 679 (1985)
L.52   B. Lengeler: Ber. Bunsenges. Phys. Chemie 90, 649 (1986)
L.53   M. Lenglet et al.: Mater. Res. Bull. 18, 935 (1983)
L.54   M. Lenglet, W. Lopitaux, J. Arsene: J. Solid State Chem. 50, 294 (1983)
        M. Lenglet et al.: J. Solid State Chem. 58, 194 (1985); Mater. Res. Bull. 20, 745 (1985)
        M. Lenglet, P. Foulatier, J. Durr, J. Arsene: Phys. Stat. Sol. (a) 94, 461 (1986)
L.55   M. Lenglet et al.: J. Phys. C 19, L363 (1986)
L.56   P. Leonard: J. Phys. F. 8, 467 (1978)
L.57   E.A. Leone: Scanning Electron Microsc. 3, 1023 (1984)
L.58   G. Leonhardt: Zur theoretischen Interpretation röntgenspektroskopischer Untersuchungen an Verbindungen der 3d-Übergangselemente. Dissertation, Karl-Marx-Universität, Leipzig (1969)
L.59   G. Leonhardt, A. Meisel: Spectrochim. Acta B 25, 163 (1970)
L.60   G. Leonhardt, A. Meisel: J. Chem. Phys. 52, 6189 (1970)
L.61   G. Leonhardt, J. Hedman, C. Nordling, A. Meisel: In [Ref.1.60, Vol.II, p.313]
L.62   G. Leonhardt, P. Pelowa, A. Meisel: Z. Anorg. Allg. Chemie 397, 209 (1973)
L.63   G. Leonhardt et al.: Phys. Scr. 16, 448 (1977)
L.64   G. Leonhardt, A. Kosakov, H. Sommer, M. Petke: In [Ref.1.62, p.57]
        A. Kosakov, H. Neumann, G. Leonhardt: Phys. Lett. A 61, 57 (1977); ibid. A 62, 95 (1977)
L.65   R.M. Levy: J. Chem. Phys. 43, 398, 1846 (1965)
L.66   R.M. Levy, J.R. Van Wazer, J. Simpson: Inorg. Chem. 5, 332 (1966)
L.67   P.H. Lewis: J. Phys. Chem. 66, 105 (1962); ibid. 67, 2151 (1963); J. Catalysis 11, 162 (1968)
L.68   G. Licheri et al.: Chem. Phys. Lett. 83, 384 (1981)
        G. Licheri, G. Pinna, G. Navarra, G. Vlaic: Z. Naturforsch. 38 a, 559 (1983)

410

G. Licheri, G. Paschina, G. Piccaluga, G. Pinna: J. Chem. Phys. **79**, 2168 (1983)

L.69 G. Licheri et al.: J. Non-Cryst. Solids **72**, 211 (1985)
L.70 R.I. Liefeld: In [Ref.1.78, p.133]
L.71 R.J. Liefeld, A.F. Burr, M.B. Chamberlain: Phys. Rev. A **9**, 316 (1974)
L.72 M. Lindroos, G. Graeffe, M. Karras: X-Ray Spectrom. **6**, 161 (1977)
L.73 R.G. Linford, P.G. Hall, C. Johnson, S.S. Hasnain: Solid State Ionics **14**, 199 (1984)
L.74 M. Linkoaho, T. Åberg, G. Graeffe, J. Utriainen: Z. Naturforsch. **24 a**, 775 (1969)
L.75 M.V. Linkoaho: In [Ref.1.93, p.146]
L.76 M. Linkoaho, J. Utriainen: Phys. Fenn. **8**, 67 (1973)
L.77 L.B. Litvinova, N.S. Pastushuk, N.E. Kasyanov: In [Ref.1.58, Vol.I, p.206]
L.78 V.A. Lobach, A.B. Sobolev, B.V. Shulgin: Zh. Strukt. Khim. **27**, Nr.6, 3 (1986)
L.79 N.R. Lokhande, A.R. Chetal: J. Phys. Soc. Jpn. **47**, 614 (1979)
L.80 N.R. Lokhande, C. Mande: Phys. Stat. Sol. (b) **102**, K11 (1980)
L.81 N.R. Lokhande: Phys. Stat. Sol. (b) **114**, K35 (1982)
      N.R. Lokhande, C. Mande: J. Phys. Chem. Solids **43**, 731 (1982)
L.82 G.G. Long, J. Kruger, D. Tanaka: J. Electrochem. Soc. **134**, 264 (1987)
L.83 R.D. Lorentz, S.S. Laderman, A.I. Bienenstock: In [Ref.1.83, p.280]
L.84 P.P. Lottici: Phys. Rev. B **35**, 1236 (1987)
L.85 M.K. Loudjani, J. Roy, A.M. Huntz, R. Cortes: J. Am. Ceram. Soc. **68**, 559 (1985)
L.86 A.J. Lowe, S.R. Elliott, G.N. Greaves: Phil. Mag. B **54**, 483 (1986)
L.87 A.R. Lubinsky, D.E. Ellis, G.S. Painter: Phys. Rev. B **11**, 3131 (1975)
L.88 A. Lucasson: Ann. Physique **5**, 509 (1960)
L.89 A.P. Lukirskii, I.A. Brytov: Fiz. Tverd. Tela **6**, 43 (1964)
L.90 A.P. Lukirskii, T.M. Zimkina: Izv. Akad. Nauk SSSR, Ser. Fiz. **28**, 765 (1964)
      A.S. Vinogradov, T.M. Zimkina, J.F. Maltsev: Fiz. Tverd. Tela **11**, 3354 (1969)
L.91 A.P. Lukirskii, I.A. Brytov: Izv. Akad. Nauk SSSR, Ser. Fiz. **28**, 841 (1968)
L.92 Lu Kunquan, Wan Jun: Solid State Commun. **63**, 797 (1987)
L.93 E.N. Luzina: Ukr. Fiz. Zh. **24**, 1755 (1979)
L.94 I.I. Lyakhovskaya, T.M. Zimkina, V.A. Fomichev: Fiz. Tverd. Tela **12**, 174 (1970); Izv. Akad. Nauk SSSR, Ser. Fiz. **36**, 393 (1972)
      T.M. Zimkina: In Proc. Colloq. Spectrosc. Int'l Invited Lect., 20th Prague 1977, Vol.2 p.73
L.95 I.I. Lyakhovskaya, T.M. Zimkina, V.A. Fomichev: Fiz. Tverd. Tela **17**, 1417 (1975)
L.96 I.I. Lyakhovskaya, V.M. Ipatov, T.M. Zimkina: Zh. Strukt. Khim. **18**, 668 (1977)
      T.M. Zimkina, A.S. Shulakov, A.P. Braiko: Fiz. Tverd. Tela **23**, 2006 (1981)
      T.M. Zimkina et al.: Fiz. Tverd. Tela **25**, 26 (1983)
L.97 F.W. Lytle: Adv. X-Ray Anal. **9**, 398 (1966)
L.98 F.W. Lytle: Acta Crystallogr. **22**, 321 (1967)
L.99 F.W. Lytle: Appl. Phys. Lett. **24**, 45 (1974)
L.100 F.W. Lytle: J. Catal. **43**, 376 (1976)
      P.S.P. Wei, F.W. Lytle: Phys. Rev. B **19**, 679 (1979)
L.101 F.W. Lytle, G.H. Via, J.H. Sinfelt: Prepr. Div. Pet. Chem. Am. Chem. Soc. **21**, 366 (1976); J. Chem. Phys. **67**, 3831 (1977); in [Ref.1.50, p.401]
L.102 F.W. Lytle et al.: J. Chem. Phys. **70**, 4849 (1979)
      G.H. Via, J.H. Sinfelt, F.W. Lytle: J. Chem. Phys. **71**, 690 (1979)

M.1 Y. Ma, E.A. Stern: Phys. Rev. B 35, 2678 (1987)
M.2 Y. Ma, E.A. Stern, F.W. Gayle: Phys. Rev. Lett. 58, 1956 (1987)
M.3 H. Maeda, T. Tanimoto, H. Terauchi, M. Hida: Phys. Stat. Sol. (a) 58, 629 (1980)
M.4 H. Maeda et al.: Jpn. J. Appl. Phys. 21, 1342 (1982)
M.5 H. Maeda: J. Phys. Soc. Jpn. 56, 2777 (1987)
M.6 K. Maeda: J. Phys. Chem. Solids 42, 981 (1981)
M.7 K. Maeda: J. Phys. Chem. Solids 43, 121 (1982)
M.8 K. Maeda, M. Uda, Y. Hayasi: Phys. Lett. A 112, 431 (1985)
M.9 T. Maekawa, N. Kikuchi, N. Fukuda, T. Yokokawa: Bull. Chem. Soc. Jpn. 51, 777 (1978)
M.10 T. Maekawa, Y. Kon, T. Yokokawa: Yogyo-Kyokai-Shi 94, 1236 (1986)
M.11 T. Magnusson: Nova Acta Reg. Soc. Sci. Upsal. IV 11, No.3 (1938)
M.12 G.D. Mahan: In [Ref.1.86, p.635]; Phys. Rev. B 11, 4814 (1975)
M.13 A. Maiste, A. Saar, M. Elango: Fiz. Tverd. Tela 16, 1720 (1974)
M.14 A. Maiste, R. Ruus: Opt. Spektrosk. 46, 197 (1979)
M.15 A. Maiste, R. Ruus, M. Elango: Zh. Eksp. Teor. Fiz. 79, 1671 (1980)
     R.E. Ruus, A.A. Maiste, Yu.A. Maksimov: Izv. Akad. Nauk SSSR, Ser. Fiz. 46, 789 (1982)
M.16 L.L. Makarov, Yu.P. Kostikov, G.P. Shabanova, A.V. Panin: Dokl. Akad. Nauk SSSR 185, 642 (1969)
M.17 L.L. Makarov et al.: Dokl. Akad. Nauk SSSR 209, 893 (1973)
M.18 L.L. Makarov et al.: Teor. Eksp. Khim. 11, 221 (1975)
M.19 L.L. Makarov, Yu.M. Zaitsev, Yu.F. Batrakov: Teor. Eksp. Khim. 12, 78 (1976)
     L.L. Makarov et al.: J. Res. Inst. Catalysis, Hokkaido Univ. 24, 102 (1976); Koord. Khim. 2, 1646 (1976)
     Yu.I. Dyachenko et al.: Teor. Eksp. Khim. 14, 44 (1978)
     Yu.M. Zaitsev, Yu.F. Batrakov, L.L. Makarov: Teor. Eksp. Khim. 14, 352 (1978)
M.20 L.L. Makarov, Yu.M. Zaitsev, L.M. Bakhmeteva: Zh. Obshch. Khim. 50, 1910 (1980)
M.21 L.L. Makarov et al.: Fiz. Khim. Stekla 10, 8 (1984)
     N.D. Aksenov et al.: Fiz. Khim. Stekla 10, 395 (1984)
     Yu.F. Batrakov et al.: Teor. Eksp. Khim. 20, 689 (1984)
M.22 S.K. Malik et al.: Phys. Rev. B 31, 4728 (1985); Phys. Rev. Lett. 55, 316 (1985)
M.23 M. Malkowska: Exp. Techn. Physik 6, 130 (1958); Thesis Prague 1963
M.24 P.J. Mallozzi, R.E. Schwerzel, H.M. Epstein, B.E. Campbell: Science 206, 353 (1979)
M.25 D. Malterre, G. Krill, J. Durand, G. Marchal: J. Physique 46, C8-199 (1985)
M.26 D. Malterre et al.: Phys. Rev. B 34, 2176 (1986)
M.27 D. Malterre et al.: J. Magn. Magn. Mater. 63-64, 521 (1987)
M.28 W. Malzfeldt: Thesis, Kiel 1985
M.29 R. Manaila, D. Macovei, R. Grigorovici, P. Pausescu: J. Non-Cryst. Solids 75, 183 (1985)
     R. Mañaila, D. Macovei, P. Pausescu, R. Grigorovici: J. Non-Cryst. Solids 77-78, 233 (1985)
M.30 C. Mande: Ann. Physique 13, 1559 (1960)
M.31 C. Mande, A.R. Chetal: Indian J. Phys. 28, 433 (1964); Curr. Sci. 33, 707 (1964); ibid. 34, 556 (1965); Proc. Indian Acad. Sci. 62, 97 (1966)
     A.R. Chetal: Phys. Fenn. 9, S 1-104 (1974)

D.K. Kulkarni, C. Mande: Indian J. Pure Appl. Phys. 12, 60 (1974)

A. De Kozak, S. Chattopadhyay, C. Mande: Indian J. Phys. A 59, 34 (1985)

M.32   C. Mande et al.: Indian J. Pure Appl. Phys. 3, 401 (1965); ibid. 4, 400 (1966)

Y.L. Rao, C. Mande: Phys. Stat. Sol. (b) 92, K61 (1979)

M.33   C. Mande, R.N. Patil, A.S. Nigavekar: Nature 211, 578 (1966)

C. Mande, A.S. Nigavekar: Proc. Indian Acad. Sci. A 67, 166 (1968); ibid. 69, 316 (1969)

M.34   C. Mande, A.S. Nigavekar, P. Chivate: Indian J. Phys. Proc. Indian Assoc. Cultivat. Sci. 41/50, 897 (1967)

M.35   C. Mande, N.V. Joshi: Indian J. Pure Appl. Phys. 7, 65 (1969)

M.36   C. Mande, D.K. Kulkarni, A.R. Chetal: Brit. J. Appl. Phys. 2, 635 (1969)

M.37   C. Mande, A.R. Chetal, V.B. Sapre: Curr. Sci. 39, 391 (1970)

C. Mande, V.B. Sapre: Indian J. Pure Appl. Phys. 17, 331 (1979)

M.38   C. Mande, A.V. Pendharkar, M.C. Chakravorti: Proc. Indian Acad. Sci. A 75, 209 (1972)

A.V. Pendharkar, C. Mande: Pramana 1, 104 (1973); Physica 66, 204 (1973); Chem. Phys. 7, 244 (1975)

M.39   C. Mande, V.B. Sapre: In [Ref.1.60, Vol.I, p.237]

B. Sapre, C. Mande: Indian J. Pure Appl. Phys. 12, 74 (1974)

Y.L. Rao, A.R. Chourasia, C. Mande: J. Non-Cryst. Solids 46, 13 (1981)

M.40   C. Mande, A.R. Chourasia: Indian J. Phys. B 60, 72 (1986)

M.41   I. Manescu: C.R. Acad. Sci. 225 537 (1947); ibid. 226, 1010 (1948)

M.42   R. Manne, J. Chem. Phys. 46, 4645 (1967)

A. Stogard, R. Manne: Chem. Phys. 8, 348 (1975)

M.43   R. Manne: J. Chem. Phys. 52, 5733 (1970)

M.44   R. Manne, M. Karras, E. Suoninen: Chem. Phys. Lett. 15, 34 (1972)

E.-K. Kortela, M. Karras: Spectrochim. Acta 29 A, 1293 (1973)

M.45   R. Manne, T.J. Arstad, J. Muller: Phys. Scr. 21, 549 (1980)

M.46   M.W.D. Mansfield, J.P. Connerade: Proc. Roy. Soc. London A 344, 303, 421 (1975)

M.47   M.W.D. Mansfield, J.P. Connerade: Proc. Roy. Soc. London A 352, 125 (1976)

M.48   M.W.D. Mansfield: Proc. Roy. Soc. London A 358, 253 (1978)

M.49   A. Mansour, S.E. Schnatterly, R.D. Carson: Phys. Rev. B 31, 6521 (1985)

M.50   A. Mansour, S.E. Schnatterly, J.J. Ritsko: Phys. Rev. Lett. 58, 614 (1987)

A. Mansour, S. Schnatterly: Phys. Scr. 35, 595 (1987)

M.51   A. Manthiram et al.: J. Phys. Chem. 84, 2200 (1980)

M.52   A. Marcelli, A Bianconi, I. Davoli, S. Stizza: In [Ref.1.83, p.52]; J. Magn. Magn. Mater. 47–48, 206 (1984)

A. Bianconi, A. Marcelli, I. Davoli, S. Stizza: In [Ref.1.66, p.403]

M.53   A. Marcelli et al.: J. Physique 46, C8-107 (1985)

M.54   M. Maret, M. Belakhovsky, F. Merdrignac, A.M. Flank: J. Physique 46, C8-211 (1985)

M.55   J.M. Mariot, R.C.Karnatak: J. Phys. F 4, L223 (1974)

M.56   J.M. Mariot, C.F. Hague, P. Oelhafen, H.-J. Güntherodt: J. Phys. F 16, 1197 (1986)

P. Oelhafen et al.: J. Non-Cryst. Solids 61-62, 1067 (1984)

M.57   E.C. Marques, D.R. Sandström, F.W. Lytle, R.B. Greegor: J. Chem. Phys. 77, 1027 (1982)

M.58   M.I. Marques, M.C. Martins, J.G. Ferreira: J. Phys. B 13, 41 (1980)

M.59   G. Martens, P. Rabe, N. Schwentner, A. Werner: Phys. Rev. Lett. 39, 1411 (1977); Phys. Rev. B 17, 1481 (1978); J. Phys. C 11, 3125 (1978)

G. Martens, P. Rabe: Phys. Stat. Sol. (a) 57, K 31 (1980)

M.60   G. Martens, P. Rabe, G. Tolkiehn, A. Werner: Phys. Stat. Sol. (a) 55, 105
       (1979)
       P. Rabe, G. Tolkiehn, A. Werner: J. Phys. C 12, L545 (1979)
M.61   G. Martens, P. Rabe, P. Wenck: Phys. Stat. Sol. (a) 88, 103 (1985)
M.62   R.M. Martin, J.B. Boyce, J.W. Allen, F. Holtzberg: Phys. Rev. Lett. 44, 1275
       (1980)
M.63   S. Maruno, S. Fujii: Jpn. J. Appl. Phys. 9, 1428 (1970)
       S. Maruno: Proc. 6th Int'l Conf. X-Ray Optics Microanal. (1971) p.579
M.64   S. Maruno, M. Mitsuda: Nagoya Kogyo Daigaku Gakuho 26, 197 (1975)
M.65   L.A. Marusak, L.L. Tongson: J. Appl. Phys. 50, 4350 (1979)
M.66   S.B. Maslenkov, A.I. Kozlenkov, S.A. Filin, A.I. Shulgin: Phys. Stat. Sol. (b)
       123, 605 (1984)
M.67   G. Materlik, J.E. Müller, J.W. Wilkins: Phys. Rev. Lett. 50, 267 (1983)
       G. Materlik, B. Sonntag, M. Tausch: Phys. Rev. Lett. 51, 1300 (1983)
M.68   Y. Mathey, H. Mercier, A. Michalowicz, A. Leblanc: J. Phys. Chem. Solids
       46, 1025 (1985)
M.69   T. Matsushita, H. Oyanagi, S. Saigo, H. Kihara: In [Ref.1.83, p.476]
M.70   T. Matsuura, T. Fujikawa, H. Oyanagi: J. Phys. Soc. Jpn. 53, 2837 (1984)
M.71   J.A.D. Matthew: Phys. Lett. A 50, 401 (1975)
M.72   J.A.D. Matthew, F.P. Netzer, C.W. Clark, J.F. Morar: Europhys. Lett. 4, 677
       (1987)
M.73   R.A. Mattson, R.C. Ehlert: Adv. X-Ray Anal. 9, 471 (1966)
M.74   R.A. Mattson, R.C. Ehlert: J. Chem. Phys. 48, 5465, 5471 (1968)
M.75   M. Maurer, J.M. Friedt, G. Krill: J. Phys. F 13, 2389 (1983)
M.76   M. Maurer, J.M. Friedt, J.P. Sanchez: J. Phys. F 15, 1449 (1985)
M.77   L.N. Mazalov, E.E. Vainshtein, V.G. Syryanov: Dokl. Akad. Nauk SSSR 164,
       545 (1965); Zh. Strukt. Khim. 7, 475 (1966)
M.78   L.N. Mazalov, S.M. Blokhin, E.E. Vainshtein: Fiz. Tverd. Tela 8, 2420 (1966)
       E.E. Vainshtein, L.N. Mazalov, E.A. Galtsova: Dokl. Akad. Nauk SSSR 168,
       376 (1966)
M.79   L.N. Mazalov, A.V. Nikolaev et al.: Izv. Sib. Otd. Akad. Nauk SSSR, Ser.
       Khim. Nauk 1971, 3
       L.N. Mazalov, A.P. Sadovskii et al.: Zh. Strukt. Khim. 14, 76 (1973)
M.80   L.N. Mazalov, A.P. Sadovskii et al.: Izv. Sib. Otd. Akad. Nauk SSSR, Ser.
       Khim. Nauk 1971, 51
M.81   L.N. Mazalov et al.: Zh. Strukt. Khim. 14, 262 (1973); ibid. 16, 262, 267
       (1975)
M.82   L.N. Mazalov, V.I. Baranovskii et al.: Zh. Strukt. Khim. 15, 51 (1974)
M.83   L.N. Mazalov et al.: Zh. Strukt. Khim. 15, 1099 (1974)
M.84   L.N. Mazalov, A.P. Sadovskii et al.: Zh. Strukt. Khim. 15, 800, 805 (1974)
       F.Kh. Gelmukhanov, L.N. Mazalov et al.: Dokl. Akad. Nauk SSSR 225, 597
       (1975)
       E.S. Gluskin, L.N. Mazalov, A.P. Sadovskii, D.A. Zhogolev: Zh. Strukt. Khim.
       16, 1061 (1075)
M.85   L.N. Mazalov, V.V. Murakhtanov, T.I. Guzhavina: Zh. Strukt. Khim. 16, 49
       (1975)
M.86   L.N. Mazalov, A.V. Kondratenko, V.V. Murakhtanov, T.I. Guzhavina: Zh.
       Strukt. Khim. 17, 174 (1976)
M.87   L.N. Mazalov, E.A. Kravtsova, S.V. Zemskov, Yu.I. Nikonorov: Zh. Strukt.
       Khim. 18, 565 (1977)
       L.N. Mazalov, A.A. Voityuk, E.A. Kravtsova: Zh. Strukt. Khim. 22, 169
       (1981)
M.88   L.N. Mazalov: Zh. Strukt. Khim. 18, 607 (1977)

M.89   L.N. Mazalov, V.V. Volkov, S.Ya. Dvurechenskaya, L. Nasonova: Zh. Neorg. Khim. **23**, 1860 (1978)

M.90   L.N. Mazalov, V.D. Yumatov, G.N. Dolenko: Zh. Strukt. Khim. **21**, 21 (1980)
V.D. Yumatov, L.N. Mazalov, E.A. Ilinchik: Zh. Strukt. Khim. **21**, Nr.5, 24 (1980)
L.N. Mazalov, G.N. Dolenko, V.D. Yumatov: Zh. Strukt. Khim. **22**, No.1,18 (1981)
L.N. Mazalov: In [Ref.1.66, p.327]

M.91   L.N. Mazalov et al.: Zh. Strukt. Khim. **25**, No.3, 34 (1984)
G.K. Parygina et al.: Zh. Strukt. Khim. **26**, No.4, 165 (1985)

M.92   L.N. Mazalov, V.D. Yumatov, A.V. Okotrub: Izv. Akad. Nauk SSSR, Ser. Fiz. **49**, 1483 (1985)

M.93   L.N. Mazalov, E.A. Kravtsova: Zh. Strukt. Khim. **27**, Nr.4, 84 (1986)

M.94   A.J. McAlister: Phys. Rev. **186**, 595 (1969)

M.95   A.J. McAlister, M.L. Williams, J.R. Cuthill, R.C. Dobbyn: Solid State Commun. **9**, 1775 (1971)
A.J. McAlister et al.: In [Ref.1.79, p.191]

M.96   A.J. McAlister, J.R. Cuthill, R.C. Dobbyn, M.L. Williams: Phys. Rev. Lett. **29**, 179 (1972); in [Ref.1.60, Vol.II, p.426]

M.97   J.L. McAtee, N.R. Smith: J. Colloid Interface Sci. **29**, 389 (1969)

M.98   J.M. McCaffrey, D.A. Papaconstantopoulos: Solid State Commun. **14**, 1055 (1974)
D.A. Papaconstantopoulos: In [Ref.1.62, p.192]

M.99   C.F. McConville et al.: Surf. Sci. **166**, 221 (1986)

M.100 D.G. McCrary. P. Richard: Phys. Rev. A **5**, 1249 (1972)

M.101 A.A. McFarlane: Proc. 6th Nat'l Conf. Electron Probe Anal. Pittsburgh (1971) p.49

M.102 A.A. McFarlane: Carbon **11**, 73 (1973)

M.103 E.J. McGuire: Scandia Res. Rep. **70**, 721 (1970); Phys. Rev. A **5**, 2313 (1972)

M.104 D.A. McKeown, G.A. Waychunas, G.E. Brown: J. Non-Cryst. Solids **74**, 325, 349 (1985)

M.105 A. Meagher et al.: Polymer **27**, 979 (1986)

M.106 G. Mechelke, K. Ulmer: Phys. Stat. Sol. (a) **19**, 573 (1973)

M.107 A.I. Medvedev et al.: Fiz. Met. Metalloved. **57**, 245 (1984)
Y.M. Yarmoshenko et al.: Solid State Commun. **55**, 19 (1985)
V.R. Galakhov et al.: J. Electron Spectrosc. **35**, 87 (1985)
V.I. Anisimov, M.A. Korotin, E.Z. Kurmaev: Fiz. Met. Metalloved. **61**, 728 (1986)

M.108 S. Mehta, K.N. Saxena, R.G. Anikhindi: Jpn. J. Appl. Phys. **17**, S 2-187 (1978)
S. Mehta, R.G. Anikhindi: Phys. Lett. **70 A**, 158 (1979)

M.109 A. Meisel, E. Döring: Z. Phys. Chemie **220**, 397 (1962)

M.110 A. Meisel: Izv. Akad. Nauk SSSR, Ser. Fiz. **28**, 811 (1964)

M.111 A. Meisel, G. Leonhardt: Z. Anorg. Allg. Chemie **339**, 1 (1965)

M.112 A. Meisel, To ba Trong: J. Prakt. Chemie **29**, 192 (1965)

M.113 A. Meisel: In [Ref.1.57, p.212]

M.114 A. Meisel, R. Szargan: Z. Phys. Chemie **236**, 113 (1967)

M.115 A. Meisel, A. Szargan: Z. Phys. Chemie **238**, 136 (1968)

M.116 A. Meisel, U. Schmidt: Z. Phys. Chemie **239**, 395 (1968)

M.117 A. Meisel, H. Sommer: In [Ref.1.58, Vol.I, p.234]

M.118 A. Meisel, R. Szargan: In [Ref.1.58, Vol.I, p.297]

M.119 A. Meisel, G. Leonhardt: In [Ref.1.92, p.150]

M.120 A. Meisel, H. Keilacker: Z. Phys. Chemie **247**, 321 (1971)

M.121　A. Meisel, R. Szargan, G. Leonhardt, H.-J. Köhler: J. Physique 32, C 4-301 (1971)

M.122　G. Melchart, H. Metzger, G. Wiech: In [Ref.1.60, Vol.I, p.437]

M.123　H. Mendel: Proc. Kon. Ned. Akad. Wetensch. 70 B, 276 (1967)

M.124　H. Mendel, A.S. Koster: J. Phys. C 3, 855 (1970)

M.125　H. Mendel: J. Phys. B 5, 1794 (1972)

M.126　A.S. Menshikov, K.M. Kolobova, E.Z. Kurmaev, S.I. Doroshek: Fiz. Met. Metalloved. 21, 374 (1966)
　　　　I.I. Sasovskaya, M.M. Noskov, A.S. Menshikov: Fiz. Met. Metalloved. 27, 272 (1969)

M.127　A.S. Menshikov: Phys. Stat. Sol. 35, 89 (1969)

M.128　A.S. Menshikov, I.A. Brytov, E.Z. Kurmaev: In [Ref.1.58, Vol.I p.115]

M.129　J. Merritt, E.J. Agazzi: Analyt. Chem. 38, 1954 (1966)

M.130　H. Merz, K. Ulmer: Z. Physik 210, 92 (1968)

M.131　H. Merz, K. Ulmer: Z. Physik 212, 435 (1968)
　　　　K. Ulmer: In [Ref.1.62, p.92]

M.132　H. Merz: In [Ref.1.79, p.543]

M.133　A. Michalowicz, J.J. Giverd, J. Goulon: Inorg. Chem. 18, 3004 (1979)

M.134　F.M. Michel-Calendini: J. Phys. Chem. Solids 35, 1163 (1974)

M.135　Yu.F. Migal: Zh. Strukt. Khim. 17, 404 (1976)

M.136　T. Mihalisin, A. Harrus, S. Raaen, R. Parks: J. Appl. Phys. 55, 1966 (1984)
　　　　R.D. Parks et al.: Phys. Rev. B 28, 3556 (1983)
　　　　S. Raaen, R.D. Parks: Solid State Commun. 48, 199 (1983)
　　　　S. Raaen, M.L. Denboer, V. Murgai, R.D. Parks: Phys. Rev. B 27, 5139 (1983)
　　　　J.M. Lawrence, M.L. Den Boer, R.D. Parks, J.L. Smith: Phys. Rev. B 29, 568 (1984)
　　　　P. Scoboria et al.: J. Appl. Phys. 55, 1969 (1984)
　　　　M. Croft et al.: Phys. Rev. B 30, 4164 (1984); J. Magn. Magn. Mater. 47-48, 115 (1985)
　　　　R.A. Neifeld et al.: Phys. Rev. B 32, 6928 (1985)
　　　　R.A. Neifeld, M. Croft: J. Magn. Magn. Mater. 47-48, 36 (1985)
　　　　H. Jhans, M. Croft: J. Magn. Magn. Mater. 47-48, 203 (1985)
　　　　E. Kemly et al.: J. Magn. Magn. Mater. 47-48, 403 (1985)

M.137　S.S. Mikhailova, G.V. Volf, I.V. Gribov, L.D. Finkelshtein: Zh. Strukt. Khim. 23, Nr.1, 83 (1982)
　　　　A.N. Gusatinskii, A.V. Soldatov, I.A. Smirnov, S.S. Mikhailova: Fiz. Tverd. Tela 25, 3188 (1983)

M.138　J.C. Mikkelsen, J.B. Boyce: Phys. Rev. B 24, 5999 (1981)
　　　　J.C. Mikkelsen, J.B. Boyce: Phys. Rev. Lett. 49, 1412 (1982); Phys. Rev. B 28, 7130 (1983); SSRL Report Nr. 84/06 (1984) p.25

M.139　A.G. Mikolaichuk, A.N. Kogut, V.I. Kovalchuk, P.I. Shevchuk: Ukr. Fiz. Zh. 23, 1393 (1978)

M.140　A. Miller, J. Phys. Chem. Solids 29, 633 (1968)

M.141　V.I. Minin, L.D. Finkelshtein et al.: Izv. Akad. Nauk SSSR, Ser. Fiz. 36, 424 (1972)

M.142　V.I. Minin, L.D. Finkelshtein et al.: In [Ref.1.94, p.95]; in [Ref.1.98, p.198]

M.143　S.L. Mironov, K.I. Tkachenko, A.A. Chuiko: Teor. Eksp. Khim. 14, 538 (1978)

M.144　G.R. Mitchell, W.W. Beeman: J. Chem. Phys. 20, 1298 (1952)
　　　　G.R. Mitchell: J. Chem. Phys. 37, 216 (1962)

M.145　M. Miyake, O. Kaji, S. Nagahara, T. Suzuki: J. Chem. Soc., Faraday Trans. II 82, 687 (1986)

M.146 M.F. Mogilevkina, E.A. Galtsova, L.N. Mazalov: Izv. Sib. Otd. Akad. Nauk SSSR, Ser. Khim. Nauk (1970) 97

M.147 L.I. Molkanov et al.: Proc. 5th Int'l Conf. Mössbauer Spectrosc., Prague (1975) Vols.1-3, p.546

M.148 L.I. Molkanov, Yu.S. Grushko, K.Ya. Mishin, V.K. Isupov: Zh. Eksp. Teor. Fiz. **78**, 467 (1980)

M.149 N.M. Molokhia: Thesis, Sheffield 1970

M.150 L.M. Monastyrskii: Fiz. Met. Metalloved. **44**, 194 (1977)

M.151 P.A. Montano, G.K. Shenoy: Solid State Commun. **35**, 53 (1980)

M.152 P.A. Montano et al.: Phys. Rev. B **30**, 672 (1984); Surf. Sci. **156**, 228 (1985); Phys. Rev. Lett. **56**, 2076 (1986)

M.153 D.W. Moon et al.: Surf. Sci. **180**, L123 (1987)

M.154 C.F. Moore, D.K. Olsen, B. Hodge, P. Richard: Z. Physik **257**, 288 (1972)

M.155 C.F. Moore, I. McWherter, D.K. Olsen, P. Richard: Lett. Nuovo Cim. **5**, 873 (1972); J. Appl. Phys. **44**, 519 (1973)

M.156 C.F. Moore, H.H. Wolter et al.: J. Phys. B **5**, L262 (1972); ibid. **6**, L124 (1973)

M.157 C.F. Moore, D.L. Matthews, H.H. Wolter: Phys. Lett. A **54**, 407 (1975)

M.158 H.R. Moore, F.C. Chalklin: Proc. Phys. Soc. A **68**, 717 (1955)

M.159 B. Moraweck, G. Clugnet, A.J. Renouprez: Surf. Sci. **81**, L631 (1978)

M.160 E. Morawitz et al.: Synth. Met. **1**, 267 (1980)

M.161 A. Morita, M. Watanabe: J. Phys. Soc. Jpn. **25**, 1060 (1968)

M.162 T.I. Morrison et al.: J. Am. Chem. Soc. **100**, 3262 (1978)

M.163 T.I. Morrison et al.: J. Chem. Phys. **72**, 6276 (1980); ibid. **73**, 4705 (1980)

M.164 T.I. Morrison et al.: Phys. Rev. B **31**, 5474 (1985)

M.165 M. Motoyama, H. Soezima, G. Hashizume: Proc. Int'l Conf. X-Ray Optics Microanal., Tokyo (1972) p.585

M. Motoyama, M. Tanaka, G. Hashizume: Jpn. J. Appl. Phys. **17**, S 2-418 (1978)

M. Motoyama, G. Hashizume: Bunko Kenkyu **29**, 92 (1980)

M.166 M. Motoyama, M. Tanaka, G. Hashizume: Yogyo Kyokai Shi **85**, 454 (1977)

M.167 D.L. Mott: Phys. Rev. **144**, 94 (1966)

M.168 N. Motta, M. De Crescenzi, A. Balzarotti: Phys. Rev. B **27**, 4712 (1983)

M.169 M.M. Motylinskaya, E.A. Zhurakovskii, B.I. Kotlyar: Dop. Akad. Nauk Ukr. SSR A **33**, 1043 (1971)

M.170 L.E. Mstibovskaya, Yu.N. Romashchenko, I.A. Brytov: Fiz. Tverd. Tela **20**, 2526 (1978)

M.171 T. Mukoyama, H. Adachi: J. Phys. Soc. Jpn. **53**, 984 (1984)

M.172 T. Mukoyama, H. Kaji, K. Yoshihara: Phys. Lett. A **118**, 44 (1986)

M.173 T. Mukoyama, K. Taniguchi, H. Adachi: Phys. Rev. B **34**, 3710 (1986)

M.174 H. Müller, H. Kirchmayr, A. Szasz, J. Kojnok: Physica **130** B+C, 59 (1985)

M.175 H. Müller et al.: Z. Physik B **67**, 193 (1987)

M.176 J. Müller, K. Feser, G. Wiech, A. Faessler: Phys. Lett. **44** A, 263 (1973)

C. Beyreuther, G. Wiech: Phys. Fenn. **9**, S 1-176 (1974)

C. Beyreuther, R. Hierl, G. Wiech: Ber. Bunsenges. Phys. Chemie **79**, 1081 (1975)

E. Tegeler, N. Kosuch, G. Wiech, A. Faessler: Phys. Stat. Sol. (b) **84**, 561 (1977); ibid. **91**, 223 (1979)

W. Burghard, M. Umeno, G. Wiech, W. Zahorowski: J. Phys. C **16**, 4243 (1983)

M.177 J.E. Müller, O. Jepsen, O.K. Andersen, J.W. Wilkins: Phys. Rev. Lett. **40**, 720 (1978)

M.178 J.E. Müller, O. Jepsen, J.W. Wilkins: Solid State Commun. **42**, 365 (1982)
J.E. Müller, W.L. Schaich: Phys. Rev. B 27, 6489 (1983)
J.E. Müller, J.W. Wilkins: Phys. Rev. B 29, 4331 (1984)

M.179 P. Müller: Thesis, Leipzig 1976
P. Müller, J. Finster, A. Meisel: Izv. Akad. Nauk SSSR, Ser. Fiz. **40**, 373 (1976)

M.180 R. Munch et al.: Phys. Rev. Lett. **50**, 1619 (1983)

M.181 M.C. Munoz, P.J. Durham, B.L. Gyorffy: J. Phys. F **12**, 1497 (1982)

M.182 K. Murakami et al.: Phys. Rev. Lett. **56**, 655 (1986)

M.183 V.V. Murakhtanov, L.N. Mazalov, I.N. Timonova: Izv. Akad. Nauk SSSR, Ser. Fiz. **49**, 1525 (1985)

M.184 V.V. Murakhtanov, L.N. Mazalov: Zh. Strukt. Khim. **27**, No.5, 40 (1986)

M.185 N.I. Murashko, A.V. Koval: Ukr. Khim. Zh. **40**, 86 (1974); ibid. **41**, 1325 (1975)

M.186 T. Murugesan, P.R. Sarode, J. Gopalakrishnan, C.N.R. Rao: J. Chem. Soc., Dalton Trans. (1980) 837

M.187 K. Myers, G. Andermann: J. Phys. Chem. **76**, 3975 (1972); ibid. **77**, 280 (1973)

N.1 E.G. Nadjakov, R.L. Barinskii: Dokl. Akad. Nauk SSSR **129**, 1279 (1959); Izv. Akad. Nauk SSSR, Ser. Fiz. **24**, 415 (1960)
R.L. Barinskii, E.G. Nadjakov: Dokl. Bolg. Akad. Nauk **13**, 31 (1960): Izv. Akad. Nauk SSSR, Ser. Fiz. **24**, 407 (1960)
R.L. Barinskii: Zh. Strukt. Khim. **1**, 200 (1960)

N.2 E. Nadjakov, D. Genchev: Dokl. Bolg. Akad. Nauk **18**, 207, 417 (1965); ibid. **22**, 643 (1969)
D.P. Genchev: C.R. Acad. Bulg. Sci. **37**, 489 (1984)

N.3 E. Nadjakov et al.: Dokl. Bolg. Akad. Nauk **24**, 167 (1971); ibid. **26**, 1169 (1973)

N.4 E. Nadjakov, D. Genchev, P. Pelova: Dokl. Bolg. Akad. Nauk **26**, 1327 (1973)

N.5 I. Nagakura: Sci. Rep. Tohoku Univ. I49, 1, 166 (1965)

N.6 I. Nagakura, V. Aita et al.: Proc. 3rd Int'l Conf. VUV Rad. Phys. Tokyo (1971)

N.7 N. Nagasima: Sci. Rep. Tohoku Univ. I49, 49, 57 (1966)

N.8 D.J. Nagel, D.A. Papaconstantopoulos, J.W. McCaffrey, J.W. Criss: In [Ref.1.60, Vol.II, p.51]

N.9 S.R. Nagel et al.: Phys. Rev. Lett. **49**, 575 (1982)

N.10 V.A. Nagorny, V.V. Nemoshkalenko: Dokl. Akad. Nauk SSSR **166**, 847 (1966)
V.V. Nemoshkalenko, V.A. Nagorny: Ukr. Fiz. Zh. **13**, 1182 (1968)

N.11 S. Nakai, H. Nakamori et al.: Phys. Rev. B 9, 1870 (1974)

N.12 S. Nakai, C. Sugiura, S. Kuni, T. Suzuki: Jpn. J. Appl. Phys. **17**, S 2-197 (1978)

N.13 S. Nakai et al.: J. Phys. Soc. Jpn. **55**, 2436 (1986)

N.14 H. Nakamori, C. Sugiura, K. Tsutsumi: J. Phys. Soc. Jpn. **35**, 1708 (1973); J. Appl. Phys. **44**, 3473 (1973)
K. Tsutsumi, H. Nakamori, K. Ichikawa: Phys. Rev. B 13, 929 (1976)

N.15 H. Nakamura, K..Ichikawa: Jpn. J. Appl. Phys. **20**, 2061 (1981)

N.16 H. Nakamura, K. Ichikawa, T. Watanabe, K. Tsutsumi: J. Phys. Soc. Jpn. **52**, 4014 (1983)

N.17 M. Nakamura: Phys. Rev. **178**, 80 (1969)
Y. Moriota, M. Nakamura et al.: J. Chem. Phys. **61**, 1426 (1974)

N.18   M. Nakamura, T. Hayashi, E. Ishiguro, M. Sasanuma: Proc. 3rd Int'l Conf.
       VUV Rad. Phys. Tokyo (1971) p.A1
N.19   Y. Nakamura et al.: Solid State Commun. 9, 2017 (1971)
       S. Sato et al.: J. Phys. Soc. Jpn. 33, 1638 (1972)
       M. Watanabe: J. Phys. Soc. Jpn. 34, 755 (1973)
N.20   T. Nakano, A. Kotani, J.C. Parlebas: J. Phys. Soc. Jpn. 56, 2201 (1987)
N.21   M.S. Nakhmanson, V.I. Baranovskii: Vestn. Leningr. Univ. (1972), Nr.10, 35
N.22   K.I. Narbutt, I.P. Laputina: Izv. Akad. Nauk SSSR, Ser. Fiz. 26, 409 (1962)
N.23   K.I. Narbutt, I.S. Smirnova: Izv. Akad. Nauk SSSR, Ser. Fiz. 36, 354 (1972)
       K.I. Narbutt: Izv. Akad. Nauk SSSR, Ser. Fiz. 40, 355 (1976); Phys. Chem.
       Miner. 5, 285 (1980)
N.24   K.I. Narbutt: Izv. Akad. Nauk SSSR, Ser. Fiz. 38, 548 (1974)
N.25   L. Natarajan, A.V. Tankhiwale, C. Mande: Pramana 26, 55 (1986)
N.26   L. Natarajan, A.V. Tankhiwale, C. Mande: Pramana 27, 275 (1986)
N.27   C.R. Natoli, D.K. Misemer, K. Doniach, P.W. Kutzler: Phys. Rev. A 22, 1104
       (1980)
N.28   A. Neckel, P. Rastl, P. Weinberger, R. Mechtler: Theor. Chim. Acta 24, 170
       (1972)
       A. Neckel et al.: Z. Naturforsch. 29 A, 107 (1974); Mikrochim. Acta (1975)
       Suppl. 6, 257
       K. Schwarz, A. Neckel: Ber. Bunsen-Ges. Phys. Chemie 79, 1071 (1975)
       P. Weinberger: Theor. Chim. Acta 41, 169 (1976)
       K. Schwarz: J. Phys. C 10, 195 (1977)
       P. Weinberger, F. Rosicky: Theor. Chim. Acta 48, 349 (1978)
       E. Wimmer et al.: J. Phys. Chem. Solids 43, 439 (1982)
N.29   H. Neddermeyer: Dissertation, Universität München (1969)
N.30   H. Neddermeyer: Phys. Lett. A 38, 329 (1972); in [Ref.1.79, p.153]; Phys.
       Fenn. 9, S 1-195 (1974); in [Ref.1.86, p.665]; Phys. Rev. B 13, 2411 (1976);
       Phys. Stat. Sol. (b) 78, 609 (1976); ibid. 80, 611 (1977)
N.31   H. Neddermeyer: Z. Physik 271, 329 (1974)
N.32   V.I. Nefedov: Izv. Akad. Nauk SSSR, Ser. Fiz. 28, 816 (1964)
N.33   V.I. Nefedov: Thesis, Moscow 1965; Zh. Strukt. Khim. 7, 549, 672, 719
       (1966)
N.34   V.I. Nefedov: Zh. Strukt. Khim. 8, 686, 1037 (1967)
N.35   V.I. Nefedov, V.A. Fomichev: Zh. Strukt. Khim. 9, 126, 217, 268, 279 (1968)
N.36   V.I. Nefedov: Zh. Strukt. Khim. 10, 691, 837, 938 (1969)
N.37   V.I. Nefedov: In [Ref.1.58, Vol.II, p.201]
N.38   V.I. Nefedov: Zh. Strukt. Khim. 11, 292, 299 (1970)
N.39   V.I. Nefedov: Zh. Strukt. Khim. 12, 303, 521 (1971)
N.40   V.I. Nefedov: A.P. Sadovskii et al.: Zh. Strukt. Khim. 12, 681 (1971)
N.41   V.I. Nefedov, V.J. Matyskin, I.B. Borovskii: Zh. Strukt. Khim. 12, 893 (1971)
N.42   V.I. Nefedov, L.N. Mazalov, A.P. Sadovskii: Zh. Strukt. Khim. 12, 1015
       (1971)
N.43   V.I. Nefedov: Zh. Strukt. Khim. 12, 1019 (1971)
       K.I. Narbutt, V.I. Nefedov et al.: Zh. Strukt. Khim. 13, 451 (1972)
N.44   V.I. Nefedov, K.I. Narbutt: Zh. Strukt. Khim. 13, 63 (1972)
N.45   V.I. Nefedov, E.Z. Kurmaev et al.: Zh. Strukt. Khim. 13, 637 (1972)
N.46   V.I. Nefedov: Phys. Fenn. 9, S 1-112 (1974)
N.47   V.I. Nefedov, A.P. Savoskii et al.: Koord. Khim. 1, 950 (1975)
N.48   V.I. Nefedov, Ya.V. Salyn, A.P. Sadovskii, L. Beyer: J. Electron Spectrosc. 12,
       121 (1977)
N.49   V.I. Nefedov et al.: Inorg. Chim. Acta 35, L343 (1979)
N.50   V.I. Nefedov, A.P. Sadovskii, Ya.V. Salyn: Koord. Khim. 5, 1204 (1979)

N.51 V.I. Nefedov, P.P. Pozdeev, V.A. Terekhov, E.V. Sotnikova: J. Electron Spectrosc. **40**, 11 (1986)

N.52 K.M. Neiman et al.: Zh. Strukt. Khim. **23**, Nr.3, 30 (1982)

N.53 K.M. Neiman et al.: Zh. Strukt. Khim. **24**, Nr.4 106 (1983)

N.54 K.M. Neiman et al.: Zh. Strukt. Khim. **26**, Nr.6, 22 (1985)

N.55 G.C. Nelson, B.G. Saunders et al.: Phys. Rev. **188**, 4 (1969); Z. Physik **235**, 308 (1970)

N.56 G.C. Nelson, B.G. Saunders: Phys. Rev. **188**, 108 (1969); ibid. A **2**, 542 (1970); J. Physique **32**, C 4-97 (1971)

N.57 S.A. Nemnonov, V.V. Klyushin: Fiz. Met. Metalloved. **6**, 951 (1958)

N.58 S.A. Nemnonov, L.D. Finkelshtein: Izv. Akad. Nauk SSSR, Ser. Fiz. **25**, 1007 (1961)

N.59 S.A. Nemnonov, K.M. Kolobova: Fiz. Met. Metalloved. **14**, 874 (1962); ibid. **22**, 680 (1966)

N.60 S.A. Nemnonov, V.G. Syryanov et al.: Fiz. Met. Metalloved. **22**, 375 (1966); J. Physique **32**, C 4-307 (1971)

N.61 S.A. Nemnonov, K.M. Kolobova: Fiz. Met. Metalloved. **23**, 456 (1967); ibid. **24**, 268 (1967); ibid. **27**, 1026 (1969)

N.62 S.A. Nemnonov, M.F. Sorokina: Fiz. Met. Metalloved. **23**, 732 (1967)
E.Z. Kurmaev et al.: Izv. Akad. Nauk SSSR, Ser. Fiz. **36**, 312 (1972)

N.63 S.A. Nemnonov et al.: Fiz. Met. Metalloved. **25**, 1064 (1968); ibid. **26**, 45 (1968)

N.64 S.A. Nemnonov, V.A. Trofimova, V.A. Trapeznikov: In [Ref.1.95, p.115]

N.65 S.A. Nemnonov: Fiz. Met. Metalloved. **27**, 949 (1969)
S.A. Nemnonov, A.N. Gusatinskii: Fiz. Met. Metalloved. **28**, 67 (1969)

N.66 S.A. Nemnonov et al.: Fiz. Met. Metalloved. **28**, 177 (1969); ibid. **30**, 199 (1970); ibid. **31**, 335 (1971)

N.67 S.A. Nemnonov, A.S. Menshikov et al.: Trans. Met. Soc. AIME **245**, 1191 (1969)
N.P. Sergushin et al.: Fiz. Met. Metalloved. **35**, 947 (1973)
V.A. Trofimova, S.A. Nemnonov: Fiz. Met. Metalloved. **39**, 215 (1975)
V.A. Trofimova et al.: Fiz. Met. Metalloved. **40**, 524 (1975); in [Ref.1.100, p.178]

N.68 S.A. Nemnonov, V.G. Syryanov, V.A. Trofimova: Fiz. Met. Metalloved. **29**, 585 (1970); ibid. **32**, 1302 (1971)

N.69 S.A. Nemnonov, E.Z. Kurmaev, S.V. Belash: Phys. Stat. Sol. (b) **39**, 39 (1970)
E.Z. Kurmaev et al.: Fiz. Met. Metalloved. **31**, 753 (1979); ibid. **33**, 578 (1972)
S.A. Nemnonov, E.Z. Kurmaev: In [Ref.1.79, p.237]

N.70 S.A. Nemnonov, V.G. Syryanov, V.I. Minin, M.F. Sorokina: Phys. Stat. Sol. (b) **43**, 319 (1971); ibid. **46**, 77 (1971)

N.71 S.A. Nemnonov, E.Z. Kurmaev et al.: Izv. Akad. Nauk SSSR, Ser. Fiz. **36**, 317 (1972)
A.Z. Menshikov, E.Z. Kurmaev: Fiz. Met. Metalloved. **41**, 748 (1976)
E.Z. Kurmaev, V.P. Belash, S.A. Nemnonov: Fiz. Met. Metalloved. **43**, 443 (1977)
E.Z. Kurmaev et al.: Solid State Commun. **29**, 59 (1979)

N.72 S.A. Nemnonov, V.A. Trofimova: Phys. Stat. Sol. (b) **52**, K 111 (1973)
V.A. Trofimova et al.: Izv. Akad. Nauk SSSR, Ser. Fiz. **38**, 649 (1974)

N.73 S.A. Nemnonov, S.S. Mikhailova: Izv. Akad. Nauk SSSR, Ser. Fiz. **38**, 493, 777 (1974); Fiz. Met. Metalloved. **39**, 1178 (1975)
K.M. Kolobova, S.A. Nemnonov: Izv. Sib. Otd. Akad. Nauk SSSR, Ser. Khim. Nauk (1975) Nr.9, H.4, 34

S.S. Mikhailova et al.: Izv. Akad. Nauk SSSR, Ser. Fiz. **40**, 439 (1976)

K.M. Kolobova et al.: Fiz. Met. Metalloved. **41**, 1201 (1976)

N.74   S.A. Nemnonov: Fiz. Met. Metalloved. **42**, 723 (1976)

N.75   V.V. Nemoshkalenko, V.V. Gorskii: Ukr. Fiz. Zh. **12**, 819 (1967); Phys. Stat. Sol. **28**, K 15 (1968); in [Ref.1.95, p.169, 177]

V.V. Nemoshkalenko, L.S. Voskrekasenko, V.P. Krivitskii, L.I. Nikolaev: Fiz. Met. Metalloved. **43**, 191 (1977)

N.76   V.V. Nemoshkalenko, M.A. Mindlina, B.P. Mamko: Phys. Stat. Sol. (b) **30**, 703 (1968)

N.77   V.V. Nemoshkalenko et al.: Ukr. Fiz. Zh. **13**, 1172, 1273, 1430 (1968)

N.78   V.V. Nemoshkalenko et al.: Ukr. Fiz. Zh. **14**, 1972 (1969); ibid. **22**, 498 (1977)

N.79   V.V. Nemoshkalenko: In [Ref.1.58, Vol.I, p.77]

N.80   V.V. Nemoshkalenko, V.P. Krivitskii: Phys. Lett. **30 A**, 44 (1969)

V.V. Nemoshkalenko, B.P. Mamko, V.Ya. Nagornyi, V.A. Yatsenko: Fiz. Met. Metalloved. **42**, 420 (1976)

N.81   V.V. Nemoshkalenko, V.Ya. Nagornyi: In [Ref.1.58, Vol.I, p.153]; Ukr. Fiz. Zh. **15**, 515 (1970)

N.82   V.V. Nemoshkalenko et al.: In [Ref.1.96, Vol.I, p.93]; Izv. Akad. Nauk SSSR, Ser. Fiz. **36**, 411 (1972)

N.83   V.V. Nemoshkalenko et al.: In [Ref.1.96, Vol.II, p.107]; Metallofizika 1970, Nr.29, 13; ibid. 1973, Nr. 46, 322; ibid. Nr.47, 57; ibid. 1974, Nr.52, 127; ibid. Nr.55, 79; J. Phys. Chem. Solids **36**, 277 (1975); Fiz. Met. Metalloved. **40**, 1191 (1975); Izv. Akad. Nauk SSSR, Neorg. Mater. **19**, 748 (1983)

N.84   V.V. Nemoshkalenko et al.: Fiz. Met. Metalloved. **31**, 444, 634 (1971); Metallofizika 1974, Nr.52, 54; ibid. 1976, Nr.68, 38

N.85   V.V. Nemoshkalenko, V.G. Aleshin: Ukr. Fiz. Zh. **15**, 849 (1971)

N.86   V.V. Nemoshkalenko, V.Ya. Nagornyi, L.I. Nikolaev: Ukr. Fiz. Zh. **17**, 690 (1972)

V.V. Nemoshkalenko, V.Ya. Nagorny, B.P. Mamko, V.A. Yatsenko: Metallofizika **68**, 38 (1977)

N.87   V.V. Nemoshkalenko, V.Ya. Nagornyi, G.P. Gigolashvili: Dokl. Akad. Nauk SSSR **203**, 79 (1972); Izv. Akad. Nauk SSSR, Ser. Fiz. **36**, 407 (1972)

N.88   V.V. Nemoshkalenko, V.V. Gorskii, R.F. Nasedkina: Ukr. Fiz. Zh. **17**, 1633 (1972).

N.89   V.V. Nemoshkalenko, V.Ya. Nagornyi, L.I. Nikolaev: Fiz. Tverd. Tela **14**, 2131 (1972)

N.90   V.V. Nemoshkalenko: In [Ref.1.60, Vol.II, p.115]

V.V. Nemoshkalenko et al.: Izv. Akad. Nauk SSSR, Neorg. Mater. **18**, 964 (1982); ibid. **19**, 606 (1983)

N.91   V.V. Nemoshkalenko et al.: Phys. Stat. Sol. (b) **57**, 321 (1973); Metallofizika 1975, Nr.58, 62; Fiz. Met. Metalloved. **40**, 837 (1975); Izv. Sib. Otd. Akad. Nauk SSSR, Ser. Khim. Nauk 1975, Nr.9, H.4, 97; J. Phys. Chem. Solids **36**, 277 (1975); Izv. Akad. Nauk SSSR, Ser. Fiz. **42**, 420 (1976); ibid. **44**, 884 (1977); Ukr. Fiz. Zh. **22**, 498 (1977); ibid. **25**, 1643 (1980); Izv. Akad. Nauk SSSR, Neorg. Mater. **14**, 1641 (1979)

N.92   V.V. Nemoshkalenko et al.: Fiz. Tverd. Tela **15**, 155, 3465 (1973)

N.93   V.V. Nemoshkalenko, V.G. Aleshin, M.T. Panchenko, A.I. Senkevich: Dokl. Akad. Nauk SSSR **214**, 543 (1974)

N.94   V.V. Nemoshkalenko, V.G. Aleshin et al.: Izv. Akad. Nauk SSSR, Ser. Fiz. **38**, 626 (1974)

K.K. Sidorin et al.: Metallofizika **60**, 48 (1975)

N.95   V.V. Nemoshkalenko et al.: Solid State Commun. **15**, 9, 104 (1974); Izv. Akad. Nauk SSSR, Ser. Fiz. **38**, 639 (1974); Izv. Sib. Otd. Akad. Nauk SSSR, Ser. Khim. Nauk 1975, Nr.9, H.4, 97; Metallofizika 2, 29 (1980)

N.96   V.V. Nemoshkalenko et al.: Dop. Akad. Nauk Ukr. SSR A **36**, 1112 (1974); Izv. Sib. Otd. Akad. Nauk, SSSR, Ser. Khim. Nauk 1975, Nr.9, H.4, 92

N.97   V.V. Nemoshkalenko, V.N. Antonov, V.G. Aleshin: Dop. Akad. Nauk Ukr. SSR A 1975, 546

N.98   V.V. Nemoshkalenko, V.Ya. Nagorny et al.: Ukr. Fiz. Zh. **20**, 1602 (1975)

N.99   V.V. Nemoshkalenko, T.B. Shashkina, V.G. Aleshin, A.I. Senkevich: J. Phys. Chem. Solids **36**, 37 (1975)

N.100   V.V. Nemoshkalenko, A.P. Shpak, V.P. Krivitskii: Fiz. Met. Metalloved. **39**, 278 (1975)

N.101   V.V. Nemoshkalenko et al.: Metallofizika **60**, 67 (1975); ibid. **66**, 51 (1976); Dokl. Akad. Nauk SSSR **228**, 837 (1976); Fiz. Met. Metalloved. **51**, 217 (1981)
V.V. Nemoshkalenko, V.P. Krivitskii, M.M. Kindrat, B.P. Mamko: Izv. Akad. Nauk SSSR, Neorg. Mater. **18**, 1596 (1982)

N.102   V.V. Nemoshkalenko et al.: Dopov. Akad. Nauk Ukr. SSR, Ser. A 1976, 169

N.103   V.V. Nemoshkalenko, V.N. Antonov, V.N. Antonov: Dopov. Akad. Nauk Ukr. SSR, Ser. A 1978, 642

N.104   V.V. Nemoshkalenko et al.: Phys. Stat. Sol. (b) **93**, 575 (1979); Wiss. Ber. Akad. Wiss. DDR **17**, 133 (1979)
V.V. Nemoshkalenko, Vl.N. Antonov, V.N. Antonov: Opt. Spektrosk. **57**, 788 (1984)

N.105   V.V. Nemoshkalenko, V.N. Antonov, Vl.N. Antonov: Phys. Stat. Sol. (b) **99**, 471 (1980); Izv. Akad. Nauk SSSR, Ser. Fiz. **46**, 731 (1982)

N.106   V.V. Nemoshkalenko, A.N. Timoshevskii, V.N. Antonov: Metallofizika 2, 34 (1980); Cryst. Res. Technol. **15**, 1429 (1980)

N.107   V.V. Nemoshkalenko, A.N. Timoshevskii, V.N. Antonov: Dokl. Akad. Nauk SSSR **253**, 1116 (1980)

N.108   V.V. Nemoshkalenko, V.N. Uvarov, E.G. Litvin, A.V. Podenezhko: Ukr. Fiz. Zh. **27**, 1696 (1982)
V.V. Nemoshkalenko et al.: Ukr. Fiz. Zh. **30**, 1220 (1985)

N.109   V.V. Nemoshkalenko et al.: Metallofizika 7, Nr.3, 22 (1985)

N.110   V.V. Nemoshkalenko et al.: Ukr. Fiz. Zh. **32**, 916 (1987)

N.111   V.S. Neshpor, L.G. Nikolaeva, S.M. Karalnik, J.I. Korolenko: Ukr. Fiz. Zh. **4**, 814 (1960)

N.112   V.S. Neshpor, I.I. Lyakhovskaya, V.P. Nikitin: Vestn. Leningr. Univ., Fiz. Khim. 1972, 55

N.113   T.P. Nguyen, M.H. Nguyen, S.S. Minn: C.R. Hebd. Sceances Acad. Sci. B **281**, 449 (1975)
M.H. Nguyen, T. Nguyen: J. Phys. C **9**, 1535 (1976)

N.114   C.J. Nicholls, E.I. Esmail, D.S. Urch: J. Chem. Soc. Chem. Commun. 1974, 39, 213

N.115   C.J. Nicholls, D.S. Urch: J. Chem. Soc., Dalton Trans. 901 (1974)

N.116   C. Nicholls, D.S. Urch: J. Mol. Struct. **31**, 327 (1976)
D.E. Haycock et al.: J. Chem. Soc., Dalton Trans. 1785, 1791 (1978)
D.S. Urch et al.: Spectrosc. Lett. **13**, 487 (1980)

N.117   W. Niederlag, W. Blau, K. Kleinstück: Phys. Stat. Sol. (b) **82**, 495 (1977)
W. Niederlag, W. Blau, A. Burenkov, K. Kleinstück: Fiz. Met. Metalloved. **43**, 972 (1977)

N.118   U. Nielsen, R. Haensel, W.H.E. Schwarz: J. Chem. Phys. **61**, 3581 (1974)

N.119   W. Niemann et al.: J. Magn. Magn. Mater. **47–48**, 462 (1985)

N.120   W. Niemann et al.: Phys. Rev. B **35**, 1099 (1987)

N.121 A.K. Nigam, M.K. Gupta: J. Phys. F **4**, 1084 (1974)
      N.N. Saxena, S.N. Gupta, R.G. Anikhindi: Chem. Phys. Lett. **29**, 551 (1974)
      N.N. Saxena, R.G. Anikhindi: Indian J. Pure Appl. Phys. **13**, 122 (1975)
N.122 A.K. Nigam, M.K. Gupta, R.K. Gupta: Phys. Stat. Sol. (b) **70**, K 173 (1975)
N.123 A.N. Nigam, Q.S. Kapoor: Chem. Phys. Lett. **20**, 219 (1973); J. Phys. B **6**, 2464 (1973)
      A.N. Nigam, R.B. Mathur: Chem. Phys. Lett. **33**, 579 (1975); Phys. Rev. A **13**, 1756 (1976); Can. J. Phys. **54**, 2193 (1976); Phys. Lett. A **73**, 159 (1979); Physica B+C **100**, 279 (1980)
N.124 A.N. Nigam, R.B. Mathur: Chem. Phys. Lett. **28**, 41 (1974); J. Phys. B **7**, 2489 (1974)
N.125 A.N. Nigam, R.B. Mathur: J. Phys. B **9**, 2613 (1976); Can. J. Phys. **55**, 1385 (1977); Chem. Phys. Lett. **55**, 149 (1978); Physica 125 B+C, 377 (1984)
N.126 A.N. Nigam, A. Kumar, T.S. Srivastava, U.C. Agarwala: Indian J. Phys. 52 A, 255 (1978)
N.127 A.N. Nigam, S.N. Soni: Nat. Acad. Sci. Lett. (India) **2**, 145 (1979)
N.128 A.N. Nigam, S. Kothari: Phys. Rev. A **21**, 1256 (1980)
      A.N. Nigam, S. Arora: Physica B+C **141**, 115 (1986)
N.129 A.N. Nigam, S.N. Soni: J. Phys. C **13**, 1567 (1980)
N.130 A.N. Nigam, S. Kothari: Can. J. Phys. **58**, 94 (1980)
N.131 A.N. Nigam, S.N. Soni: Can. J. Phys. **63**, 1418 (1985)
N.132 H.L. Nigam, U.C. Srivastava: Inorg. Chim. Acta Res. **5**, 338 (1971)
      U.C. Srivastava, H.L. Nigam, A.N. Vishnoi: Indian J. Pure Appl. Phys. **10**, 61 (1972); Curr. Sci. **41**, 251 (1972)
N.133 H.L. Nigam, U.C. Srivastava: Z. Naturforsch. **26** b, 997 (1971)
      A.K. Nigam, M.K. Gupta: J. Phys. F **3**, 1251 (1973)
N.134 A.S. Nigavekar: Thesis, Poona 1967
N.135 A.S. Nigavekar, S. Bergwall: Phys. Lett. **28A**, 232 (1968)
N.136 I.Ya. Nikiforov: Thesis, Rostov/Don (1966)
N.137 I.Ya. Nikiforov: In [Ref.1.57, p.241]; in [Ref.1.58, Vol.I, p.147]
N.138 I.Ya. Nikiforov, I.I. Geguzin, M.A. Blokhin, G.I. Alperovich: In [Ref.1.98, p.53]
N.139 I.Ya. Nikiforov, Yu.F. Maltsev: Fiz. Met. Metalloved. **33**, 422 (1972)
N.140 I.Ya. Nikiforov, E.V. Shtern: Fiz. Met. Metalloved. **38**, 273 (1974)
N.141 I.Ya. Nikiforov, A.T. Kozakov: Fiz. Tekh. Poluprov. **8**, 1384 (1974)
N.142 A.V. Nikolaev, L.N. Mazalov et al.: Dokl. Akad. Naük SSSR **181**, 119 (1968); ibid. **189**, 784 (1969); Izv. Sib. Otd. Akad. Nauk, Ser. Khim. Nauk 1972, 7
N.143 A.V. Nikolaev, L.N. Mazalov et al.: Dokl. Akad. Nauk SSSR **190**, 1113 (1970); ibid. **191** 144 (1970)
N.144 A.V. Nikolaev, L.N. Mazalov et al.: Izv. Sib. Otd. Akad. Nauk SSSR, Ser. Khim. Nauk 1970, 3
      V.V. Murakhtanov, T.I. Guzhavina et al.: Izv. Sib. Otd. Akad. Nauk SSSR, Ser. Khim. Nauk 1971, 8
N.145 A.V. Nikolaev, V.N. Andrievskii et al.: Zh. Neorg. Khim. **15**, 1336 (1970)
      A.V. Ablov, N.N. Poroskina et al.: Zh. Neorg. Khim. **17**, 2849 (1972)
      A.P. Sadovskii, L.N. Mazalov et al.: Zh. Strukt. Khim. **14**, 667 (1973)
N.146 A.V. Nikolaev et al.: Izv. Sib. Otd. Akad. Nauk SSSR, Ser. Khim. Nauk 1975, Nr.9, H.4, P.79
      Yu.A. Zhdanov et al.: Zh. Strukt. Khim. **18**, 677, 681,804 (1977); ibid. **19**, 779 (1978)
      L.N. Mazalov, V.D. Yumatov, G.N. Dolenko: Zh. Strukt. Khim. **21**, 21 (1980)
      G.N. Dolenko et al.: Zh. Neorg. Khim. **25**, 761 (1980)

N.147 L.I. Nikolaev, I.D. Shcherba, N.V. German, T.V. Koba: Metallofizika 7, Nr.1, 106 (1985)

N.148 A.P. Nikolskii, E.A. Zhurakovskii: Dokl. Akad. Nauk SSSR, 182, 313 (1968)
L.N. Sharanevich, E.A. Zhurakovskii, P.P. Kirichok: Dopov. Akad. Nauk Ukr. SSR A 1979, 390

N.149 P.O. Nilsson, T. Jarlborg, C.G. Larsson: Jpn. J. Appl. Phys. 17, S 2-135 (1978)

N.150 P.O. Nilsson, C.G. Larsson: Jpn. J. Appl. Phys. 17, S 2-144 (1978)

N.151 N. Nishimiya et al.: J. Less-Common Met. 88, 263 (1982)

N.152 S. Nishiyama, H. Fujimoto: J. Phys. Soc. Jpn. 36, 1614 (1974)

N.153 F. Niskauen: Can. Met. Quart. 8, 263 (1969)

N.154 V.J. Nithianandam, S.E. Schnatterly: Phys. Rev. B 36, 1159 (1987)

N.155 M. Nomura et al.: Chem. Phys. Lett. 122; 538 (1985)

N.156 B. Nordfors: Ark. Fysik 10, 279 (1956)

N.157 J. Nordgren, L.O. Werme, H. Ågren, C. Nordling, K. Siegbahn: J. Phys. B 8, L18 (1975)

N.158 J. Nordgren, H. Ågren, L. Selander, C. Nordling, K. Siegbahn: Phys. Scr. 16, 280 (1977)
H. Ågren, J. Nordgren, L. Selander, C. Nordling, K. Siegbahn: Phys. Scr. 18, 499 (1978); Chem. Phys. 37, 161 (1979)
J. Nordgren et al.: J. Chem. Phys. 76, 3928 (1982)
J.E. Rubensson et al.: J. Chem. Phys. 82, 4486 (1985)
L. Pettersson et al.: J. Phys. B 18, L 125 (1985)

N.159 J. Nordgren et al.: Phys. Scr. 27, 169 (1983)
J. Nordgren et al.: Chem. Phys. 84, 333 (1984)
R. Brammer et al.: Chem. Phys. Lett. 106, 425 (1984)
R. Brammer, N. Wassdahl, J.-E. Rubensson, J. Nordgren: UUIP 1148 (1986)
R. Brammer, J.-E. Rubensson, N. Wassdahl, J. Nordgren: UUIP 1149 (1986)

N.160 C. Nordling: Jpn. J. Appl. Phys. 17, S 2-7 (1978)
J. Nordgren et al.: Phys. Scr. 20, 623 (1979)

N.161 E. Noreland, B. Ekstig, J. Auleytner: Ark. Fysik 25, 1 (1963)

N.162 D. Norman, K.B. Garg, P.J. Durham: Solid State Commun. 56, 895 (1985)

N.163 D. Norman et al.: Phys. Rev. Lett. 58, 519 (1987)

N.164 P.R. Norris: Phys. Lett. A 45, 387 (1973)

N.165 P.R. Norris, R.S. Crisp, R.K. Dimond: In [Ref.1.79, p.229]
A. Dev, A. Breeze, P.R. Norris, L.M. Watson: Phys. Fenn. 9, S 1-428 (1974)
J.C. Fuggle, L.M. Watson, P.R. Norris, D.J. Fabian: J. Phys. F 5, 590 (1975)

N.166 K. Norrish, D.G.W. Smith: X-Ray Spectrom. 4, 65 (1975)

O.1 M. Obashi: Sci. Rep. Coll. Gen. Educat. Osaka Univ. 10, 1 (1961); ibid. 16, 1 (1967)

O.2 M. Obashi, T. Nakamura: Jpn. J. Appl. Phys. 10, 1437 (1971); Sci. Rep. Osaka Univ. 20, 31 (1972)

O.3 M. Obashi: Phys. Fenn. 9, S1-148 (1974)

O.4 M. Obashi: Jpn. J. Appl. Phys. 16, 167 (1977)

O.5 M. Obashi: Jpn. J. Appl. Phys. 17, 563 (1978)

O.6 M. Obashi, T. Matsukawa: J. Phys. Soc. Jpn. 52, 1071 (1983); ibid. 53, 4420 (1984)

O.7 H.M. O'Bryan, H.W.B. Skinner: Phys. Rev. 45, 370 (1934); Proc. Roy. Soc. A 176, 229 (1940)

O.8 Y. Ohmura, K. Ishikawa: J. Phys. Soc. Jpn. 49, 1829 (1980)

O.9 K. Ohno: Bunseki Kagaku 20, 308 (1971)

O.10 M. Ohno: Phys. Scr. 21, 589 (1980); J. Phys. C 13, 447 (1980)

O.11 M. Ohno: J. Phys. B **15**, 513 (1982)
O.12 M. Ohno, A. Kaakkronen, A. Vuoristo, G. Graeffe: Phys. Scr. **34**, 146 (1986)
O.13 M. Ohno, G. Wendin: Z. Physik D **5**, 233 (1987)
O.14 Y. Ohno et al.: Phys. Rev. B **27**, 3811 (1983); J. Phys. C **16**, 6695 (1983)
      Y. Ohno, K. Kaneda, S. Okada, K. Hirama: J. Solid State Chem. **54**, 170 (1984)
      Y. Ohno, K. Kaneda, K. Hirama: Phys. Rev. B **30**, 4648 (1982)
O.15 Y. Ohno, S. Nakai: J. Phys. Soc. Jpn. **54**, 3591 (1985)
O.16 Y. Ohno, K. Hirama: J. Solid State Chem. **63**, 258 (1986)
O.17 N. Oikawa, M. Yokota, H. Fujimoto, Y. Siota: In [Ref.1.60, Vol.I, p.163]
O.18 H. Oizumi et al.: J. Phys. Soc. Jpn. **54**, 4027 (1985)
O.19 H. Oizumi et al.: Jpn. J. Appl. Phys. **24**, 1475 (1985)
O.20 E. Ojala: Phys. Stat. Sol. (b) **119**, 269 (1983)
O.21 N. Okamoto, M. Kajikawa, K. Hasegawa: J. Chem. Soc. Jpn., Pure Chem. Sect. **87**, 363 (1966); ibid. **88**, 165 (1967)
O.22 T. Okamoto, Y. Fukushima: J. Non-Cryst. Solids **61-62**, 379 (1984)
O.23 A.V. Okotrub, L.N. Mazalov, V.D. Yumatov: Zh. Strukt. Khim. **25**, Nr.6, 66 (1984)
O.24 M. Okuno et al.: J. Non-Cryst. Solids **87**, 312 (1986)
O.25 M. Okusawa et al.: Jpn. J. Appl. Phys. **17**, S2-161 (1978)
O.26 M. Okusawa, K. Ichikawa, O. Aito, K. Tsutsumi: Phys. Rev. B **35**, 478 (1987)
O.27 J.S. Olsen et al.: Physica B+C **144**, 56 (1986)
O.28 C.G. Olson, D.W. Lynch: Solid State Commun. **31**, 601 (1979); ibid. **36**, 513 (1980); ibid. **33**, 849 (1980)
O.29 C.G. Olson, D.W. Lynch: J. Opt. Soc. Am. **72**, 88 (1982)
O.30 C.G. Olson et al.: J. Phys. C **16**, 3813 (1983)
O.31 C.G. Olson, D.W. Lynch: Phys. Rev. B **35**, 7658 (1987)
O.32 R.K. O'Nions, D.G.W. Smith: Nature **231**, 130 (1971)
O.33 Y. Onodera: J. Phys. Soc. Jpn. **39**, 1482 (1975)
O.34 K. Onoue, T. Suzuki: Jpn. J. Appl. Phys. **17**, S2-439 (1978)
O.35 M. Onozuka et al.: IRCS Med. Sci. Libr. Compend. **8**, 545 (1978)
O.36 H. Oppolzer, E. Wolfgang: Mikrochim. Acta, Suppl. **6**, 311 (1975)
O.37 E. Orgaz, M. Gupta: J. Less-Common Met. **130**, 293 (1987)
O.38 E.G. Orlova: Deposited Doc. 1975 (VINITI 1345-76) 12
O.39 B.R. Orton, G.K. Malra, A.T. Steel: J. Phys. F **17**, L45 (1987)
O.40 J. Osterwalder: Z. Physik B **61**, 113 (1985)
O.41 H.R. Ott, Y. Baer, K. Andres: In *Valence Fluctuations in Solids*, ed. by L.M. Falicov, W. Hanke, M.B. Maple (North-Holland, Amsterdam 1981) p.297
O.42 D. Ottewell, E.A. Stewardson, J.E. Wilson: J. Phys. B **6**, 2184 (1973)
O.43 D.A. Outka, R.J. Madix, J. Stöhr: Surf. Sci. **164**, 235 (1985)
O.44 D.A. Outka et al.: Phys. Rev. B **35**, 4119 (1987)
O.45 D.A. Outka et al.: Surf. Sci. **185**, 53 (1987)
O.46 D.A. Outka et al.: Phys. Rev. Lett. **59**, 1321 (1987)
O.47 I.A. Ovsyannikova, E.E. Vainshtein, G.V. Samsonov: Fiz. Met. Metalloved. **18**, 637 (1964)
O.48 I.A. Ovsyannikova: Izv. Sib. Otd. Akad. Nauk SSSR, Ser. Khim. Nauk 1966, 151
O.49 I.A. Ovsyannikova, S.S. Batsanov et al.: Izv. Akad. Nauk SSSR, Ser. Fiz. **31**, 922 and 936 (1967)
O.50 I.A. Ovsyannikova, S.S. Batsanov: In [Ref.1.58, Vol.II, p.255]
O.51 I.A. Ovsyannikova, L.I. Nasonova: Zh. Strukt. Khim. **11**, 548 (1970)
O.52 I.A. Ovsyannikova: In [Ref.1.93, p.141]

O.53   H. Oyanagi, K. Tanaka, S. Hosoya, S. Minomura: J. Physique 42, C4-221 (1981)

O.54   H. Oyanagi et al.: J. Non-Cryst. Solids 77-78, 1023 (1985)

O.55   H. Oyanagi et al.: In *Gallium Arsenide and Related Compounds*, ed. by M. Fujimoto (Hilger, Bristol 1986) p.295

T. Sasaki, T. Onda, R. Ito, N. Ogasawara: Jpn. J. Appl. Phys. 25, 231 (1986); ibid. 25, 640 (1986)

O.56   H. Oyanagi, M. Tokumoto, T. Ishiguro: Synth. Met. 17, 491 (1987)

H. Oyanagi et al.: Synth. Met. 18, 59 (1987)

O.57   H. Oyanagi et al.: Jpn. J. Appl. Phys. Part 2 26, L488, L638, L828, L1233, L1561 (1987)

P.1   Z. Paal et al.: Appl. Surf. Sci. 14, 101 (1982)

P.2   B.D. Padalia, C.S. Gupta, A. Paigankar, B.C. Halder: Curr. Sci. 38, 490 (1969)

P.3   B.D. Padalia, V. Krishnan: Indian J. Pure Appl. Phys. 9, 813 (1971)

P.4   B.D. Padalia, S.N. Gupta: J. Phys. F 2, 189 (1972)

P.5   B.D. Padalia, S.N. Gupta, V. Krishnan: J. Chem. Phys. 58, 2084 (1973)

P.6   B.D. Padalia, V. Krishnan et al.: J. Phys. Chem. Solids 34, 1173 (1973); Phys. Stat. Sol. (a) 25, K177 (1974); Phys. Rev. B 12, 443 (1975); J. Phys. Chem. Solids 36, 199 (1975)

M.N. Ghatikar, B.D. Padalia, R.M. Nayak: J. Phys. C 10, 4173 (1977)

R.M. Nayak, B.D. Padalia: Phys. Stat. Sol. (b) 96, 259 (1979)

P.7   B.D. Padalia, S.N. Gupta et al.: Chem. Phys. Lett. 27, 224 (1974); J. Phys. F 4, 938 (1974)

P.N. Koul, T.K. Hatwar, M.N. Ghatikar, B.D. Padalia: Phys. Stat. Sol. (b) 93, 223 (1979)

P.8   M.A. Paesler, D.E. Sayers: Phys. Rev. B 28, 4550 (1983)

P.9   V.N. Pak, Yu.P. Kostikov: Zh. Prikl. Spektrosk. 26, 755 (1977)

P.10   P. Palmieri, C. Zauli: Theoret. Chim. Acta 7, 89 (1967)

P.11   H.K. Pan et al.: J. Chem. Phys. 82, 1529 (1985)

P.12   A.J. Panson, M. Kuriyama: Rev. Sci. Instr. 36, 1488 (1965)

P.13   S.T. Pantelides, R.M. Martin, P.N. Sen: In [Ref.1.86, p.387]

P.14   S.T. Pantelides: Phys. Lett. A 48, 433 (1974)

S.T. Pantelides, F.C. Brown: Phys. Rev. Lett. 33, 298 (1974); Solid State Commun. 16, 95 (1975)

P.15   S.T. Pantelides, W.A. Harrison: Phys. Rev. B 13, 2667 (1976)

P.16   G. Paolucci et al.: Phys. Rev. B 34, 1340 (1986)

P.17   D.A. Papaconstantopoulos: Phys. Rev. Lett. 31, 1050 (1073); Phys. Rev. B 11, 4801 (1975); in [Ref.1.62, p.192]

P.18   D.A. Papaconstantopoulos, D.J. Nagel: Int'l J. Quantum Chem. 12, 497 (1978)

P.19   L.G. Parratt, E.L. Jossem: J. Phys. Chem. Solids 2, 67 (1957)

P.20   K. Parthasaradhi et al.: Nucl. Instrum. Methods Phys. Res. A 255, 54 (1987)

P.21   N.J. Patel, B.C. Haldar: J. Inorg. Nuclear Chem. 29, 1037 (1967)

P.22   B.A. Patki, G.S. Agrawal, V.G. Kher: J. Phys. F 5, 2208 (1975)

P.23   P. Pausescu et al.: Lucr. Simp. Natl. Fiz. Solidului 1978, 154

D. Macovei et al.: J. Appl. Crystallogr. 15, 39 (1982)

P.24   M. Pavicevic, P. Ramdohr, A. El Goresy: Proc. 3rd Lunar Science Conf. (1972) Vol.I, p.295

P.25   M.K. Pavicevic: In *Analyse extraterrestischen Materials*, ed. by W. Kiesl, H. Malissa jun. (Wien/New York 1974) p.289

P.26   A.A. Pavlychev et al.: Fiz. Tverd. Tela 20, 3671 (1978); Opt. Spektrosk. 48, 192 (1980)

A.A. Pavlychev, A.S. Vinogradov, T.M. Zimkina: Opt. Spektrosk. **52**, 231 (1982)

P.27   A.A. Pavlychev, A.S. Vinogradov, D.E. Onopko, S.A. Titov: Opt. Spektrosk. **46**, 821 (1979)

        A.A. Pavlychev et al.: Opt. Spektrosk. **47**, 73 (1979)

P.28   A.A. Pavlychev et al.: Fiz. Tverd. Tela **22**, 260 (1980)

P.29   A.A. Pavlychev, I.V. Kondrateva: Fiz. Tverd. Tela **28**, 837 (1986)

        A.A. Pavlychev, A.S. Vinogradov, I.V. Kondrateva: Fiz. Tverd. Tela **28**, 2881 (1986)

P.30   D.M. Pease, L.V. Azaroff: J. Appl. Phys. **44**, 3419 (1973)

        D.M. Pease: Phys. Rev. B **7**, 3568 (1973)

        D.M. Pease, L.V. Azaroff: J. Appl. Phys. **50**, 6605 (1979)

        B. Cordt, D.M. Pease, L.V. Azaroff: Phys. Rev. B **22**, 4692 (1980)

P.31   D.M. Pease, T.K. Gregory: Solid State Commun. **18**, 1133 (1976)

P.32   D.M. Pease et al.: J. Non-Cryst. Solids **61-62**, 1359 (1984)

        M. Choi et al.: Phys. Rev. B **32**, 7670 (1985)

P.33   C.J. Peimann, M. Skibowski: Phys. Stat. Sol. **46**, 655 (1971)

P.34   J. Pelka, J. Auleytner: Acta. Phys. Pol. A **71**, 901 (1987)

P.35   J. Perel: Phys. Rev. **147**, 463 (1966)

P.36   J. Perel, R.D. Deslattes: Phys. Rev. B **2**, 1317 (1970)

P.37   R.C.C. Perera, B.L. Henke: Jpn. J. Appl. Phys. **17**, S2-112 (1978); J. Chem. Phys. **70**, 5398 (1979); X-Ray Spectrom. **9**, 81 (1980)

P.38   R.C.C. Perera, J. Barth, R.E. LaVilla, C. Nordling: In [Ref.1.83, p.501]

P.39   R.C.C. Perera et al.: Phys. Rev. A **32**, 1489 (1985)

P.40   R.C.C. Perera, R.E. LaVilla, G.V. Gibbs: J. Chem. Phys. **86**, 4824 (1987)

P.41   B.I. Peshchevitskii et al.: Koord. Khim. **5**, 1838 (1979)

        A.P. Sadovskii, S.V. Tsemskov, E.A. Kravtsova, V.N. Mitkin: Koord. Khim. **6**, 1727 (1980)

P.42   M. Pessa: Phys. Lett. A **40**, 407 (1972)

        M. Pessa, E.-K. Kortela, A. Suikkanen, E. Suoninen: Phys. Rev. A **8**, 48 (1973)

P.43   V.M. Pessa: X-Ray Spectrom. **2**, 169 (1973)

P.44   M. Pessa, E. Suoninen, T. Valkonen: Phys. Fenn. **8**, 71 (1973)

P.45   M. Pessa, R. Uusitalo: Solid State Commun. **13**, 1703 (1973)

P.46   V.M. Pessa, R.R.O. Uusitalo: Phys. Stat. Sol. (b) **63**, 691 (1974)

P.47   V.M. Pessa: J. Phys. C **8**, 1769 (1975)

P.48   V.M. Pessa: Phys. Rev. B **15**, 1223 (1977)

P.49   H. Peterson, C. Kunz: Phys. Rev. Lett. **35**, 863 (1975)

P.50   H. Peterson, K. Radler, B. Sonntag, R. Haensel: J. Phys. B **8**, 31 (1975)

        H. Peterson: Phys. Stat. Sol. (b) **72**, 591 (1975)

        K. Radler, B. Sonntag: Chem. Phys. Lett. **39**, 371 (1976)

        K. Radler, B Sonntag, H.W. Wolff: In [Ref.1.62, p.54]

P.51   J. Petiau et al.: Phys. Rev. B **34**, 7350 (1986)

P.52   E.V. Petrov: Zh. Fiz. Khim. **46**, 2129 (1972)

P.53   E.V. Petrov, V.P. Tsvetkov: Metallofizika **40**, 98 (1972); ibid. **44**, 83 (1972); Zh. Fiz. Khim. **47**, 1039 (1973)

P.54   E.V. Petrovich, O.I. Sumbaev et al.: Zh. Eksp. Teor. Fiz. **53**, 796 (1967); ibid. **55**, 745 (1968); in [Ref.1.58, Vol.II, p.179]

P.55   E.V. Petrovich, Yu.P. Smirnov et al.: Zh. Eksp. Teor. Fiz. **61**, 1756 (1971)

P.56   J. Petru: Phys. Stat. Sol. (b) **135**, 233 (1986)

P.57   L. Pettersson et al.: J. Phys. B **17**, L 279 (1984)

        J.-E. Rubensson, N. Wassdahl, R. Brammer, J. Nordgren: UUIP 1152 (1986)

P.58   R.F. Pettifer, P.W. McMillan: Phil. Mag. **35**, 871 (1977)
R.F. Pettifer: Trends Phys. Pap. Gen. Conf. Eur. Phys. Soc., 4th 1978 (publ. 1979) 522
S.J. Gurman, R.F. Pettifer: Phil. Mag. B **40**, 345 (1979)
P.59   R.F. Pettifer, C. Hermes: J. Appl. Crystallogr. **18**, 404 (1985)
P.60   R.F. Pettifer, A.J. Bourdillon: J. Phys. C **20**, 329 (1987)
P.61   R.A. Phillips, F.P. Larkins: Aust. J. Phys. **39**, 717 (1986)
P.62   M. Piacentini et al.: Solid State Commun. **51**, 467 (1984)
V. Grasso, R. Girlanda: Helv. Phys. Acta **58**, 234 (1985)
P.63   A. Pimpale, C. Mande: Pramana **1**, 147 (1973)
A.V. Barhate, V.P. Sapre, C. Mande: Indian J. Phys. **56** A, 81 (1982); J. Phys. Chem. Solids **43**, 761 (1982)
P.64   B. Pitault, E. Belin, D. Boutouaba, C. Senemaud: Chem. Phys. Lett. **81**, 123 (1981)
P.65   R. Podloucky, R. Laesser, E. Wimmer, P. Weinberger: Phys. Rev. B **19**, 4999 (1979)
P.66   G.P. Polovina: Ukr. Fiz. Zh. **28**, 422 (1983).
G.P. Polovina, L.V. Shevchenko: Izv. Akad. Nauk SSSR, Neorg. Mater. **19**, 1020 (1983)
P.67   S.G. Porutsky, E.A. Zhurakovskii: J. Less-Common Met. **120**, 273 (1986)
P.68   A.V. Postnikov, S.N. Shamin, V.A. Trofimova, E.Z. Kurmaev: Fiz. Met. Metalloved. **58**, 288 (1984)
P.69   A.V. Postnikov, V.R. Galakhov, V.P. Antropov, E.Z. Kurmaev: Izv. Akad. Nauk SSSR, Ser. Fiz. **49**, 1505 (1985)
P.70   V.I. Potorocha, V.A. Tskhai, P.V. Geld, E.Z. Kurmaev: Fiz. Met. Metalloved. **33**, 666 (1972); Dokl. Akad. Nauk SSSR **203**, 1118 (1972)
P.71   B. Poumellec, F. Lagnel, J.F. Marucco, B. Touzelin: Phys. Stat. Sol. (b) **133**, 371 (1986)
P.72   B. Poumellec, J.F. Marucco, B. Touzelin: Phys. Stat. Sol. (b) **137**, 519 (1986)
P.73   B. Poumellec, J.F. Marucco, B. Touzelin: Phys. Rev. B **35**, 2284 (1987)
P.74   L. Powers, P. Eisenberger, J. Stamatoff: Ann. N.Y. Acad. Sci. **307**, 113 (1978)
P.75   L. Powers et al.: Biochim. Biophys. Acta **546**, 520 (1979)
P.76   V. Prabhawalkar, B.D. Padalia: Phys. Stat. Sol. (b) **110**, 659 (1982)
B.D. Padalia, T.K. Hatwar, M.N. Ghatikar: J. Phys. C **16**, 1537 (1983)
V. Prabhawalkar et al.: J. Less-Common Met. **91**, 217 (1983)
V. Prabhawalkar, B.D. Padalia: Phys. Stat. Sol. (b) **121**, K 65 (1984)
B. Darshan, B.D. Padalia: In [Ref.1.83, p.452]
B. Darshan, B.D. Padalia, C. Prakash: Phys. Stat. Sol. (b) **122**, K 59 (1984)
B. Darshan, B.D. Padalia: J. Phys. C **17**, L 281 (1984)
B. Darshan, B.D. Padalia, R. Nagarajan, R. Vijayaraghavan: J. Phys. C **17**, L 445 (1984)
B.D. Padalia, B.D. Darshan: J. Phys. C **18**, 1087 (1985)
B.D. Padalia, V. Prabhawalkar: J. Less-Common Met. **105**, 321 (1985)
T.K. Hatwar, B.D. Padalia, M.N. Ghatikar, D.R. Chopra: Phys. Stat. Sol. (b) **126**, 279 (1984)
P.77   J. Prasad, H.L. Nigam, U. Agarwala: J. Phys. C **9**, 4349 (1976)
J. Prasad, V. Krishna, H.L. Nigam: J. Chem. Soc., Dalton Trans. (1976) 241; indian J. Chem. A **15**, 303 (1977); J. Phys. Chem. Solids **38**, 1149 (1977)
H.L. Nigam, B.D. Srivastava, J. Prasad: Solid State Commun. **28**, 1001 (1978)
P.78   J. Prasad, M.C. Sham, B.D. Srivastava: Indian J. Chem. A **21**, 1040 (1982)
P.79   P. Putila, H. Juslén, M. Pessa, G. Graeffe: Phys. Scr. **20**, 41 (1979)

P.80   P. Putila-Mäntylä, M. Ohno, G. Graeffe: J. Phys. B **17**, 1735 (1984)
       M. Ohno, P. Putila-Mäntylä, G. Graeffe: J. Phys. B **17**, 1747 (1984)
P.81   P. Putila-Mäntylä, G. Graeffe: Phys. Rev. A **35**, 673 (1987)

R.1   P. Rabe, B. Sonntag, T. Sagawa, R. Haensel: Phys. Stat. Sol. (b) **50**, 559 (1972)
R.2   P. Rabe: Thesis Hamburg 1974; DESY F 41-74/2 (1974)
R.3   P. Rabe, G. Tolkiehn, A. Werner: Jpn. J. Appl. Phys. **17**, S2-215 (1978)
R.4   P. Rabe, G. Tolkiehn, A. Werner: J. Phys. C **12**, 899 (1979)
R.5   P. Rabe, G. Tolkiehn, A. Werner: J. Phys. C **12**, 1173 (1979)
R.6   P. Rabe, G. Tolkiehn, A. Werner, R. Haensel: Z. Naturforsch. **34 a**, 1528 (1979)
R.7   P. Rabe, G. Tolkiehn, A. Werner: Nucl. Instrum. Methods **171**, 329 (1980)
R.8   P. Rabe, G. Tolkiehn, A. Werner: J. Phys. C **13**, 1857 (1980)
R.9   S.I. Radautsan et al.: Dokl. Akad. Nauk SSSR **234**, 575 (1977)
R.10   K. Radler, B. Sonntag, T.C. Chang, W.H.E. Schwarz: Chem. Phys. **13**, 363 (1976)
R.11   E.R. Radtke: J. Phys. B **12**, L77 (1979)
R.12   A.M. Radwan, M. Taut: Phys. Stat. Sol. (b) **76**, 605 (1976)
       M. Taut, A.M. Radwan: Phys. Stat. Sol. (b) **82**, 507 (1977)
R.13   S. Rai, P.P. Varma, R.B. Singh: J. Phys. B **3**, 1186 (1970)
R.14   S. Rai: Acta Phys. Pol. A **46**, 631 (1974)
R.15   S. Rai, G.B. Deodhar: Acta. Phys. Pol. A **48**, 835 (1975)
R.16   S. Rai, S.D. Rai: Acta. Phys. Pol. A **49**, 289 (1976)
R.17   K.N. Ramachandran, C.D. Cox: J. Vacuum Sci. Technol. **10**, 1068 (1973)
R.18   B.G. Ramesh et al.: Physica B+C **138**, 201 (1986)
R.19   B.G. Ramesh et al.: Phys. Rev. A **36**, 386 (1987)
R.20   A. Ramos, M. Gandais, J. Petiau: J. Physique **46**, C8-491 (1985)
       T. Dumas, J. Petiau: J. Non-Cryst. Solids **81**, 201 (1986)
R.21   L. Ramqvist, B. Ekstig, E. Källne: J. Phys. Chem. Solids **30**, 1849 (1969)
       L. Ramqvist: J. Appl. Phys. **42**, 2113 (1971)
R.22   L. Ramqvist, R. Manne: J. Phys. Chem. Solids **32**, 149 (1971)
R.23   C.N.R. Rao et al.: Chem. Phys. Lett. **76**, 413 (1980)
R.24   J.B. Rao, A.R. Chetal: J. Phys. C **15**, 6281 (1982); J. Phys. Stat. Sol. (b) **113**, 727 (1982); ibid. (b) **118**, 551 (1983)
R.25   J.B. Rao, A.R. Chetal: J. Phys. Chem. Solids **44**, 677 (1983)
R.26   K.J. Rao, J. Wong: J. Chem. Phys. **81**, 4832 (1984)
       K.J. Rao, J. Wong, M.W. Shafer: J. Solid State Chem. **55**, 110 (1984)
       K.J. Rao, J. Wong, B.G. Rao: Phys. Chem. Glasses **25**, 57 (1984)
R.27   K.J. Rao, J. Wong, S. Hemlata: Proc. Indian Acad. Sci. Chem. Sci. **94**, 449 (1985)
R.28   D. Raoux et al.: Rev. Phys. Appl. **15**, 1079 (1980)
       G. Calas et al.: Rev. Phys. Appl. **15**, 1161 (1980)
R.29   D. Raoux, A. Fontaine, P. Lagarde, A. Sadoc: Phys. Rev. B **24**, 5547 (1981)
R.30   D. Ravot, C. Godart, J.C. Achard, P. Lagarde: In *Valence Fluctuations in Solids*, ed. by L.M. Falicov, W. Hanke, M.B. Maple (North-Holland, Amsterdam 1981) p.423
R.31   J. Reed, P. Eisenberger, B.-K. Teo, B.M. Kincaid: J. Am. Chem. Soc. **99**, 5217 (1977); ibid. **100**, 2374 (1978)
       J. Reed, P. Eisenberger: J. Chem. Soc., Chem. Commun. 1977, 628; Acta Crystallogr. B **43**, 344 (1978)
R.32   J. Reed, P. Eisenberger, J. Hastings: Inorg. Chem. **17**, 481 (1978)

R.33 J.J. Rehr, E.A. Stern: Phys. Rev. B **14**, 4413 (1976)
E. Sevillano, H. Meuth, J.J. Rehr: Phys. Rev. B **20**, 4908 (1979)

R.34 J.J. Rehr, R.C. Albers, C.R. Natoli, E.A. Stern: Phys. Rev. B **34**, 4350 (1986)

R.35 B. Reihl, R.R. Schlittler, H. Neff: Phys. Rev. Lett. **52**, 1826 (1984); Phys. Rev. B **29**, 2267 (1984)

R.36 H. Rempp: Z. Physik **267**, 181, 187 (1974)

R.37 H. Renner, G. Brauer, A. Faessler: Z. Naturforsch. **10 a**, 171 (1955)

R.38 P. Rennert, Th. Dorre, U. Gläser: Phys. Stat. Sol. (b) **87**, 221 (1978)
P. Rennert, H. Schelle, U.-H. Gläser: Phys. Stat. Sol. (b) **121**, 673 (1984)

R.39 C.G. Ribbing: Solid State Commun. **13**, 1717 (1973)
E. Källne, C.G. Ribbing: J. Phys. C **8**, 2953 (1975)

R.40 T.J. Ribble: Thesis Temple 1971; Phys. Stat. Sol. (a) **6**, 473 (1971)

R.41 P. Richard, J. Bolger, D.K. Olsen, C.F. Moore: Phys. Lett. A **269** (1972)

R.42 P. Richard, D.K. Olsen et al.: Phys. Rev. A **7**, 1437 (1973)

R.43 J. Richter, V.P. Sachenko, V.V. Krivitskii: Fiz. Met. Metalloved. **33**, 843 (1972)

R.44 R. Rieger et al.: Phys. Rev. B **34**, 7295 (1986)

R.45 F. Riehle, K. Ulmer: Phys. Stat. Sol. (a) **32**, K23 (1975)

R.46 F. Riehle: Jpn. J. Appl. Phys. **17**, S2-314 (1978)

R.47 F. Riehle: Phys. Stat. Sol. (b) **98**, 245 (1980)

R.48 J.R. Riter, Jr.: Acta Crystallogr. A **36**, 330 (1980)

R.49 J.J. Ritsko, S.E. Schnatterly, P.C. Gibbons: Phys. Rev. Lett. **32**, 671 (1974); Phys. Rev. B **10**, 5017 (1974)

R.50 G. Rittmayer: Thesis München 1960

R.51 K.J. Ro, J. Wong: J. Chem. Phys. **78**, 6228 (1983)

R.52 J. Robertson: Phys. Rev. B **28**, 3378 (1983)

R.53 A. Roche et al.: J. Microsc. Spectrosc. Electron **4**, 3 (1979)

R.54 A. Roche et al.: J. Microsc. Spectrosc. Electron **4**, 351 (1979)
R. Bador, M. Romand, M. Charbonnier, A. Roche: Adv. X-Ray Anal. **24**, 351 (1981)

R.55 A. Roche et al.: Appl. Surf. Sci. **9**, 227 (1981)

R.56 B.M. Rode: Chem. Phys. Lett. **27**, 264 (1974)

R.57 A.L. Roe et al.: J. Am. Chem. Soc. **106**, 1676 (1984)

R.58 J. Röhler, J.P. Kappler, G. Krill: Nucl. Instrum. Meth. Phys. Res. **208**, 647 (1983)
J. Röhler: J. Magn. Magn. Mater. **47–48**, 175 (1985)
A. Slebarski et al.: J. Magn. Magn. Mater. **47–48**, 595 (1985)
P. Weidner et al.: J. Magn. Magn. Mater. **47–48**, 75 (1985); ibid. **47–48**, 599 (1985)
B. Bittins et al.: Z. Physik B **62**, 21 (1985)

R.59 J. Röhler: Physica B+C **144**, 27 (1986)

R.60 M. Romand, J.S. Solomon, W.L. Baun: X-Ray Spectrom. **1**, 147 (1972); Spectrochim. Acta B **28**, 17, 45 (1973)

R.61 M. Romand, J.S. Solomon, W.L. Baun: X-Ray Spectrom. **2**, 7 (1973)

R.62 M. Romand, J.S. Solomon, W.L. Baun: J. Phys. Chem. Solids **34**, 1765 (1973)

R.63 M. Romand, R. Bador, A. Roche, G. Bouyssoux: J. Microsc. Spectrosc. Electron **2**, 627 (1977)

R.64 M. Romand et al.: J. Microsc. Spectrosc. Electron **9**, 95 (1984)
M.J. Romand et al.: J. Physique **45**, C2-371 (1984)

R.65 M. Romand, R. Bador, M. Charbonnier, F. Gaillard: X-Ray Spectrom. **16**, 7 (1987)

R.66 G.N. Ronami, V.P. Berezina: In [Ref.1.93, p.153]

R.67 G.A. Rooke: J. Phys. Chem. **1**, 776 (1968; in [Ref.1.78, p.185]

R.68   G.A. Rooke: In [Ref.1.101, p.287]
R.69   T. Rose, G.L. Borchert, O.W.B. Schult: Z. Naturforsch. A 39, 924 (1984)
R.70   R.A. Rosenberg et al.: Phys. Rev. B 28, 3026 (1983)
R.71   R.A. Rosenberg et al.: Phys. Rev. B 28, 6083 (1983)
R.72   R.A. Rosenberg, P.J. Love, V. Rehn: Phys. Rev. B. 33, 4034 (1986)
R.73   G. Rossi et al.: Phys. Rev. B 27, 5154 (1983)
       G. Rossi, A. Barski: Solid State Commun. 57, 277 (1986)
       G. Rossi, P. Roubin, D. Chandesris, J. Lecante: Surf. Sci. 168, 787 (1986)
R.74   G. Rossi, D. Chandesris, P. Roubin, J. Lecante: Phys. Rev. B 34, 7455 (1986)
R.75   P.N. Roul, B.D. Padalia: X-Ray Spectrom. 12, 128 (1983)
R.76   J.-E. Rubensson, H. Agren, R. Manne: J. Electron Spectrosc. 36, 307 (1985)
R.77   L. Rudstrom: Ark. Fysik 12, 287 (1957)
R.78   A.R. Ruffa: J. Appl. Phys. 43, 4263
R.79   K.C. Rule: Phys. Rev. 68, 246 (1945)
R.80   A.S. Rylnikov et al.: Pisma Zh. Eksp. Teor. Fiz. 12, 128 (1970); Zh. Eksp. Teor. Fiz. 63, 53 (1972)
R.81   M.V. Ryzhkov et al.: Physica B+C 101, 364 (1980)

S.1   A.M. Saar, M.A. Elango: Fiz. Tverd. Tela 13, 3532 (1971)
S.2   V.P. Sachenko et al.: Phys. Fenn. 9, S 1-129 (1974); Phys. Lett. 48 A, 169 (1974)
S.3   A. Sadoc, D. Raoux, P. Lagarde, A. Fontaine: J. Non-Cryst. Solids 50, 331 (1982)
      A. Sadoc et al.: J. Non-Cryst. Solids 65, 109 (1984)
      A. Sadoc, R. Krishnan, P. Rougier: J. Phys. F 15, 241 (1985)
      A. Sadoc, J.C. Lasjaunias: J. Phys. F 15, 1021 (1985)
      A. Sadoc, J.C. Lasjaunias: J. Physique 46, C8-505 (1985)
      A.M. Flank, D. Raoux, A. Naudon, J.F. Sadoc: J. Non-Cryst. Solids 61-62, 445 (1984)
S.4   A. Sadoc, P. Lagarde, G. Vlaic: J. Phys. C 18, 23 (1985)
S.5   A. Sadoc et al.: J. Physique 47, 873 (1986)
S.6   A. Sadoc, Y. Calvayrac: J. Non-Cryst. Solids 88, 242 (1986)
S.7   A.P. Sadovskii, A.V. Belyaev: Izv. Sib. Otd. Akad. Nauk SSSR, Ser. Fiz. 1967, 57
S.8   A.P. Sadovskii: Thesis Rostov/Don 1968
      A.P. Sadovskii, L.N. Mazalov, V.M. Bertenev: Teor. Eksp. Khim. 6, 502 (1970)
S.9   A.P. Sadovskii, S.V. Larionov, E.E. Vainshtein: Zh. Strukt. Khim. 8, 1043 (1967)
      A.P. Sadovskii, S.V. Larionov: Zh. Strukt. Khim. 9, 533 (1968)
      R.L. Barinskii, I.M. Kulikova: Izv. Akad. Nauk SSSR, Ser. Fiz. 38, 444 (1974)
S.10   A.P. Sadovskii, V.M. Bertenev, S.M. Blokhin: Teor. Eksp. Khim. 4, 533 (1968)
S.11   A.P. Sadovskii, V.A. Kogan, I.N. Lobanov: Zh. Strukt. Khim. 11, 681 (1970)
S.12   A.P. Sadovskii, G.N. Dolenko et al.: Izv. Akad. Nauk SSSR, Ser. Fiz. 38, 606 (1974)
      G.N. Dolenko et al.: Izv. Sib. Otd. Akad. Nauk SSSR, Ser. Khim. Nauk 1974 157
S.13   A.P. Sadovskii, E.A. Kravtsova et al.: Izv. Sib. Otd. Akad. Nauk SSSR, Ser. Khim. Nauk 1975, Nr.9, H.4, 30
S.14   A.P. Sadovskii, E.A. Kravtsova, L.N. Mazalov: Izv. Sib. Otd. Akad. Nauk SSSR, Ser. Khim. Nauk 1975, Nr.9, H.4, 62

S.15   A.P. Sadovskii, L.N. Mazalov, E.A. Kravtsova, L.I. Nasonova: Izv. Sib. Otd. Akad. Nauk SSSR, Ser. Khim. Nauk 1975, Nr.9, H.4, 113, 128
A.P. Sadovskii, E.A. Kravtsova: Koord. Khim. 4, 418 (1978)

S.16   A.P. Sadovskii, E.A. Kravtsova, L.N. Mazalov: Izv. Sib. Otd. Akad. Nauk SSSR, Ser. Khim. Nauk 1975, 116, 128
L.N. Mazalov: Zh. Strukt. Khim. 18, 607 (1977)

S.17   A.P. Sadovskii, L.I. Nasanova: Zh. Strukt. Khim. 18, 673 (1977)
L.N. Mazalov et al.: Zh. Strukt. Khim. 23, 1860 (1978)
A.P. Sadovskii, E.A. Kravtsova: Koord. Khim. 5, 197, 890 (1979)

S.18   A.P. Sadovski, S.V. Zemskov, E.A. Kravtsova, V.N. Mitkin: Koord. Khim. 6, 1727 (1980)
A.A. Voityuk, L.N. Mazalov, E.A. Kravtsova: Zh. Strukt. Khim. 24, Nr.6, 26 (1983)

S.19   T. Sagawa, Y. Iguchi et al.: J. Phys. Soc. Jpn. 21, 2602 (1966)

S.20   T. Sagawa: J. Physique 32, C4-186 (1971)

S.21   V. Saile, N. Schwentner et al.: Phys. Lett. A 46, 245 (1973)

S.22   K. Saito, T. Sagawa: J. Phys. Soc. Jpn. 50, 1660 (1981)

S.23   P.U. Sakellaridis: J. Physique 16, 422 (1955); C.R. Acad. Sci. 247, 876 (1958)

S.24   P.U. Sakellaridis: Chim. Chron. (Athen) 27, 27 (1962)

S.25   A.I. Sakharov, V.A. Fomichev: Fiz. Tverd. Tela 15, 3434 (1973)

S.26   S. Sakka: Bull. Inst. Chem. Res., Kyoto Univ. 49, 349 (1971)

S.27   S. Sakka: Yogyo Kyokai Shi 85, 168, 299 (1977)

S.28   S. Sakka, A. Senga: J. Mater. Sci. 13, 505 (1978)
S. Sakka, H. Hotta: J. Mater. Sci. 14, 2335 (1979)

S.29   S. Sakka, K. Kamiya, M. Hayashi: Bull. Inst. Chem. Res. Kyoto Univ. 59, 172 (1982)
K. Kamiya, T. Yoko, S. Sakka: J. Mater. Sci. 20, 906 (1985)

S.30   A. Sakuma, Y. Kuramoto, T. Watanabe: Phys. Rev. B 34, 2231 (1986)

S.31   V.R. Salaneck, N.O. Lipari et al.: Solid State Commun. 15, 1453 (1974)

S.32   S.I. Salem, B.L. Scott: Phys. Rev. A 9, 690 (1974)
S.I. Salem, P.L. Lee: Phys. Rev. A 10, 2033 (1974)

S.33   S.I. Salem: Phys. Fenn. 9, S 1-307 (1974)

S.34   S.I. Salem, F. Boehm, P.L. Lee: Nucl. Instrum. Methods 140, 511 (1977)

S.35   S.I. Salem, C.N. Chang, T.J. Nash: Phys. Rev. B 18, 5168 (1978)

S.36   S.I. Salem: Phys. Rev. A 21, 858 (1980)

S.37   S.I. Salem, B. Dev, P.L. Lee: Phys. Rev. A 22, 2679 (1980)

S.38   S.I. Salem, V.L. Hall: J. Phys. F 10, 1627 (1980)
S.I. Salem, A. Kumar: J. Phys. B 19, 73 (1986)

S.39   S.I. Salem, A. Kumar, B.L. Scott: Phys. Lett. A 97, 100 (1983)
S.I. Salem, A. Kumar: Phys. Rev. A 28, 2245 (1983)

S.40   S.I. Salem, B.L. Scott: Phys. Rev. A 35, 1607 (1987)

S.41   S.V. Salvi, P.H. Umadikar, V.S. Darshane, N.V. Bhat: Spectrochim. Acta B 37, 965 (1982)
S.V. Salvi et al.: Spectrochim. Acta B 39, 965 (1984)

S.42   E.V. Sampathkumaran et al.: Phys. Rev. B 29, 5702 (1984); J. Magn. Magn. Mater. 47-48, 212 (1985); Phys. Rev. B 31, 3185 (1985); Phys. Rev. Lett. 54, 1067 (1985)

S.43   W.C. Sander, J.A. Huddle, J.D. Wilson, R.E. LaVilla: Phys. Lett. 63 A, 313 (1977)

S.44   D.R. Sandström, H.W. Dodgen, F.W. Lytle: J. Chem. Phys. 67, 473 (1977)
D.R. Sandström: J. Chem. Phys. 71, 2381 (1979)

S.45   D.R. Sandström, F.W. Lytle: Ann. Rev. Phys. Chem. 30, 215 (1979)

S.46   D.R. Sandström: In [Ref.1.83, p.409]

432

S.47 D.R. Sandström et al.: Phys. Rev. B **32**, 3541 (1985)

S.48 M. Sano, T. Maruo, H. Yamatera: Chem. Phys. Lett. **101**, 211 (1983); Chem. Phys. **84**, 66 (1986)

S.49 V.B. Sapre: Thesis, Nagpur (1972)
V.B. Sapre, C. Mande: J. Phys. Chem. Solids **34**, 1351 (1973)
V. Kondawar, C. Mande: Phys. Stat. Sol. (b) **75**, 79 (1976)
C. Mande, V.B. Sapre: Indian J. Pure Appl. Phys. **17**, 331 (1979)

S.50 A.C. Sarma, W.G. Bos: J. Phys. Chem. Solids **32**, 1423 (1971); ibid. **33**, 935 (1972)

S.51 D.D. Sarma et al.: Z. Physik B **59**, 159 (1985)

S.52 P.R. Sarode, A.R. Chetal: J. Phys. F **6**, L163 (1976)

S.53 P.R. Sarode, A.R. Chetal: Jpn. J. Appl. Phys. **15**, 1383, 2459 (1976)

S.54 P.R. Sarode, A.R. Chetal: Proc. 12th Rare Earth Res. Conf. (1976) Vol.2, p.562
A.R. Chetal, P.R. Sarode: In [Ref.1.62, p.163]

S.55 P.R. Sarode, A.R. Chetal: J. Phys. Soc. Jpn. **40**, 1637 (1976); J. Phys. F **7**, 745, 1103 (1977); ibid. C **10**, 153 (1977)
P.R. Sarode, A.R. Chetal, C. Mande: J. Physique C5-86 (1979)
R. Parthasarathy, R.V. Prasad, R. Sarode, K.J. Rao: Proc. Indian Acad. Sci. Chem. Sci. **91**, 201 (1982)

S.56 P.R. Sarode: Phys. Stat. Sol. (b) **88**, K35 (1978)

S.57 P.R. Sarode: Z. Naturforsch. **33** a, 946 (1978)

S.58 P.Sarode, A.V. Pendharkar: Chem. Phys. **28**, 455 (1978)

S.59 P.R. Sarode: J. Chem. Soc., Dalton Trans. 1979, 993

S.60 P.R. Sarode, S. Ramasesha, W.H. Madhusudan, C.N.R. Rao: J. Phys. C **12**, 2439 (1979)
P.R. Sarode, G. Sankar, C.N.R. Rao: Proc. Indian Acad. Sci. Chem. Sci. **92**, 527 (1983)
G. Sankar, P.R. Sarode, C.N.R. Rao: Chem. Phys. **76**, 435 (1983)
G. Sankar, S. Vasudevan, C.N.R. Rao: Chem. Phys. Lett. **127**, 620 (1986); J. Chem. Phys. **85**, 2291 (1986)

S.61 P.R. Sarode, K.J. Rao, M.S. Hegde, C.N.R. Rao: J. Phys. C **12**, 4119 (1979)
C.N.R. Rao, P.R. Sarode, R. Parthasarathy, K.J. Rao: Phil. Mag. B **41**, 581 (1980)

S.62 P.R. Sarode et al.: J. Phys. C **15**, 6655 (1982)

S.63 P.R. Sarode: Phys. Stat. Sol. (a) **98**, 391 (1986)

S.64 S. Sato et al.: J. Phys. Soc. Jpn. **30**, 459 (1971)
S. Sato: J. Phys. Soc. Jpn. **41**, 913 (1976)

S.65 T. Sato, Y. Takahashi, K. Yabe: Bull. Chem. Soc. Jpn. **40**, 298 (1967)
Y. Takahashi, K. Yabe, T. Sato: Bull. Chem. Soc. Jpn. **42**, 2707 (1969)

S.66 N.D. Savchenko, P.V. Gel, V.P. Tsvetkov: Ukr. Fiz. Zh. **15**, 835 (1970)
N.D. Savchenko, V.P. Tsvetkov: In [Ref.1.98, p.211]

S.67 M. Sawada, K. Taniguchi, H. Nakamura: In [Ref.1.58, Vol.II, p.122]; J. Phys. Soc. Jpn. **33**, 1496 (1972)

S.68 M. Sawada, K. Taniguchi, H. Nakamura: In [Ref.1.60, Vol.I, p.477]

S.69 K.N. Saxena, N.N. Saxena, R.G. Anikhindi: Chem. Phys. Lett. **30**, 152 (1975)

S.70 N.N. Saxena, K.N. Saxena, R.G. Anikhindi: Phys. Lett. A **50**, 181 (1974)
N.N. Saxena: J. Phys. C **8**, 1450 (1975)
K.N. Saxena, C.P. Saxena, R.G. Anikhindi, A.S. Kaveeshwar: Phys. Lett. A **78**, 325 (1980)

S.71 R.R. Saxena, R.H. Bragg: Carbon **12**, 210 (1974)

S.72 S.G. Saxena, K.B. Garg: Phys. Stat. Sol. (a) **87**, K 25 (1985)

S.73   D.E. Sayers, E.A. Stern, F.W. Lytle: Phys. Rev. Lett. 27, 1204 (1971)
        E.A. Stern, D.E. Sayers: Phys. Rev. Lett. 30, 174 (1973)
S.74   D.E. Sayers, F.W. Lytle, E.A. Stern: J. Non-Cryst. Solids 8-10, 401 (1972);
        Trans. Am. Crystallogr. Assoc. 10, 45 (1974)
S.75   D.E. Sayers, F.W. Lytle, M. Weissbluth, P. Pianetta: J. Chem. Phys. 62, 2514
        (1975)
S.76   D.E. Sayers, E.A. Stern, J.R. Herriott: J. Chem. Phys. 64, 427 (1976)
S.77   O. Schaaber, H. Vetters: Mikrochim. Acta S5-47 (1974)
S.78   O. Schaaber, H. Vetters: Härterei-Techn. Mitteil. 30, 359 (1975)
S.79   W.L. Schaich: Phys. Rev. 14, 4420 (1976)
S.80   H. Scheidt, M. Globl, V. Dose, J. Kirschner: Phys. Rev. Lett. 51, 1688 (1983)
S.81   M. Schluter, J.R. Chelikowsky: Solid State Commun. 21, 1123 (1977)
        J.R. Chelikowsky, M. Schluter: Phys. Rev. B 15, 4020 (1977)
S.82   P.M. Schneider, W.B. Fowler: Phys. Rev. Lett. 36, 425 (1976)
S.83   E. Schnell: Monatsh. Chemie 94, 703 (1963)
S.84   H.W. Schnopper: In [Ref.1.57, p.303]
S.85   H.W. Schnopper, L.G. Parratt: In [Ref.1.57, p.314]
S.86   H.W. Schnopper: Phys. Rev. 154, 118 (1967)
S.87   P. Schreiber: Thesis Hamburg 1970
        R. Haensel, N. Kosuch, U. Nielsen, B. Sonntag, U. Rössler: Phys. Rev. B 7,
        1577 (1973)
S.88   H. Schrenk: Thesis München 1969
S.89   W. Schwander, K. Ulmer: Phys. Stat. Sol. (a) 41, K167 (1977)
S.90   K. Schwarz, A. Neckel, J. Nordgren: J. Phys. F 9, 2509 (1979)
S.91   K. Schwarz, E. Wimmer: J. Phys. F 10, 1001 (1980)
S.92   K. Schwarz, H. Ripplinger, A. Neckel: Z. Physik B 48, 79 (1982)
S.93   W.H.E. Schwarz: Chem. Phys. 9, 157 (1975); ibid. 11, 217 (1975)
S.94   W.H.E. Schwarz, L. Mensching, K.-H. Hallmeier, R. Szargan: Chem. Phys. 82,
        57 (1983)
S.95   N. Schwentner, M. Skibowski, W. Steinmann: Phys. Rev. B 8, 2965 (1973)
S.96   W. Seka, H.P. Hanson: J. Chem. Phys. 50, 344 (1969)
S.97   H. Semat: Phys. Rev. 46, 688 (1934)
        S. Kawata: Proc. Phys. Math. Soc. Jpn. 17, 89 (1935)
S.98   C. Senemaud, C. Hague: J. Physique 32, C 4-193 (1971)
        C. Senemaud: J. Physique 32, 89 (1971); Phys. Rev. B 18, 3929 (1978)
S.99   C. Senemaud: In [Ref.1.60, Vol.I, p.216]
S.100   C. Senemaud, M.T. Costa Lima: Phys. Lett. 47 A, 395 (1974); J. Non-Cryst.
        Solids 33, 141 (1979)
        M.T. Costa Lima, C. Senemaud: Chem. Phys. Lett. 40, 157 (1976)
S.101   C. Senemaud, M.-T. Costa Lima, J.A. Roger: Chem. Phys. Lett. 26, 431
        (1974)
        C.F. Hague, C. Senemaud, H. Ostrowiecki: J. Phys. F 10, L 267 (1980)
        C. Senemaud, B. Pitault, B. Bourdon: Solid State Commun. 43, 483 (1982)
S.102   C. Senemaud, C. Cardinaud, G. Villela: Solid State Commun. 50, 643 (1984)
S.103   I.P. Sergushin et al.: Zh. Strukt. Khim. 18, 698 (1977)
S.104   F. Sette et al.: Phys. Scr. T17, 209 (1987)
        F. Sette, S.J. Pearton, J.M. Poate, J.E. Rowe: Nucl. Instrum. Methods Phys.
        Res. B 19-20, 408 (1987)
S.105   F. Sette, C.T. Chen, J.E. Rowe, P.H. Citrin: Phys. Rev. Lett. 59, 311 (1987)
S.106   P.B. Sewell, D.F. Mitchell: Surf. Sci. 55, 367 (1976)
S.107   S.A. Shabalovskaya, V.E. Panin, S.F. Tyumentseva, N.I. Frese: Izv. Vyssh.
        Uchebn. Zaved., Fiz. 17, 119 (1974)

S.108 I.N. Shabanova, N.P. Sergushin, K.M. Kolobova, V.A. Trapeznikov, V.I. Nefedov: Fiz. Met. Metalloved. **34**, 1187 (1972)

S.109 V.A. Shaburov, A.E. Sovestnov, O.I. Sumbaev: Pisma Zh. Eksp. Teor. Fiz. **18**, 425 (1973); Phys. Lett. **49** A 83 (1974)
       A.E. Sovestnov, A.S. Rylnikov, O.I. Sumbaev, V.A. Shaburov: Zh. Eksp. Teor. Fiz. **71**, 1119 (1976)

S.110 M.C. Shah, B.D. Srivastava, J. Prasad: Natl. Acad. Sci. Lett. (India) 2, 444 (1979)

S.111 S.M. Shah, K. Das Gupta: J. Phys. Soc. Jpn. **37**, 1069 (1974)

S.112 T.K. Saito, T. Sagawa: J. Phys. Soc. Jpn. **50**, 1660 (1981)

S.113 T.K. Sham, J.B. Hastings, M.L. Perlman: Chem. Phys. Lett. **83**, 391 1981)

S.114 T.K. Sham: J. Am. Chem. Soc. **105**, 2269 (1983)
       T.K. Sham: J. Chem. Phys. **79**, 1116 (1983); ibid. **83**, 3222 (1985)

S.115 T.K. Sham, S.M. Heald: J. Am. Chem. Soc. **105**, 5142 (1983)

S.116 T.K. Sham, R.A. Holroyd: J. Chem. Phys. **80**, 1026 (1984)
       T.K. Sham, R.G. Carr: J. Chem. Phys. **83**, 5914 (1985)

S.117 T.K. Sham: Phys. Rev. B **31**, 1888 and 1903 (1985); J. Chem. Phys. **84**, 7054 (1986)

S.118 T.K. Sham, R.G. Carr: J. Chem. Phys. **84**, 4091 (1986)

S.119 V.T. Sharai, O.I. Paskovskii: In [Ref.1.94, p.103]

S.120 O.P. Sharkin: In [Ref.1.57, pp.272, 288]

S.121 O.P. Sharkin, S.B. Stepchenko, R.F. Nasedkina: Konst. Svoistva Miner. **10**, 18 (1976)

S.122 T.B. Shashkina, B.I. Kotlyar et al.: In [Ref.1.96, Vol.I, p.121]

S.123 T.B. Shashkina: Phys. Stat. Sol. (b) **44**, 571 (1971)

S.124 T.B. Shashkina, I.A. Brytov, Yu.N. Romashchenko, V.I. Minin: In [Ref.1.98, p.182]

S.125 C.H. Shaw: Phys. Rev. **57**, 877 (1940)

S.126 S.G. Shevchenko et al.: Zh. Strukt. Khim. **23**, Nr.3, 43 (1982)

S.127 N.J. Shevchik, J. Tejeda, M. Cardona: Phys. Rev. B **9**, 2627 (1974)

S.128 O. Shimomura et al.: Jpn. J. Appl. Phys. **17**, S 2-221 (1978)

S.129 V.T. Shipatov, Yu.P. Kostikov, P.P. Seregin: Vestn. Leningr. Univ., Fiz. Khim. 1971, 148
       L.L. Makarov, Yu.P. Kostikov, G.P. Kostikova: Zh. Teor. Eksp. Khim. **8**, 403 (1972)

S.130 Yu.S. Shorikov et al.: Izv. Akad. Nauk SSSR, Neorg. Mater. **22**, 974 (1986)

S.131 A.I. Shponko, L.N. Mazalov et al.: Izv. Sib. Otd. Akad. Nauk SSSR, Ser. Khim. Nauk 1975, Nr.9, H.4, 138

S.132 B.D. Shrivastava, R.K. Jain, V.S. Dubey: Phys. Lett. A **54**, 299 (1975); J. Phys. B **8**, 2948 (1975)
       B.D. Shrivastava, D. Singh, V.S. Dubey: Chem. Phys. Lett. **37**, 521 (1976); Phys. Lett. **56** A, 263 (1976)

S.133 B.D. Shrivastava, R.K. Jain, V.S. Dubey: Phys. Lett. **59** A, 323 (1976); Physica B+C **84**, 281 (1976)
       B.D. Shrivastava, D.C. Gaur: Natl. Acad. Sci. Lett 5, 431 (1982); ibid. 6, 279 (1983); Acta Phys. Pol. A **65**, 197 (1984); Indian J. Pure Appl. Phys. 22, 426 (1984)
       B.D. Shrivastava, A. Mishra: Acta Phys. Pol. A **67**, 1153 (1985)
       B.D. Shrivastava, A. Mishra, D. Singh: Phys. Lett. A **110**, 323 (1985)
       B.D. Shrivastava, P.R. Landge: J. Less-Common Met. **114**, 329 (1985)

S.134 B.D. Shrivastava, R.K. Jain, V.S. Dubey: Can. J. Phys. **55**, 1521 (1977)
       B.D. Shrivastava, D. Singh: Naturwissensch. **65**, 387 (1978)

S.135 B.D. Shrivastava, N.K. Mahajan: Proc. Nucl. Phys. Solid State Phys. Symp. 21 C, 86 (1978)

S.136 B.D. Shrivastava, P.R. Landge: Nuovo Cim. 49 B, 118 (1979)

S.137 B.D. Shrivastava, P.R. Landge: Phys. Lett. 75 A, 507 (1980)

S.138 B.D. Shrivastava, P.R. Landge: Phys. Stat. Sol. (a) 59, 133 (1980)

S.139 B.D. Shrivastava, P.R. Landge: Acta Phys. Pol. A 62, 143 (1982)

S.140 B.D. Shrivastava, R.K. Jain, A. Mishra: J. Phys. B 19, 3839 (1986)

S.141 R.L. Shrivastava, K.S. Srivastava, O.K. Harsh, V. Kumar: Indian J. Pure Appl. Phys. 15, 517 (1977)
K.S. Srivastava, R.L. Shrivastava: Indian J. Pure Appl. Phys. 17, 476 (1979)

S.142 V.C. Shrivastava, H.L. Nigam, A.N. Vishnoi: J. Pure Appl. Phys. 9, 63 (1970)
H.L. Nigam, U.C. Srivastava, K.B. Pandeya: Indian J. Chem. 11, 62 (1973)

S.143 A.S. Shulakov, V.A. Fomichev, T.M. Zimkina: Fiz. Tverd. Tela 14, 3088 (1972)
T.M. Zimkina et al.: Fiz. Tverd. Tela 26, 1981 (1984); in [Ref.1.66, p.263]; Izv. Akad. Nauk SSSR, Ser. Fiz. 46, 720 (1982)

S.144 A.S. Shulakov. I.I. Lyachovskaya, V.A. Fomichev: Fiz. Tverd. Tela 15, 2246 (1973)
I.I. Lyachovskaya, T.M. Zimkina, V.A. Fomichev: Fiz. Tverd. Tela 17, 1417 (1975)

S.145 A.S. Shulakov, T.M. Zimkina et al.: Fiz. Tverd. Tela 15, 3598 (1973)

S.146 A.S. Shulakov, T.M. Zimkina, V.A. Fomichev, N.S. Neshpor: Fiz. Tverd. Tela 16, 401 (1974); Izv. Akad. Nauk SSSR, Ser. Fiz. 41, 216 (1977)

S.147 A.S. Shulakov, T.M. Zimkina, V.A. Fomichev, V.S. Neshpor: Fiz. Tverd. Tela 18, 793 (1976)

S.148 A.S. Shulakov et al.: Fiz. Tverd. Tela 22, 442 (1980)

S.149 A.S. Shulakov, T.M. Zimkina, A.P. Stepanov: Fiz. Tverd. Tela 27, 112 (1985)
A.S. Shulakov, T.M. Zimkina, A.P. Braiko, A.P. Stepanov: Opt. Spektrosk. 58, 776 (1985)

S.150 A.S. Shulakov, T.M. Zimkina, V.A. Fomichev: Izv. Akad. Nauk SSSR, Ser. Fiz. 49, 1495 (1985)

S.151 A.S. Shulakov et al.: Fiz. Met. Metalloved. 62, 1136 (1986)
A.S. Shulakov, A.P. Stepanov, D. Atiekh: Fiz. Tverd. Tela 29, 241 (1987)

S.152 R.G. Shulman et al.: J. Mol. Biol. 124, 305 (1978); Ann. Rep. Biophys. Bioeng. 7, 559 (1978)
R.G. Shulman: Trends Biochem. Sci. 3, N282 (1980)

S.153 S.V. Shutov, Yu.M. Yarmoshenko, E.Z. Kurmaev: Fiz. Met. Metalloved. 63, 187 (1987)

S.154 A.T. Shuvaev: Izv. Akad. Nauk SSSR, Ser. Fiz. 25, 986, 992 (1961)

S.155 A.T. Shuvaev, G.M. Kulyabin: Izv. Akad. Nauk SSSR, Ser. Fiz. 27, 322 (1963)
A.T. Shuvaev: Thesis, Rostov/Don (1963)

S.156 A.T. Shuvaev: Izv. Akad. Nauk SSSR, Ser. Fiz. 28, 758, 934 (1964)

S.157 A.T. Shuvaev, V.G. Syryanov, V.V. Gorskii: Izv. Akad. Nauk SSSR, Ser. Fiz. 28, 823 (1964)

S.158 A.T. Shuvaev et al.: Izv. Akad. Nauk SSSR, Ser. Fiz. 31, 898 (1967)

S.159 A.T. Shuvaev, M.A. Blokhin, E.A. Israilevich: Izv. Akad. Nauk SSSR, Ser. Fiz. 31, 919 (1967)

S.160 A.T. Shuvaev, V.V. Krivitskii, A.P. Semlyanov: Izv. Akad. Nauk SSSR, Ser. Fiz. 36, 259 (1972)

S.161 A.T. Shuvaev, I.A. Sarubin: Izv. Akad. Nauk SSSR, Ser. Fiz. 36, 367 (1972)

S.162 A.T. Shuvaev, A.P. Semlyanov et al.: Zh. Strukt. Khim. 15, 433 (1974)

S.163 A.T. Shuvaev et al.: Koord. Khim. 3, 690 (1977)
B.Yu. Khelmer et al.: Izv. Akad. Nauk SSSR, Ser. Fiz. 40, 348 (1976)

A.T. Shuvaev et al.: Zh. Strukt. Khim. **25**, Nr.2, 172 (1984)

S.164   A.T. Shuvaev et al.: Zh. Strukt. Khim. **20**, 736 (1979)
A.V. Nefedov et al.: Zh. Strukt. Khim. **21**, 68 (1980)

S.165   A.T. Shuvaev, S.A. Prosandeev, I.A. Zarubin: Izv. Akad. Nauk SSSR, Ser. Fiz. **46**, 753 (1982)

S.166   A.T. Shuvaev, M.M. Tatevosyan, V.M. Kopylov, N.N. Karabaev: Zh. Teor. Eksp. Khim. **20**, 369 (1984)

S.167   A.T. Shuvaev et al.: Izv. Akad. Nauk SSSR, Ser. Fiz. **49**, 1471 (1985)

S.168   I.G. Shveitser, M.A. Blokhin: Izv. Akad. Nauk SSSR, Ser. Fiz. **31**, 947 (1967)
I.G. Shveitser, V.P. Sachenko, I.Ya. Nikiforov: Izv. Akad. Nauk SSSR, Ser. Fiz. **31**, 949 (1967)

S.169   M. Siegbahn, T. Magnusson: Z. Physik **96**, 1 (1935)

S.170   J. Siivola: Bull. Geol. Soc. Finl. **43**, 1 (1971); Phys. Fenn. **9**, 111 (1974)

S.171   A. Simunek: Czech. J. Phys. **24**, 942, 1063 (1974)

S.172   A. Simunek: Proc. Int'l Conf. Radiat. Recomb. Relat. Phen. in III–V Comp. Semicond. Prague (1979) p.223

S.173   A. Simunek, G. Wiech: Phys. Rev. B **30**, 923 (1984)

S.174   A.K. Singh, B.M.S. Kashyap: J. Phys. F **5**, 822 (1975)
A.K. Singh: Indian J. Pure Appl. Phys. **14**, 498 (1976)

S.175   R.B. Singh, S.K. Händel, B. Stenerhag: Z. Physik **249**, 241 (1972); ibid. **252**, 1 (1972)

S.176   S.P. Singh: Indian J. Pure Appl. Phys. **14**, 856 (1976); ibid. **15**, 449 (1977)

S.177   S.P. Singh: Curr. Sci. **47**, 896 (1978)

S.178   S.P. Singh: Indian J. Pure Appl. Phys. **16**, 637, 715 (1978); ibid. **17**, 405, 783 (1979)

S.179   S.P. Singh: Phys. Lett. **74 A**, 137 (1979)

S.180   S.P. Singh, N.C. Tiwari, R.C. Joshi: Indian J. Pure Appl. Phys. **23**, 480 (1985)

S.181   V.P. Singh, B.K. Agarwal: J. Phys. Chem. Solids **35**, 465 (1974); J. Phys. C **7**, 831 (1974)

S.182   V.N. Sivkov, V.N. Akimov, A.S. Vinogradov, T.M. Zimkina: Opt. Spektrosk. **57**, 265 (1984)

S.183   H.W.B. Skinner, J.E. Johnston: Proc. Roy. Soc. A **161**, 420 (1937); Proc. Cambridge Phil. Soc. **34**, 109 (1938)

S.184   H.W.B. Skinner: Phil. Trans. Roy. Soc. A **239**, 95 (1940)

S.185   H.W.B. Skinner, T.G. Bullen, J.E. Johnston: Phil. Mag. **45**, 1070 (1954)

S.186   I.S. Slabkovskii, V.G. Sinyushko, R.F. Kripyakevich: Navodor. Met. Borba Vodorodn. Khrupk. 1968, 37

S.187   R.A. Slater, D.S. Urch: J. Chem. Soc. Chem. Commun. 564 (1972); in [Ref.1.79, p.655]

S.188   A. Slebárski, K. Lawniczak, J. Auleytner: Phys. Stat. Sol. (a) **54**. 79 (1979)

S.189   A. Slebarski, J. Auleytner: Phys. Stat. Sol. (b) **109**, 125 (1982)

S.190   A. Slebarski, W. Zahorowski: J. Phys. F **14**, 1553 (1984)

S.191   A. Slebarski: J. Magn. Magn. Mater. **66**, 107 (1987)

S.192   J.H. Slowik, F.C. Brown: Phys. Rev. Lett. **29**, 934 (1972)

S.193   Yu.P. Smirnov, O.I. Sumbaev et al.: Zh. Eksp. Teor. Fiz. **57**, 1139 (1969)

S.194   I.S. Smirnova, K.I. Narbutt: Izv. Akad. Nauk SSSR, Ser. Fiz. **21**, 1375 (1957); in [Ref.1.58, Vol.I, p.262]

S.195   D.A. Smith, M.J. Heeg, W.R. Heineman, R.C. Elder: J. Am. Chem. Soc. **106**, 3053 (1984)

S.196   D.G.W. Smith, R.K. O'Nions: J. Phys. D **4**, 147 (1971)
R.K. O'Nions, D.G.W. Smith: Am. Mineral. **56**, 1452 (1971); Chem. Geol. **9**, 145 (1972)
D.G.W. Smith, R.K. O'Nions, K. Norrish: Spectrochim. Acta **29 B**, 63 (1974)

S.197 T.A. Smith et al.: J. Am. Chem. Soc. 107, 5945 (1985)

S.198 L. Smrcka: Czech. J. Phys. 21, 683 (1971)

S.199 T.M. Snyder: Phys. Rev. 59, 168 (1941)

S.200 E. Sobczak, K. Persy: Phys. Scr. 22, 88 (1980)

S.201 A.I. Sokolenko, V.I. Sokolenko, I.B. Stary: Izv. Akad. Nauk SSSR, Ser. Fiz. 38, 646 (1974); Ukr. Fiz. Zh. 19, 2 (1974); Dopov. Akad. Nauk Ukr. SSR A 37, 449 (1975); Ukr. Fiz. Zh. 22, 833 (1977)
V.I. Sokolenko, O.I. Sokolenko, E.A. Zhurakovskii: Dopov. Akad. Nauk Ukr. SSR A 1977, 57, 1979, 209
A.I. Sokolenko, V.I. Sokolenko: In [Ref.1.100, p.143]; Fiz. Tverd. Tela 22, 3720 (1980)

S.202 V.I. Sokolenko, I.M. Kungurov: Issled. po Molekulyar. Fiz. i Fiz. Tverdogo Tela (1976) 90
A.I. Sokolenko, V.I. Sokolenko: Ukr. Fiz. Zh. 22, 833 (1977)

S.203 V.I. Sokolov et al.: Fiz. Tverd. Tela 27, 2118 (1985)

S.204 J.S. Somers et al.: Surf. Sci. 183, 576 (1987)

S.205 H. Sommer, V.F. Volkov: Izv. Vyssh. Uchebn. Zaved., Fiz. 16, 153 (1973)

S.206 H. Sommer, A. Meisel: Izv. Akad. Nauk SSSR, Ser. Fiz. 40, 379 (1976)

S.207 H. Sommer, G. Leonhardt, A. Meisel, D. Hirsch: Jpn. J. Appl. Phys. 17, S 2-278 (1978)
E. Sobczak, H. Sommer: Acta. Phys. Pol. A 60, 71 (1981)
H. Neumann et al.: Phys. Stat. Sol. (b) 121, 641 (1984)

S.208 D. Sondericker, Z. Fu, D.C. Johnston, W. Eberhardt: Phys. Rev. B 36, 3983 (1987)

S.209 B. Sonntag, G. Zimmerer, T. Tuomi: DESY SR - 72/9 (1972)

S.210 B. Sonntag, T. Tuomi, G. Zimmerer: Phys. Stat. Sol. (b) 58, 101 (1973)

S.211 B.F. Sonntag: Phys. Rev. B 9, 3601 (1974)

S.212 B. Sonntag, F.C. Brown: Phys. Rev. B 10, 2300 (1974)

S.213 S.N. Sony: J. Phys. B 13, 2859 (1980)

S.214 B. Sorkin, A. Saar, M. Elango: Izv. Akad. Nauk Est. SSR 22, 105 (1973)

S.215 H. Sorum: Phys. Stat. Sol. (b) 113, 197 (1982)

S.216 H. Sorum, J. Bremer: J. Phys. F 12, 2721 (1982)

S.217 H. Sorum: J. Phys. F 17, 417 (1987)

S.218 W. Speier et al.: Phys. Rev. Lett. 55, 1693 (1985)

S.219 K.S. Srivastava, S.P. Singh, R.L. Shrivastava: Phys. Stat. Sol. (b) 63, K 25 (1974); Phys. Lett. 47 A, 305 (1974)

S.220 K.S. Srivastava, R.L. Shrivastava, S.P. Singh: Ind. J. Pure Appl. Phys. 14, 134 (1976)

S.221 K.S. Srivastava, S.P. Singh, R.L. Shrivastava: Phys. Rev. B 13, 3213 (1976)
K.S. Srivastava, R.L. Shrivastava, O.K. Harsh: Indian J. Pure Appl. Phys. 15, 516 (1977)
R.L. Shrivastava, K.S. Srivastava: Indian J. Pure Appl. Phys. 16, 125 (1978)
S.P. Singh: Indian J. Pure Appl. Phys. 16, 560 (1978); ibid. 17, 406 (1979)

S.222 K.S. Srivastava, R.L. Shrivastava, O.K. Harsh, V. Kumar, S.S. Sangal: Curr. Sci. 46, 251 (1977)

S.223 K.S. Srivastava et al.: Phys. Lett. A 64, 263 (1977); indian J. Pure Appl. Phys. 17, 54 (1979)

S.224 K.S. Srivastava, O.K. Harsh, V. Kumar: Phys. Stat. Sol. (b) 91, K169 (1979)

S.225 K.S. Srivastava, R.L. Shrivastava, O.K. Harsh, V. Kumar: J. Phys. Chem. Solids 40, 489 (1979)
V. Kumar, K.S. Srivastava, A.R. Chetal: Indian J. Pure Appl. Phys. 21, 552 (1983)

M.C. Shah, B.D. Shrivastava, K.B. Pandeya, K. Srivastava: J. Phys. C 16, 3601 (1983)

S.226   K.S. Srivastava, R.L. Shrivastava, O.K. Harsh, V. Kumar: Phys. Rev. B 19, 4336 (1979)

S.227   K.S. Srivastava et al.: Acta Phys. Pol. A 64, 639 (1983)

S.228   K.S. Srivastava et al.: Acta Phys. Pol. A 65, 531 (1984)

S.229   U.C. Srivastava, H.L. Nigam: Indian J. Chem. 10, 751 (1972)
        U.C. Srivastava: Nuovo Cim. 11 B, 68 (1972)
        M.N. Srivastava et al.: J. Inorg. Nucl. Chem. 28, 1897 (1976)

S.230   U.C. Srivastava: Indian J. Pure Appl. Phys. 16, 114 (1978)

S.231   U.C. Srivastava: Indian J. Pure Appl. Phys. 18, 258 (1980)

S.232   W. Starke, O. Brümmer, G. Dräger, I.Ya. Nikiforov, V.P. Sachenko, J. Richter: Izv. Akad. Nauk SSSR, Ser. Fiz. 36, 237 (1972)

S.233   G. Stegemann: Report Jul-2075, Kernforschungsanlage Jülich (1986)

S.234   S. Stephenson: Phys. Rev. 58, 873 (1940)

S.235   E.A. Stern, D.E. Sayers: Phys. Rev. Lett. 30, 174 (1973); Phys. Rev. B 10, 3027 (1974)
        F.W. Lytle, D.E. Sayers, E.A. Stern: Phys. Rev. B 11, 4825 (1975)
        E.A. Stern, D.E. Sayers, F.W. Lytle: Phys. Rev. B 11, 4836 (1975)
        D.E. Sayers, E.A. Stern, F.W. Lytle: Phys. Rev. Lett. 35, 584 (1975)

S.236   E.A. Stern, D.E. Sayers, F.W. Lytle: Phys. Rev. Lett. 37, 298 (1976)

S.237   E.A. Stern et al.: J. Magn. Magn. Mater. 7, 188 (1978)

S.238   E.A. Stern et al.: Phys. Rev. Lett. 38, 767 (1977)
        J.J. Rehr, E.A. Stern, R.L. Martin, E.R. Davidson: Phys. Rev. B 17, 560 (1978)
        S.M. Heald, E.A. Stern: Phys. Rev. B 17, 4069 (1978)
        E.A. Stern, S.M. Heald, B. Bunker: Phys. Rev. Lett. 42, 1372 (1979)

S.239   E.A. Stern, B.A. Bunker, S.M. Heald: Phys. Rev. B 21, 5521 (1980)
        E.A. Stern, Phys. Rev. Lett. 49, 1353 (1982)
        K.-Q. Lu, E.A. Stern: Nucl. Instrum. Meth. Phys. Res. 212, 475 (1983)

S.240   E.A. Stern et al.: Phys. Rev. Lett. 54, 905 (1985)

S.241   E.A. Stern, Y.Ma, K. Bauer, C.E. Bouldin: J. Physique 47, C3-371 (1986)

S.242   S. Stizza et al.: In [Ref.1.83, p.331];J. Physique 46, C8-255 (1985); J. Non-Cryst. Solids 80, 175 (1986)
        M. Benfatto et al.: J. Non-Cryst. Solids 77-78, 1325 (1985)

S.243   J. Stöhr et al.: In [Ref.1.87, p.43]
        D. Denley: Phys. Rev. B 21, 2267 (1980)
        R.S. Williams, D. Denley, D.A. Shirley, J. Stöhr: J. Am. Chem. Soc. 102, 5717 (1980)
        J. Stöhr, R. Jaeger: Phys. Rev. B 26, 4111 (1982)
        R.J. Koestner, J. Stöhr, J.L. Gland, J.A. Horsley: Chem. Phys. Lett. 105, 332 (1984)
        J. Stöhr, F. Sette, A.L. Johnson: Phys. Rev. Lett. 53, 1684 (1984)
        J. Stöhr et al.: Phys. Rev. Lett. 53, 2161 (1984)
        J.A. Horsley, J. Stöhr, R.J. Koestner: J. Chem. Phys. 83, 3146 (1985)
        F. Sette et al.: Phys. Rev. Lett. 54, 935 (1985)
        A.L. Johnson, E.L. Muetterties, J. Stöhr, F. Sette: J. Phys. Chem. 89, 4071 (1985)

S.244   J. Stöhr, D. Denley, P. Perfetti: Phys. Rev. B 18, 4132 (1978)
        L.I. Johansson, J. Stöhr: Phys. Rev. Lett. 43, 1882 (1979)
        J. Stöhr et al.: Phys. Rev. B 22, 4052 (1980)

S.245   J. Stöhr: Jpn. J. Appl. Phys. 17, S 2-217 (1978); J. Vac. Sci. Technol. 16, 37 (1979)

439

J. Stöhr, L. Johansson, I. Lindau, P. Pianetta: Phys. Rev. B **20**, 664 (1979)

S.246 J. Stöhr et al.: Appl. Opt. **19**, 3911 (1980)

D. Norman et al.: Phys. Rev. Lett. **51**, 2052 (1983)

S.247 J. Stöhr, R. Jaeger: J. Vac. Sci. Technol. **21**, 619 (1982)

S.248 J. Stöhr et al.: Phys. Rev. Lett. **55**, 1468 (1985)

F. Sette et al.: Phys. Rev. Lett. **56**, 2637 (1986)

S. Brennan, J. Stöhr, R. Jaeger: Phys. Rev. B **24**, 4871 (1981)

S.249 J. Stöhr et al.: Phys. Rev. B **36**, 2976 (1987)

S.250 M.J. Stott, N.H. March: In [Ref.1.78, pp.283, 303]

S.251 H. Sugawara, K. Sato, Y. Siota, H. Fujimoto: Ann. Rep. Res. Inst. Sci. Educat. **5**, 1 (1969)

S.252 C. Sugiura et al.: Techn. Rep. Tohoku Univ. **33**, 119, 205, 331 (1968); ibid. **34**, 125 (1969)

S.253 C. Sugiura, Y. Fujino, S. Kiyono: Techn. Rep. Tohoku Univ. **34**, 107, 307 (1969)

S.254 C. Sugiura, S. Kiyono: Techn. Rep. Tohoku Univ. **35**, 61, 249 (1970)

S.255 C. Sugiura, S. Kiyono: Techn. Rep. Tohoku Univ. **35**, 243 (1970); J. Phys. Soc. Jpn. **32**, 494 (1972); ibid. **33**, 455 (1972); Phys. Rev. B **6**, 1709 (1972); ibid. B **8**, 823 (1973); ibid. B **9**, 2679 (1974); J. Chem. Phys. **58**, 3527 (1973)

S.256 C. Sugiura: Jpn. J. Appl. Phys. **10**, 1120 (1971)

C. Sugiura et al.: In [Ref.1.62, p.168]

T. Matsukawa, M. Obashi, S. Nakai, C. Sugiura: Jpn. J. Appl. Phys. **17**, S 2-184 (1978)

C. Sugiura, Y. Gohshi: J. Chem. Phys. **74**, 4204 (1981)

S.257 C. Sugiura, O. Aita, S. Kiyono: Techn. Rep. Tohoku Univ. **36**, 77 (1971)

S.258 C. Sugiura, S. Kiyono: Proc. 6th Int'l Conf. X-Ray Opt. Microanal. Tokyo (1972) p.567

C. Sugiura, S. Muramatsu: Phys. Stat. Sol. (b) **129**, K 157 (1985)

S.259 C. Sugiura, Y. Hayasi: Jpn. J. Appl. Phys. **11**, 327, 598 (1972)

S. Muramatsu, C. Sugiura: Phys. Rev. B **26**, 3092 (1982)

S.260 C. Sugiura: J. Chem. Phys. **59**, 4907, 5444 (1973)

C. Sugiura, T. Suzuki: J. Chem. Phys. **75**, 4357 (1981)

C. Sugiura: J. Chem. Phys. **77**, 681 (1982)

C. Sugiura, M. Obashi: J. Chem. Phys. **78**, 88 (1983)

C. Sugiura: J. Chem. Phys. **79**, 3645 and 4811 (1983)

C. Sugiura, S.I. Nakai: Phys. Rev. B **28**, 1088 (1983)

C. Sugiura, S. Muramatsu: J. Phys. Chem. Solids **46**, 1215 (1985)

C. Sugiura, M. Kitamura, S. Muramatsu: Phys. Stat. Sol. (b) **135**, K 57 (1986)

S.261 C. Sugiura, Y. Gohshi, I. Suzuki: Phys. Rev. B **10**, 338 (1974)

C. Sugiura, I. Suzuki, J. Kashiwakura, Y. Gohshi: J. Phys. Soc. Jpn. **40**, 1720 (1976)

S.262 C. Sugiura: J. Chem. Phys. **62**, 1111 (1975); Jpn. J. Appl. Phys. **14**, 1691 (1975)

S.263 C. Sugiura, S. Nakai: Jpn. J. Appl. Phys. **17**, S 2-190 (1978)

S.264 C. Sugiura: J. Chem. Phys. **80**, 1047 (1984)

S.265 C. Sugiura, S. Muramatsu: J. Chem. Phys. **82**, 2191 (1985)

S.266 C. Sugiura, M. Kitamura, S. Muramatsu: J. Chem. Phys. **84**, 4824 (1986)

S.267 C. Sugiura, M. Kitamura, S. Muramatsu: J. Chem. Phys. **85**, 5269 (1986)

S.268 C. Sugiura, M. Kitamura, S. Muramatsu: Phys. Stat. Sol. (b) **140**, 631 (1987)

S.269 C. Sugiura, M. Kitamura, S. Muramatsu: Phys. Stat. Sol. (b) **141**, K173 (1987)

S.270 S.L. Suib et al.: J. Chem. Phys. **80**, 2203 (1984)

S.271 Yu.V. Sukhetskii et al.: Phys. Stat. Sol. (b) **132**, 103 (1985)

S.272 Yu.V. Sukhetskii et al.: Litov. Fiz. Sb. **26**, 31 (1986)

S.273 V.L. Sukhorukov, V.F. Demekhin, V.V. Timoshevskaya, S.V. Lavrentev: Opt. Spektrosk. **47**, 407 (1979)

S.274 V.L. Sukhorukov, L.A. Demekhina, V.A. Yavna, V.F. Demekhin: Fiz. Tverd. Tela **21**, 2976 (1979)

S.275 V.L. Sukhorukov, V.A. Yavna, V.F. Demekhin: Izv. Akad. Nauk SSSR, Ser. Fiz. **46**, 763 (1982)

S.276 A.S. Suleimanov, F.A. Namatsova, S.M. Gusein-Tsade: Isl. Obl. Neorg. Fiz. Khim. (1971) 370

S.277 O.I. Sumbaev, E.V. Petrovich et al.: Zh. Eksp. Teor. Fiz. **53**, 1545 (1967)

S.278 O.I. Sumbaev et al.: Zh. Eksp. Teor. Fiz. **56**, 536 (1969); ibid. **57**, 1716 (1969); Phys. Lett. A **30**, 129 (1969)
A.I. Grushko et al.: Zh. Eksp. Teor. Fiz. **74**, 501 (1978)

S.279 O.I. Sumbaev: Mod. Phys. Chem. **1**, 31 (1976)

S.280 E. Suoninen, V. Lantto, V. Polvi: Ann. Acad. Sci. Fenn. A **296**, 1 (1968)

S.281 E. Suoninen, M. Pessa: Phys. Scr. **7**, 89 (1973)

S.282 E. Suoninen, T. Valkonen, M. Pessa: Phys. Fenn. **9**, S 1-184 (1974)
E.J. Suoninen, T.V.O. Valkonen: J. Phys. F **5**, 837 (1975)

S.283 Y. Suwa, S. Naka, T. Noda: Kogyo Kagaku Zasshi **74**, 845 (1971)

S.284 S. Sužuki et al.: J. Magn. Magn. Mater. **52**, 458 (1985)

S.285 Y. Suzuki, E. Asada: J. Chem. Soc. Jpn., Ind. Chem. Sect. **70**, 652 (1967)

S.286 V.N. Svechnikov, V.V. Nemoshkalenko et al.: Metallofizika **52**, 50 (1974); in [Ref.1.98, p.328]

S.287 N. Swanson, K. Codling: J. Opt. Soc. Am. **58**, 1192 (1968)

S.288 C.A. Swarts, J.D. Dow, C.P. Flynn: Phys. Rev. Lett. **43**, 158 (1979)

S.289 K. Syassen et al.: Phys. Rev. B **26**, 4745 (1982)

S.290 V.G. Syryanov, S.A. Nemnonov, M.F. Sorokina: In [Ref.1.95, p.53]

S.291 V.G. Syryanov, S.A. Nemnonov: Fiz. Met. Metalloved. **31**, 515 (1971)

S.292 R. Szargan: Thesis Leipzig 1970
R. Szargan, A. Meisel: Wiss. Z. Karl-Marx-Univ. Leipzig, Math.-Naturwiss. R. **20**, 41 (1971)

S.293 R. Szargan, H.-J. Köhler, A. Meisel: Spectrochim. Acta **27** B, 43 (1972)

S.294 R. Szargan, G. Leonhardt, F.-U. Flöther, A. Meisel: Spectrochim. Acta **28** B, 359 (1973)

S.295 R. Szargan, E. Suoninen, M. Lähdeniemi, M. Pessa: Spectrochim. Acta **33** A, 129 (1977)

S.296 R. Szargan, A. Meisel, E. Hartmann, G. Brunner: Jpn. J. Appl. Phys. **17**, S 2-174 (1978)
E. Hartmann, R. Szargan: Chem. Phys. Lett. **68**, 175 (1979)

S.297 A. Szasz et al.: Vacuum **33**, 107 (1983)

S.298 A. Szasz, J. Kojnok: Appl. Surf. Sci. **24**, 34 (1985)

S.300 F. Szmulowicz, B. Segall: Phys. Rev. B **21**, 5628 (1980)

T.1 J.A. Tagle, E.T. Arakawa, T.A. Callcott: Phys. Rev. B **21**, 4552 (1980)

T.2 J.A. Tagle, E.T. Arakawa, T.A. Callcott: Phys. Rev. B **22**, 2716 (1980)

T.3 S. Takagi, G. Yamaguchi: Yogyo Kyokai Shi **85**, 268 (1977)
S. Takagi, G. Yamaguchi, K. Shirakusa: Sekko To Sekkai **148**, 119 (1977)

T.4 Y. Takahashi, K. Yabe: Bull. Chem. Soc. Jpn. **42**, 3064 (1969)

T.5 Y. Takahashi, Bull. Chem. Soc. Jpn. **44**, 587 (1971); ibid. **46**, 2039 (1973)

T.6 Y. Takahashi, K. Yabe. T. Sato, T. Takahashi: Proc. 6th Int'l Conf. X-Ray Optics Microanal. (1972) p.553

T.7 T. Takigushi, E. Asada: J. Chem. Soc. Jpn. **70**, 214 (1967)

T.8   Y. Tamaki, T. Omori, T. Shiokawa: Jpn. J. Appl. Phys. 17, S2-425 (1978)

T.9   K. Tanaka, M. Matsumoto, S. Maruno, A. Hiraki: Appl. Phys. Lett. 27, 529 (1975)

K. Tanaka, A. Hiraki: Jpn. J. Appl. Phys. 17, S-2121 (1978)

K. Tanaka, T. Saito, M. Yasuda: J. Phys. Soc. Jpn. 52, 1718 (1983)

T.10   K. Tanaka, N. Hamasaka, M. Yasuda, Y. Fukai: Solid State Commun. 30, 173 (1979)

K. Tanaka, M. Higatani, K. Kai, K. Suzuki: J. Less-Common Met. 38, 317 (1982)

T.11   K. Tanaka, C. Sugiura, S. Nakai, Y. Ohno: Jpn. J. Appl. Phys. 20, 41 (1981)

T.12   K. Tanaka, M. Yoshino, K. Suzuki: J. Phys. Soc. Jpn. 51, 3882 (1982)

K. Tanaka, Y. Yamada, K. Kai, K. Suzuki: J. Phys. Soc. Jpn. 53, 1783 (1984)

K. Tanaka, T. Saito, K. Suzuki, R. Hasegawa: Phys. Rev. B 32, 6853 (1985)

T.13   J.G. Tandberg: Ark. Mat. Astr. Fysik 24, Nr.6 (1933)

T.14   C. Tang et al.: Phys. Rev. B 31, 1000 (1985)

T.15   K. Taniguchi: Adv. X-Ray Anal. 23, 193 (1980)

T.16   K. Taniguchi: Jpn. J. Appl. Phys. 23, 358 (1984); Bull. Chem. Soc. Jpn. 57, 909, 915, and 921 (1984)

T.17   H. Tanino, H. Oyanagi, M. Yamashita, K. Kobayashi: Solid State Commun. 53, 953 (1985)

T.18   H. Tanino, K. Takahashi: Solid State Commun. 59, 825 (1986)

T.19   S. Tanuma et al.: Synth. Met. 12, 371 (1985)

T.20   J.M. Tarascon et al.: J. Physique 41, 1135, 1141 (1980)

T.21   M. Tavernier, P. Chevallier, P. Briand, V. Kostroun: Jpn. J. Appl. Phys. 17, S2-147 (1978)

T.22   D.A. Taylor et al.: Chem. Phys. Lett. 121, 482 (1985)

T.23   E. Tegeler, N. Kosuch, G. Wiech, A. Faessler: Phys. Stat. Sol. (b) 84, 561 (1977); ibid. 91, 223 (1979)

T.24   E. Tegeler, N. Kosuch, G. Wiech, A. Faessler: Jpn. J. Appl. Phys. 17, S2-97 (1978)

E. Tegeler: Dissertation, Universität München (1978)

E. Tegeler, G. Wiech, A. Faessler: J. Phys. B 13, 4771 (1980)

E. Tegeler, M. Iwan, E.E. Koch: J. Electron Spectrosc. 22, 297 (1981)

T.25   E. Tegeler, N. Kosuch, G. Wiech, A. Faessler: J. Electron Spectrosc. 18, 23 (1980)

T.26   R.G. Teller, M.R. Antonio, J.F. Brazdil, R.K. Grasselli: J. Solid State Chem. 64, 249 (1986)

T.27   D.H. Templeton, L.K. Templeton: Acta Crystallogr. A 36, 237 (1980)

T.28   B.-K. Teo et al.: J. Am. Chem. Soc. 99, 3854 (1977); ibid. 100, 621, 3225 (1978)

T.29   B.-K. Teo, P. Eisenberger, B.M. Kincaid: J. Am. Chem. Soc. 100, 1735 (1978)

T.30   B.-K. Teo, R.G. Shulman, G.S. Brown, A.E. Meixner: J. Am. Chem. Soc. 101, 5624 (1979)

B.-K. Teo: Accounts Chem. Res. 13, 412 (1980)

N.R. Antonio, B.-K. Teo, W.E. Cleland, B. Averill: J. Am. Chem. Soc. 105, 3477 (1983)

B.-K. Teo, M.R. Antonio, B.A. Averill: J. Am. Chem. Soc. 105, 3751 (1983)

M.R. Antonio et al.: J. Am. Chem. Soc. 107, 3583 (1985)

T.31   B.-K. Teo, H.S. Chen, R. Wang, M.R. Antonio: J. Non-Cryst. Solids 58, 249 (1983)

T.32   H. Terauchi et al.: J. Phys. Soc. Jpn. 53, 3286 (1984)

T.33   V.A. Terekhov et al.: Izv. Akad. Nauk SSSR, Ser. Fiz. 46, 749 (1982)

T.34   V.A. Terekhov et al.: Fiz. Tverd. Tela 25, 2482 (1983)

T.35 V.A. Terekhov et al.: Fiz. Tekh. Poluprovodn. **18**, 1897 (1984); Poverkhn. Fiz. Khim. Mekh. 1984, Nr.6, 91

T.36 V.A. Terekhov et al.: Fiz. Tekh. Poluprovodn. **20**, L658 (1986)

T.37 E.C. Theil, D.E. Sayers, M.A. Brown: J. Biol. Chem. **254**, 8132 (1979)

T.38 R. Theisen: Mikrochim. Acta, Suppl. II, 211 (1967)

T.39 M.L. Theye, A. Gheorghiu, H. Launois: J. Phys. C **13**, 6569 (1980)

T.40 B.T. Thole et al.: Phys. Rev. B **31**, 6856 (1985); ibid. B **32**, 5107 (1985)
B.T. Thole, G. Van der Laan, G.A. Sawatzky: Phys. Rev. Lett. **55**, 2086 (1985)

T.41 J.S. Thomsen: J. Phys. B **16**, 1171 (1983)

T.42 F.R. Thornley, N.T. Barrett, G.N. Greaves, G.M. Antonini: J. Phys. C **19**, L563 (1986)

T.43 W. Thulke, R. Frahm, R. Haensel, P. Rabe: Phys. Stat. Sol. (a) **75**, 501 (1983)

T.44 W. Thulke, P. Rabe: J. Phys. C. **16**, L 955 (1983)

T.45 J. Tilgner, I. Topol, G. Leonhardt, A. Meisel: J. Phys. Chem. Solids **36**, 27 (1975)

T.46 K. Tohji, Y. Udagawa: Rev. Sci. Instrum. **54**, 1482 (1983)

T.47 M. Tokumoto et al.: Solid State Commun. **48**, 861 (1983)
H. Oyanagi et al.: J. Phys. Soc. Jpn. **53**, 4044 (1984)
M. Tokumoto et al.: Mol. Cryst. Liq. Cryst. **117**, 139 (1985)

T.48 M. Tomellini, D. Gozzi, A. Bianconi, I. Davoli: J. Chem. Soc., Faraday Trans. I **83**, 289 (1987)

T.49 T.P. Tooman: Thesis, Las Cruces, New Mexico 1975

T.50 I. Topol, E. Hess: Phys. Stat. Sol. (b) **56**, K9 (1973); Ann. Physik **30**, 211 (1973)

T.51 I.A. Topol, G. Leonhardt, K. Unger, E. Hess: Phys. Stat. Sol. (b) **61**, 285 (1974)
G. Leonhardt: Fiz. Tverd. Tela **17**, 3 (1975)

T.52 I. Topol, J. Tilgner, G. Leonhardt, A. Meisel: J. Phys. Chem. Solids **35**, 1657 (1974)

T.53 I.A. Topol, A.V. Kondratenko, L.N. Mazalov: Opt. Spektrosk. **50**, 494 (1981); Izv. Akad. Nauk SSSR, Ser. Fiz. **46**, 776 (1982)

T.54 J.A. Tossell: Geochim. Cosmochim. Acta **37**, 583 (1973); J. Phys. Chem. Solids **34**, 307 (1973)
D.J. Vaughan, J.A. Tossell: Am. Mineral. **58**, 765 (1973)
J.A. Tossell, J. Phys. Chem. Solids **36**, 1273 (1975); ibid. **37**, 1043 (1976); Chem. Phys. **15**, 303 (1976)
J.A. Tossell, G.V. Gibbs: Phys. Chem. Miner. **2**, 21 (1977)
J.A. Tossell: Am. Mineral. **62**, 136 (1977)

T.55 J.A. Tossell: Inorg. Chem. **19**, 3328 (1980)

T.56 J.A. Tossell, J.W. Davenport: J. Chem. Phys. **80**, 813 (1984)

T.57 J.A. Tossell: Am. Mineral. **71**, 1170 (1986)

T.58 J.A. Tossell: Phys. Chem. Miner. **14**, 320 (1987)

T.59 G. Tourillon et al.: Surf. Sci. **156**, 536 (1985)

T.60 G. Tourillon, H. Dexpert, P. Lagarde: J. Electrochem. Soc. **134**, 327 (1987)

T.61 G. Tourillon et al.: Phys. Lett. A **121**, 251 (1987)

T.62 G. Tourillon et al.: Surf. Sci. **184**, L345 (1987)

T.63 G. Tourillon et al.: Phys. Rev. B **35**, 9863 (1987)

T.64 H. Toyuki: Yogyo Kyokai Shi **85**, 263 (1977)

T.65 J.M. Tranquada, R. Ingalls: Phys. Rev. B **34**, 4267 (1986)

T.66 J.M. Tranquada, C.Y. Yang: Solid State Commun. **63**, 211 (1987)

T.67 V.A. Trofimova et al.: Fiz. Met. Metalloved. **63**, 880 (1987)

T.68   M. Tronc, G.C. King, F.H. Read: J. Phys. B **9**, L555 (1976); ibid. **12**, 137 (1979)

T.69   N.V. Troneva: Izv. Akad. Nauk SSSR, Ser. Fiz. **27**, 403 (1963); ibid. **28**, 809 (1964)

T.70   K.L. Tsang et al.: Phys. Rev. B **35**, 8374 (1987)

T.71   Y.F. Tsay, A. Vaidyanathan, S.S. Mitra: Phys. Rev. B **19**, 5422 (1979)

T.72   M.E. Tsimbler, S.M. Karalnik, A.V. Koval, S.M. Tsimbler: Ukr. Khim. Zh. **39**, 646 (1973)
P.B. Mikhelson, S.M. Karalnik, A.V. Koval, V.D. Dobrovolskii: Ukr. Khim. Zh. **41**, 604 (1975)

T.73   K. Tsutsumi: J. Phys. Soc. Jpn. **14**, 1696 (1959); in [Ref.1.57, p.336]

T.74   K. Tsutsumi, H. Nakamori: J. Phys. Soc. Jpn. **25**, 1418 (1968); in [Ref.1.60, Vol.I, p.100]

T.75   K. Tsutsumi, M. Obashi: In [Ref.1.58, Vol.I, p.65]

T.76   K. Tsutsumi, H. Nakamori, K. Ichikawa: Phys. Rev. B **13**, 929 (1976)

T.77   K. Tsutsumi, O. Aita, K. Ichikawa: Phys. Rev. B **15**, 4638 (1977)

T.78   K. Tsutsumi et al.: J. Phys. Soc. Jpn. **47**, 1920 (1979)

T.79   V.P. Tsvetkov, A.V. Kalenichenko, O.M. Porolenko: In [Ref.1.58, Vol.I, p.227]; in [Ref.1.96, Vol.I, p.147]

T.80   V.P. Tsvetkov, A.V. Kalenichenko: Ukr. Fiz. Zh. **16**, 1299 (1971); Vopr. Obshch. Prikl. Fiz. 1972, 39

T.81   V.P. Tsvetkov, A.V. Kalenichenko, A.P. Ganzha: Izv. Akad. Nauk SSSR, Ser. Fiz. **36**, 287 (1972)

T.82   V.P. Tsvetkov, N.D. Savchenko, A.V. Kalenichenko: In [Ref.1.94, p.45]

T.83   M.H. Tuilier, J. Dexpert-Ghys, H. Dexpert, P. Lagarde: J. Solid State Chem. **69**, 153 (1987)

T.84   J. Tulkki, T. Åberg: J. Phys. B **15**, L 435 (1982)

T.85   T.D. Tullius, D.M. Kurtz, Jr., S.D. Conradson, K.O. Hodgson: J. Am. Chem. Soc. **101**, 2776 (1979)

T.86   T. Tullius, P. Frank, K.O. Hodgson: Proc. Natl. Acad. Sci. USA **75**, 4069 (1980)

T.87   T.D. Tullius, W.O. Gillum, R.M.K. Carlson, K.O. Hodgson: J. Am. Chem. Soc. **102**, 5670 (1980)

T.88   I.I. Tupitsyn, I.I. Lyakhovskaya et al.: Fiz. Tverd. Tela **16**, 3117 (1974)

T.89   F. Tyrén: Nova Acta Reg. Soc. Sci. Upsal. **12**, Nr.1 (1940)

U.1   R. Ubgade, P.R. Sarode: Phys. Stat. Sol. (a) **99**, 295 (1987)

U.2   H. Uchikawa, M. Numata: Onoda Kenkyu Hokoku **25**, 18 (1972); Yogyo Kyokai Shi **81**, 189 (1973)

U.3   A. Udris, L.L. Makarov et al.: Liet. Fiz. Rinkinys **19**, 442 (1979)

U.4   K. Ulmer: In [Ref.1.58, Vol.II, p.79]
K. Böhm, J. Kieser: Z. Physik **259**, 365 (1973)

U.5   K. Ulmer: In [Ref.1.79, p.521]

U.6   K. Ulmer: Jpn. J. Appl. Phys. **17**, S2-154 (1978)
S. El-Kholy, K. Ulmer: Z. Physik B **38**, 1 (1980)

U.7   M. Umeno, G. Wiech: Phys. Stat. Sol. (b) **59**, 145 (1973)
G. Wiech: In [Ref.1.62, p.195]

U.8   P.H. Umadikar, S.G. Mestry, N.V. Bhat, B.C. Haldar: Spectrochim. Acta B **31**, 411 (1976)

U.9   D. Urch: Adv. X-Ray Anal. **14**, 250 (1971)

U.10   D.S. Urch, S. Webber: X-Ray Spectrom. **6**, 64 (1977)

U.11 D.S. Urch: In *Electron Spectroscopy-Theory, Techniques and Applications*, Vol.3, ed. by G.R. Brundle, A.D. Baker (Academic, London 1979) p.1
U.12 V.S. Urusov: Dokl. Akad. Nauk SSSR **166**, 660 (1966)
U.13 J. Utriainen, M. Linkoaho, T. Åberg: Comment. Phys. Math. Soc. Sci. Fenn. **42**, 296 (1972)
U.14 J. Utriainen: In [Ref.1.60, Vol.I, p.382]
U.15 J. Utriainen, J. Valjakka, M. Linkoaho: J. Phys. C **8**, 3710 (1975)

V.1   A.S. Vaingankar, B.V. Khasbardar, R.N. Patil: J. Phys. F **10**, 1615 (1980)
      V.S. Sahasrabudhe, A.S. Vaingankar: Solid State Commun. **43**, 299 (1982)
      V.S. Sahasrabudhe: Solid State Commun. **46**, 697 (1983)
      V.S. Sahasrabudhe: Indian, J. Pure. Appl. Phys. **23**, 168 (1985)
V.2   E.E. Vainshtein: *X-Ray Spectra of Atoms in Molecules, Chemical Compounds and Alloys* (in Russian), Moscow 1950; Zh. Eksp. Teor. Fiz. **20**, 442, 446 (1950)
V.3   E.E. Vainshtein, Yu.F. Kopelev: Zh. Strukt. Khim. **3**, 448 (1962)
V.4   E.E. Vainshtein et al.: Dokl. Akad. Nauk SSSR **151**, 120, 1360 (1963)
V.5   E.E. Vainshtein, R.M. Ovrutskaya et al.: Fiz. Tverd. Tela **5**, 2955 (1963); **7**, 2120 (1965)
V.6   E.E. Vainshtein, S.M. Bokhin et al.: Fiz. Tverd. Tela **6**, 2909 (1964); Zh. Neorg. Khim. **10**, 121 (1965)
V.7   E.E. Vainshtein, L.N. Mazalov: Fiz. Tverd. Tela **7**, 1099 (1965)
V.8   E.E. Vainshtein, V.I. Chirkov et al.: Dokl. Akad. Nauk SSSR **163**, 63 (1965); Izv. Sib. Otd. Akad. Nauk SSSR, Ser. Khim. Nauk 1966, 12
      V.I. Chirkov et al.: Dokl. Akad. Nauk SSSR **165**, 1354 (1965); Fiz. Tverd. Tela **9**, 1116 (1967); Izv. Akad. Nauk SSSR, Neorg. Mater. **3**, 1017, 1022 (1967)
V.9   E.E. Vainshtein, A.P. Sadovskii, S.V. Larionov: Zh. Strukt. Khim. **7**, 623 (1966)
V.10 E.E. Vainshtein, S.M. Blokhin, V.M. Bertenev: Izv. Sib. Otd. Akad. Nauk SSSR, Ser. Khim. Nauk 1966, 59
V.11 E.E. Vainshtein, M.N. Bril et al.: Izv. Akad. Nauk SSSR, Neorg. Mater. **3**, 644, 1685 (1967)
V.12 P.P. Vaishnava et al.: Phys. Rev. B **34**, 4599 (1986)
V.13 J. Valasek: Phys. Rev. **52**, 250 (1937); ibid. **53**, 274 (1938)
V.14 J. Valjakka, J. Utriainen, T. Åberg, J. Tulkki: Phys. Rev. B **32**, 6892 (1985)
V.15 G. van der Laan et al.: Solid State Commun. **56**, 673 (1985); Phys. Rev. B **33**, 4253 (1986)
V.16 G. van der Laan et al.: J. Phys. Chem. Solids **47**, 413 (1986)
V.17 G. van der Laan: In *Giant Resonances in Atoms, Molecules and Solids*, ed. by J.P. Connerade, J.M. Esteva, R.C. Karnatak (Plenum, New York 1987) p.447
V.18 M.J. Van der Wiel, R.B. Kay, P.E. Van der Leeuw: In [Ref.1.87, p.101]
V.19 J.P. Van Dyke: Phys. Rev. B **5**, 4206 (1972)
V.20 R.A. Van Nordstrand: Adv. Catalysis **12**, 149 (1960)
V.21 R.A. Van Nordstrand: In [Ref.1.57, p.255]
V.22 H.F.J. Van't Blik et al.: J. Phys. Chem. **87**, 2264 (1983)
      J.B.A.D. Van Zon et al.: J. Chem. Phys. **80**, 3914 (1984)
      J.B.A.D. Van Zon, D.C. Koningsberger, H.F.J. Van't Blik, D.E. Sayers: J. Chem. Phys. **82**, 5742 (1985)
V.23 V.M. Vdovenko, L.L. Makarov et al.: Dokl. Akad. Nauk SSSR **202**, 868 (1972)

L.L. Makarov et al.: Vestn. Leningr. Univ. 1975, Nr.16, 87; Radiokhimiya 20, 116 (1978); Zh. Neorg. Khim. 24, 1014 (1979)

V.24   R.V. Vedrinskii, V.P. Sachenko et al.: Vopr. Obshch. Prikl. Fiz. (1972) 50; Izv. Akad. Nauk SSSR, Ser. Fiz. 38, 434 (1974)

R.V. Vedrinskii, V.L. Kraizman: Zh. Eksp. Teor. Fiz. 74, 1215 (1978)

V.25   R.V. Vedrinskii, S.A. Prosandeev, V.V. Krivitskii, A.N. Pavlov: Zh. Strukt. Khim. 21, 83 (1980)

V.26   R.V. Vedrinskii, S.A. Prosandeev, A.N. Pavlov, A.P. Kovtun: Teor. Eksp. Khim. 16, 19 (1980)

V.27   R.V. Vedrinskii, V.L. Kraizman: Zh. Neorg. Khim. 25, 2858 (1980)

V.28   R.V. Vedrinskii, L.A. Bugaev: Fiz. Tverd. Tela 28, 2516 (1986)

V.29   E.T. Verkhovtseva, P.S. Pogrebnyak, Ya.M. Fogel: In [Ref.1.87, p.82]; Phys. Lett. A 65, 106 (1978)

V.30   E.T. Verkhovtseva, P.S. Pogrebnyak: Opt. Spektrosk. 45, 866 (1978); ibid. 48, 858 (1980); J. Phys. B 13, 3535 (1980)

E.T. Verkhovtseva, P.S. Pogrebnyak, A.A. Tkachenko: Opt. Spektrosk. 59, 457 (1985)

E.T. Verkhovtseva, E.V. Gnatchenko, P.S. Pogrebnyak: J. Phys. B 19, 2089 (1986)

V.31   E.T. Verkhovtseva, E.V. Gnatchenko, A.A. Tkachenko: Opt. Spektrosk. 61, 43 (1986)

V.32   L.P. Verma, B.K. Agarwal: Indian J. Pure Appl. Phys. 5, 241 (1967)

A.M. Vishnoi, B.K. Agarwal: Indian J. Pure Appl. Phys. 9, 1074 (1971)

V.33   L.P. Verma: Curr. Sci. 45, 17 (1976)

V.34   S.P. Vernon, M.B. Stearns: Phys. Rev. B 29, 6968 (1984)

V.35   R.C. Vickery, R. Sedlacek, A. Ruben: J. Chem. Soc. (1959) 505

V.36   A.S. Vinogradov, T.M. Zimkina: Fiz. Tverd. Tela 12, 1492 (1970); Opt. Spektrosk. 32, 33 (1972)

A.A. Pavlychev, A.S. Vinogradov: Fiz. Tverd. Tela 23, 3564 (1981)

V.37   A.S. Vinogradov, T.M. Zimkina et al.: Izv. Akad. Nauk SSSR, Ser. Fiz. 38, 508 (1974); Izv. Sib. Otd. Akad. Nauk SSSR, Ser. Khim. Nauk 1975, Nr.9, H.4, 88

V.N. Akimov, A.S. Vinogradov, A.Yu. Dukhnyakov: Fiz. Tverd. Tela 24, 169 (1982)

V.N. Akimov, A.S. Vinogradov, T.M. Zimkina: Opt. Spektrosk. 53, 109 (1982); ibid. 53, 476 and 918 (1982)

V.38   A.S. Vinogradov et al.: Fiz. Tverd. Tela 22, 2602 (1980)

V.39   A.S. Vinogradov et al.: Fiz. Tverd. Tela 24, 1417 (1982); ibid. 25, 400 (1983)

V.N. Sivkov, A.S. Vinogradov: Fiz. Tverd. Tela 25, 897 (1983)

A.Yu. Dukhnyakov, A.S. Vinogradov: Opt. Spektrosk. 53, 841 (1982)

V.40   A.N. Vishnoi, B.K. Agarwal: Phys. Lett. 29, 105 (1969); indian J. Pure Appl. Phys. 7, 812 (1969); ibid. 9, 1074 (1971); Lett. Nuovo Cim. I 4, 771 (1970)

V.41   A.N. Vishnoi: J. Phys. C 2, 227 (1970)

V.42   G. Vlaic, J.C.J. Bart, W. Cavigiolo, S. Mobilio: Chem. Phys. Lett. 76, 453 (1980)

V.43   S.V. Vlasov, O.V. Farberovich: Fiz. Tverd. Tela 24, 941 (1982)

V.44   S.V. Vlasov et al.: Fiz. Tverd. Tela 27, 2173 (1985)

V.45   S.V. Vlasov, O.V. Farberovich, G.P. Nizhnikova, R.S. Dagis: Litov. Fiz. Sb. 26, 408 (1986)

V.46   Vo Chiong Ki et al.: Fiz. Tverd. Tela 26, 3521 (1984)

V.47   A.A. Voityuk et al.: Zh. Strukt. Khim. 25, Nr.3, 12 (1984)

V.48   V.F. Volkov et al.: Fiz. Met. Metalloved. 25, 1134 (1968); ibid. 26, 376 (1968)

V.49   D. Vollath: Mikrochim. Acta, Suppl. **3**, 1 (1968)
R. Brückner, W. Poch, D. Vollath: Glastechn. Ber. **42**, 322 (1969)

V.50   L.S. Voskrekasenko, V.V. Didyk, L.I. Nikolaev: Doprov. Akad. Nauk Ukr. SSSR, Ser. A (1977) 459

V.51   D.D. Vvedensky, J.B. Pendry, U. Döbler, K. Baberschke: Phys. Rev. B **35**, 7756 (1987)

V.52   Ts. Vylov, B.S. Dzhelepov, R.B. Ivanov, M.A. Mikhailova, V.O. Sergeev: Izv. Akad. Nauk SSSR, Ser. Fiz. **36**, 2136 (1972)

W.1   W.G. Waddington, P. Rez, I.P. Grant, C.J. Humphreys: Phys. Rev. B **34**, 1467 (1986)

W.2   A. Wagner, G. Wiech: In [Ref.1.60, Vol.I, p.225]

W.3   M. Wakagi, M. Chigasaki, M. Nomura: J. Phys. Soc. Jpn. **56**, 1765 (1987)
M. Wakagi, I. Ohno, M. Chigasaki, M. Nomura: J. Phys. Soc. Jpn. **56**, 2413 (1987)

W.4   S. Wakoh, Y. Kuvo: Jpn. J. Appl. Phys. **17**, S2-193 (1978)

W.5   P.F. Walch, D.E. Ellis: Phys. Rev. B **8**, 5920 (1973)

W.6   S.S. Wald, S.K. Cheng: Adv. X-Ray Anal. **21**, 241 (1978)

W.7   Wang Wen-Cai, Ge Sen-Lin, Chen Yu: Acta Phys. Sin. **35**, 1164 (1986)

W.8   H. Watanabe, T. Hayasi: Sci. Rep. Tohoku Univ. I. **50**, 157 (1967)

W.9   M. Watanabe, K. Nishioka: Jpn. J. Appl. Phys. **17**, S2-201 (1978)

W.10   T. Watanabe, C. Horie: In [Ref.1.62, p.341]

W.11   L.M. Watson, R.K. Dimond, D.J. Fabian: In [Ref.1.78, p.45]; in [Ref.1.58, Vol.II, p.56]

W.12   L.M. Watson, V.V. Nemoshkalenko et al.: Izv. Akad. Nauk SSSR, Ser. Fiz. **36**, 415 (1972)
L.M. Watson et al.: In [Ref.1.62, p.205]

W.13   L.M. Watson, Q.S. Kapoor, D. Hart: In [Ref.1.60, Vol.II, p.135]
C.M. Sayers et al.: J. Phys. F **5**, L207 (1975)

W.14   L.M. Watson, C.A.W. Marshall, C.P. Cardoso: J. Phys. F **14**, 113 (1984)

W.15   R.L. Watson, T. Chiao, F.E. Jenson: Phys. Rev. Lett. **35**, 254 (1975)
R.L. Watson et al.: Phys. Rev. A **15**, 914 (1977)
J.A. Demarest, R.L. Watson: Phys. Rev. A **17**, 1302 (1978)

W.16   R.L. Watson, A. Langenberg, F.E. Jenson, R.M. Hedges: Jpn. J. Appl. Phys. **17**, S2-93 (1978)

W.17   G.A. Waychunas: Am. Mineral. **72**, 89 (1987)

W.18   J.H. Weaver, C.G. Olson: In [Ref.1.87, Vol.2, p.52]

W.19   W.M. Weber: Thesis, Groningen (1972); in [Ref.1.60, Vol.I, p.143]; Phys. Fenn. **9**, S1-77 (1974)

W.20   W.M. Weber: Phys. Rev. B **11**, 2744 (1975)

W.21   W. Weber, J. Peisl: Phys. Rev. B **28**, 806 (1983)

W.22   T. Wehara, K. Koto, S. Emura, F. Kanamaru: Solid State Ionics Diffus. React. **23**, 331 (1987)

W.23   P. Weightman, M. Davies, P.T. Andrews: Phys. Rev. B **30**, 5586 (1984)

W.24   P. Weinberger: Ber. Bunsenges. Phys. Chemie **81**, 459 (1977)

W.25   P. Weinberger: Phys. Stat. Sol. (b) **97**, 565 (1980); ibid. **98**, 207,591 (1980)

W.26   P. Weinberger, J. Staunton, B.L. Gyorffy: J. Phys. F **12**, 199 (1982)

W.27   J. Weisbach, W. Blau, G. Merz: Phys. Fenn **9**, S1-181 (1974)
W. Niederlag, W. Blau, A. Burenkov, K. Kleinstück: Fiz. Met. Metalloved. **43**, 972 (1977)
W. Blau, J. Weisbach, G. Merz, K. Kleinstück: Phys. Stat. Sol. (b) **93**, 713 (1979)

	A. Himsel et al.: Phys. Stat. Sol. (b) 100, 179 (1980)
	C. Müller, W. Blau, P. Ziesche: Phys. Stat. Sol. (b) 116, 561 (1983)
W.28	H.E. Welch: Thesis Texas 1970
W.29	A. Wenger, G. Burri, S. Steinemann: Solid State Commun. 9, 1125 (1971)
W.30	A. Wenger, S. Steinemann: Helv. Phys. Acta 47, 321 (1974)
W.31	F. Werfel, G. Dräger, O. Brümmer, M. Jurisch: Phys. Stat. Sol. (b) 80, K95 (1977)
	F. Werfel, O. Brümmer, M. Jurisch: In [Ref.1.100, p.267]
W.32	F. Werfel, O. Brümmer, H. Oppermann: In [Ref.1.100, p.273]
W.33	L.O. Werme et al.: Phys. Lett. A 41, 113 (1972); C.R. Acad. Sci. B 279, 119 (1974)
W.34	L.O. Werme, B. Grennberg, J. Nordgren, C. Nordling, K. Siegbahn: Nature 242, 453 (1973)
W.35	L.O. Werme et al.: J. Electron Spectrosc. 2, 435 (1973); Z. Physik A 272, 131 (1975)
	H. Ågren, J. Müller: Phys. Scr. 20, 627 (1979); J. Chem. Phys. 72, 4078 (1980)
W.36	D. Wesner, W. Eberhardt: Phys. Rev. B 28, 7087 (1983)
W.37	E.W. White, H.A. McKinstry, T.F. Bates: Adv. X-Ray Anal. 2, 239 (1960); ibid. 9, 376 (1966)
W.38	E.W. White, R. Roy: Solid State Commun. 2, 151 (1964)
	E.W. White, G.V. Gibbs: Am. Mineral. 52, 985 (1967)
W.39	E.W. White, G.V. Gibbs: Am. Mineral. 54, 931 (1969)
W.40	H.C. Whitehead, G. Andermann: J. Phys. Chem. 77, 721 (1973); ibid. 78, 2592 (1974)
W.41	G. Wiech: Dissertation, Universität München (1964)
W.42	G. Wiech: Z. Physik 193, 490 (1966)
	G. Wiech, E. Zöpf: J. Physique 32, C4-200 (1971)
W.43	G. Wiech: Z. Physik 207, 428 (1967)
W.44	G. Wiech: In [Ref.1.78, p.59]
W.45	G. Wiech: Z. Physik 216, 472 (1968)
W.46	G. Wiech, E. Zöpf: In [Ref.1.101, p.335]
W.47	G. Wiech, E. Zöpf: Z. Physik 244, 94 (1971)
W.48	G. Wiech, W. Koeppen, D.S. Urch: Inorg. Chem. Acta 6, 376 (1972)
W.49	G. Wiech, E. Zöpf: In [Ref.1.79, p.173]
W.50	G. Wiech, E. Zöpf: In [Ref.1.79, p.629]
	E.Z. Kurmaev, G. Wiech: J. Non-Cryst. Solids 70, 187 (1985)
W.51	G. Wiech, C. Beyreuther, E. Hieke: Monatsh. Chemie 105, 302 (1974)
W.52	G. Wiech, E. Zöpf, H.U. Chun, R. Brückner: J. Non-Cryst. Solids 21, 251 (1976)
	A. Alter, G. Wiech: Jpn. J. Appl. Phys. 17, S2-288 (1978)
W.53	G. Wiech: Solid State Commun. 52, 807 (1984)
	G. Wiech, E.Z. Kurmaev: J. Phys. C 18, 4393 (1985)
W.54	D.W. Wilbur: Thesis Livermore 1965
	D.W. Wilbur, J.W. Gofman: Adv. X-Ray Anal. 9, 354 (1966)
W.55	M.L. Williams, R.C. Dobbyn, J.R. Cuthill, A.J. McAlister: In [Ref.1.101, p.303]
W.56	E. Wimmer, P. Weinberger, A. Kosakov, G. Leonhardt: Phys. Stat. Sol. (b) 89, 619 (1978)
W.57	J. Woicik, P. Mahowald, R. List, P. Pianetta: SSRL Report Nr. 84/06, p.24 (1984)
W.58	A. Wolberg, J.F. Roth: J. Catalysis 15, 250 (1969)
W.59	H.-W. Wolff, K. Radler, B. Sonntag, R. Haensel: Z. Physik 257, 353 (1972)
W.60	H.W. Wolff, R. Bruhn, K. Radler, B. Sonntag: Phys. Lett. A 59, 67 (1976)

W.61   J. Wong, F.W. Lytle: J. Appl. Phys. **51**, 280 (1980)
W.62   J. Wong, H.H. Liebermann: Phys. Rev. B **29**, 651 (1984)
W.63   J. Wong, F.W. Lytle, R.P. Messmer, D.H. Maylotte: Phys. Rev. B **30**, 5596 (1984)
        J. Wong, G.A. Slack: J. Solid State Chem. **61**, 203 (1986)
W.64   J. Wong, S.H. Lamson, K.J. Rao: In [Ref.1.83, p.49]
W.65   P.R. Wood, D.S. Urch: J. Chem. Soc., Dalton Trans. (1976) 2472; Chem. Phys. Lett. **37**, 13 (1976); J. Phys. F **8**, 543 (1978)
        D.S. Urch, P.R. Wood: X-Ray Spectrom. **7**, 9 (1978)
W.66   G. Wortmann et al.: J. Magn. Magn. Mater. **49**, 325 (1985)
W.67   F. Wuilleumier: In [Ref.1.58, Vol.II, p.327]
W.68   F. Wuilleumier: J. Physique **32**, C4-88 (1971)
W.69   F. Wuilleumier, M.O. Krause: In [Ref.1.60, Vol.I, p.397]
W.70   R.K. Wyrick, T.A. Cahill: Phys. Rev. A **8**, 2288 (1973)

Y.1   O. Yagci, J.E. Wilson: J. Phys. C **16**, 383 (1983)
Y.2   O. Yagci: J. Phys. C **19**, 3487 (1986)
Y.3   S. Yamaguchi, H. Hanyu, H. Koike: In [Ref.1.87, Vol.2, p.58]
Y.4   T. Yamaguchi, O. Lindqvist, T. Claeson, J.B. Boyce: Chem. Phys. Lett. **93**, 528 (1982)
Y.5   T. Yanagase, K. Morinaga, S. Sumita: J. Physique **43**, C9-51 (1982)
Y.6   E.A. Yanchuk: Mineral. Sb. **25**, 259 (1971)
Y.7   B.X. Yang, J. Kirz, T.K. Sham: Phys. Lett. A **110**, 301 (1985)
        B.X. Yang et al.: Phys. Lett. A **113**, 283 (1985)
Y.8   B.X. Yang, J. Kirz: Phys. Rev. B **35**, 6100 (1987); ibid. B **36**, 1361 (1987)
Y.9   C.Y. Yang, M.A. Paesler, D.E. Sayers: Phys. Rev. B **36**, 980 (1987)
Y.10   T.C. Yao, J.J. Holst: Spectrochim. Acta B **23**, 19 (1967)
Y.11   J.A. Yarmoff et al.: Phys. Rev. B **36**, 3967 (1987)
Y.12   Yu.M. Yarmoshenko et al.: Fiz. Met. Metalloved. **62**, 932 (1986)
Y.13   S. Yasuda, H. Kakiyama: X-Ray Spectrom. **7**, 23 (1978); Bunseki Kagaku **27**, 183 (1978)
Y.14   S. Yasuda, H. Kakiyama: Spectrochim. Acta A **35**, 485 (1979); Bunseki Kagaku **29**, 447 (1980); Rep. Gov. Ind. Res. Inst. Kynshu **25**, 1553 (1980)
Y.15   S. Yasuda, H. Kakiyama: X-Ray Spectrom. **10**, 85 (1981)
Y.16   B.N. Yatsyk: Ukr. Fiz. Zh. **20**, 851 (1975)
Y.17   V.A. Yavna et al.: Phys. Met. **6**, 1286 (1986)
Y.18   Yu.I. Yermakov et al.: React. Kinet. Catal. Lett. **7**, 309 (1977); ibid. **8**, 377 (1978)
Y.19   C.D. Yin et al.: J. Non-Cryst. Solids **69**, 97 (1984); ibid. **74**, 237 (1985)
Y.20   K.L. Yip, A.B. Kunz, W.S. Williams: Phys. Stat. Sol. (b) **75**, 533 (1976)
Y.21   T. Yoshiyama: Kyoto Sangyu Daigaku Ronshu **9**, 58 (1980)
Y.22   J.T. Yue, S. Doniach: Phys. Rev. B **8**, 4578 (1973)
Y.23   V.D. Yumatov, G.N. Dolenko et al.: Izv. Sib. Otd. Akad. Nauk SSSR, Ser. Khim. Nauk 1975, Nr.9, H.4, 124
Y.24   V.D. Yumatov et al.: Zh. Strukt. Khim. **26**, Nr.4, 59 (1985)
Y.25   V.D. Yumatov et al.: Zh. Strukt. Khim. **27**, Nr.1, 169 (1986)
Y.26   Yu.A. Yurakov, E.P. Domashevskaya, Ya.A. Ugai, L.Ya. Tverdokhlebov: Fiz. Khim. Obrab. Mater. (1978) 122

Z.1   I.I. Zalyubovskii et al.: Fiz. Tekh. Poluprovodn. **14**, 2081 (1980)
Z.2   I.A. Zarubin, A.T, Shuvaev et al.: Izv. Sib. Otd. Akad. Nauk SSSR, Ser. Khim.

Nauk 1975, Nr.9, H.4, 37; Izv. Akad. Nauk SSSR, Ser. Fiz. **40**, 340 (1976)

A.T. Shuvaev, I.A. Zarubin et al.: Izv. Akad. Nauk SSSR, Ser. Fiz. **40**, 333 (1976)

Z.3   A.P. Zemlyanov, A.T. Shuvaev, V.P. Krivitskii, M.G. Voronkov: Izv. Akad. Nauk SSSR, Ser. Fiz. **36**, 255 (1972)

A.T. Shuvaev, M.A. Blokhin et al.: Phys. Fenn. **9**, S1-156 (1974)

M.M. Tatevosyan, A.T. Shuvaev et al.: Zh. Strukt. Khim. **18**, 684 (1977)

Z.4   G. Zhang, M. Boudart: SSRL Report Nr. 84/06, p.61 (1984)

Z.5   A.Z. Zhmudskii et al.: Fiz. Tverd. Tela **12**, 1527 (1970); Ukr. Fiz. Zh. **17**, 856 (1972)

Z.6   A.Z. Zhmudskii: Metallofizika 1971, Nr.37, 6

Z.7   A.Z. Zhmudskii, V.F. Surzhko, V.I. Shiyanovskii: Ukr. Fiz. Zh. **24**, 1057 (1979)

Z.8   I.I. Zhukova, V.A. Fomichev, T.M. Zimkina: Izv. Akad. Nauk SSSR, Ser. Fiz. **31**, 952 (1967)

Z.9   I.I. Zhukova, V.A. Fomichev, A.S. Vinogradov, T.M. Zimkina: Fiz. Tverd. Tela **10**, 1383 (1968)

Z.10   E.A. Zhurakovskii, A.A. Vladimirova, V.P. Dseganovskii: Dokl. Akad. Nauk SSSR **170**, 548 (1966)

Z.11   E.A. Zhurakovskii, V.V. Gorskii: Dopov. Akad. Nauk Ukr. SSR 1428 (1966); ibid. 1127 (1967)

Z.12   E.A. Zhurakovskii, V.P. Dseganovskii: Poroshk. Metallurg. **6**, 70, 95 (1966); Ukr. Fiz. Zh. **15**, 1477 (1970)

I.N. Frantsevich et al.: Izv. Akad. Nauk SSSR, Neorg. Mater. **3**, 8 (1967)

V.P. Dseganovskii et al.: Ukr. Fiz. Zh. **15**, 1488 (1970)

Z.13   E.A. Zhurakovskii, V.I. Kotlyar, T.B. Shashkina: Izv. Akad. Nauk Ukr. SSR A 1969, 654

Z.14   E.A. Zhurakovskii, N.N. Vasilenko: Dokl. Akad. Nauk SSSR **187**, 562 (1969)

I.N. Frantsevich, E.A. Zhurakovskii, N.N. Vasilenko: Dokl. Akad. Nauk SSSR **198**, 1066 (1971)

N.N. Vasilenko, M.D. Lyutaya, E.A. Zhurakovskii, I.N. Frantsevich: Teor. Eksp. Khim. **8**, 347 (1972)

Z.15   E.A. Zhurakovskii, L.V. Nikitin: Izv. Akad. Nauk SSSR, Neorg. Mater. **5**, 1526 (1969)

Z.16   E.A. Zhurakovskii: Dokl. Akad. Nauk SSSR **184**, 1317 (1969); ibid. **194**, 312 (1970)

Z.17   E.A. Zhurakovskii, V.P. Dseganovskii, T.N. Bondarenko: In [Ref.1.58, Vol.I, p.184]; Tugopl. Karb. (1970) 119; Izv. Sib. Otd. Akad. Nauk SSSR, Ser. Khim. Nauk (1975) Nr.9, H.4, 102

I.Ya. Dekhtyar, E.A. Zhurakovskii, V.I. Shevchenko, D.I. Podyalovskii: Ukr. Fiz. Zh. **21**, 1714 (1976)

E.A. Zhurakovskii, L.V. Shevchenko: Dopov. Akad. Nauk Ukr. SSR, Ser. A 1978, 561

E.A. Zhurakovskii: In [Ref.1.100, p.98]

L.V. Shevchenko, E.A. Zhurakovskii, K.S. Pzoskurka: Ukr. Fiz. Zh. **27**, 1659 (1982)

E.A. Zhurakovskii et al.: Dokl. Akad. Nauk SSSR **283**, 1347 (1985)

Z.18   E.A. Zhurakovskii et al.: Izv. Akad. Nauk SSSR, Ser. Fiz. **4**, 50 (1971); Izv. Akad. Nauk SSSR, Neorg. Mater. **8**, 708 (1972); Teor. Eksp. Khim. **12**, 274 (1976)

Z.19   E.A. Zhurakovskii, P.V. Gel, V.V. Sokolenko, A.I. Sokolenko: Ukr. Fiz. Zh. **18**, 843 (1973)

P.V. Gel: Dopov. Akad. Nauk Ukr. SSSR B **36**, 53 (1974)

E.A. Zhurakovskii et al.: Ukr. Fiz. Zh. **22**, 65 (1977); ibid. **23**, 1477 (1978)

Z.20 E.A. Zhurakovskii, V.S. Neshpor, T.N. Bondarenko, V.P. Nikitin: Poroshk. Metallurg. **13**, 75 (1973)

Z.21 E.A. Zhurakovskii, P.V. Gel, A.I. Sokolenko, V.I. Sokolenko: Ukr. Fiz. Zh. **23**, 1477 (1978); in [Ref.1.100, p.135]

Z.22 E.A. Zhurakovskii, V.I. Trefilov, Ya.V. Zaulichnyi, G.I. Savvakin: Dokl. Akad. Nauk SSSR **284**, 1360 (1985)

Z.23 E.A. Zhurakovskii, Ya.V. Zaulichnyi: Fiz. Tverd. Tela **27**, 3452 (1985)

Z.24 P. Ziesche et al.: Wiss. Ber. Akad. Wiss. DDR, Zentralinst. Festkörperphys. Werkstofforsch. **17**, 125 (1979)

Z.25 T.M. Zimkina, O.A. Ershov, A.P. Lukirskii: Izv. Akad. Nauk SSSR, Ser. Fiz. **28**, 836 (1964)

Z.26 T.M. Zimkina, A.P. Lukirskii: Fiz. Tverd. Tela **7**, 1462 (1965)

O.A. Ershov, V.M. Burtseva: Opt. Spektrosk. **28**, 167 (1970)

Z.27 T.M. Zimkina et al.: Fiz. Tverd. Tela **9**, 1447 (1967)

Z.28 T.M. Zimkina, V.A. Fomichev: Fiz. Tverd. Tela **10**, 1392 (1968)

Z.29 T.M. Zimkina, A.S. Vinogradov: In [Ref.1.58, Vol.II, p.229]

Z.30 T.M. Zimkina, A.S. Vinogradov: J. Physique **32**, C4-3, 278 (1971)

A.S. Vinogradov, T.M. Zimkina: Zh. Strukt. Khim. **31**, 685 (1971)

Z.31 T.M. Zimkina, A.S. Vinogradov: Izv. Akad. Nauk SSSR, Ser. Fiz. **36**, 248 (1972)

A.S. Vinogradov et al.: Fiz. Tverd. Tela **22**, 2602 (1980)

A.A. Pavlychev et al.: Opt. Spektrosk. **52**, 506 (1982)

A.S. Vinogradov, E.O. Filatova, T.M. Zimkina: Fiz. Tverd. Tela **25**, 1120 (1983)

Z.32 T.M. Zimkina, V.A. Fomichev, S.A. Gribovskii: Fiz. Tverd. Tela **15**, 1629, 2685 (1973)

S.A. Gribovskii, T.M. Zimkina: Opt. Spektrosk. **35**, 179 (1973)

V.A. Fomichev, S.A. Gribovskii, T.M. Zimkina: In [Ref.1.60, Vol.II, p.191]

Z.33 E. Zöpf: Thesis, Universität München (1972): In [Ref.1.98, p.188]

Z.34 E. Zschech et al.: Phys. Stat. Sol. (a) **86**, 117 (1984)

Z.35 E. Zschech et al.: J. Non-Cryst. Solids **86**, 336 (1986)

Subject Index